AF443073

Further Understanding of the Human Machine
The Road to Bioengineering

SERIES ON BIOENGINEERING AND BIOMEDICAL ENGINEERING

Series Editor: John K-J Li *(Department of Biomedical Engineering, Rutgers University, USA)*

Series on Bioengineering & Biomedical Engineering – Vol. 7

Further Understanding of the Human Machine

The Road to Bioengineering

Max E. Valentinuzzi Editor

National University of Tucumán (UNT)
National Scientific and Technical Research
Council (CONICET), Argentina

NEW JERSEY · LONDON · SINGAPORE · BEIJING · SHANGHAI · HONG KONG · TAIPEI · CHENNAI · TOKYO

Published by

World Scientific Publishing Co. Pte. Ltd.

5 Toh Tuck Link, Singapore 596224

USA office: 27 Warren Street, Suite 401-402, Hackensack, NJ 07601

UK office: 57 Shelton Street, Covent Garden, London WC2H 9HE

Library of Congress Cataloging-in-Publication Data

Names: Valentinuzzi, Max E., 1932–   author.

Title: Further understanding of the human machine : the road to bioengineering /
   Max E. Valentinuzzi, National Scientific and
   Technical Research Council (CONICET), Argentina.

Description: New Jersey : World Scientific, 2016. | Series: Series on bioengineering and
   biomedical engineering ; volume 7 | Includes bibliographical references and index.

Identifiers: LCCN 2016023757 | ISBN 9789813147256 (hardcover : alk. paper)

Subjects: LCSH: Human physiology. | Bioengineering. | Biomedical engineering.

Classification: LCC QP34.5 .V344 2016 | DDC 610.28--dc23

LC record available at https://lccn.loc.gov/2016023757

British Library Cataloguing-in-Publication Data

A catalogue record for this book is available from the British Library.

Typeset by Stallion Press
Email: enquiries@stallionpress.com

Printed in Singapore

"O my body!
I dare not desert the likes of you in other
men and women,
nor the likes of the parts of you.
I believe the likes of you are to stand or fall
with the likes of the soul,
and that they are the soul.
I believe the likes of you shall stand or fall
with my poems,
and that they are my poems."
Walt Whitman
(1819–1892)
I sing the Body Electric

It appeared in the 1860 third edition of his revolutionary volume of
poetry, *Leaves of Grass,* as a poem of the *"Enfans d'Adam"*.

(see http://whitmanarchive.org/published/LG/1860/poems/57)

DEDICATION

To Tropezón (Tropi), our family Labrador, for his unbreakable love, faithfulness and so many moments of company, waiting patience and joy, and through him, a tribute to all those less happy and Heroic Dogs that silently and quietly offered their lives to better understand physiology and health.

Max Valentinuzzi IV, Max Valentinuzzi III & Tropi

However, let us not forget the Frog and Toad, wonderful and harmless characters of the Animal Kingdom; they also meekly gave us essential background scientific knowledge about our bodily functions . . . and funny amazement, too, when getting dormant after gently scratching his belly (only males so respond). Do not leave out, neither, the bewitched Prince who happily broke out of the evil trance after a sweet Princess' kiss!

Argentine toad, formerly *Bufo arenarum*, later reclassified as *Rhinella arenarum*. Alert and dormant. Picture by MEV.

ACKNOWLEDGMENTS

This second updated edition took quite a while, with effort, perspiration, and other secretions, too. No doubt, the contributors are the first to receive my full recognition. Boy, they sweated out! However, there are always people behind the curtain whose names remain well hidden in the backstage. The editorial house, WSP, in Singapore, must be acknowledged because several of its officials believed in my word. Among them, I want to underline Ms Yubin Zhai's name, stationed in the US office, and Ms Catherine Yeo, our patient and helpful editorial assistant, in Singapore. To our devoted artist, Gustavo Idemi, who generously in San Juan City, Argentina, produced, drew or redrew pictures of different kind, serious and funny ones. Lastly, my recognition, too, to Dr. Natalia Martina López Celani, also in San Juan, former student of mine, for taking care of the DROP-BOX, where the files were securely saved. Other editorial details were handled by my daughter, Verónica S. Valentinuzzi, co-author of one chapter. To all them, my appreciation and sincere thanks, along with my apologies for harassing and annoying them perhaps too much, especially during this last early 2016 period!

Argentina, by Gustavo Idemi

Max E. Valentinuzzi
Yerba Buena, Province of Tucumán, Argentina
April 20, 2016

CONTRIBUTORS

Gisele Akemi **Oda**, gaoda@ib.usp.br

Gisele is an assistant professor at the Department of Physiology, Institute of Biosciences, University of São Paulo, Brazil since 2007 and a Corresponding Researcher of CONICET, Argentina, since 2014. She completed her undergraduate formation in the Institute of Physics, University of São Paulo, in 1990, and got the master and doctorate degrees at the same institute (1991–1995). She worked as a Postdoctoral Fellow at the Center for Biological Timing, University of Virginia, USA (1998–2000), and at the Institute of Physics, University of São Paulo (2001–2005). Her research deals with chronobiology, circadian rhythms, synchronization, and mathematical modeling.

Pedro D. **Arini**, pedro.arini@conicet.gov.ar

Born in Buenos Aires, Argentina, he obtained in 1995 the MS degree in electronics engineering from the National Technological University (UTN), Buenos Aires. Afterwards, in 2001, he got the MSc in BME from the Favaloro University, Buenos Aires. Subsequently, Pedro carried out doctoral studies at Zaragoza University, Zaragoza, Spain, in 2007, reaching the PhD level. He has worked in several areas of biomedical and electronics engineering (communications, video systems, control engineering, speech signal processing, neuroscience, cardiac electrophysiology, and digital processing of biomedical signals. Since 2008, he has been an investigator of the National Research Council of Argentina (CONICET) carrying out activities at the Institute of Mathematics *Alberto P. Calderón.* Besides, since 2010, he holds the position of associate professor at the Engineering School of the University of Buenos Aires (UBA). Dr. Arini has published over 30 scientific papers in international journals, about 50 communications to conference proceedings and one book chapter. His current research interests include biomedical signal processing, with main focus on cardiovascular signals.

Alfredo **Coviello**, dr.acoviello@gmail.com

A physiologist specialized in renal physiology and hypertension, he was a full professor of physiology at the Medical School of the National University of Tucumán, Argentina. He was, during his young training years, a research fellow at the Catholic University of Louvain, Belgium, under Professor Jean Crabbé. In that laboratory, he began studies about the role of angiotensin II in sodium and water transport in isolated toad kidney, bladder, and skin. After returning to Argentina, he obtained the doctoral degree supervised Dr. Julia Uranga, a disciple of Nobel Prize Bernardo Houssay. Alfredo demonstrated for the first time that angiotensin II increases sodium and water transport in the renal proximal tubule and that the octapeptide stimulates osmotic water permeability mediated by cyclic AMP. He is at present an active investigator dealing with the effects of psychological stress in markers of prothrombotic factors in healthy young volunteers.

Rosana **Chehín**, rosana@fbqf.unt.edu.ar; rosanachehin@gmail.com

Rosana Chehín was born in Tucumán, Argentina, in 1967, getting a biochemistry BSc degree in 1992 and a PhD in 1995 from the Universidad Nacional de Tucumán (UNT). She performed postdoctoral studies in protein biophysics in Spain and now she holds the positions of assistant professor, UNT, and CONICET Career Investigator. She has national and international grants as well as leads international cooperation programs with France, Spain, and Brazilian universities. Dr. Chehín is the team leader of the *Instituto Superior de Investigaciones Biológicas* (INSIBIO) Biophysical and Bioinformatics Group (IBBG), which is composed of four young researchers, two postdoctoral fellows, four doctoral, and three undergraduate students. IBBG has published significant contributions in high-impact journals within the protein biophysics area.

Ting-Hsuan **Chen** (Cecil), thchen@cityu.edu.hk

Ting-Hsuan Chen is an assistant professor in the Department of Mechanical and Biomedical Engineering at City University of Hong Kong since 2012. He received his BS degree (2003) and MSc degree (2005) from the National Tsing Hua University, Taiwan, and PhD degree from Mechanical and Aerospace Engineering at the University of California, Los Angeles, USA (2012). His research interests are mainly applying micro/nanotechnology for diverse fields such as tissue morphogenesis, cell mechanics, and biosensors. Dr. Chen has been spearheading an array of multidisciplinary projects addressing the pattern formation of self-organizing biological systems. He has made a fundamental discovery of cellular left–right symmetry breaking of vascular stem cells and successfully exploited this finding to engineer intricate patterns of tissue architecture. His recent researches include the *in-vitro* development of cardiac tissue culture, nanowire platforms for measuring cell mechanics, and nanomaterials for development of assays for intracellular detection or point-of-care applications.

Esteban **Cynowiec**, ecynowiec@fi.uba.ar

Esteban was born in Buenos Aires in 1972, getting a biology degree at the University of Buenos Aires in 2000. Currently, he holds an assistant professorship position at the Institute of Biomedical Engineering in the same university, where he is in charge of the Biological Systems course. His interests reside within the neurosciences, especially working with neural networks, but also crossing to the epistemological value of animal models often applied to human psychiatric illnesses. He is interested in philosophy of neuroscience and science in general and has written a literary book; besides, he publishes short stories in several magazines, too. Esteban is also a scientific writer and journalist having published a great deal of articles in general scientific magazines and newspapers.

Sibel **Ertek**, sibelertek@yahoo.it

Sibel is an associate professor of Endocrinology and Metabolic Diseases at the Memorial Ankara Hospital, Turkey, since 2014. She completed her medical education in Hacettepe University, in 1999, and specialized on internal medicine in Ankara University (2000–2005). She worked at the Nephrology and Hypertension Departments of Ankara and Baskent Universities. Besides, she worked at the Internal Medicine, Hypertension, Atherosclerosis, and Metabolic Disease Departments of Bologna University, Italy, in 2006. In Hacettepe University Medical School, she proceeded at the Pharmacology Department, in 2007, and had Endocrinology and Metabolic Diseases education (2008–2011) at Ufuk University, Ankara. She has worked on kidney, liver, and pancreas-transplanted patients in Medicalpark Antalya Hospital and Transplantation Center, Turkey, in 2013. She has published in the fields of metabolic syndrome, hypertension, dyslipidemia, insulin resistance, diabetes, and clinical endocrinology. She is an active member of the Endocrine Society, the European Thyroid Association, and the Society of Endocrinology and Metabolism of Turkey (SEMT).

Ignacio **Ghersi**, ign.ghersi@gmail.com

Born in Buenos Aires, Argentina (1985), Ignacio is a researcher at the Biomechanics and Health Engineering Laboratory (LaBIS), Faculty of Engineering, Pontifical Catholic University (FI-UCA). He received an Engineering Degree in Electronics and Communications in 2009 from FI-UCA, and a PhD in Design, Architecture School, Universidad de Buenos Aires (UBA), in 2016. Between 2010 and 2015, he received a research scholarship from UCA, to work in methods for signal processing, human-balance assessments and medical technologies and devices. His doctoral dissertation was developed at the Center for Research in Industrial Design of Complex Products (CIDI-FADU-UBA) and LaBIS-FI-UCA, and concerned with updated mechatronics and interfaces for a novel medical-bed design.

Federico N. **Guerrero**, federico.guerrero@ing.unlp.edu.ar

Federico Nicolás Guerrero was born in Comodoro Rivadavia, southern Argentina, in 1986. He received the Engineering degree in electronics from La Plata National University (UNLP), La Plata, Argentina, in 2011. Since 2012, he has been working towards the PhD degree at the Institute of Industrial Electronics, Control and Instrumentation (LEICI), UNLP and

CONICET, La Plata. He is also a teaching assistant at the UNLP Engineering Department. His research has been concerned with instrumentation and control for biopotential measurement systems.

Anthony **Guiseppi-Elie**, guiseppi@tamu.edu

Professor and department head of biomedical engineering at Texas A&M University (TAMU). Tony is also a founder and scientific director at ABTECH Scientific, Inc. Holds the ScD in materials science and engineering from MIT, the MSc in chemical engineering from UMIST and the BSc (First Class Honors) in Applied, Analytical, and Biochemistry from UWI. He has spent 15 years in intrapreneurial industrial research and product development before becoming a full professor of chemical and life science engineering and professor of emergency medicine at VCU/MCV in 1998. In 2006, he joined Clemson as the Dow Chemical Professor, later moving to Texas A&M in 2015. His research interests are in **engineered bioresponsive systems in the service of human health and medicine** (145 papers, 4324 citations, h-index = 34). Tony has been a 2015–2017 Fulbright Specialist Award recipient, the 2014–2015 Visiting Distinguished Professor of Industrial Bioelectronics at *l'Ecole des mines d'Alès*, France, the 2013 Avis Professor in Pharmaceutics at the University of Tennessee Health Sciences Center, a 2012–2013 IEEE-EMBS Distinguished Lecturer and a 2012 Microsystems Distinguished Lecturer at the University of Maryland. Besides, he is an Editor-in-Chief of *Bioengineering* and Associate Editor of *Biomedical Microdevices*. AIMBE Fellow, Fellow of the Royal Society of Chemistry, IEEE Fellow, lifetime member of AIChE and holds memberships to other societies.

Gustavo **Idemi**, gustavo.idemi@gmail.com

Our draftsman and artist was born in 1965, in the city of San Juan, west side of Argentina, by the *Cordillera de los Andes*. A professional of caricatures and drawings, he describes himself as an illustrator, graphic designer using digital techniques, and publicist. He likes the Creative Arts, but humor is his main specialty. Simultaneously, Gustavo carried out several developments in the area of strategic communications, political marketing, packaging design (that is, planning and fashioning the complete form and structure of a product's package), and traditional 2D animations. Currently, his activities place him between visual communications and editorial illustrations intended for different types of publications (newspapers and magazines). See http://gusidemi.com/blog2/

Alberto J. **Kohen**, ajkohen@yahoo.com

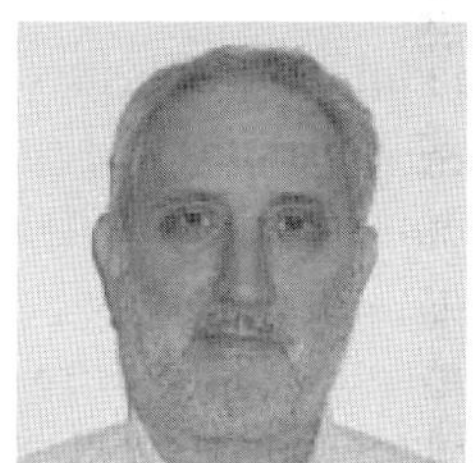

Alberto José Kohen (66) obtained his engineering degree in electromechanics and electronics at UBA in 1974. He is currently a professor at the *Instituto de Ingeniería Biomédica* and Secretary to the Electronics Department of the University of Buenos Aires (UBA). Besides, he holds also a professorship of BME at the *Universidad Nacional de San Martín*, Argentina. Coauthor of more than 20 papers and one patent, he has had active participation in industry for over 40 years and in the development of Medical Instrumentation, Clinical Engineering and in several education programs.

Rossana E. **Madrid**, rmadrid@herrera.unt.edu.ar

She is with the *Laboratorio de Medios e Interfases*, Bioengineering Department, Faculty of Exact Sciences and Technology, University of Tucumán (UNT) and *Instituto Superior de Investigaciones Biológicas* (INSIBIO) — CONICET. Rossana obtained the EE degree and PhD in bioengineering, both at UNT. She holds the positions of assistant professor and investigator, the latter since 2004, having published R&D papers in national and international journals, and two chapters in books. Besides, she holds four patents. She leads research national grants and international cooperation grants with two German Universities. She currently supervises five doctoral dissertations in the area of biosensors and microfluidics. She is a professor of Biomedical Transducers and Biosensors and Microsystems of the Biomedical Engineering Program, and also at the doctoral level. Her main research fields include sensors, biosensors, and microfluidic systems for biomedical and environmental applications.

Mónica **Miralles**, mmiralles@gmail.com.

Mónica is a professor of physics at the Department of Industrial Design since 2004, and professor in the Career of Specialization in Biodesign and Mechatronic Products (Biomechanics and Bionics modules, respectively), since 2007, at the Faculty of Architecture, Design and Urban Planning (FADU), of the National University of Buenos Aires (UBA, *Universidad de Buenos Aires*, Argentina). She is also a professor of fluid mechanics, since 2006, at the Engineering School of the Pontifical Catholic University (UCA). She obtained the PhD degree in physical sciences at the Faculty of Exact and Natural Sciences of UBA (FCEyN-UBA) in 1998, with specialization in materials, at the National Atomic Energy Commission of Argentina (CNEA), where she earned doctoral and post-doctoral fellowships (1993–2001). Since 2004, she is a researcher for the Center of Research in Industrial Design of Complex Products (CIDI-FADU-UBA). Since 2009, she is founder and director of the Biomechanics and Health Engineering Laboratory (LaBIS-UCA), conducting research projects in the fields related to aspects of human biomechanics and, in particular, human balance of older adults and ataxic patients (risk of falls). She has presented and published extensive work regarding nuclear materials, human balance, biomechanics, and motor control of human movements (signal processing), in national and international journals.

Chi-Sang **Poon**, cpoon@mit.edu

Chi-Sang Poon is with the Department of Health Sciences and Technology as Principal Research Scientist, Division of Health Sciences and Technology, of the Massachusetts Institute of Technology (MIT). He got the BSc in electrical engineering, University of Hong Kong 1975, master in Bioelectronics, Chinese University of Hong Kong, 1977, and PhD in bioengineering, UCLA, 1981. He spent a period in France, 1994, and is a fellow member of IEEE and of the American Institute for Medical and Biological Engineering (AIMBE). His research takes a multidisciplinary approach, combining biology with engineering and computational science, to investigate the mechanisms of neural control at both the cellular and systems levels. Work on respiration focuses on mechanisms of control in neural and neuronal plasticity, in which physiological changes during development and in response to environmental stimuli produce adaptive behavior. One critical problem in cardiac rhythm is distinguishing random fluctuations from background noise, so that we can identify subtle patterns in heart-rate variability that may be correlated with specific types of arrhythmia or sudden cardiac death.

Enrique **Spinelli**, spinelli@ing.unlp.edu.ar

Enrique Mario Spinelli was born in Balcarce, Province of Buenos Aires, Argentina, in 1964. He received the engineering degree in electronics and the MS and PhD degrees from La Plata National University (UNLP), La Plata, Argentina, in 1989, 2000, and 2005, respectively. Since 1990, he has been with the Institute of Industrial Electronics, Control and Instrumentation (LEICI), UNLP, and CONICET, La Plata, where he is currently a professor of control systems in the Engineering Department, working on scientific instrumentation. He is also a Career Investigator of the National Scientific and Technical Research Council (CONICET), Argentina. His research interests are analog signal processing and brain control interfaces.

Max E. **Valentinuzzi**, maxvalentinuzzi@arnet.com.ar; maxvalentinuzzi@ieee.org

 Born in Buenos Aires 1932. National College Buenos Aires (1950), Telecommunications Engineer, University of Buenos Aires (1956). He received a PhD in physiology & biophysics, Baylor College of Medicine (Houston, TX, USA, 1969). Assistant Prof. Baylor (1969–1973). Prof. Bioengineering & Head of Lab (1972—retired 2001), *Univ Nacional Tucumán* (UNT) & Career Investigator (1977—retired 2000) of *Consejo Nacional Investigaciones Científicas y Técnicas* (CONICET), Argentina. Emeritus Investigator CONICET. Cofounder & first director (1980–7) of *Instituto Superior Investigaciones Biológicas* (INSIBIO). Professor of BME at the *Instituto de Ingeniería Biomédica* (IIBM), University of Buenos Aires (UBA) (2007–2014). 1973: shared Nightingale Prize Bioengineering (IFMBE). 1980: Houssay Prize of Biology (*Sociedad Argentina Biología*). 1984: Golden Route Prize Science, Buenos Aires. 1985: Catalina B. de Barón Accesit Prize Cardiology. 1996: IEEE/EMBS Career Achievement Award. Life Fellow Member IEEE, Emeritus Member American Physiol Soc, member of other societies. Inducted into several academies. 2004: Bernardo Houssay Prize. 2005: Felix Cernuschi Prize of Bioengineering. 2008: UNT Emeritus Professor. 2013: Platinum KONEX Prize. Author/coauthor over 135 scientific and technical papers and more than 30 teaching and/or general articles, textbooks, and chapters. Guest editor of special issues. Lectures at several US and Latin American Universities. He has been editorial board member of some journals. Worked in bioimpedance, cardiovascular system, fibrillation–defibrillation, impedance microbiology, numerical deconvolution and BME education. He loves his

students, children, animals, and music. Being is much better than having. Editor of *IEEE/EMB Pulse Magazine's* RETROSPECTRO-SCOPE SECTION (since 2011).

Veronica S. **Valentinuzzi**, veroval@hotmail.com; vvalentinuzzi@crilar-conicet.gob.ar

Born in Bahia Blanca, Argentina, in 1963. She received the degree in Animal Husbandry at the National University of Tucumán (UNT), Argentina, in 1989. Completed an MSc and PhD in Physiology–Biology at the State University of Campinas-UNICAMP, Brazil, in 1993 and 1999, respectively. She was a visiting predoctoral researcher at the Center for Biological Timing, Circadian and Seasonal Rhythms, Northwestern University, USA (1994–1997). Between 2000 and 2003, she had a postdoctoral position at the National University of São Paulo-USP, Brazil, and, from 2004 to 2006, a visiting professorship at the National University of *Rio Grande do Norte*-UFRN, in Natal, Brazil. Since 2007, she is a CONICET Investigator at the Regional Research Center CRILAR (*Centro Regional de Investigaciones Científicas y Transferencia Tecnológica de La Rioja*), in Anillaco, province of La Rioja, Argentina, being in charge of the Chronobiology Laboratory.

Silvano **Zanutto**, silvano.z@gmail.com.

B. Silvano Zanutto received the electronics engineering degree from the University of Buenos Aires (UBA) in 1980 and the PhD degree in Biology from UBA, too, in 1993. He was a research associate at the Department of Psychology and Neuroscience of Duke University from 1993 to 1997, and an assistant professor at the electronics department (UBA) from 1998 to 2002. Currently, and since 2002, he holds a full professorship and the Direction of the *Instituto de Ingeniería Biomédica* (IIBM) at UBA. He has been a researcher of IBYME-CONICET since 1998. Author/coauthor over 85 scientific papers and book chapters, and more than 100 congress meetings. He is interested in bioengineering, behavioral brain research, theoretical neuroscience, and bio-robotics. Topics of his interest are computational theories of operant learning and decision making based on psychological and neurobiological data.

CONTENTS

PREFACE

*Nothing ends ... or perhaps this is what we inwardly would like to happen
...*

This book is a continuation of a previous one published in 2004 (Valentinuzzi, 2004), which was the result of over 35 years of teaching bioengineering and physiology at the undergraduate and graduate levels, in this country and abroad, in engineering and in medical schools. This book is from the perspective of a simple teaching assistant under the supervision of great teachers like Hebbel E. Hoff and Leslie A. Geddes at Baylor College of Medicine in Houston, Texas, USA, to the level of full professorship and directorship of a graduate program at the *Universidad Nacional de Tucumán* (UNT), in Argentina; a result of collecting notes and material of different kinds, of actual research and technological development activities, of close interaction with students and collaborators alike, of joyful, and sometimes painful work. This older text acts as launching platform of the present one, which is not a **Primer** anymore, as it was stated in its former title, but it still intends to be a **Road to Bioengineering** bringing **Further News**. Thus, the former did not end, or I may have not wanted it to end.

The newer is not comprehensive and not a handbook, either. It just is what its title suggests. It is multi-author and the main scriptwriter of the preceding textbook takes now the position of Editor and Principal Author as well. Current knowledge has become so wide and complex that rarely, if ever, a single mind is able to manage it all, no matter how experienced the person may be. Several contributors were invited (see the list), but not too many, otherwise the text would lose continuity or steadiness, but each is proficient in his or her specific subject. Most of them are well set in Argentina, while the rest develop their activities abroad.

Perhaps the book is somewhat biased in many respects, or lacking subjects, especially because always when a book is made public . . . well . . . it is already old or outdated. My apologies, but remember that often things depend on the color of the lens through which we look, as centuries ago was apparently stated one way or another by renowned poets like Jorge Manrique (1440–1479), William Shakespeare (1564–1616), Pedro Calderón de la Barca (1600–1681), and Ramón de Campoamor (1817–1901).

The main audience aimed at should be formed by relatively advanced students, say, near graduation or already in the graduate road, including young researchers deepening on established lines or perhaps opening new tracks. The well-set investigator may also pick up a few hints. Let me insist: The intent is to bring, not all, but **Further News**.

The style is light, interspersed with historical side notes and, why not, even spicing the text with a little good mood to make reading easier. I hope such feature does not bother the reader. A well-intended smile never hurts, and eventually, it opens doors.

Perhaps, a good leitmotiv for the book could be phrased as,

remembering slightly the past,
strongly living the present,
while trying to look if possible into the future.

This is a good place to quote from an old book by Postman and Weingartner (1969):

> Once you have learned how to ask questions, relevant and appropriate and substantial questions, you have learned how to learn and no one can keep you from learning whatever you want or need to know.

Another relevant oldie is due to Anatole France (1844–1924), quoted by Tilden (1967):

> Do not try to satisfy your vanity by teaching a great many things. Awaken instead your students' curiosity.

Most important, keep your mind active, and look at examples such as William Hansel's, who at 96, is still a professor of animal physiology at Louisiana State University, Pennington Biomedical Research Center, in Baton Rouge, LA, a position he has held since his "retirement" from Cornell University in 1990, and now devoting his efforts to cancer biology and therapy (Musch, 2014). Another example is 1977 Nobel Prize winner of Medicine and Physiology Roger Guillemin, who at 92, keeps going to his Salk Institute office in San Diego, CA. Or look at Stephen Hawking, 74 years old, dragging amyotrophic lateral sclerosis (ALS) and writing: *It has been a glorious time to be alive and doing research in theoretical physics. I'm happy if I have added something to our understanding of the universe* (Hawking, 2013).

Aesthetics should be pursued, too, as once wrote in a personal letter Ludwig van Beethoven (Solomon, 1998; Cobra, 2000). That is the way to reach full freedom, finding the aesthetic content of our actions, and Arts, Science and Education, which call for a lot of love, are no doubt masters in this respect.

Our current hectic World, where the concept of freedom appears nebulous or even lost (Valentinuzzi, 2015), needs more than ever before high doses of love, understanding, tolerance, mutual respect, and smiling faces, even in the face of pain, disease, and suffering. No, no ... the statement does not mean at all trying to return to the old aestheticism movement or its endorsement. It simply aims in the end at offering tools to help life and find a better future for humankind as many problems still batter population sectors in large geographical areas of the world, in spite of tremendous scientific and technological advances.

Godlee (2013), for example, editor of the *British Medical Journal* (BMJ), pointed out that death in children under 5 years of age have fallen impressively in the past quarter of a century; excellent news, indeed, but not for all, since they still do it rather slow for the world at large. Progress is especially slow in the poorest countries. Where should countries invest for the quickest and biggest gains in life expectancy? There is agreement that several factors are associated with childhood mortality with no time delay, such as national income, access to sanitation, and/or human immunodeficiency virus (HIV) prevalence, while others show delayed associations, such as urbanization, health spending, corruption, and political instability. Women's education is also associated with lower mortality, but the benefits seem to level off after only a few years of schooling. Surprisingly, it was found no link between childhood mortality and undernourishment [*and I, the author of this PREFACE, say, is it really so? I personally doubt it. Visit areas where undernourishment hits and you will have a small sample*]. Expectations do not lead to answers that can be applied worldwide. The world could, **if it chose to**, eliminate health inequalities between nations over the next 20 years by doubling investment in new drugs, vaccines, and health technologies, saving perhaps 10 million lives in low and middle income countries in the year 2035 alone ... however, and unfortunately, there are and there will be natural and **man-made disasters**. Should you be a doubter, just check the news and take note of the craving for power, richness, and fame of too many human beings and some countries, too. None the less, it sounds healthy to remember Francisco when he says ***never yield to pessimism***. Let us take this simple thought, whether you are or not a believer, as our Olympic Torch carried by Caloi's Clemente, who was declared the Paradigmatic XXIst Century Bioengineer in the preceding version of this book, while he continues as the reader's guide, always reminding:

> You will begin to respect and love your neighbor
> the very moment you learn how to laugh at yourself.

Copyright comic by Caloi, 1983; later on by WSP, 2004

Max E. Valentinuzzi, EE, PhD
maxvalentinuzzi@arnet.com.ar
maxvalentinuzzi@ieee.org
maxvalentinuzzi071@gmail.com

Yerba Buena, Tucumán, Argentina, June 21, 2013, Winter Solstice in the Southern Hemisphere, a nice time coincidence. The *Intiwatana*, known as the "hitching post of the sun," at Machu Picchu, Perú, is an altar the *Inca* devised to connect with the Divine Source. On it, the longest winter night and the shortest winter day of the year, *Inti* (the Sun), who seems to be weakening and dying, is reborn, and begins to grow in radiance and renewal. Rebirth and regeneration represent the essence of this concept, and along with wickedness the human being carries always an enormous rebirth and regeneration capacity. Why not taking this beautiful metaphor as a good omen for bioengineering, science, and technology, and finally, for our so-many-times-badly-beaten Earth with all its inhabitants?

References

Cobra, M. From Ode to Hymn-From Schiller to Beethoven. HODIE™ Productions, 2000. http://hodie-world.com/cms_fr_full/component/option,com_docman/task,cat_view/gid,34/Itemid,32/lang,es/?mosmsg.

Godlee, F. Editorial Note. *Br Med J* 347:f7236, 2013. Published 04 Dec; http://www.bmj.com/content/347/bmj.f7236?etoc=.

Hawking S. My brief history. Bantan Books, New York, 127 (see p 126), 2013.

Musch T. Discipuli supra omni praeferentur: Above all, the students come first. *Physiologist* 57(1):48–50, 2014.

Postman N, Weingartner C. Teaching as a subversive activity. Dell Publishing, New York, 23, 1969.

Solomon, M. Beethoven. 2nd ed. Schirmer Trade Books, New York, 554, 1998.

Tilden F. Interpreting our heritage. University of North Carolina Press, Chapel Hill, NC, 1967.

Valentinuzzi ME. Understanding the human machine: A primer to bioengineering. World Scientific Publisher, New Jersey & Singapore, (xii + 396), 8 chapters. Series Bioeng & Biomed Eng, Vol. 4, 2004. ISBN 981-238-930-X & ISBN 981-256-043-2.

Valentinuzzi ME. Technological acceleration: A few musings. A short break to think over: To be free or not to be free?. *IEEE Pulse Mag* 6(6):39, 2015.

CHAPTER 1

INTRODUCTION

Max E. Valentinuzzi

*What is first goes first, what is last goes last, and often there is something
in between ... as a sandwich.*

Abstract

Bioengineering has evolved steadily and increasingly over the last two
or three decades, but especially when it crossed the centuries dividing
line, as if the millennium transition meant a unique omen leap. In the
beginning, it was a simple mixture of a few ingredients, and people talked
of Medical Electronics (ME) as a *multidiscipline*. Step by step, the link
became more intimate, and it was called an *interdiscipline*, because there
was free crossing from one area into the other. Nowadays, new disciplines
show up, reminding of transgenic organisms, where the very deoxyribonucleic acid (DNA) is modified. Thus, we should speak of *transdiscipline*. Far off is the naïve 1950s idea of putting together in a room a
bunch of engineers, biologists, and physicians to produce ME!

1.1 The Seven Lamps of Bioengineering

They sure enlighten its road, as a rainbow

It is said that Saint Mathesis, patron of mathematics, used to
be surrounded by seven lamps eternally lit: *lampas utilitatis* (usefulness), *lampas imaginationis* (imagination), *lampas poesis* (creativity), *lampas infinitatis* (infinite), *lampas decoris* (beauty), *lampas
mysterii* (mystery), and *lampas religionis* (religion). Bioengineering,
strongly supported by mathematics, searches the quantification of
the biomedical sciences. No doubt, these lamps also illuminate its

1

course because this interdiscipline is certainly useful, for it applies directly or indirectly to the human being. It requires tremendous imagination and creativity and extends unlimited in all directions branching off here and there. It is as beautiful as an enchanted fairy while it shines with some degree of mystery when penetrating the darkness of the unknown, so spurring curiosity, and shows also even mystical and mythical[1] edges when human, ethical, and religious aspects are touched on (Smith, 1947; Valentinuzzi, 2004, 2010).

1.2 Current Definition of Bioengineering

Oh, definitions... are they really necessary? Some people become obsessed with them!

This book refers to the use and application of principles, laws, techniques, and general knowledge, taken from the physical and engineering sciences, to the better understanding and solution of biological and medical problems at large. No one argues this anymore. We just realize how wide and ambitious the intent is. Of the several terms that have danced around for a long time, Bioengineering and Biomedical Engineering seem to stay with us, showing no differences between the two, the former being shorter than the latter, if a hard eye looks for one. *Clinical Engineering* though, the newest of those words, is clearly linked to Health Care Systems since it deals with problems found in hospitals and emergency services, and works side by side with medicine. It shows a well-defined personality with publications of its own. Bioengineering has become more biological in some respects and more engineer-like in others, quite medical on occasions and very technological in many specific areas, such as ophthalmology or anesthesiology, or when getting into agricultural or ecological problems, which in the end support life on Earth. Boundaries are

---

[1] *Mystical*: Something having a spiritual reality not apparent to the intelligence or senses, or related to, or stemming from direct communion with a divinity. Also related to rites or strange practices. *Mythical*: Something existing in an imaginary or fictitious idea, as the legendary unicorn, or sirens, or elves. Representing the first one, there is a famous fresco, probably by Domenico Zampieri, circa 1602, in the Palazzo Farnese, Rome.

not well delimited, nor they need be, and information and activities must flow freely from one disciplinary compartment to another in a constant exchange if results are to be fruitful. Interestingly enough, the two biggest international organizations do not officially use either of the two current terms, preferring longer combinations of words: The *Institute of Electrical and Electronics Engineers/Engineering in Medicine and Biology Society* (IEEE/EMBS) and the *International Federation for Medical and Biological Engineering* (IFMBE). However, Bioengineering or Biomedical Engineering, and their much more bounded sub-discipline Clinical Engineering, constitutes common language terms in their respective publications and meetings as well.

Definitions are always controversial and, in this particular case, perhaps more than ever for there is still a state of development and evolution of concepts, making terms obsolete in a relatively short time. Thus, *a flexible and open mind becomes mandatory* (quite an attractive characteristic of Bioengineering, indeed!). Definitions are not absolute, they are secondary; they can and should be changed as required. They are simply convenient. Look with a serious frown at those people who stick stubbornly to definitions (and even worse, to standards!). Imagine what it would be like had we fastened ourselves to the definitions (and the concepts supporting them) of the Middle Age or, without going as far back, of the more recent last two centuries.

1.3 Multidiscipline, Interdiscipline and Transdiscipline

Discipline, among other meanings, for us is a field of study

Engineers and physicists measure, they measure to describe magnitudes in terms of numbers attaching them to units, too, so that they better know about a given event; they measure to design and adjust, to quantify, as the case is, for example, with the amount of cement and number of bricks to build a wall. Physicians also measure, parameters such as body temperature, body weight, blood pressure, and other physiological events. Nutritionists have a similar stand with meals and their caloric content. What should we eat and how much? Dietetics, strongly based on nutrition concepts had to draw upon physics and physiology.

Bioengineering, as meals, developed first as a simple mixture of biological and technological ingredients (a salad). Early in its modern life, say during the 1950s, it barely was electronic instrumentation used in biology and medicine. Not much was the knowledge of the intervening engineers and physicians about the other guy's area. The book by Donaldson (1958, 1959) stands as an almost bibliographic relic, to be highly cherished and kept by those who still hold a copy, for its content well offers a clear description of those instruments and their conceptual philosophy. That kind of bioengineering was, yes, a simple mixture, with little or no interaction of the parts, but nicely set the route of the biological stimulators and amplifiers plus other more or less successful gadgets. At one point, Biological and Medical Electronics popped shyly up as a *multidiscipline*, including some physics and math, too, but not much, for it took considerable time to add more elaborated methods (Valentinuzzi, 2014; Valentinuzzi and Kohen, 2013). No doubt, much water has run under the bridge, as wisely stated by an old motto.

Slowly but steadily, the relationship became wider and deeper, as a kind of integrative cooking action (Fig. 1.1). The informational content increased and the concept of *interdiscipline* came up, because there is free information transfer from one discipline to the other, and as in dancing parties, girls and boys integration takes place, or as in stews, each ingredient during the cooking process delivers its own set of components, with a deeper effect than in the simple former mixture.

Foods, meals, and their associated numerical caloric values recognize equipment, procedures, theoretical and experimental models, observations, in both animals and clinical tests, involved in obtaining the values that a dietician or nutritionist needs to recommend a given diet within safety margins. The dietician may not realize, but this is Bioengineering, too, in the end.

Figure 1.1: Bioengineer, as a cook, mixes and combines several ingredients to come out with a potent product, as a goulash to face a hard Nordic winter night. Drawn by Gustavo Idemi.

New disciplines are being generated because interactions get deeper and stronger so that *transdiscipline* appeared as another word, in an evolving phenomenon somewhat similar to the transgenic organisms (transgenic rats, transgenic seeds), holding in their germinal line an exogenous deoxyribonucleic acid (DNA) experimentally introduced. Thus, several bioengineering areas are now on the scene, such as nanoscience, neuroengineering, biorobotics, and biomechatronics, the latter attempting to reach human biology with the idea of assisting or even replacing members, senses, or injured organs caused by trauma, disease, or congenital defects. Closer we get to the science fictional Bionic Man or Woman leaving quite behind the modest XIXth Century Frankenstein Monster. And all this transgenic science, or transgenic engineering, is *transdiscipline*! How much of such turmoil will remain . . . we do not know, but exciting it is, no doubt.

The word above — *transdiscipline* — brings us to the prefix *trans-* or *tran-*, from Latin, which means *across, beyond,* or *crossing,* according to the Collins English Dictionary (2009). Hence, the act of translating refers usually to rendering from one language into another, as a kind of transformation. Its adjective — *translational* — qualifies a given activity or discipline, as *translational research*. The latter, for example, and getting now closer to our area, helps to make findings from basic science useful for practical applications that enhance human health and well-being. It is practiced in the medical, behavioral, and social sciences. In medicine, it is used to "translate" findings in basic research quickly into medical practice and meaningful health outcomes. Applying knowledge from basic science often is a major stumbling block. *Translational research* is another term for *translative* research and *translational science*, while *translational medicine* is research that aims at moving "from bench to bedside" or from laboratory experiments through clinical trials to point-of-care patient applications (i.e., near the site of patient's care). Following this trend of new terms, there is a journal — the *Journal of Translational Engineering in Health and Medicine* (JTEHM) — launched by IEEE/EMBS, which supports the movement of biotechnological innovations from idea to clinical trials and ultimately to commercialization; it also fosters global conversation and interdisciplinary collaborations on translational research. It intends to bridge the clinical and engineering worlds, along with a variety of interactive contents, all aimed at a better global healthcare, and dealing with Medical Devices, Nano-biosensors and Biosystems, Medical Imaging, Cardiovascular, Neurovascular and Rehabilitation Devices, Surgical and Interventional Devices, Bio and Surgical Robotics, Wearable Sensors and Health Monitoring, Wireless and Communication Technologies for Bio-Information, Electronic Medical Records, Global Health, Harsh Environments, Social Determinants of Health, Law, Ethics and Policy, all appealing emerging areas, indeed. How many out of this list will survive? How many even newer will appear? We do not know, and it is not easy to foresee, perhaps in a way resembling a kind of survival of the fittest and of natural selection, as introduced in biology by Charles Darwin in the XIX century (Darwin, 1859) after his long and famous voyage (1831–1836) on board of His/Her Majesty Ship *HMS Beagle*, under the orders of Vice-Admiral Robert FitzRoy (1805–1865).

1.4 Frames of Reference and Humor

Two women meet at the supermarket. One looks cheerful, the other depressed. The former said: "What's eating you? Nothing's eating me. Death in the family? No, God forbid! Worried about money? No ... no ... Trouble with the kids? Well, sort of, it's my little Jimmy. What's wrong with him? Nothing really, his teacher advised me to see a psychiatrist. PAUSE. Well ... what's wrong with that? Nothing ... just the psychiatrist says he's got an Oedipus complex. ANOTHER PAUSE. Well ... Oedipus, Shmoedipus, what the heck, I wouldn't worry so long as he's a good boy and loves his momma"

In the days of the Austro-Hungarian Empire, a dashing but penniless young officer tried to obtain the favors of a beautiful courtesan during a dance in the romantic Vienna palace, both caressed by an orchestra's charming waltz. To shake off this unwanted suitor, she coldly said that her heart was not free. Recovering from the unexpected shock, the chap replied politely: "Mademoiselle, I never aimed as high as that". Both stories from Koestler (1964).

The creative act of the humorist consists in bringing about a momentary fusion between two habitually incompatible matrices. In the little first story above, the cheerful woman's statement is ruled by the logic of common sense: if Jimmy is a good boy and loves his momma, there cannot be much wrong. But in the context of Freudian psychiatry, the relationship to the mother carries entirely different associations (Fig. 1.2). The second story does not require an explanation for it is too commonplace and is good to tell during a social evening.

Scientific discovery can be described, in very similar terms, as the permanent fusion of matrices of thought previously believed to be incompatible. A problem, when looked at from two or more points of view, may be understood and solved for more easily. One reference frame may be, say, strictly biological, while another may be purely mathematical. They tend to enlighten each other so favoring the intellectual act of creation. It is like taking photographs of a landscape from different positions. Some may show aspects that others may not. The creative act, by connecting previously unrelated dimensions of experience, enables the scientist to attain a higher level of mental evolution. It is an act of liberation, it is aesthetic, the defeat

Figure 1.2: Alice and Martha discussing Jimmy's Oedipus Complex. Good collision of two reference frames. Drawn by Gustavo Idemi.

of the habit by originality. This is a theory put forward by Arthur Koestler — an Austrian journalist, also philosopher and thinker — in one of his several books (Koestler, 1964). In his Theory of Humor, Koestler states that a joke would lead to laugh when two reference frames collide. An expert joke-teller should carefully prepare the first frame to suddenly exit into a much unexpected situation, causing the collision and, thus, triggering laughter; in short, the stronger the collision, the higher the laugh intensity. Bioengineering, by its very definition and nature, favors all kinds and sorts of intellectual collisions. It brings together the improbable and the apparently unrelated, biology and mathematics, physiology and physics, medicine and engineering. Thus, through such cross-fertilization, it supplies a good culture broth or scenario for comprehension, discovery and invention. Good light from the *Lampas Poesis* this is, indeed, so much that somehow, even music gets entangled in such view (Valentinuzzi, 2013; Valentinuzzi *et al.*, 2015).

Hence, regular seminars and meetings organized with people from different disciplines appear as fertile, powerful ground for the generation of new ideas, such as stews — a goulash, for example — full of nutrients, calories, and spicy taste (Fig. 1.1).

1.5 Guiding Philosophy

A map, a router, a GPS, are good helpers to find the right road... even a good sniffing.

This book follows the following simple philosophy: Our source is the biological system (be it a vegetal, a whole animal, the human being, an isolated organ, a single cell, or even an organelle). We must get to know it fully, and Biology and Physiology at large are absolutely necessary. This is why most of its chapters are mainly devoted to the Physiological Systems, although using an approach that differs from the traditional one found in the traditional physiology textbooks. Their respective authors (listed in the front page) cover a range of wide knowledge and expertise, and each has put the best of his/her previous years of dedicated research. The advice given to the readers, obviously already aware of it for they are already in this *metier*, insists on stating:

> *Know the ground thou set thy foot... do not forget it, even if being carried on by technology.*

Let us recall that ever since the appearance of man on the surface of the Earth, diseases along with natural phenomena and catastrophes have interfered with his life. Sorcerers, witch-doctors, and their witchery were the only way to fight them off. As knowledge slowly and painfully advanced, the former practice freely interlaced with charlatanism or quackery. In between, we find the "old mothers", "barefoot doctors", medicine men, and healers of all sorts — many times applying surprisingly good and efficient concoctions or procedures, so indicating keen observational abilities — and still acting all over the world. Supernatural mysterious actions have always attracted people, so that, do not be surprised, because,

> *From the witch-doctor to the biomedical engineer, when suffering hits, people seek relief wherever and from any hand: From the real and from the illusion, from the truth and from lies, in spells and in magic, in*

praying and faith. They're hopes and beliefs of so many that few, very few, just barely are able to grasp.

The latter words need to be remembered by all professionals involved in health delivery, and with due respect, not flippantly or irreverently because they refer to human pain. Understanding is required from one side while simple clear teaching is a must. All these concepts are part of ethics and empathy.

1.6 The Recording Channel Extended

Verba volant, scripta manent... even better if the script is on a chip or in a cell

One of the objectives of Bioengineering is to bring quantification to the biological and biomedical sciences. Above we underlined the need to measure, now we add, *we need to record what is measured,* and this simple concept has been emphasized often either in research or in teaching activities (Valentinuzzi, 2015a,b). Johannes Gutenberg's printing machine (1398–1468) with movable types represented an essential step forward in the advancement of culture for it permitted permanent records of the human thought. Before 1847, the physiological message given off by an animal or by man himself could not be graphically registered and Carl Ludwig (1816–1895) introduced the first machine to do that (Valentinuzzi, Beneke, and González, 2012). Before, Stephen Hales (1677–1761) had to describe in words the ups and downs of the blood column (as it slowly and blandly coagulated), when he measured arterial blood pressure for the first time, back in 1728. Later, James Hope (1801–1841), in 1830, called for the presence of witnesses to certify (thus, "record") his description of the heart sounds, to prove their valvular nature during experiments made in donkeys.

To pick up biological signals is essential because they carry, if not all, most of the information needed to understand the behavior of the system and, eventually, to take a decision (say, a therapy). Thus, the *recording channel* appears as a second system (a technological one) to implement, coupled to the biological system; *this book, as in its*

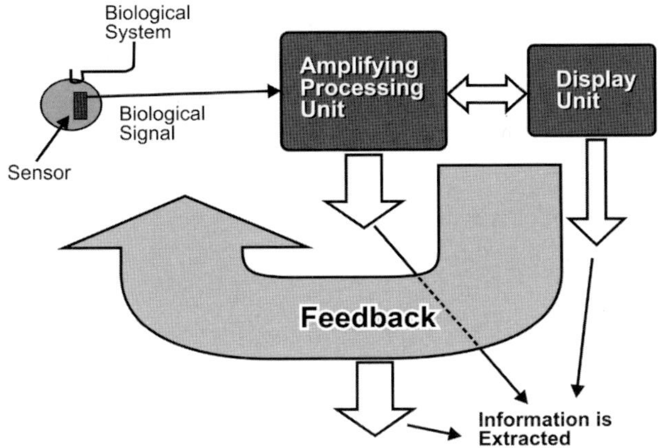

Figure 1.3: The recording channel.

predecessor version, still takes it as basic blueprint, but now refreshing, updating, and expanding the concept (Fig. 1.3). The biological system produces signals picked up by specific sensors, signals thereafter amplified processed and displayed. Eventually, feedback signals modify different input levels according to information obtained from other parts; however, several technological novelties can be introduced or are conceivable after relatively recent advancements and perhaps newer to come about. The link between the source and the recorder proper can be telemetric, as in the case of studies on free animals or ambulatory patients. Why not, then, digitally inscribing the record in a miniature memory to be later retrieved by a computer? The processor becomes much more than an amplifying unit because signals can be mathematically handled to extract hidden information or to produce evaluating parameters. The signal can be transmitted from inside the body by a miniature chip to an external receptor-processing unit. In fact, like an updated and by far more versatile old Meccano, which was invented in 1901, in England, by Frank Hornby (1863–1936), becoming the delight and perhaps inspiration of millions of children the world over, the bioengineer is only limited by his/her ingenuity and creativity.

The biological system produces signals that are picked up by specific sensors. They are amplified, processed, and displayed.

Eventually, feedback signals modify different input levels according to information obtained from the different outputs; however, several technological novelties are possible or are conceivable after relatively recent advancements.

There are very many biological signals; however, their nature comes always down to a few kinds: electrical, mechanical, chemical, and thermal. Accordingly, electrodes and transducers represent a key bottleneck, permanently opened to research and new developments. Most of the time, if not always, the signal requires amplification and processing. This basic unit needs to fulfill unique characteristics not covered by the amplifier usually applied in other fields. This is why the chapter devoted to the biological amplifier needed full revamp. After faithful display or store, we have to read the signal aiming at a meaningful interpretation, that is, and in simple words, analyzing the signal. The information produced by the analysis of the biological signals hopefully offers a variety of messages leading to one or more decisions. For example, the signal pick up needs improvement, signal processing has to be changed, stimulation of some kind is advisable to isolate a given response, a therapy is instituted, a relationship is uncovered or discovered, or novel evaluation parameters are defined. Interpretation of the signals, by means of from simple visual inspection to sophisticated techniques, is essential for the understanding of normal phenomena and diagnosis of pathologies. In other words, signals must be perused. In the end, the development of mathematical models leads to *prediction*, which is the highest level of quantification. Perhaps a mathematical model can be set, thus starting a fruitful feedback loop. Feedback is meant here in a very wide sense and context. As such, astrophysics or atomic physics are well ahead of any other discipline. Recall, for example, how accurately the movements and positions of astronomical phenomena can be anticipated. After all, what is a physician trying to do when he/she examines a patient? To determine what the disease is and, above all, to *predict its most probable course.* A veterinarian, an ecologist, and other biologically related activities take a similar stand. A still distant and well-yearned objective is to anticipate disease, as much in advance and as much quantifiably as possible, based on the current known condition of a given individual. For the time being, even with the tools nowadays at hand of the medical profession, that prediction is still far from being exact and precise. The most we can do is obtaining a risk probability, as for example in cardiology (Valentinuzzi *et al.*, 2013). Genetics and the sequencing of the Human Genome in 2000 certainly brought us closer to that possible prediction. The final chapter tries to put things together, looking into the Crystal Ball in a Nostradamus-like modern fashion and including, also some of those subjects that were forgotten.

1.7 Objectives of the Book

Set always the goal. It will be easier to find the way. Be a Pathfinder. Remember the sniffing dog above!

This introductory chapter is being closed establishing clearly the book objectives, so that the reader or the potential reader knows where he/she is getting into and what is to be found in the text:

- To revise critically what Bioengineering is about.
- To add new material, both in concepts and techniques.
- To give a historical feeling in some of the subjects.
- To stimulate one self's learning capacity.

Motivated work, some inspiration, a lot of sweat and, perhaps, some light from the *Seven Lamps*, would do the rest.

References

Collins English Dictionary — Complete & Unabridged. 10th ed. William Collins Sons & Co. Ltd. 1979, 1986; HarperCollins Publishers 1998, 2000, 2003, 2005, 2006, 2007, 2009. http://dictionary.reference.com/browse/trans.

Darwin C. The origin of species. Bantam edition published in 2008, Random House, New York, 495 pp., 1859.

Donaldson PEK. Electronic apparatus for biological research. Butterworths Scientific Publications, London, 718 pp., 1958 and 1959.

Koestler A. The act of creation. Pan Books Limited, London, 491 pp., 1964.

Smith DE. The poetry of mathematics. Scripta Mathematica, New York, 90 pp., 1947.

Valentinuzzi ME. Understanding the human machine: A primer to bioengineering. In: Series bioengineering & biomedical engineering, Vol 4, chap.8, pp. xii+396. World Scientific, New Jersey & Singapore, 2004. ISBN 981-238-930-X; ISBN 981-256-043-2.

Valentinuzzi ME. The seven lamps of bioengineering. *IEEE Pulse* 1(1):15–16, 2010.

Valentinuzzi ME. Bioingeniería: La ciencia de medir, definir, modelar, predecir... y cocinar. (In Spanish, Bioengineering: Science of Measuring, Defining, Modelling, Predicting... and Cooking.) 1st ed. Siglo Veintiuno Editors, Buenos Aires, 128, 2013. Serie Ciencia que Ladra, director Diego Golombek, ISBN 978-987-629-352-5.

Valentinuzzi ME. Máximo Valentinuzzi (1907–1985): Perhaps the first Latin American biophysicist, biomathematician and bioengineer. *IEEE Pulse* 5(3):66–75, 2014.

Valentinuzzi ME. The good old blackboard and chalk. *IEEE Pulse* 6(3):58–62, 2015a.

Valentinuzzi ME. Physiological records projected on a screen. *IEEE Pulse* 6(4):64–69, 2015b.

Valentinuzzi ME, Arini PD, Laciar E, Bonomini MP, Correa RO. Cardiac risk assessment: When and who? *IEEE Pulse* 4(4):38–48, 2013.

Valentinuzzi ME, Beneke K, González GE. Ludwig: The bioengineer, *IEEE Pulse*, 3(4):68–78, 2012.

Valentinuzzi ME, Kohen AJ. The mathematization of biology and medicine: Who, when, how? *IEEE Pulse* 4(1):50–56, 2013.

Valentinuzzi ME, López Celani NM, Hortt F, Idemi G. Música, Fisiología, Bioingeniería y Algo Más. (In Spanish, Music, Physiology, Bioengineering and Something Else.) Editorial Académica Española OmniScriptum GmbH & Co, Saarbrücken, Germany, 120, 2015. ISBN-10: 3659093602; ISBN-13: 978-3659093609.

CHAPTER 2

ONE ESSENTIAL SOURCE: ELECTROPHYSIOLOGY

Max E. Valentinuzzi and Alberto J. Kohen

Tiny short electric pulses run downwards and upwards signaling what to do and when.

Abstract

The basic concepts are briefly reviewed, for nerves, muscles, and heart (resting potential, action potential, and conduction velocity), as they are essential source of information. The ion channels and their most recent concepts are also covered succinctly since, for transport efficiency, channels have an advantage over carriers in that up to 100 million ions can pass through one open channel each second, a rate 10^5 times greater than the fastest rate of transport mediated by any known carrier protein. Besides, the relationship to the Maxwell and Kirchhoff equations is given as the strongest theoretical ground electrophysiology may have.

2.1 Basic Concepts

2.1.1 *Excitable tissues*

Nerves, skeletal, cardiac, and smooth muscles — all excitable tissues — arc characterized by a resting and stable electrical state, E_1. In fact, such a condition is so important that, when it disappears, it flags the death of the cell, offering an unmistakable criterion for the experimental researcher to decide whether his/her preparation is still viable (e.g., when working with microelectrodes). When an *adequate stimulus* is applied to the cell, say, at time t_0 having a finite duration $(t_1 - t_0)$, the response is a change in the electrical state to another level, E_2, remaining at it for a specific length of time, $t_d = (t_3 - t_2)$,

15

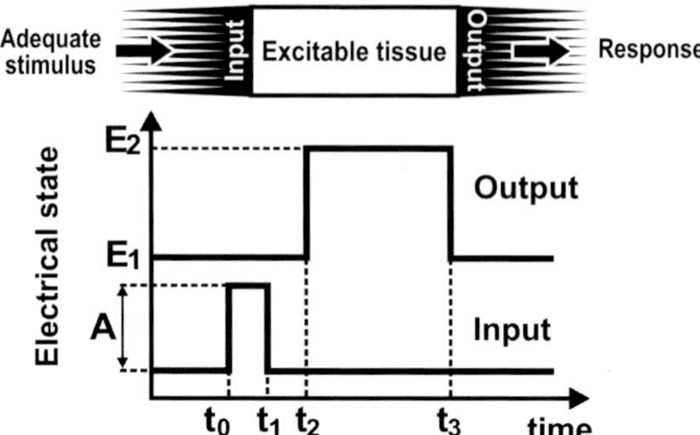

Figure 2.1: Idealization of an excitable tissue. An *adequate stimulus* applied to the cell, say, at time t_0, with a duration $(t_1 - t_0)$, triggers a change in the cell electrical state from E_1 to E_2, remaining at the latter for a specific duration, $t_d = (t_3 - t_2)$, which depends on the tissue. This behavior reminds of an electronic monostable multivibrator.

which depends on the tissue (Fig. 2.1). The time difference between t_2 and t_0 is called the *latency*, that is, it measures how long it takes the cell to react and it is also a characteristic of the tissue. The amplitude of the response, the *action potential*, is obviously the difference in the two electrical levels, E_2 and E_1. In the laboratory, the stimulus comes from a biological stimulator, whereas in the physiological situation, other action potentials act as stimuli (Weiss, 1901; Lapicque, 1909; Geddes and Bourland, 1985).

2.1.2 *Resting potential*

Most of the excitable cells are long and cylindrical in shape surrounded by a membrane, about 100 Å thick, which separates the intracellular fluid (ICF) and its contents from the extracellular fluid (ECF). When a microelectrode, connected to one of the inputs of a high impedance amplifier, is introduced into the ICF puncturing the thin cell membrane while the other terminal is hooked to a return electrode immersed in the ECF, the recording instrument shows a displacement of the base line from the zero level to about -80 to $-90\,\mathrm{mV}$. This is the resting membrane potential (the stable state), E_1. The internal side of the membrane is negative with respect to its external counterpart.

The ECF contains a high concentration of sodium ions and a low level of potassium ions. Conversely, the ICF shows a low level of sodium and a high concentration of potassium. Besides, chloride ions accompany the two previously mentioned ions. The membrane is relatively permeable to all these charged carriers; however, it does not permit the passage of large proteic anions, which abound within the cell. Moreover, the membrane is a good insulator constituted by oriented proteins and phospholipids, roughly containing one ion per 5000 water molecules. ECF and ICF, instead, have in the order of 1 ion per 175 water molecules, meaning that these fluids are by far better electrical conductors.

The ionic theory of excitation states: Within the cell membrane, an *active ionic pump* extrudes sodium from the cell and carries potassium into it. For the time being, let us assume that the ionic exchange is 1:1, or, for each sodium out, one potassium is brought in. It is *active* because the pump consumes energy supplied by the cell itself. Slowing it down or inactivating it tends to level concentrations off and, in the end, kills the cell. Skou (1957), in Denmark, discovered Na+/K+-ATPase, identified as the sodium–potassium pump; he shared in 1997 the Nobel Prize in Chemistry. This pump is an enzyme located in the plasma membrane of all animal cells, and hence, what originally was the first postulate of the theory became a fact.

Skou (1918–) completed his medical doctorate in 1954. His dissertation was published in Danish under the title *Local Anaesthetics*. In Aarhus, however, he did not have access to nerves from squid, and Skou decided to use nerves from crabs, instead. Crab nerves have the same properties as squid nerves because they also contain a protein substance with a high level of enzymatic activity. Unfortunately, they are not the same size, so a considerable number of crabs had to be used. He made an agreement with a fisherman and, in the following years, a distinct odour invaded the Department of Physiology, where approximately 20,000–25,000 crabs had to be boiled to study the threadlike fibers. *These studies led to Skou's description of the sodium–potassium pump: An ion-transporting enzyme (NaK-ATPase) that maintains the balance between sodium ions (Na^+) and potassium ions (K^+) in the cells.* Skou did not initially use the term *pump* about the enzyme. The paper published in 1957 and referred to above, was the article that 40 years later earned him the Nobel Prize (see http://www.iucr.org/people/nobel-prize/skou). Its file can be obtained from the INTERNET and even it has much later been reissued including enlightening comments by Dr. Skou himself (see http://jasn.asnjournals.org/content/9/11/2170.full.pdf). The prize was shared and it stated: *The Royal Swedish Academy of Sciences has decided to award the 1997 Nobel Prize in Chemistry with one half to Prof. Paul D. Boyer, University of California, Los Angeles, USA, and Dr. John E. Walker, Medical Research Council Laboratory of Molecular Biology, Cambridge, United Kingdom, for their elucidation of the enzymatic mechanism underlying the synthesis of adenosine triphosphate (ATP) and with one half to Prof. Jens C. Skou, Aarhus University, Denmark, for the first discovery of an ion-transporting enzyme, Na^+, K^+-ATPase* (see http://www.nobelprize.org/nobel_prizes/chemistry/laureates/1997/press.html).

The other postulate of the Ionic Theory (it should be now called its *First Postulate*) says: *At rest, the ratio between the membrane permeability's to sodium and potassium ions is much larger to the first ion than to the latter (about 50- to 75-fold)*. Due to the concentration gradients, there is a passive (with no energy expenditure) back diffusion of both ions; one tends to return sodium to the ICF and the other pushes potassium to the ECF. However, because of the large permeability difference, for each Na^+ getting back inside the cell, there are 50–75 K^+ leaving it, so accumulating positive charges on the ECF side, depleting the ICF of them and, hence, creating a charge imbalance with the outer face positive to the inner. Thus, the membrane resting potential, E_m, builds up, an event that obviously breaks neutrality (Fig. 2.2). Here, it is underlined what in a slightly different way has been said above. The reference by Heitler (to be found in the Web) is highly recommended because it proceeds slowly step by step allowing also modification of some parameters to visualize their respective effects (William J. Heitler, The Origin of the Resting Membrane Potential, School of Biology, University of Saint Andrews, UK; see http://www.st-andrews.ac.uk/~wjh/neurotut/mempot.swf).

The membrane is impermeable to the large proteic anions and they stay within the cell, but chloride can pass through it passively following the electric field sustained by the resting potential to reach an equilibrium state. Due to osmotic forces that depend on the concentrations of the osmotically active particles (largely the small ions), water also moves across the membrane so determining cell volume.

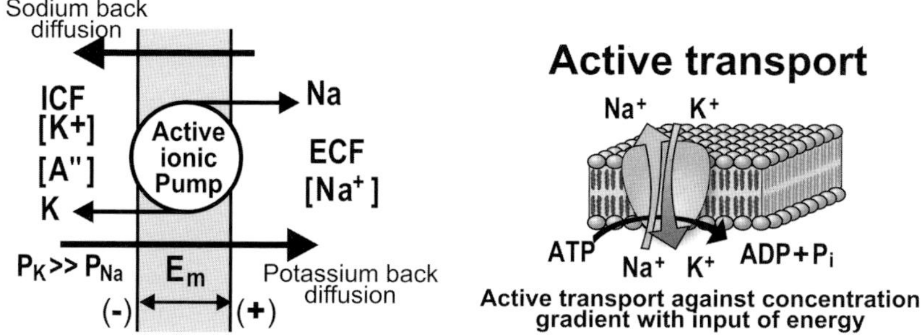

Figure 2.2: The sodium–potassium pump and membrane resting potential. Left: The two vertical lines represent, respectively, the internal (left) and external (right) sides of the cell membrane delimiting its thickness. The active ionic pump maintains the ionic concentration gradients (potassium: higher in ICF than in ECF; sodium: the opposite). Besides, in the resting condition, the *permeability of the membrane to potassium is much higher than its permeability to sodium*, thus, potassium back diffuses easier to the ECF than sodium does it to the ICF. As both carry positive electric charges, the net result is an accumulation of positive charge on the ECF side, breaking neutrality and so giving rise to the resting potential, E_m, positive outside and negative inside. Right: adenosine triphosphate (ATP) degrades to adenosine diphosphate (ADP) liberating a phosphate to pump sodium and potassium ions out and in, respectively. Typical concentrations on both sides of the membrane are: in ICF, 140 for K^+ and 14 for Na^+; in ECF, 5 for K^+ and 150 for 150 for Na^+, these figures expressed in mEq (or mM) per liter of water. Usually, for three sodium ions out, two potassium are pumped in, that is, there is a 3:2 ratio. A'' above at the left side stands for negative anions concentration. Redrawn by Gustavo Idemi.

In real life, very rarely, if ever, the pump is electrically neutral (1:1 ratio). The ionic exchange ratio varies from tissue to tissue and, even within the same tissue, histologically different fibers may show different ratios (typical example is the heart). In fact, the determination of the sodium–potassium pump exchange ratios is still a subject of research. Therefore, the pump contributes one portion (usually less than one-third) of E_m. Ionic pumps, which sustain potentials, are called *electrogenic*. In other words, *the resting membrane potential recognizes two sources, one (the largest) comes from the large difference in permeabilities via back diffusion, in turn dependent on concentration gradients sustained by the sodium–potassium pump, and the other (much smaller) directly produced by the pump due to a not neutral pumping sodium–potassium ratio, different from 1:1*

(Kimura, 2001; PhysiologyWeb, 2000–2013). In the latter reference, full details can be found regarding mechanisms and electric potential calculations.

Transfer of sodium and potassium through the membrane involves the hydrolysis of ATP to provide energy; it is a process in which the enzyme Na^+/K^+-ATPase plays a significant role moving usually three Na^+ to the outside and two K^+ ions to the inside. The charge imbalance contributes to the separation of charge across the membrane. The pump is called a P-type ion pump because the ATP interaction phosphorylates the transport protein and causes a change in its conformation (Skou, 1957).

2.1.3 *The action potential*

It was said above that an *adequate stimulus* applied to the cell triggers a response characterized by a change in the electrical stable state to another level, E_2, remaining at the second one for a specific length of time, t_d; thus, it can be called *metastable*. When charges are removed from the external side of the membrane at rest, i.e., if the membrane is *depolarized*, the potential shifts towards zero (Fig. 2.3). If depolarization continues, the decrease in E_m (less negative inside) eventually reaches a level called the *critical*, the *firing* or the *threshold potential*. At the latter level, the membrane takes command of its own depolarization and rapidly proceeds to a maximum, E_2, with reversed polarity, to thereafter and somewhat slower, returning to the initial resting state (Fig. 2.3). Beyond the critical or firing level, the applied stimulus loses control completely. Figure 2.4 shows an actual nerve action potential. Notice the conceptual difference, often mixed up in the daily jargon, between *critical* (or *firing*) *threshold potential* and *threshold stimulus*.

The first one is just the negative potential measured from the zero line (in the figure, it lies in the order of $-70\,\mathrm{mV}$), above the resting potential E_m (about $-90\,\mathrm{mV}$), while the latter is *defined as the threshold potential minus the membrane resting potential*, or $S_{th} = E_{th} - E_m$ which, using the numerical values stated above, would yield $(-70) - (-90) = 20\,\mathrm{mV}$. Hence, the applied voltage must

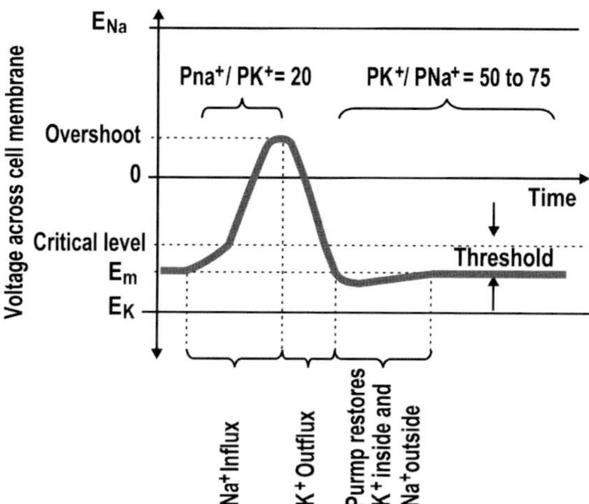

Figure 2.3: Action potential and its boundaries. The action potential is always bounded by the potassium potential (negative inside), near the membrane potential, and the sodium potential above (positive outside), which is a limiting unreachable value. Drawn by Gustavo Idemi.

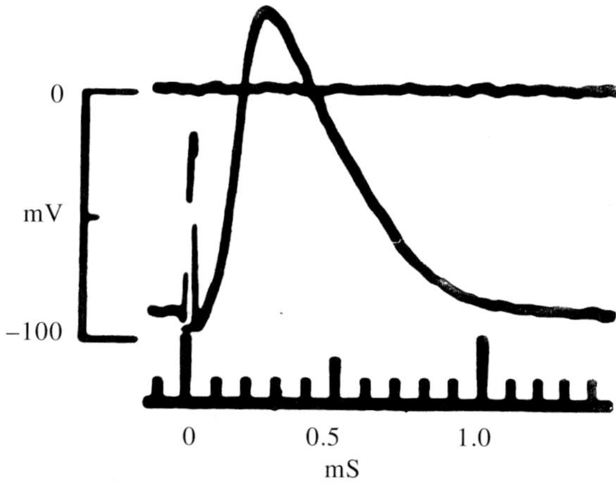

Figure 2.4: Actual action potential. Recorded with microelectrode from a cat dorsal root nerve fiber. Observe the stimulus artifact just before the depolarization upstroke, where the zero time is placed. The overall duration of the event is shorter than 1 ms. The positive overshoot appears clearly depicted. The resting potential lays around −90 mV. Reproduced from Ruch and Patton, Physiology and Biophysics, Copyright 1966, Chapter 2 by Walter Woodbury, Fig. 2.2C, p. 30, with permission from Elsevier.

overcome the difference separating the resting level from the threshold potential and, as it is applied via an internal microelectrode, it has to be a positive pulse.

A Few Exercises

Suppose the permeability ratio potassium to sodium is one-half of the stated value, would the resulting membrane potential increase, stay the same or decrease? Suppose both permeabilities were equal, what resting potential would result? Suppose the pumping rate drops to one-half of its normal value, would it affect the membrane potential? Suppose a toxin stops the pump, what would E_m be? Does the pump directly contribute to the resting membrane potential? For the latter question, consider a 1:1 pump ratio. Note: Do not confuse membrane ionic permeability ratio with ionic pumping ratio. This exercise is important for the full understanding of changes in membrane potential due to permeability changes.

Suppose the pump ratio changed to $2\mathrm{Na}^+/3\mathrm{K}^+$ and both permeabilities were equal, would the pump contribute to the membrane potential? If the answer were affirmative, which side would be positive to the other? Suppose now the ratio changed to $3\mathrm{Na}^+/2\mathrm{K}^+$, keeping always both permeabilities equal, which side would be positive?

An *adequate* stimulus was mentioned above, implying some conditions to be met by the stimulating rectangular pulse. Two conditions were introduced. There is a third, which is easily fulfilled by any electronic stimulator.

What important concepts are derived from the strength–duration curve? This experimental relationship was introduced first by Weiss (1901) and perfected in 1909 by Lapicque (1909).

The following link is a good one to visit, for it is didactic and it includes exercises: http://www2.derby.ac.uk/ostrich/intro_to_bio_psych/The%20Neuron,%20Neuronal%20Conduction%20&%20Synaptic%20Transmission/page_14.htm. Besides, the textbooks by Wickens (2009) and Pinel (2011) are also recommended. The latter author is with the University of British Columbia.

So far, we merely described the action potential. The *Ionic Theory* has what can now be considered its *Second Postulate*, i.e., *depolarization produces an increase in the membrane permeability to sodium ions*, which, due to the outside–inside concentration gradient, leads to an increase in sodium influx with more positive charges toward the internal face and more depolarization to the membrane. Recall that depolarization means either negative charges on the external side or positive on the inside. Beyond the critical level, however, this process is greatly and irreversibly accelerated, so that the external stimulating front is no longer needed because more depolarization brings about higher permeability to sodium with a rapid positive ionic entrance and even more depolarization. A positive feedback loop, called *Hodgkin activation cycle*, characterized by a *fast inward sodium current*, is established (Fig. 2.5). However, after a certain delay, the effect of depolarization on sodium permeability slows down and eventually there is a reversal while, simultaneously, the permeability of the membrane to potassium ions increases. Thus, less sodium gets into the cell and a rapid and steady potassium outflow takes place, both leading to depolarization slowing down and reversal to repolarization. The latter inactivation cycle is characterized by potassium extrusion (Hodgkin and Huxley, 1952). Actually, this was the last of a series of papers where the Ionic Theory was developed. A good overall reference is Concha (1997), from the University of Texas Medical School at Houston.

A previous experimental finding associated the action potential with a temporal membrane transverse impedance change. It was a superb biomedical engineering accomplishment carried out by Cole and Curtis (1939), with the cell membrane forming one arm of an impedance bridge and a well-designed vacuum tube electronics. As an impulse passed between the impedance electrodes after the axon had been stimulated at one end, the oscillograph line first broadened into a band, indicating a bridge unbalance, and then narrowed down to balance during recovery. Hodgkin and his group, later on, broke up the impedance change in its two sodium and potassium ionic components. Between Cole's experiments and Hodgkin's, World War II took place sadly forcing a long parenthesis in this line of research.

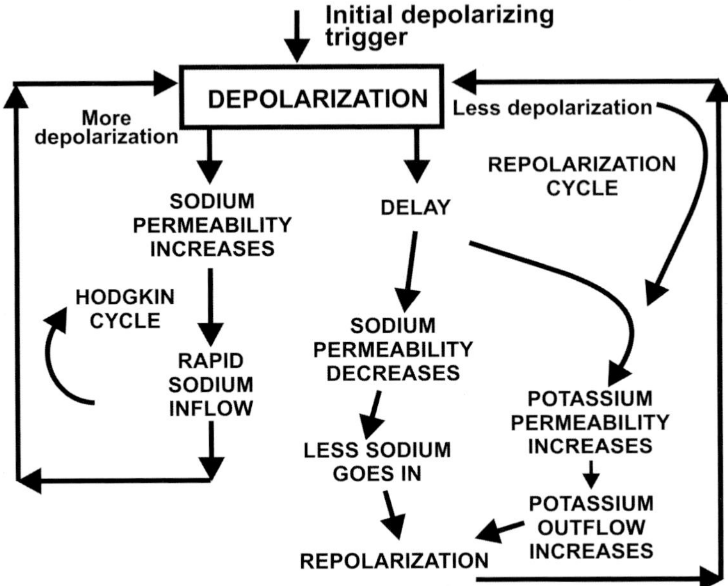

Figure 2.5: Hodgkin activation cycle. Depolarization produces an increase in the membrane permeability to sodium leading to a positive feedback loop. With some delay, depolarization starts an increase in the permeability to potassium so initiating repolarization.

Ionic membrane permeabilities to both ions represent main actors in the resting state and during excitation. Changes are rapid and dramatic, from stable potassium-to-sodium ratio of 50:75, depending on the tissue, to a full reversal of sodium to potassium in the order of 20, at the overshoot; in other words, *permeabilities act as true controllers*. Associated with these changes, there are ionic currents: very fast sodium one (depolarization) and a somewhat slower but also fast potassium efflux (repolarization). Ionic exchange has become an important physiological tool and biophysicists talk in terms of *ionic channels*. Whole conferences are being devoted to the subject discussing concepts such as channel gating, channel pore, channel associated proteins, assembly of channels, channel regulation, and ionic channel dysfunction associated with disease (PhysiologyWeb, 2000–2013).

2.1.4 *Quick Nernst equation refreshment*

The mathematics involved in the Nernst equation was developed in detail in the previous version of this book (Valentinuzzi, 2004). When such an equation is applied to the potassium and sodium concentrations found in the ECF and ICF compartments, the following typical values can be calculated out:

$$E_{K^+} = RT/F \times \ln[K_o]/[K_i] = 61.5 \log_{10} 31 = -93\text{mV}$$

$$E_{Na^+} = RT/F \times \ln[Na_o]/[Na_i] = 61.5 \log_{10}(1/24) = 84\text{mV}$$

$$(2.1)$$

The two latter describe numerically the potassium and sodium equilibrium Nernst–Gibbs–Donnan potentials, if the system were a passive one dominated either by the first or by the second ion. Hence, they are only theoretical potentials impossible to coexist. They are different and of different signs. In the resting state, the membrane potential, E_m, is rather close to the potassium potential, E_K, say, $-90\,\text{mV}$ versus $-92\,\text{mV}$, or thereabouts, depending on the specific tissue. Thus, it is said that the resting membrane potential is dominated by potassium. On the other hand, the sodium potential is way up (about $+84\,\text{mV}$). The positive overshoot (in the order of $+20\,\text{mV}$) tends to the sodium potential but it falls rather short of reaching it. As stated before, *both theoretical equilibrium potentials mark the lower and upper limits of the action potential operating band.*

Exercises

The reader might check the numerical values of the equilibrium potentials applying concentrations quoted in the literature for nerve and muscle tissues (PhysiologyWeb, 2000–2013). The temperature is, by and large, body temperature, but other physiological values are valid as well (as in amphibians or *in vitro* preparations). Many experiments have been reported modifying sodium and potassium concentrations and measuring the possible influence on the resting potential and on the overall amplitude of the action potential. Consider the following situations:

If the concentration of potassium ions were increased in the ECF,

(a) Would E_m change?
(b) If the answer above is yes, in what direction?
(c) Is the amplitude of the action potential affected?

If the concentration of potassium were decreased in the ECF,

(a) Would the resting membrane potential change?
(b) Would the stimulus threshold be different?
(c) If the answer is yes, would a larger or smaller stimulus be needed?

If the concentration of sodium were decreased in the ECF,

(a) Would the resting potential be affected?
(b) Would the action potential overshoot change?
(c) If the answer is yes, would it be smaller or larger?

Actually, the Nernst equation does not describe neither the resting nor the maximum activity of the membrane potential. A modified empirical version was proposed by Hodgkin's group, called the *Goldman–Hodgkin–Katz* or the *GHK equation*, which takes into account both permeabilities and ionic concentrations and, thus, it is applicable at any moment of the action potential,

$$E_m = -61.5 \, \log_{10} \frac{[\text{K}^+]_i + \text{P}_{\text{Na}}/\text{P}_{\text{K}}[\text{Na}^+]_i}{[\text{K}^+]_o + \text{P}_{\text{Na}}/\text{P}_{\text{K}}[\text{Na}^+]_o} \tag{2.2}$$

It is illustrative to insert actual numbers into the equation in order to test it and see how the different components affect the membrane potential. A better exercise is to plot the membrane potential as a function of each parameter in the equation keeping the others at fixed values (PhysiologyWeb, 2000–2013).

2.1.5 *Electric pulse conduction*

Excitability *per se* is a local phenomenon, but of limited use for an organism. The characteristic that really enhances bioelectricity to

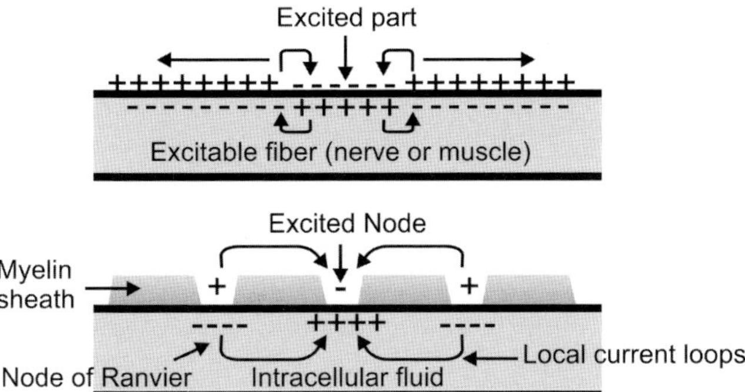

Figure 2.6: Propagation of the action potential. Schematic representation of a longitudinal section. Upper section — An excited portion acts as the stimulus to its adjacent right and left sections via the generation of depolarizing local current loops from positive to negative (curved arrows). This mechanism repeats itself to both sides resulting in the propagation of an action potential to the right and another to the left. By the time the two pulses reach their respective ends, the originally excited part is already recovered and ready to accept a second stimulus, if any. Lower section — It shows an important type of nerve fibers, covered by discontinuous sleeves of an insulating material called myelin. The cell membrane shows up only in-between, at 1μm gaps, the nodes of Ranvier. The local current loops only can be established between the excited one and the two immediate non-depolarized neighbors. Hence, there is some kind of jump (*saltatory conduction*). Drawn by Gustavo Idemi.

its full power is the propagation (or conduction) of the action potential to adjacent parts of the excitable tissue and to areas distant from the original stimulus, traversing complex networks, and eliciting responses essential for life. Once a given portion of an excitable fiber is fully depolarized to maximum activity (sign reversal), small potential gradients appear between the depolarized region and the yet resting neighbor areas to each direction, creating local current loops which depolarize them and, thus, generate new action potentials, one to the left and another to the right (Fig. 2.6, upper part). This is the basic *smooth conduction* mechanism in nerves and muscles.

A discontinuous sheath of an electrically insulating fat-like substance called myelin, however, covers an important part of nerve fibers. It forms sleeves approximately $1 - 2\,\mathrm{mm}$ long and about 2 μm thick. In between, the cell membrane shows up, at the nodes of Ranvier. They are only 1 μm wide. Hence, electrical charges concentrate around them, both outside and inside the cell. The electrical resistance across myelin is much higher than across the naked membrane (about 200 times). Conversely, the electrical capacitance across the membrane is about 200 times that across the sheath. Figure 2.6 (lower part) illustrates the conduction mechanism, similar to the one explained before, except that the current local loops "jump" or "leap" from node to node, for which reason it is named *saltatory conduction* (from Latin, jumping). It has been demonstrated that a myelinated nerve can elicit a response only when the stimulus is applied to a node. If a node is blocked (as with cold or with an anesthetic), the local depolarizing loop can reach the next adjacent one and still propagate the action potential, but if two nodes are blocked, then propagation cannot proceed. The velocity of propagation greatly increases by the leaping trick. The velocity of conduction of the action potential, in any excitable tissue, is one aspect of concern with definite patho-physiological and clinical implications.

Nerve conduction studies in combination with needle electromyography are crucial to understanding neuropathies. Detection of demyelization (conduction blocks), for example, leads to the diagnosis of multifocal neuropathy and to proper treatment. Such studies provide essential information on the spatial pattern of the neuropathy, can distinguish between primarily axonal and demyelinating pathology, and the severity of the damage. Neuropathies can be generalized, focal, or multifocal, symmetric or asymmetric, distally predominant or proximal and distal. Primarily axonal neuropathies mainly affect sensory nerve and compound muscle action potential amplitudes, whereas demyelinating neuropathies lead to slowing of nerve conduction and to increased temporal dispersion or conduction block (Franssen and van den Bergh, 2006; Padua *et al.*, 2007; Dhavalikar *et al.*, 2009; Chi *et al.*, 2010; Huynh and Kiernan, 2011).

2.1.6 *And what about the electrical behavior of the heart?*

Cardiac cells differ from skeletal myofibers in that cardiac myocytes have as a rule only one nucleus per cell. Myocardial fibers are shorter than the fibers of skeletal muscle and rarely exceed 100 μm in length,

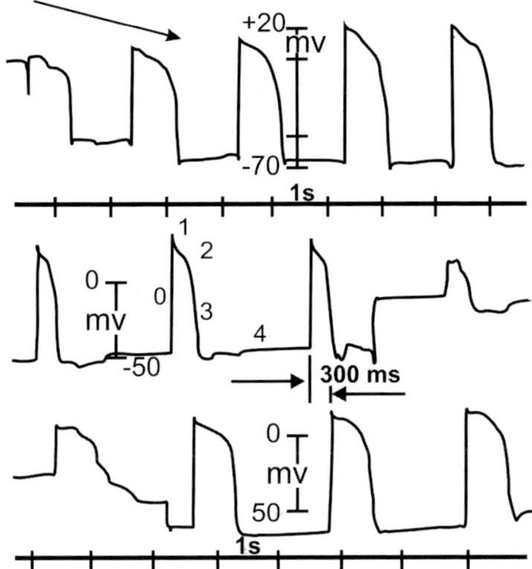

Figure 2.7: Cardiac action potentials. The upper channel comes from a floating glass microelectrode piercing the cell membrane of a ventricular frog ventricle. Slanted arrow above on the left indicates the piercing process, with a first beat when the electrode was on the surface, and a second beat with the electrode almost inside. The three last beats display full action potentials. Diastolic potential was approximately −70 mV with an overshoot of +20 mV; thus, the total amplitude was 90 mV. Lower two records: Two floating glass microelectrodes recorded simultaneously from the left atrium (above) and from the ventricle (below). The time difference of about 300 ms measures essentially the A–V delay. After the third beat (atrial channel), the electrode came off. The lowest record shows also penetration of the electrode on the left side. Adapted from Max E. Valentinuzzi, in 1967, at the Department of Physiology of Baylor College of Medicine. Redrawn by Gustavo Idemi.

with a diameter in the order of 30–50 μm. The shape is rather irregular, showing projections to contact other cells in such a way that can be described as an *anastomosing network*. Such arrangement, very likely, tends to facilitate the propagation of the electric impulse. The electrical phenomena associated with cardiac activity find their background in the general concepts of electrophysiology, but they show a number of peculiarities that have established *cardiac electrophysiology* as a separate chapter. As any excitable tissue, myocytes have a resting membrane potential and, when stimulated, trigger an action potential. Figure 2.7 shows atrial and ventricular action potentials recorded with microelectrodes. Roughly, these action potentials look like distorted rectangles and, in principle, have been modeled by a rectangular pulse. The two nodes (S–A and A–V), however, greatly deviate from that shape, displaying smooth rounded waveforms. All show a depolarization upstroke (phase 0, a standardized denomination in cardiac electrophysiology (Fig. 2.7, atrial channel), from a minimum (the resting or diastolic potential) to a maximum (the overshoot). The overshoot is not present in nodal fibers. The first and short repolarization return is usually called phase 1. Many papers are available to explain it. The overshoot is followed by a plateau (phase 2), well marked in Purkinje and ventricular fibers and slowly falling off in atrial and His bundle fibers. It is as if the repolarization process were somewhat held on for a given time; thereafter, repolarization speeds up (phase 3) until the membrane returns to the initial resting state (phase 4). The plateau does not appear in nodal fibers. Nerve and skeletal muscle action potentials are always short in duration, usually less than 1 ms. Cardiac action potentials, instead, last between 200 and 400 ms, depending on the species and on the type of fiber. In other words, it is a very long metastable period. The duration of the contraction elicited by this action potential is of the same order of magnitude. Conversely, in skeletal muscle a very short action potential triggers a longer contraction.

The action potential (AP) is generated by transport of ions through transmembrane ion channels. Rate dependence of AP repolarization is a fundamental property of cardiac cells, and its modification by disease or drugs can lead to fatal arrhythmias. Using a computational biology approach, the gating kinetics of the rapid (I_{Kr}) and slow (I_{Ks}) K^+ currents during the AP have been investigated trying to provide insights into the molecular basis of their role. Results show that I_{Kr} intensifies during the late AP plateau by progressively recovering from inactivation and generating a pronounced late peak of open-state occupancy. The delayed peak makes I_{Kr} an effective determinant of AP repolarization. I_{Ks} builds an available reserve of channels in closed states near the open state that can open rapidly to generate current during the AP repolarization phase. By doing so, I_{Ks} can provide repolarizing current when other currents

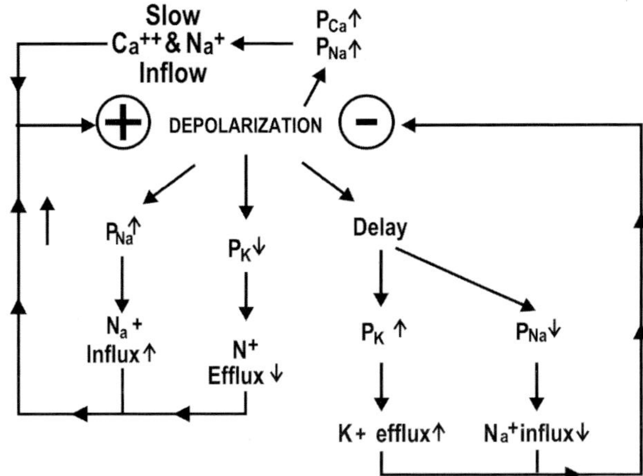

Figure 2.8: Noble's activation cycle. Compare this figure with Hodgkin's activation cycle (Fig. 2.5) for nerve and skeletal muscle. The long duration of the cardiac action potential is due to the effects of Ca^{2+} and Na^+ getting into the myocyte, but mainly to the former.

are compromised by disease or drugs, thus preventing excessive AP prolongation and arrhythmic activity (Yoram, 2008; Grant, 2009).

2.1.7 *Noble's cycle*

The activation process in heart cells is somewhat different from in either nerve or skeletal muscle. Figure 2.8 modifies Hodgkin's cycle and summarizes the chain of events (Noble, 1962). Depolarization, after started, increases the permeability to sodium ions and simultaneously decreases the permeability to potassium. Due to the existing concentration gradients maintained by the sodium/potassium pump, an increment in Na^+ influx and a decrement in K^+ efflux take place, both events so reinforcing depolarization. Thus, a positive feedback loop is established leading to the fast depolarizing upstroke (phase 0). Slower Ca^{2+} and Na^+ channels due to increases in their respective permeabilities allow entrance of these ions into the cell and, hence, as they carry positive charges, depolarization is held on (plateau or phase 2). After some delay, the effect of depolarization on the permeabilities reverses leading to the repolarization process in a braking negative loop. Noble did not consider these latter calcium and sodium channels in his first 1962 paper. They were reported years later, by McAllister *et al.* who developed a Purkinje fiber model in 1975 and which, thereafter, led to the Beeler–Reuter mammalian ventricular model in 1977. Finally, Di Francesco and Noble (1985) constructed a new and now well-recognized model of cardiac electrical activity. Thus, experimentally based models of the heart emerged in 1960, starting with the discovery of potassium channels as extensions of the Hodgkin–Huxley nerve impulse equations. The first models, including calcium balance and signaling, came up in the 1980s and reached a high degree of detail. During the 1990s, such cell models were incorporated into tissues, which led to the first virtual organ, the Virtual Heart. As a product of interaction between simulation and experiment, these models are now sufficiently refined to be useful in drug development (Noble, 2007). The latter paper well reviews the subject.

2.1.8 *Smooth muscles*

The electrophysiology of this kind of tissue is still a rather difficult and fuzzy subject because of experimental challenges, the variety of tissue types (alimentary canal, circulatory vessels, respiratory system, and urogenital apparatus), and probably, too, because of the relative scarcity of researchers devoted to this area as compared to other physiology areas. Nonetheless, and more specifically, the role of gastrointestinal electrophysiology in relation to digestive abnormalities and the ability of managing those diseases via electrical stimulation have been an area of research for the past decade or so (Du and Wu, 2009).

Normal gastrointestinal motility results from coordinated smooth muscle contractions originating in two patterns of electrical activity, i.e., *slow waves* and *spike potentials*. As other excitable tissues, gastrointestinal smooth muscle cells maintain a resting membrane potential between -50 and -60 mV, that is, somewhat lower in absolute value than the usual -80 to $-90\,\mathrm{mV}$ found in either skeletal muscle or nerve cells. Besides, this membrane potential *fluctuates spontaneously.* As these cells are electrically coupled, such fluctuating potentials spread to adjacent sections, leading to what are called

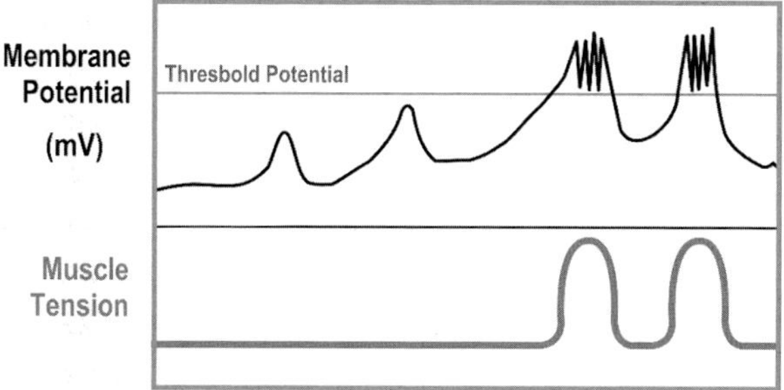

Figure 2.9: Intestinal activity. Upper trace: slow electric waves (left) and faster action potentials (right), expressed in mV. Lower trace: contractions, expressed in grams, triggered by action potentials. Slow waves do not elicit mechanical activity. Drawn by Gustavo Idemi, from Internet.

"slow waves" (waves of partial depolarization that travel along the digestive tube for long distances). These partial oscillating depolarizations are in the order of 5–15 mV. Moreover, the frequency of the slow waves depends on the section of the digestive tube. The small intestine shows rates of 10–20 times/minute, while in the stomach and large intestine, they are slower (3–8 times/minute). Slow wave activity does not seem to depend on nervous stimuli. To underline: *Slow waves do not elicit contractions*, they synchronize contractions in the gut by controlling the appearance of a second type of depolarization, spike potentials, which take place only at the crests of slow waves (Bowen, 1996).

Spike potentials are true action potentials that elicit muscle contraction (Fig. 2.9). They result when a slow wave passes over an area of smooth muscle previously exposed, primarily to neurotransmitters released by neurons of the enteric nervous system. These neurotransmitters appear as response to a variety of local stimuli, such as mechanical distension of the digestive tube wall. When a bolus of food gets into the small intestine, the following events occur:

Walls stretch, stretching stimulates nerves in the wall that release neurotransmitters, the membrane potential of that section of muscle becomes more depolarized, when a slow wave passes over this area of sensitized smooth muscle, spike potentials form and contraction results, the contraction moves around and along the gut since muscle cells are electrically coupled via gap junctions.

Contraction or relaxation of arterial smooth muscles determines their lumen, hence controlling blood flow and so playing a role in systemic blood pressure. The contraction process is mainly regulated by cytosolic (or ICF) calcium concentration ($[Ca^{2+}]_{cyt}$), which depends on several ionic channels. Such channels are intermediates in signal transduction mechanisms activated by vasoactive hormones to act on vasoconstriction/dilatation. Furthermore, ion channels are often the target of therapeutic agents (e.g., calcium blockers to induce vasodilatation and lower blood pressure).

The so denominated Kv7 voltage-activated potassium channels seem to intervene in the regulation of smooth muscle contraction (Brueggemann *et al.*, 2012). Elucidation of the roles of these channels in both physiological signal transduction and in the actions of therapeutic agents requires knowledge of their modulation activity at the cellular level as well as to evaluate their contribution in intact arteries. Potassium channels classification usually follow the primary source of energy for their gating (i.e., transition from closed to open). Three most common classes have been described, i.e.,

- voltage-gated channels,
- ligand-gated channels, and
- mechanosensitive channels.

Many K^+ channels are *voltage-gated* and are being divided into several K_v subfamilies, designated as $K_v1.x$ to $K_v12.x$ (Gutman *et al.*, 2003; Xiong *et al.*, 2007). A comprehensive view of the subject, although relatively old, is to be found in Bolton (1989). The current notation referring to these channels is complex and not easy for the uninitiated. For example, potassium voltage-gated channel subfamily KQT-member 4, also known as voltage-gated potassium channel subunit $K_v7.4$, is a protein that in humans is encoded by the KCNQ4 gene (Gutman *et al.*, 2005). The protein encoded by this gene forms a potassium channel thought to play a critical role in the regulation of neuronal excitability, particularly in sensory cells of the cochlea. *Ligand-gated channels*, instead, are a group of transmembrane ion channel proteins, which open to allow ions to pass through the membrane in response to the binding of a chemical messenger (a ligand). *Mechanosensitive channels* are membrane proteins capable of responding to mechanical stress over a wide dynamic range of external mechanical stimuli. They appear in prokaryotes and eukaryotes.

Magnetic methods for assessing the electrical activity of the gastrointestinal tract have been tried with good results. While the cutaneous electrogastrogram may reveal the frequency dynamics of gastric electrical activity, other parameters are not available from these records. Recent studies on the electroenterogram are promising, but low-conductivity abdominal layers complicate the identification of small electrical rhythms. The magnetogastrogram (MGG) and magnetoenterogram (MENG) are able to characterize gastric and intestinal electrical activity noninvasively in terms of its frequency, power, and propagation characteristics. Superconducting QUantum Interference Device (SQUID) magnetometers can detect the minute magnetic fields associated with electrical activity of the gastrointestinal syncytium. Pathological conditions appear in MGG and MENG signals, showing clinical promise (Bradshaw, 2004; Cordova *et al.*, 2004).

2.2 Ion Channels and Ion Selectivity

Do you remember "Maxwell's Demon"? It opens a gate to allow only the faster than average molecules to flow through to one side of a compartment, and only the slower than average molecules to slip to the other side, causing the first side to gradually heat up while the other side cools down, thus decreasing entropy. Well... not any more fictional! It exists!
Rather often, too, other kind of human demons pop up... unfortunately!

2.2.1 *Ion channels*

The sections above somehow anticipated what is dealt with here in more detail regarding this complex topic. Repetition is unavoidable, but it is helpful from a didactic viewpoint. Specific cations are essential for many biological processes during which, for many of them, protein actions discriminate between different ionic species, indeed, a fascinating mechanism because it calls for delicate balance of strong interactions. Identification and quantification of microscopic factors are not easy, and as several others, they are not amenable to experimental measurement. Theory and computation can provide virtual routes to help in the task, though. It represents a biophysical and biochemical area, greatly assisted by bioengineering that has shown very important advancements and where efforts in many laboratories are being invested (Rasband, 2010; Roux *et al.*, 2011; Lin *et al.*, 2013).

Main actors are potassium, sodium, and calcium, all of them carrying positive charges, in close and complex relationship somehow mastering many functions of the living systems, with selectivity — a highly complex issue yet — playing a significant role. Thus, it requires searching in supposedly simpler structures, as the case is with potassium ($\underline{K}$) $\underline{c}$rystallographically $\underline{s}$ited $\underline{A}$ctivation channel (KcsA), a prokaryotic potassium ion channel found in the soil bacteria *Streptomyces lividans*, activated by changes in pH; it became the most extensively studied K-channel and is widely used as template in research. There are figures in the literature intending to depict the probable mechanisms, as some kind of a selectivity filter at the sub-angstrom level. For structural and free energy reasons differences, smaller sodium ions cannot get through (when the sodium–potassium free energy difference is positive, selectivity to K^+ occurs, and the opposite takes place when such a difference is negative). Notice the thermodynamics involved, so pointing out to the importance of local energetics in flexible binding sites.

We should recall that in Thermodynamics, the energy in a system that can be converted into work is called free energy (FE). In particular,

- Helmholtz FE, or $A = U - TS$, is the energy that can be converted into work at a constant temperature and volume, expressed in joules or ergs, U stands for internal energy of the system, T for absolute temperature in Kelvin, and S for entropy, in joules or ergs/K;
- Gibbs FE, or $G = H - TS$, is the energy that can be converted into work at isothermic and isobaric conditions throughout a system (typical in biological systems, and is the energy referred to

previously), with H representing enthalpy. The rest reads the same as above.

Perhaps this is a good point for the reader to refresh these basic concepts and the mathematical relationships involved in them. An old but relevant and related paper is that by Lindemann (1910), where an index is defined to quantify thermally driven disorder in atoms or molecules, obviously a practical thermodynamic idea.

2.2.2 *Selectivity*

Roux *et al.* (2011), in their exhaustive review, point out that ion selectivity can mean different things, depending on the molecular system under consideration (ion channel or transporter), and on the experimental conditions to probe the system (equilibrium binding or non-equilibrium flux and ionic current measurements). Traditionally, selectivity characterization made use of permeability ratios. Computational as well as experimental studies have been applied to quantitatively characterize selectivity, especially for highly selective channels, as the potassium one (Lockless *et al.*, 2007; Cuello *et al.*, 2010; Egwolf and Roux, 2010). Incidentally, Lockless and his collaborators belong to the Laboratory of Molecular Neurobiology and Biophysics, Rockefeller University, New York, while Cuello is part of the prominent group headed by the Canadian Benoît Roux, at the Department of Biochemistry and Molecular Biology, University of Chicago. These authors showed that the ion radius dependence of selectivity stems from the channel's recognition of ion size (or volume) rather than charge density. Size recognition is a function of the channel's ability to adopt a very specific conductive structure with larger ions, concluding that ion selectivity in a K^+ channel is a property of size-matched ion-binding sites created by the protein structure. Even though K^+, Na^+, and Ca^{2+} channels display different versions and particular characteristics, they are interrelated and cannot be separated fully out from each other; hence, when discussing one, often the other species must be mentioned, too (Alberts *et al.*, 2002).

2.3 Potassium Channels

A potassium channel is sensitive to specific stimuli, yielding a transient period of ion conduction until the selectivity filter spontaneously undergoes a conformational change towards inactivation. Removal of the stimulus closes the gate and allows the selectivity filter to return to its conductive situation (recovery). The process takes up to several seconds while structural differences between both states are very small. The bacterial channel KcsA mentioned above has been used to elucidate doubts regarding inactivation-recovery at the atomic level (Ostmeyer *et al.*, 2013). The simplest view of conductive-state selectivity assumes the existence of binding sites that can be occupied by either K^+ or Na^+, and this leads to the relative free energy computation of K^+ and Na^+ in the binding sites. However, conductive-state selectivity can also result from kinetic factors due to differences in free energy barriers. Remember that under physiological conditions, the high selectivity of K^+ channels is primarily to prevent the entry of Na^+ ions from the extracellular side.

Theoretical studies based on simplified models dealing just with the ion and the nearest ligands can provide insight when selectivity is concerned with thermodynamic binding in specific sites. For example, interactions from distant parts of the protein or the solvent are not expected to directly affect relative free energy differences if, for instance, the charge of the ion remains unchanged, as the case is with Na^+ and K^+. Reduced models only treat the ion and the surrounding coordinating ligands explicitly. The effect of the missing atoms from the protein or the solvent must be incorporated indirectly via spatial constraints that are acting on the ion and ligands. Two distinct and idealized limiting mechanisms have been found as to produce ion selectivity in protein binding sites. In one limit, the geometric forces are dominant and the binding site is nearly rigid. Then, it can be shown that the free Na–K energy difference is set by the difference between the mean ion–ligand interactions, and that ion selectivity is controlled by the cavity size, according to the *snug-fit* mechanism, which is perhaps an idealized mechanism. Electrostatic (dipole, charge, and polarizability) as well as non-electrostatic (shape, radius, and size) factors play a role in these mechanisms. Selectivity in reduced models, where the optimal geometry is adapted to best fit K^+ or Na^+, a binding site is K^+-selective when the Na–K free energy difference is positive and Na^+-selective when it is negative. Ligand is an ion or molecule that binds to a central metal atom to form a complex called coordination complex. To bind a ligand, a protein must have a binding site complementary to that of the ligand. A key fitting into a lock is the classic analogy. During binding, each part adjusts its structure to the presence of the other, and the protein is said to *snuggle* around the ligand, thus the jargon term *snug-fit* (Petsko and Ringe, 2004; Asthagiri *et al.*, 2006; Yu *et al.*, 2010; Merz *et al.*, 2010; Ziervogel and Roux, 2013).

As partial summary:

> Early molecular dynamic simulations of the KcsA channel brought up the idea that ion selectivity could be achieved, leading to reexamination of microscopic factors by means of computational experiments. Classical notions about conductive-state selectivity (such as the *snug-fit* mechanism, field strength, and free energy profiles are still important), but they are now sharing the stage with other concepts relevant to flexible and loosely restricted systems. To underline and closing the section:

> *The notion that ion selectivity in a flexible binding site can emerge from local energetics has now better grounds.* The relative success of computer simulations suggests that they provide an acceptable representation of these molecular systems, which clearly indicates how intimate the exchange with the physical sciences has become.

What about cardiac K^+ channels? They fall into three broad categories, somehow reminding of the classification given in Section (2.1.8), but not quite,

- Voltage gated,
- Inward rectifier channels, with several specific subsets, and
- Background K^+ currents that stabilize the resting membrane potential.

Variation in the level of expression of these channels accounts for regional differences of the action potential in the atria, ventricles, and across the myocardial wall (from endocardium to epicardium).

Besides, K^+ channels are highly regulated and relate to variations in the heart rate. The structure of voltage-gated K^+ channels is similar to the voltage-gated Na^+ and Ca^{2+}. Chronic α-adrenergic stimulation and angiotensin II reduces channel expression. The channel is highly expressed in the SA and AV nodes, and atria, but low in ventricle. Mutations of the genes encoding cardiac K^+ channels are the principal causes of arrhythmias due to abnormal repolarization. Cardiac K^+ channels are the targets for class III antiarrhythmic drugs. The potent K^+ channel blocking action of quinidine and procainamide account for their QT prolonging action and occasionally *torsade de pointes*. Amiodarone is exceptional in that it produces K^+ channel blockade. Antiarrhythmic drugs play an important role in the prevention of atrial fibrillation recurrence.

2.4 Voltage-Gated Sodium Channels

Sodium channels are the arch-type of voltage-gated ion channels. They are responsible for initiating electrical signaling in excitable cells, as explained before. The structural basis for their voltage-dependent activation, ion selectivity, and drug block is still a research subject. In *Arcobacter butzleri* (a bacterium with a wide range of habitats, even as pathogen in human and animal), the arginine gating charges make multiple hydrophilic interactions within the voltage sensor, including hydrogen bonds to the protein backbone. It has a short selectivity filter. This structure provides the template for understanding electrical behavior in excitable cells and the actions of drugs used for pain, epilepsy, and cardiac arrhythmias (Payandeh *et al.*, 2011). This gated channel is a protein embedded in the plasma membrane, typical of nerve and muscle cells. It has a polypeptide chain of more than 1800 amino acids showing non-polar side chains forming segment coils with a length approximately the width of the membrane (about 100 Å). The side chains face outward, where they interact with the membrane lipids. The peptide bonds, instead, which are polar, face inward. This type of channels has a crucial role in neurons, synapses, and muscles, highly significant in rapid depolarization (Yellen, 1998; see also http://courses.washington.edu/conj/membrane/nachan.htm). They generally are composed of several subunits arranged in such a way that there is a central pore through which ions can travel down their electrochemical gradients.

Channels tend to be ion-specific, although similarly sized and charged ions may sometimes travel through them (see http://en. wikipedia.org/wiki/Voltage-gated_ion_channel). Sodium channels, as well as calcium channels, are made up of a single polypeptide with four homologous domains. Each contains six membrane spanning alpha-helices (the latter group is a common secondary right-handed coiled structure of proteins, where every backbone N–H pair donates an H bond to the backbone $C = O$ group. One of these is the voltage sensing helix. It has many positive charges such that a high positive charge outside the cell repels the helix, keeping the channel in its closed state. Depolarization of the cell interior causes the helix to move, inducing a conformational modification such that ions may flow through the channel (open state). Potassium channels function in a similar way, with the exception that they are composed of four separate polypeptide chains.

One portion of the channel determines its selectivity and even potassium (chemically similar) cannot traverse the channel. Another portion of the channel serves as a gate. For many ion channels, the gate opens in response to regulatory molecules that specifically bind to either the inside or outside side, but in the case of the voltage-gated sodium channel, the gate is controlled by a voltage sensor, which responds to the level of the membrane potential (Figs. 2.10 and 2.11). Cells, in general, have a small net excess of negative ions clustered under the plasma membrane. Inactivation sets a limit to the period of time the channel remains open, despite steady stimulation. Other types of ion channels, however, may not have an inactivation gate. At resting membrane potential (say, $-70\,\mathrm{mV}$), the channel is closed. Then, should any factor depolarize the membrane sufficiently (e.g., to $-50\,\mathrm{mV}$), the voltage sensor moves outward and the gate opens. Much of this knowledge has been obtained by patch fluorometry/fluorimetry, which permits conformational changes in ion channel protein with channel function, by simultaneous recordings of

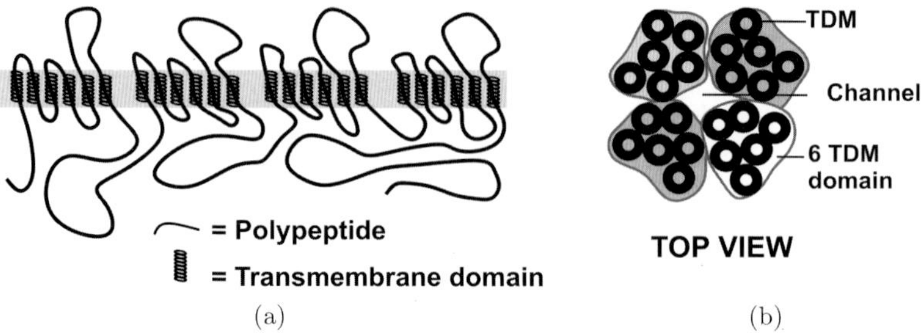

Figure 2.10: Voltage-gated sodium channel scheme. The alpha-helices are shown spread out and in a row. Also, they are shown divided into four groups. Each of these is a *homologous transmembrane domain* (TMD) with a similar sequence of amino acids. (a) Longitudinal membrane cut. (b) Viewing the membrane from above. At the center of the four domains is the channel through which the sodium ions move separated from the lipid environment of the membrane. In the case of the voltage-gated sodium channel, there are 24 such transmembrane domains in the polypeptide chain. Redrawn by Gustavo Idemi, from INTERNET.

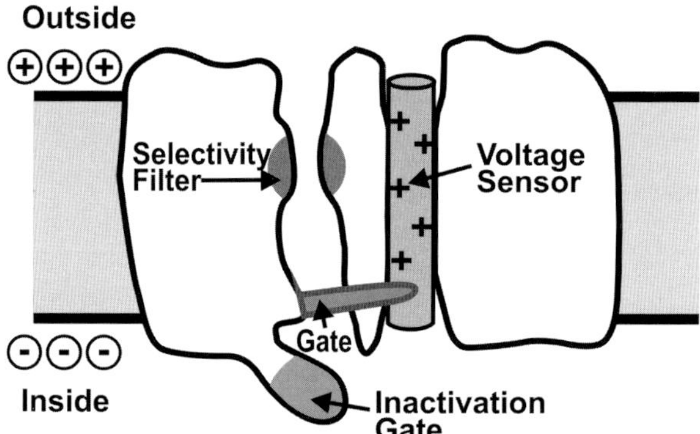

Figure 2.11: Simplified model of a voltage-gated ion channel showing the membrane thickness (positive on its outside), the selectivity filter, the voltage sensor, and the on–off gate. Ion channels have moving parts that perform useful functions. The channel proteins contain an aqueous, ion-selective pore that crosses the plasma membrane, and they use a number of distinct gating mechanisms to open and close this pore in response to biological stimuli such as the binding of a ligand or a change in the transmembrane voltage. Redrawn by Gustavo Idemi, from INTERNET.

fluorescence and current signals from ion channels in isolated membrane patches.

Patch fluorometry is an analytical technique for identifying and characterizing minute amounts of a substance by excitation of the substance with a beam of ultraviolet light. Besides, it detects and measures the characteristic wavelength of fluorescent light emitted. By this means, calcium and voltage dependence of conformational changes of the gating-ring region of BK channels has been studied (Miranda *et al.*, 2013). BK channels (in the technical jargon, "Big Potassium" or "Maxi-K") are ion channels characterized by their large conductance of K^+ through cell membranes. They open by changes in membrane electrical potential and/or by increases in concentration of intracellular Ca^{2+}. FRET (Fluorescence Resonance Energy Transfer) is the used technique.

When open, voltage-gated sodium channels allow an inward flux of sodium that makes the transmembrane potential more positive; voltage-gated potassium channels then open, allowing outward flux of potassium to restore the membrane potential to its negative resting value. Gating of these channels is carefully arranged to produce the variable electrical signals required by the nervous system for signal transmission. A review by Armstrong and Hille (1998) treated the entire voltage-gated channels subject area in historical perspective, including the seminal contributions of Hodgkin and Huxley to the original conception of gate. It must be underlined that the voltage-activated Na^+, Ca^{2+}, and K^+ channels constitute a family of structurally related integral membrane proteins. In fact, gating involves several activation–inactivation mechanisms. In response to a positive change in the transmembrane voltage, the channel opens rapidly. Immediate return of the potential to the resting level reverses the process, closing the channel. If, after activation, the positive potential is maintained, the channel closes despite the maintained activating stimulus. Yellen (1998) has also discussed all this in detail.

The highly potent neurotoxin tetrodotoxin (TTX) and the systematic mutation of residues have enabled the tentative identification of the amino acid residues that are critical for ion permeation; these residues include aspartate, glutamate, lysine, and alanine. Each sodium channel opens very briefly (< 1 ms). The channel occasionally shows alternative gating modes consisting of isolated brief openings occurring after variable and prolonged latencies and bursts of openings during which the channel opens repetitively for hundreds of milliseconds. The isolated brief openings are the result of the occasional return from the inactivated state. The bursts of openings are the result of occasional failure of inactivation. Sodium channel mutations that favor these slow gating modes are the basis of the long QT syndromes. Sodium channel inactivation is a complex process that may occur within milliseconds, seconds, or even tens of seconds, depending on the duration of the preceding depolarization. In response to depolarization lasting tens of milliseconds, the process is fast. Intermediate and slow inactivation develops over hundreds of millisecond.

2.5 Calcium Channels

These channels display selective permeability to that particular ionic species. Occasionally, it is synonymous of voltage-dependent Ca^{2+} channel although there are also ligand-gated calcium channels (Catterall, 2000). This kind of channels mediates Ca^{2+} entry into cells in response to membrane depolarization. Electrophysiological studies reveal different calcium currents that have been characterized as complexes of a pore-forming alpha subunit of approximately 190–250 kDa (Grant, 2009). The molecular basis of inherited cardiac arrhythmias has been a strong force to identifying ion channels. The genes encoding all the major ion channels have been cloned and sequenced, showing greater complexity than ever before imagined. These channels function as part of macromolecular complexes in which many components are assembled at specific sites in the membrane. The generation of the action potential and the regional differences that are observed throughout the heart are the result of the selective permeability distributed on the cell membrane. The ion channels reduce the activation energy required for ion movement across the lipophilic membrane. During the action potential, the permeability of ion channels changes. Electrochemical gradients determine whether an ion moves into the cell (depolarizing current for cations) or out of the cell (repolarizing current for cations). Homeostasis of the intracellular ion concentrations is maintained by active and coupled transport processes linked directly (or indirectly) to ATP hydrolysis. Most ion channels have a nonlinear current–voltage relationship. For the same absolute potential difference, the magnitude of the current depends on the direction of ion movement into or out of the cells. This property is termed rectification and is an important property of K^+ channels; they pass little outward current at positive (depolarized) potentials. The molecular mechanism of rectification varies with ion channel type. Block by internal Mg^+ and polyvalent cations is the mechanism of the strong inward rectification demonstrated by many K^+ channels. Gating is the mechanism of opening and closing of ion channels and is their second major property. Ion channels sometimes are sub-classified by their mechanism of gating: voltage-dependent, ligand-dependent, and mechano-sensitive gating. Voltage-gated ion channels change their conductance in response to variations in membrane potential. Voltage-dependent gating is the commonest gating mechanism. A majority of ion channels open in response to depolarization.

Ion channels have two mechanisms of closure. Certain channels such as the Na^+ and Ca^{2+} channels enter a closed inactivated state during maintained depolarization. To regain their ability to open, the channel must undergo a recovery process at hyperpolarized potentials. The inactivated state may also be accessed from the closed state. Inactivation is the basis for refractoriness in cardiac muscle and is fundamental for the prevention of premature re-excitation. If the membrane potential is abruptly returned to its hyperpolarized (resting) value while the channel is open, it closes by deactivation, a reversal of the normal activation process.

Ligand-dependent gating is the second major gating mechanism of cardiac ion channels. The most thoroughly studied channel of this class is the acetylcholine (Ach)-activated K^+ channel. The mechanosensitive or stretch-activated channels are the least studied. They belong to a class of ion channels that can transduce a physical input, such as stretch into an electric signal through a change in channel conductance. Acute cardiac dilatation is a well-recognized cause of cardiac arrhythmias. Calcium ions are the principal intracellular signaling ions. They regulate excitation–contraction coupling, secretion, and the activity of many enzymes and ion channels. Ca^{2+} is highly regulated despite its marked fluctuation between systole and diastole. Calcium channels are the principal portal of entry of calcium into the cells; a system of intracellular storage sites, and transporters such as the sodium-calcium exchanger play important roles in calcium regulation. In cardiac muscle, two types of Ca^{2+} channels, the L-low threshold type and T-type (transient-type), transport Ca^{2+} into the cells. The L-type channel shows up in all cardiac cells. The T-type channel is found principally in pacemaker, atrial, and Purkinje cells. This may be the basis for early after-depolarizations and polymorphic VT in LQTS. The overall kinetics of the Ca channel is important in controlling contractility. A sudden death syndrome that combines the features of Brugada syndrome, including the characteristic ECG pattern, and a short QT interval has been described. Calcium channel is the target for the interaction with class IV antiarrhythmic drugs. The principal class IV drugs are the phenylalkylamine, verapamil, and diltiazem. These drugs block open and inactivated calcium channels; they may cause block of conduction in cells such as those in the SA and AV nodes and slow the sinus node rate. However, the hypotensive effects of verapamil may cause an increase in sympathetic tone and increase the heart rate. A third class of Ca^{2+} channel blockers, the dihydropyridines, block open calcium channels. As can be seen, the subject area of ion channels is complex, highly specialized and far yet of having been fully explained. The literature abounds and, herein, we cannot be exhaustive. Some papers are essential, even though they are relatively old (Zilberter *et al.*, 1994; Bennett *et al.*, 1995; Richmond *et al.*, 1998; Rohl *et al.*, 1999; Antzelevitch and Dumaine, 2002; Antzelevitch *et al.*, 2005; Clancy and Kass, 2005; Splawski *et al.*, 2005; Antzelevitch *et al.*, 2007; Zhang *et al.*, 2007; Rosen *et al.*, 2008).

In short, and to keep in mind:

Calcium channel blockers slow heart rate and lower blood pressure. These channel blockers slow heart rate by blocking the number of electrical impulses that cause the heart muscle to contract and pump blood. Besides, calcium channel blockers help lower blood pressure by relaxing the muscle tissue in the blood vessels. This makes it easier for blood to flow through the vessels.

2.6 Maxwell and Kirchhoff Equations: In the Back of Electrophysiology

Maxwell enriched Newton's inheritance, consolidated Faraday's work, and gave the launching platform to Einstein while Kirchhoff started what now is circuit theory.

Dass die Elecktrodynamik Maxwells — wie dieselbe gegenwärtig aufgefasst pflegt — in ihrer Anwendung auf bewegte Körper zu Asymmetrien führt, welche den Phänomenen nicht anzuhaften scheinen, ist bekannt. [*That Maxwell's elecktrodynamics — same as currently conceived maintains — in their application to moving bodies leads to asymmetries that the phenomena do not seem to follow, is known*]. First sentence of Albert Einstein's Special Relativity paper, Zur Elecktrodynamik bewegter Körper, *Annalen der Physik*, 1905, 17(Series 4):891–921.

He that would enjoy life and act with freedom must have the work of the day continually before his eyes. James Clerk Maxwell (1831–1879).

The objective of this section is to review the derivation of the Maxwell equations, as a didactic and less common exercise because it follows a path different from the traditional line, which is set on the principles of conservation of charge and conservation of energy. Moreover, and less known, Maxwell explored the Biological Sciences, too. A second objective is to show how Kirchhoff's laws project to Physiology, perhaps indirectly, as several examples attest so. Both acted in a way as predecessors of Bioengineering (Valentinuzzi and Kohen, 2013).

2.6.1 *From M-Equations to K-Laws*

The complete set of Maxwell's equations, originally defined for 20 field variables, is known since 1865. Later, Oliver Heaviside (1850–1925) and Josiah Willard Gibbs (1839–1903) transformed them into the today's most used notation with vectors. They are the cornerstone in electrodynamics (Maxwell, 1865, 1873). These equations (or M-Equations) are usually described in two different mathematical ways,

I — As differential equations,

$$\nabla \times \vec{E} = -\frac{\partial \vec{B}}{\partial t} \quad \text{describes Faraday's Law} \tag{2.3}$$

$$\nabla \times \vec{H} = \vec{J} + \frac{\partial \vec{D}}{\partial t} \quad \text{describes Ampere's Law} \tag{2.4}$$

$$\nabla \cdot \vec{B} = 0 \quad \text{divergence of vector } B \text{ describes Gauss' Law}$$
$$\text{for magnetic fields} \tag{2.5}$$

$$\nabla \cdot \vec{D} = \rho_v \quad \text{divergence of vector } D \text{ describes Gauss's Law}$$
$$\text{for electric fields} \tag{2.6}$$

II — As integral–differential equations

$$\oint_L \vec{E} \cdot \mathrm{d}\vec{l} = -\frac{\partial}{\partial t} \iint_S \vec{B} \cdot \mathrm{d}\vec{s} \quad \text{also Faraday's Law} \tag{2.7}$$

$$\oint_L \vec{H} \cdot \mathrm{d}\vec{l} = \iint_S \left(\vec{J} + \frac{\partial \vec{D}}{\partial t} \right) \cdot \mathrm{d}\vec{s} \quad \text{also Ampere's Law} \tag{2.8}$$

$$\oiint_S \vec{B} \cdot \mathrm{d}\vec{s} = 0 \quad \text{also Gauss' Law for magnetic fields} \tag{2.9}$$

$$\oiint_S \vec{D} \cdot \mathrm{d}\vec{s} = \iiint_V \rho_v \mathrm{d}v = Q \quad \text{also Gauss' Law}$$
$$\text{for electric fields} \tag{2.10}$$

In Eqs. (2.1)–(2.4), the symbol ∇, called either *nabla* or *del*, is defined as the vector operator

$$\nabla = i\frac{\partial}{\partial x} + j\frac{\partial}{\partial y} + k\frac{\partial}{\partial z} \tag{2.11}$$

with i, j, and k standing for the unit vectors in each direction of the coordinate axes, respectively. Equations (2.5) and (2.6) describe the *divergence* of a vector, say $\vec{F}$, i.e.,

$$\nabla \cdot \vec{F} = \frac{\partial F_x}{\partial x} + \frac{\partial F_y}{\partial y} + \frac{\partial F_z}{\partial z} \tag{2.12}$$

The vector (or cross) product of two vectors is symbolized by ($\times$), and ($\cdot$) stands for scalar (or dot product), also of two vectors. The former is also a vector and the latter is a scalar magnitude. The *divergence* measures the magnitude of a vector field sustained by a source or a sink at a given point. An example: Suppose that air is heated, thus air will expand in all directions. The velocity field points outwardly and the divergence of the velocity vector would be positive, as the region acts as a source. The opposite would happen if the air were cooled. The first two equations, then, are vectors, and the second pair represents scalar magnitudes. It must be underlined that $\vec{F}$ represents a vector function with respect to position and time, i.e., $\vec{F}(\vec{r}, t)$ while $\hat{F}$ stands for a vector function depending on position and in the frequency domain, that is, $\hat{F}(\vec{r}, \omega)$. So far, so much for this brief refreshing paragraph. Let us now state in simple words the four M-Equations (2.3)–(2.6):

A time changing magnetic field $\vec{B}$ sustains an electric field $\vec{E}$ (Eq. (2.3)) and a time variable electric field $\vec{D}$ plus a current density produces a spatial magnetic field (Eq. (2.4)). The overall flow of a magnetic field is null (Eq. (2.5)) while a similar magnitude of an electric field is sustained by an electric charge (Eq. (2.6)).

If we deal just with sinusoids, then phasors become useful, and the two first equations would be written as

$$\nabla \times \hat{E} = -j\omega\hat{B} = -j\omega\mu\hat{H} \tag{2.13}$$

$$\nabla \times \hat{H} = \hat{J} + j\omega\hat{D} = \hat{J} + j\omega\varepsilon\hat{E} \tag{2.14}$$

Based on the two sets shown above, two avenues leading to the K-Equations are available:

A — First, when the frequency is relatively low, say, not above 500 MHz, a quasi-stationary approximation is applicable, offering a way to Electric Circuit Theory. This point is critical and must be underlined because it marks the link between the two subject areas, electromagnetic equations and electric circuits. However, for biomedical applications, the low frequency limit is much lower (say, 50–100 Hz at the most), a condition that makes the approximation by far easier.

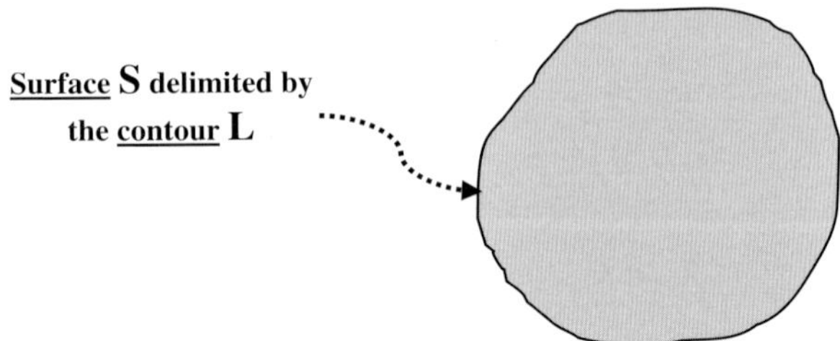

Figure 2.12: Stokes' theorem. The contour L, where the closed linear integral operates, surrounds the surface S (dashed), which represents the place of the double integral.

Thus, $j\omega\mu\hat{H}$ and $j\omega\varepsilon\hat{E}$ in (2.13) and (2.14) are negligible as compared to the other terms, so that the latter equations simplify to,

$$\nabla \times \hat{E} \cong 0 \qquad (2.15)$$

$$\nabla \times \hat{H} \cong \hat{J} \qquad (2.16)$$

Let us now consider the first of these two simpler equations. For that matter, the Stokes theorem must be recalled, i.e., the surface integral of the *curl* of a vector field (*curl* is the vector product within the integral) over a surface S in space equals the line integral over its boundary L, that is (Fig. 2.12),

The following equation holds: $\iint_S (\nabla \times \hat{E})\mathrm{d}\vec{s} = \oint_L \hat{E} \cdot \mathrm{d}\vec{l} = 0$

Since by Eq. (2.15) the above left-hand integrand is null, accordingly, the right-hand closed integral is also zero. Besides, the integrand units are (volt/meter $\times$ meter) = (volts), so that finally we end up with

$$\sum_i V_i = 0 \qquad (2.17)$$

The contour L surrounding the surface S represents an electric circuit, say, a network, and the latter equation is nothing else than the well-known first Kirchhoff's Law of Electrical Engineering. It must be underlined that $\hat{E}, \hat{D}, \hat{B}, \hat{H}$, and $\hat{J}$ are all vectors, while V (in 2.17) and I (in 2.20) are scalar magnitudes. From Eq. (2.14), and also by considering Stokes' Theorem, the left side of the equation can be replaced by the closed integral, in turn equal to a surface integral, or

$$\oint_L \hat{H} \cdot \mathrm{d}\vec{l} = \iint_S (\hat{J} + j\omega\varepsilon\hat{E}) \cdot \mathrm{d}\vec{s} \qquad (2.18)$$

If the contour tends to zero, the linear integral tends to zero, meaning that the surface integral also tends to zero. Hence,

$$\oint_L \hat{H} \cdot \mathrm{d}\vec{l} = 0 \Rightarrow \iint_S (\hat{J} + j\omega\varepsilon\hat{E}) \cdot \mathrm{d}\vec{s} = 0 \qquad (2.19)$$

Furthermore, we said before that the term $j\omega\varepsilon\hat{E}$ is negligible at low enough frequencies and, since $\hat{J}$, as current density, is measured in amps per square cm, the integrand of the former equation represents current in the usual amperes. In the end,

$$\sum_i I_i = 0 \qquad (2.20)$$

which is the second Kirchhoff's Law usually applied to nodes in electric networks. The current density $\hat{J}$ generates the magnetic field $\hat{H}$ (neglecting the minor effect due to $j\omega\varepsilon\hat{E}$). Such $\hat{H}$, in turn, creates an electric field $\hat{E}$, however, the latter does not affect the original $\hat{H}$. It is so that the induction expressed by Eq. (2.13) is only limited to "coils" and the current represented by $j\omega\varepsilon\hat{E}$ manifests itself just in "capacitors". In other words, there is lack of radiation from electric circuits (fortunately, otherwise our technological-electric-dependent lives would suffer quite a bit!). When ω is high, there is interaction between Eqs. (2.13) and (2.14) and they must be simultaneously solved for in order to find $\hat{E}$ and $\hat{H}$. Say, in a series LCR circuit with a sinusoidal generator, in C, $j\omega\varepsilon\hat{E}$ creates a magnetic $\hat{H}$ field restricted to the capacitor region. The electric field $\hat{E}$ is sustained by the electric charges in the capacitor's plates, according to the fourth Maxwell's equation, $\mathrm{div}\vec{D} = \nabla \cdot \vec{D} = \rho_1$; instead, in the inductor L, $j\omega\mu\hat{H}$ generates an electric field $\hat{E}$ limited to its region. The magnetic field $\hat{H}$ is due to the conductive electric current, in charges per second or en amps.

B — Another line is possible to reach the K-Equations. Starting from Eq. (2.4), Ampere's Law, and considering that the divergence of a rotor is null (recall that $\nabla \cdot \nabla \times \vec{F} \equiv 0$), applying the *nabla* operator to both sides, we obtain

$$\nabla \cdot \nabla \times \vec{H} = \nabla \cdot \left(\vec{J} + \frac{\partial \vec{D}}{\partial t} \right) \equiv 0 \qquad (2.21)$$

which is equivalent to

$$\nabla \cdot \vec{J} + \nabla \cdot \frac{\partial \vec{D}}{\partial t} = \nabla \cdot \vec{J} + \frac{\partial \left(\nabla \cdot \vec{D} \right)}{\partial t} \quad \text{and, recalling Eq. (2.6)}$$

$$\nabla \cdot \vec{J} + \frac{\partial \rho_v}{\partial t} = 0 \tag{2.22}$$

The latter Eq. (22) is also called Continuity Equation, which, if ρ_v is constant (as the electron charge in electric circuits or the density in an incompressible fluid are), simply becomes Eq. (2.24), or the first Kirchhoff's Law, that is, the algebraic sum of currents to and from a node is null. Remember that *nabla* is the partial derivatives sum of current density such that, on integration, becomes the common current or flow.

Now, beginning from the integral–differential Eq. (2.7) and knowing that $\iint_S \vec{B} \cdot d\vec{s} = \Phi_m$, where Φ_m stands for magnetic flux,

$$\oint_L \vec{E} \cdot d\vec{l} = -\frac{\partial \Phi_m}{\partial t} \tag{2.23}$$

If no temporal variations take place, Eqs. (2.22) and (2.23) simplify into,

$$\nabla \cdot \vec{J} = 0 \tag{2.24}$$

and

$$\oint_L \vec{E} \cdot d\vec{l} = 0 \tag{2.25}$$

Equation (2.24) is the continuity one already reached to before. By application of the divergence theorem (also called Green–Gauss–Ostrogradsky) to the same equation (Eq. (2.24)), we get

$$\oiint_S \vec{J} \cdot d\vec{s} = \iiint_V (\nabla \cdot \vec{J}) dv = 0 \tag{2.26}$$

Besides, and somewhat repeating what was said above, since current can be expressed as $I_k = \iint_{S_k} \vec{J}_k \cdot d\vec{s}$, we end up with,

$$\oiint_S \vec{J} \cdot d\vec{s} = \sum_k I_k = 0 \tag{2.27}$$

which is obviously the First Kirchhoff's Law (the reader must be patient, but there are more than one road to Rome). Meanwhile,

Eq. (2.25) speaks of a conservative vector field, i.e., the gradient of a scalar potential. Conservative fields are path independent. Thus,

$$\oint_L \vec{E} \cdot \mathrm{d}\vec{l} = -\oint_L \nabla V \cdot \mathrm{d}\vec{l} = -\oint_L \mathrm{d}V = 0 \Rightarrow$$

$$\oint_L \mathrm{d}V = \int_1 \Delta V_1 + \int_2 \Delta V_2 + \cdots + \int_{m-1} \Delta V_{m-1} + \int_m \Delta V_m$$

$$= \sum_1^m \Delta V_m = 0 \tag{2.28}$$

In other words, Eq. (2.25) says that the sum of all potential drops over a closed pathway is zero, and that is Second Kirchhoff's Law. The closed path L was divided in a series of m pieces and the sum of potential differences is null. Nice and well received surprise!

It is hoped that the above mathematical linking between the four M-Equations and the two K-Equations appears as clear enough to the reader. Obviously, some more mathematics is required than that customarily offered in initial engineering courses, but it shows beauty, that is, it has aesthetic value, keeping those philosophical essential concepts laid down by Schiller and Beethoven over two centuries ago (Valentinuzzi, 2012).

2.6.2 *What do they have to do with Physiology?*

Both Kirchhoff's Laws find a place in Physiology and, hence, become good and neat examples of the Bioengineering concept (application of the Physical Sciences to Biomedical Problems). The so-called Fick Principle is nothing else than a hemodynamic version of the Second Law (Valentinuzzi *et al.*, 1968). A similar application takes place when speaking of regional blood flows (say, coronary, hepatic, or renal).

The First Law directly leads to the two Einthoven's Laws (or E-Laws) of electrocardiography, i.e.,

$$V_{\mathrm{I}} + V_{\mathrm{III}} = V_{\mathrm{II}} \tag{2.29}$$

Any physiology textbook describes the standard leads located on the frontal plane of an individual. An electrocardiograph records the differences of potential from leads I, II, and III. These leads form

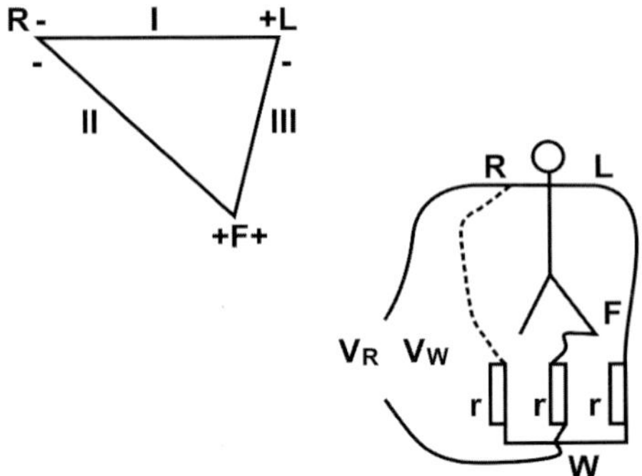

Figure 2.13: Einthoven's ECG leads I, II, and III to measure the differences of potential between left and right arm, left foot and right arm, and left foot and left arm. Resistors **r** act as equalizers and may take practical values between 5 and 10 kohm.

a triangular circuit (Einthoven's Triangle), amenable to the First Kirchhoff's Law. As in a loop, going from the negative to the positive or R to L (in branch I), from L to F (in branch III), to equate the potential difference found between F and R, the algebraic sum of all the electrical potential differences must be zero (Eq. (2.29)). Each term V stands, respectively, for the potential difference among L–R, F–L, and F–R. Such magnitudes are scalars; however, there is more, because the Second Kirchhoff's Law can be applied to the node W, where the sum of the currents should equal zero (Fig. 2.13), or

$$\frac{V_{\mathrm{R}} - V_{\mathrm{W}}}{r} + \frac{V_{\mathrm{F}} - V_{\mathrm{W}}}{r} + \frac{V_{\mathrm{L}} - V_{\mathrm{W}}}{r} = 0 \qquad (2.30)$$

$$V_{\mathrm{R}} + V_{\mathrm{F}} + V_{\mathrm{L}} = 3V_{\mathrm{W}} \qquad (2.31)$$

where each V represents the electrical potential (not the difference) of the right arm, left arm, and left foot. The last potential is called Wilson's terminal and works as a virtual reference for the arrangement. Either of the latter two equations (2.30–2.31) describes Einthoven's Second Law. By joining the wires from the right arm, left arm, and left foot with 5000 ohm resistors, Frank N. Wilson (1890–1952) defined an *indifferent electrode* later called the *Wilson Central Terminal*. The combined lead acts as a virtual earth and is attached to the negative terminal of the ECG recorder. Jin *et al.* (2012) published a good didactic paper on Einthoven's Triangle. The heart, muscles, nerves, and brain generate magnetic fields and the E-Equations are right there, too, perhaps less prominent, but still backing them up. Magnetocardiography, magnetoencelography, and magnetomyography are neat examples. Electromagnetic fields are inseparable phenomena of the biological systems at large. Somehow, Maxwell stands behind them.

By and large, Kirchhoff's equations are not applicable to time variable systems, unless relatively short intervals are considered, when changes are not too marked. In biology, for example, all signals (usually of low levels) are always contaminated by electromagnetic noise, so that they are significantly affected, and this is an aspect better helped out in its interpretation by the M-Equations, we might say, bringing us closer to reality. There are different notations to the M-equations. Depending on the application, one or another can be more useful, but in the end such a variety is not fully satisfactory for, in all likelihood, the correct final form has not been yet found, so that the lack of symmetry in Maxwell's equations pointed out by Einstein still must be considered as a drawback.

Maxwell went into Physiology, Biophysics, and/or Bioengineering, too. In 1854, Maxwell wrote from Trinity College to his aunt Cay:

> I have made an instrument for seeing into the eye through the pupil. The difficulty is to throw light in at that small hole and look in at the same time; but that difficulty is overcome, and I can see a large part of the back of the eye quite distinctly with the image of the candle on it. People find no inconvenience in being examined, and I have got dogs to sit quite still and keep their eyes steady. Dog's eyes are very beautiful behind, a copper-colored ground, with glorious bright patches and networks of blue, yellow, and green, with blood-vessels great and small.

Only Helmholtz before, in 1851, published an article on an ophthalmoscope, made mainly of brass. Was Maxwell aware of Helmholtz earlier contribution? Perhaps, we do not know, but evidently, he is to be considered a contributor and a pioneer in that subject, too. Also the same year of 1854, on November, James wrote again to his aunt describing trials to obtain certain colours by combining blue, red, and yellow and regarding his interest in checking people in Edinburgh "who do not see colours", that is, color-blind people. Another letter, now of May 15, 1855, indicates tests with different persons to perfect his instrument "for looking into the eye". Later on, 27 February 1857, always to Cay, he says: "I have just been getting cods' and bullocks' eyes, to refresh my memory and practice dissection. The size of the cod and the ox eye is nearly the same". Thus, Maxwell not only experimented in Physics, but also in Physiology and Anatomy. Even he showed interest in pain and sensations (Valentinuzzi and Kohen, 2013).

As bioengineers, we may find some kind of peace of mind for our feet stand on solid grounds with an outstanding intellectual ancestry. It enhances the importance of theoretical studies, even though they may occasionally look as impossible dreams. Both Maxwell and Kirchhoff were young men when they developed their respective theories; thus, the importance of stimulating young students becomes apparent. With some prudishness, embarrassment, or even lack of modesty, we repeat almost verbatim, only slightly adapted, the closing words written by Hawking (2013) in his famous and appealing book:

> *It has been a glorious time to be alive and doing research in bioengineering. We are happy if we have added something to our understanding of this multi–inter–transdiscipline.*

2.7 Summary

We went over the basics of general electrophysiology and cardiac electrophysiology to delve thereafter into ionic channels and their different variants. Maxwell equations are finally developed demonstrating their importance in this essential physiology chapter. Mathematics appears in almost all these subjects, in greater or lesser degree, as some kind of key helping in opening doors for comprehension.

References

Alberts B, Johnson A, Lewis J, Raff M, Roberts K, Walter P. Ion channels and the electrical properties of membranes. In: Molecular biology of the cell. 4th ed. Garland Science, New York, 2002. http://www.ncbi.nlm.nih.gov/books/NBK26910/.

Antzelevitch C, Brugada P, Borggrefe M, Brugada J, Brugada R, Corrado D, Gussak I, LeMarec H, Nademanee K, Riera ARP, Shimizu W, Schulze-Bahr E, Tan H, Wilde A. Brugada syndrome: Report of the second consensus conference. *Heart Rhythm* 2:429–440, 2005.

Antzelevitch C, Dumaine R. Electrical heterogeneity in the heart: Physiological, pharmacological and clinical implications. In: Handbook of physiology, Chap. 17, Oxford University Press, New York, 2002.

Antzelevitch C, Pollevick GD, Cordeiro JM, Casis O, Sanguinetti MC, Aizawa Y, Guerchicoff A, Pfeiffer R, Oliva A, Wollnik B, Gelber P, Bonaros EP, Jr Burashnikov E, Wu Y, Sargent JD, Schickel S, Oberheiden R, Bhatia A, Hsu, L-F, Haïssaguerre M, Schimpf R, Borggrefe M, Wolpert C. Loss-of-function mutations in the cardiac calcium channel underlie a new clinical entity characterized by ST-segment elevation, short QT intervals, and sudden cardiac death. *Circulation* 115:442–449, 2007.

Armstrong CM, Hille B. Voltage-Gated ion channels and electrical excitability. *Neuron* 20:371–380, 1998.

Asthagiri D, Pratt LR, Paulaitis ME. Role of fluctuations in a *snug-fit* mechanism of KcsA channel selectivity. *J Chem Phys* 125:24701, 2006.

Bennett PB, Yazawa K, Makita N, George AL Jr. Molecular mechanism for an inherited cardiac arrhythmia. *Nature* 376:683–685, 1995.

Bolton TB. Electrophysiology of the intestinal musculature. In Handbook of physiology, the gastrointestinal system, motility and circulation Suppl. 16, 1989. Online Jan 2011; 10.1002/cphy.cp060106.

Bowen R. Control of Digestive System Function, 1996. http://arbl.cvmbs. colostate.edu/hbooks/pathphys/digestion/basics/slowwaves.html. Nov 23. See also Alexandra Valderrama Bonilla, Physiological and molecular mechanisms contribute to altered smooth muscle contractility during dextran sodium sulfate-induced colitis in mice. University of Calgary, Alberta, Canada, 2008.

Bradshaw LA. Biomagnetic techniques for assessing gastric and small bowel electrical activity. *AIP Conference Proceedings* 724(1):8, 2004.

Brueggemann LI, Mani BK, Haick J, Byron KL. Exploring arterial smooth muscle Kv7 potassium channel function using patch clamp electrophysiology and pressure myography. *J Vis Exp* 67:e4263, 2012. doi: 10.3791/4263; http:// www.jove.com/video/4263/exploring-arterial-smooth-muscle-kv7-potassium-channel-function-using.

Catterall WA. Structure and regulation of voltage-gated Ca^{2+} channels. *Annu Rev Cell Dev Biol* 16:521–55, 2000.

Clancy CE, Kass RS. Inherited and acquired vulnerability to ventricular arrhythmias: Cardiac Na^+ and K^+ Channels. *Physiol Rev* 85:33–47, 2005.

Cole KS, Curtis HJ. Electric impedance of the squid giant axon during activity. *J Gen Physiol* 22(5):649–670, 1939. doi: 10.1085/jgp.22.5.649.

Concha JD. Department of neurobiology and anatomy, University of Texas Medical School at Houston, 1997. http://neuroscience.uth.tmc.edu/s1/index.htm.

Cordova T, Bradsha LA, Adilton A, Sosa M. Biomagnetic signals of the large intestine. *AIP Conf Proc* 1032(1):37, 2004.

Cuello LG, Jogini V, Cortes DM, Pan AC, Gagnon DG, Dalmas O, Cordero-Morales JF, Chakrapani S, Roux B, Perozo E. Structural basis for the coupling between activation and inactivation gates in K^+ channels. *Nature* 466:272–275, 2010. doi:10.1038/nature09136.

Dhavalikar M, Narkeesh A, Gupta N. Effect of skin temperature on nerve conduction velocity and reliability of temperature correction formula in Indian females. *JESP* 5(1):24–29, 2009.

Di Francesco D, Noble DA. A model of the cardiac electrical activity incorporating ionic pumps and concentration changes. Simulations of ionic currents and concentration changes. *Philos Trans R Soc Lond* B307:353–398, 1985.

Egwolf B, Roux B. Ion selectivity of the KcsA channel: A perspective from multi-ion free energy landscapes. *J Mol Biol* 401:831–842, 2010. doi:10.1016/j.jmb.2010.07.006.

Franssen H, van den Bergh PYK. Nerve conduction studies in polyneuropathy: Practical physiology and patterns of abnormality. *Acta Neurol Belg* 106:73–81, 2006.

Geddes LA, Bourland JD. The strength–duration curve. *IEEE Trans Biomed Eng* 32(6):458–459, 1985.

Grant AO. Basic Science for the clinical electrophysiologist: Cardiac ion channels. *Circ Arrhythm Electrophysiol* 2:185–194, 2009. print ISSN 1941-3149; online ISSN 1941-3084; doi 10.1161/CIRCEP.108.789081; http://circep.ahajourna lsorg/content/2/2/185; http://circep.ahajournals.org/content/2/2/185.full. pdf+html; grant007@mc.duke.edu.

Gutman GA, Chandy KG, Adelman JP, Aiyar J, Bayliss DA, Clapham DE, Covarrubias M, Desir GV, Furuichi K, Ganetzky B, Garcia ML, Kurachi Y, Lazdunski M, Lesage F, Lester HA, Mckinnon D, Nichols CG, O'Kelly I, Robbins J, Robertson GA, Rudy B, Sanguinetti M, Seino S, Stuehmer W, M. Tamkun MM, Vandenberg CA, Wei A, Wulff H, Wymore RS. International union of pharmacology. XLI. Compendium of voltage-gated ion channels: Potassium channels. *Pharmacol Rev* 55(4):583–586, 2003.

Gutman GA, Chandy KG, Grissmer S, Lazdunski M, McKinnon D, Pardo LA, Robertson GA, Rudy B, Sanguinetti MC, Stuhmer W, Wang X. International union of pharmacology. LIII. Nomenclature and molecular relationships of voltage-gated potassium channels. *Pharmacol Rev* 57(4):473–508, 2005.

Hawking S. A Brief History of Time. My Brief History. Random House, Inc., New York, 127, 2013.

Hodgkin AL, Huxley AF. A quantitative description of membrane current and its application to conduction and excitation in nerve. *J Physiol* (London) 117:500–544, 1952.

Huynh W and Kiernan MC. Nerve conduction studies. *Aust Fam Phys* 40(9):693–697, 2011. http://www.neura.edu.au/sites/neura.edu.au/files/Nerve%20Con duction%20Studies.pdf.

Jin BE, Wulff H, Widdicombe JH, Zheng J, Bers DM, Puglisi JL. A simple device to illustrate the Einthoven triangle. *Adv Physiol Educ* 36:319–324, 2012. doi: 10.1152/advan.00029.2012.

Kimura J. Electrodiagnosis in diseases of nerve and muscle: Principles and practice. Oxford University Press, New York, 2001. ISBN 0-19-512977-6.

Lapicque L. Définition expérimentale de l'excitabilité [Experimental definition of excitability]. *Comptes Rendus des Séances de la Société de Biologie*, 67:280–283, 1909.

Lin YL, Meng Y, Jiang W, Roux B. Explaining why *Gleevec* is a specific and potent inhibitor of Abl kinase. *Proc Natl Acad Sci USA* 110:1664–1669, 2013. Note: *Gleevec* or *Glivec* is the commercial name of *imatinib*, a drug used in certain types of cancer.

Lindemann FA. The calculation of molecular vibration frequencies. *Z Phys* 11:609–612, 1910.

Lockless SW, Zhou M and MacKinnon R. Structural and thermodynamic properties of selective ion binding in a K+ channel. *PLoS Biol* 5(5, May):e121, 2007.

McAllister RE, Noble D, Tsien RW. Reconstruction of the electrical activity of cardiac Purkinje fibres. *J Physiol* 251(1):1–59, 1975.

Maxwell J. A dynamical theory of the electromagnetic field. *Roy Soc Trans* 155:459–512, 1865.

Maxwell JC. A treatise on electricity and magnetism. Dover Publications, New York, 1873. Two volumes; ISBN 0-486-60636-8 & 0-486-60637-6.

Merz KM Jr, Ringe D, Reynolds CH (editors). Drug design: Structure and ligand-based approaches. Cambridge University Press, New York, 287, 2010. online ISBN: 9780511730412; hardback ISBN:9780521887236; paperback ISBN:9781107475915; book doi: 10.1017/CBO9780511730412.

Miranda P, Contreras JE, Plested AJR, Sigworth FJ, Holmgren M, Giraldez T. State-dependent FRET reports calcium- and voltage-dependent gating-ring motions in BK channels. *Proc Nat Acad Sci*, 2013. Online before print March 11; 10.1073/pnas.1219611110.

Noble DA. A modification of the Hodgkin–Huxley equations applicable to Purkinje fibre action and pacemaker potentials. *J Physiol* (London) 160:317–352, 1962.

Noble DA. From the Hodgkin–Huxley axon to the virtual heart. Hodgkin–Huxley–Katz Prize Lecture. *J Physiol* 580(1):15–22, 2007.

Ostmeyer J, Chakrapani S, Pan AC, Perozo E, Roux B. Recovery from slow inactivation in K^+ channels is controlled by water molecules. *Nature* 501(Sept): 121–124, 2013. doi:10.1038/nature12395; online 28 July.

Padua L, Caliandro P, Erik Stalberg E. A novel approach to the measurement of motor conduction velocity using a Single Fibre EMG electrode. *Clin Neurophysiol* 118:1985–1990, 2007.

Payandeh J, Scheuer T, Zheng N, Catterall WA. The crystal structure of a voltage-gated sodium channel. *Nature* 475(21 July):353–358, 2011. doi:10.1038/nature10238.

Peng Du, Alan Wu. Smooth Muscle Electrophysiology and Siggraph 2008, Asian Conference, Bioeng Institute, Auckland University, 2009. http://www.abi.auckland.ac.nz/uoa/home/about/events/abi-events-seminars/events/template/event_item.jsp?cid=214973.

Petsko GA, Ringe D. Protein function and structure. New Science Press, London, 195, 2004.

PhysiologyWeb. Secondary Active Transport, 2000–2013. www.physiologyweb. com; http://www.physiologyweb.com/lecture_notes/membrane_transport/ secondary_active_transport.html.

Pinel John PJ. Biopsychology. 8th ed. Allyn & Bacon, Boston, 608, 2011. ISBN-10:0205832563; ISBN-13:9780205832569.

Rasband MN. Ion channels and excitable cells. *Nat Educ* 3(9):41, 2010. http:// www.nature.com/scitable/topicpage/ion-channels-and-excitable-cells-1440 6097.

Richmond JE, Featherstone DE, Hartmann HA, Ruben PC. Slow inactivation in human cardiac sodium channels. *Biophys J* 74:2945–2952, 1998.

Rohl CA, Boeckman FA, Baker C, Scheuer T, Catterall WA, Klevit RE. Solution structure of the sodium channel inactivation gate. *Biochemistry* 38:855–861, 1999.

Rosen MR, Brink PR, Cohen IS, Robinson RB. Basic science for the clinical electrophysiologist. Cardiac pacing: From biological to electronic . . . to biological? *Circ Arrhythm Electrophysiol* 1:54–61, 2008.

Roux B, Bernèche S, Egwolf B, Lev B, Noskov SY, Rowly CN, Yu H. Ion selectivity in channels and transporters. *J Gen Physiol* 137(5):415–426, 2011. http://www.physiol.gy/content/137/5/415.full; roux@uchicago.edu.

Skou JC. The influence of some cations on an adenosine triphosphatase from peripheral nerves. *Biochim Biophys Acta* 23(2):394–401, 1957.

Splawski I, Timothy KW, Decher N, Kumar P, Sachse FB, Beggs AH, Sanguinetti MC, Keating MT. Severe arrhythmia disorder caused by cardiac L-type calcium channel mutations. *Proc Natl Acad Sci USA* 102:8089–8096, 2005.

Valentinuzzi ME. Understanding the human machine: A primer to bioengineering. World Scientific Publisher, NJ & Singapore, xii + 396, 8 chapters. Series on bioengineering & biomed engineering, vol. 4, 2004. ISBN 981-238-930-X & ISBN 981-256-043-2.

Valentinuzzi ME. 50 Years a biomedical engineer: Remembering a long and fascinating journey. *BioMed Eng OnLine* 11:1, 2012. doi:10.1186/1475-925X-11-1.

Valentinuzzi ME, Geddes LA, Baker LE, Hoff HE. The node equation and its applications in physiology. *Med Biol Eng* 6(4):387–397, 1968.

Valentinuzzi ME, Kohen AJ. James Clerk Maxwell's, Kirchhoff's Laws and their implications in modelling physiology. *IEEE Pulse* 4(2, March–April):40–46, 2013.

Weiss G. Sur la possibilité de rendre comparable entre eux les appareils servant à l'excitation électrique [On the possibility of making direct comparisons between equipment used to create electrical excitation]. *Arch Ital Biol* 35: 413–446, 1901.

Wickens AP. Introduction to biopsychology. 3rd ed. Harlow, Essex, England and Pearson Education, New York (xxxvii + 626), 2009.

Xiong Q, Gao Z, Wang W, Li M. Activation of Kv7 (KCNQ) voltage-gated potassium channels by synthetic compounds. *Trends Pharmacol Sci* 29(2):99–107, 2007.

Yellen G. The moving parts of voltage-gated ion channels. *Quartely Rev Biophys* 31(3):239–295, 1998.

Yoram R. Molecular basis of cardiac action potential repolarization. In control and regulation of transport phenomena in the cardiac system. *Ann N Y Acad Sci* 1123:113–118, 2008.

Yu H, Noskov SY, Roux B. Two mechanisms of ion selectivity in protein binding sites. *Proc Natl Acad Sci* 107(47):20329–20334, 2010.

Yu Mike Chi, Tzyy-Ping Jung, Cauwenberghs G. Dry-contact and noncontact biopotential electrodes: Methodological review. *IEEE Rev BME* 3:106–119, 2010.

Zilberter YI, Starmer CF, Starobin J, Grant AO. Late Na channels in cardiac cells: The physiological role of background Na channels. *Biophys J* 67:153–160, 1994.

Zhang, Z-S, Tranquillo J, Neplioueve V, Bursac N, Grant A. Sodium channel kinetic changes that produce Brugada syndrome or progressive cardiac conduction system disease. *Am J Physiol Heart Circ Physiol* 292:H399–H407, 2007.

Ziervogel BK, Roux B. The binding of antibiotics in OmpF Porin. *Structure* 21:76–87, 2013.

CHAPTER 3

ANOTHER SOURCE: CARDIAC MECHANICAL ACTIVITY

Max E. Valentinuzzi and Alberto J. Kohen

It doesn't matter how good the electrical signal is... if no mechanical contraction shows up

And now a bold and pretentious poetic summary about the cardiovascular system,

With rhythmic pulses running down its road	(electrical activity)
the heart and blood move restlessly abroad.	(mechanical activity, blood reach)
The magic of its numbers is astounding	(measurements, heart rate...)
no matter how they're handled or transformed.	(signal processing)
Its reddish elixir is wonder	(blood)
that flows through those palpitant channels and yonder	(hemodynamics)
against oscillating resistance	(peripheral resistance)
and generously crossing fine barriers.	(capillaries)
We read all its messages, hesitatingly	(recording)
we think we understand them, expectantly	(diagnosis, interpretation)
while taking decisions, vacillatingly.	(therapeutics)
Oh! Circuit revered and much feared,	(still there are unknowns)
by probes and by catheters inspected,	(more studies are needed)
from inside our destiny thou markest!	(but in the end, we all die)

Abstract

Muscle contraction is the mechanical event, triggered by an electrical pulse, which in the end produces useful force, be it to lift a weight or to push a burst of blood forward. Basic laws, principles, and concepts govern such mechanical activity and they are revisited attempting to offer better mathematical background.

3.1 Brief Review

The previous poetic and pretentious account briefly summarizes what the system is about. From it, the variables and components of the cardiovascular system (CVS) can be anticipated:

Blood volume (V_b), an almost intuitive concept closely related to *mass* and *density*, is a volume that does not stay put but moves, so becoming *blood flow* (F_b), measured in terms of mass per unit time, to reach every single piece of tissue and, in so doing, guaranteeing *perfusion*. The total resting outflow, either from the left or from the right ventricle, has an average value of about 5–6 L/min (*cardiac output*, CO). Each vascular bed takes only a portion of this total value. When the pulsating action of the heart is considered, and as each ventricle ejects during each contraction the *stroke volume*, SV (about 70–80 mL/beat, at a frequency of 70/min), the CO value given above is easily verified. Without blood, the tissues die. *Ischemia* describes the condition characterized by a diminished perfusion. If it happens in the brain or in the ventricles, it may end up in serious cerebral or cardiac injury. It may take place also in any tissue, as, for example, in the gut. The existence of pressure does not necessarily assure a good passage of blood. An obstructed vessel, that is, with no flow, usually has good (and obviously useless) hydraulic pressure at the arterial side. How knowledge about CO developed along the decades, a good saga, indeed, is a nice chapter in the history of physiology that places Adolph Fick (1829–1901) as central actor (Valentinuzzi and Leder, 2013).

Blood pressure (P_b) is a clear physical concept (force exerted over a unit surface). Maintenance of a head pressure is what keeps the blood moving. Except for friction losses, arterial blood pressure at the entrance of any bed is the same. Its mean value is somewhat lower than 100 mmHg. Application of Poiseuille's law leads to the definition of a fourth (and this time compound) variable, p*eripheral resistance* R_p. If applied to a specific region, the concept of *regional resistance* R_r can also be defined uncovering vascular beds able to display wide variations (such as the skin or the guts) in opposition to the brain, heart, or kidneys, which always require patent vessels with low resistance.

The heart clearly is a muscle divided in two sections, left and right, each with an atrium and a ventricle. Hence, it has four chambers, two are mainly collecting ones (with a relatively minor propelling function), while the other two are essentially ejecting structures. The heart *in toto* acts as two hydraulic pumps connected in series. The unidirectionality of the circuit is manifested by the four cardiac valves (kind of hydraulic diodes). Moreover, the heart can be considered as a periodic pulse generator that, on the average, maintains an *equivalent constant flow*. Such flow, according to the requirements, can be adjusted (up to a maximum of about 30 L/min during heavy exercise, in a healthy young adult). However, this continuous equivalent concept is merely a practical simplification and it must be emphasized that, by its nature, *the circulatory system is pulsatile* (recall the concept of stroke volume). Even more, there is evidence indicating that life would not be viable, were the pump of the true constant type. Thus, apparently, pulsations are necessary, although some people question such stand. The bibliography is abundant and has been well reviewed (Gourlay and Taylor, 2000). In the cardiac arrest condition, for example, pulsatile extracorporeal circulation provides more blood flow, higher flow velocity and less resistance to coronary artery than non-pulsatile circulation, as contended by Son *et al.* (2005), a research group from Korea University. Their study was carried out in swines. Besides, it is now well recognized that the vascular system shows reflecting waves and a nonlinear behavior.

The problem of blood flow relies greatly on how *laminar* or *turbulent* is and, certainly, it varies with the vessel, the instant in the cardiac cycle and with the vascular bed, too, when away from the heart. If the fluid moves slowly or if it is very viscous (something like thickness or stickiness), it tends to be laminar, that is, it passes in layers and without mixing (when Reynolds number $Re < 2000$). Conversely, at high velocities, mixing takes place and the flow is called turbulent (when $Re > 4000$). A *transition* or *critical region* is encompassed by the $2000 < Re < 4000$ range. For example, in the descending aorta, Re is about 1000, and in the brain, it is in the order of 100. The situation appears more complex at the ventricular outflow and around the aortic arch. This number is defined as

$$Re = \text{convective inertial force/shear force} = \text{CIF/SF} \qquad (3.1)$$

after Osborne Reynolds (1842–1912). Another parameter significant in fluid dynamics is Womerseley number Wo, after John R. Womersley (1907–1958), and defined as

$$Wo = \text{transient inertial force/shear force} = \text{TIF/SF} \qquad (3.2)$$

When Re is large, flow is dominated by *convective* inertial effects (a motion by *convection* is caused by an external force, as, for example, gravity or temperature gradient); instead, if Re is small, flow is dominated by *shear* effects (*shear stress* arises from the force vector component parallel to the cross section, say, of a vessel). A large Wo (10 or more) means that oscillatory inertial forces predominate and that the velocity profile is flat. If Womersley parameter is low, viscous forces tend to dominate the flow, velocity profiles are parabolic in shape, and the centerline velocity oscillates in phase with the driving pressure gradient.

We must briefly now remind the Navier–Stokes (N–S) equation referred to a Cartesian system (Fung, 1997), that is,

$$\rho(\partial u/\partial t) + [\rho u(\partial u/\partial x) + \rho v(\partial u/\partial y) + \rho w(\partial u/\partial z)]$$
$$= \rho g - \partial p/\partial x + [\mu(\partial^2 u/\partial x^2) + \mu(\partial^2 v/\partial y^2) + \mu(\partial^2 w/\partial z^2)]$$
$$(3.3)$$

which, expressed in words, says that the transient inertial forces (TIF) plus the convective inertial forces (CIF) — first and second terms, respectively, on the left side of the equation above — are balanced out (see right hand side) by the sum of the gravitational force (GF), the pressure force (PF), and the viscous force (VF), or, in short,

$$\text{TIF} + \text{CIF} = \text{GF} + \text{PF} + \text{VF} \qquad (3.4)$$

Let us define each symbol for the sake of clarity, even if there is some repetition, including their dimensions, while underlining that the use of density makes everything referred to volume:

ρ is the fluid density (M/L^3); for blood it is $1.060\,\text{kg/m}^3$ at $37°\text{C}$;

u is the fluid velocity (M/T); for blood, it depends on the vessel;
$\rho(\partial u/\partial t)$ is the mass per unit volume times acceleration, an inertial force;
u, v, and w are the x, y, and z Cartesian components of the fluid velocity, each in (M/T);
$[\rho u(\partial u/\partial x) + \rho v(\partial v/\partial y) + \rho w(\partial u/\partial z)]$ are convective forces, where each partial derivative describes the velocity gradient in that direction. Observe that dimensions are consistent with force.

The three terms on the right-hand side of Eq. (3.3) — see also Eq. (3.4) — are identified as

(1) ρg, the gravitational component GF (M L^{-2} T^{-2});
(2) $\partial p/\partial x$, the pressure force that opposes the GF, also given in (M L^{-2} T^{-2}), and
(3) $[\mu(\partial^2 u/\partial x^2) + \mu(\partial^2 v/\partial y^2) + \mu(\partial^2 w/\partial z^2)]$, the viscous component VF, with μ, say, in (N s m^{-2}).

Besides, μ is the dynamic viscosity of the fluid, measured, for example, in (Pa s), equivalent to (N s/m^2). The viscosity of blood at body temperature is normally about 3.5×10^{-3} [Pa.s], or 3.5 centipoise (cP), the latter in the old CGS system [1 cP = 0.001 Pa s].

This important equation for the dynamics of fluids was developed by Claude-Louis Navier (1785–1836), and further improved later on by George Gabriel Stokes (1819–1903). From a philosophical viewpoint, all this can be criticized as a somewhat *reductionist* approach. Good point to remember René Descartes (1596–1650) who held that non-human animals could be explained as automata (Fig. 3.1), an interesting view followed by many others thereafter. However, and without downgrading such view that in many respects has been quite helpful, research proceeds and new knowledge is constantly added, sometimes correcting or even disproving a previous one.

Arterial stenosis constitutes a current circulatory pathology. One question that may be posed leads to how the mechanical concepts briefly stated above can help in its understanding. It has been found, for example, that the instantaneous streamlines of the physiological

Figure 3.1: Mechanical duck. Descartes's reductionist view of a non-human being. Amusing, indeed. Redrawn by Gustavo Idemi after download from https://en. wikipedia.org/wiki/Digesting_Duck. This kind of representation was often criticized by other scientists.

flow show considerable variation throughout the cardiac cycle, characterized by the formation of a vortex just downstream of a stenosed zone during the accelerating phase and the vortex elongation and formation of a second, also rotating vortex, during the decelerating phase. Moreover, the magnitude of *wall shear stress* (WSS) seems to be the largest at the peak systolic time, that is, when the flow reaches a maximum in the cycle. The unsteady flow produces a higher pressure drop than the steady flow, while the stenosed area leads to a net increase in the *oscillatory shear index* (OSI) at the far-downstream side of the artery. At higher Womerseley numbers, the instantaneous flow field shows a stronger influence of the flow profiles. With an increase in Womersley number, the peak value of time-averaged WSS decreases, while the OSI values exhibit an increasing trend (Banerjee *et al.*, 2012).

3.2 Characteristics and Properties of the Heart

The heart is simultaneously a pump and a muscle and, as any muscle, it has contractile elements (*sarcomeres*) that change their length.

The net result is a change in the cross-sectional horizontal ventricular diameter, relatively much larger than the change in the longitudinal (or *basal–apical*) length. Wall thickness also changes during contraction, and all this occurs in synchronism, both in the left and in the right ventricle. The increment in intraventricular pressures opens the arterial output valves, as these pressures become higher than the arterial blood pressures, so starting the ejection phase. Hence, valve opening is a *passive* phenomenon, with a pressure gradient being the *only* driving mechanism (no muscle or tiny motor is involved). In the aortic root, pressure reaches a maximum of about 120 mmHg (or 160 hPa, *systolic pressure*) and, because of its pulsatile characteristic, shows a minimum in the order of 80 mmHg (or 106 hPa, *diastolic pressure*). In the pulmonary artery, the maximum and minimum are, respectively, 25 and 10 mmHg (or 33.3 and 13.3 hPa) with a mean value of about 15 (or 20 hPa). The difference between the two is called the *differential systemic arterial pressure* or the *differential pulmonary pressure*, both with significant clinical implications. *Pulse pressure* is another term in use for the same concept. Within the ventricles, the numbers, instead, are rather different. In both ventricles, the minimum is always zero, which offers a splendid criterion to know when a catheter is within the chamber, going up to a maximum of 120 and 25 mmHg, respectively, for the left and for the right side, obviously coincident with the maxima given above (remember that the ventricular and the nearby arterial cavities become temporarily a single one when the valves are opened during the ejection phase). A unique characteristic of any muscle is its *contractility*, somewhat elusive in its definition for researchers still argue about which is the best. Nonetheless, as intraventricular pressures rise from zero to a maximum because of the contraction, the velocity of doing it, that is, the temporal derivative of the intraventricular pressure provides an acceptable means for estimating how good the muscle contracts. Hence, contractility, either of the left or of the right ventricle, can be defined as (dP_{iv}/dt), where P_{iv} stands for the corresponding intraventricular pressure. Usually, the maximum of this derivative, $(dP_{iv}/dt)_{max}$, is chosen as the proper descriptor. Cardiac contractility is an extremely important physiological and clinical parameter. Any of the four cardiac chambers possesses its own. However, the left ventricle is what people are mostly concerned about for in heart disease, very commonly, ventricular contraction decreased strength leads to cardiac insufficiency and failure. Nervous or hormonal influences can deeply affect cardiac contractility (and, therefore, the ability to develop pressure). For example, sympathetic activity will increase it (the so-called *positive inotropic effect*, producing stronger contractions and higher pressure. Other agents may elicit *negative inotropic responses* (i.e., weaker contractions). Cardiac biomechanics deals precisely with the subjects quickly overviewed in this paragraph, especially recalling two basic laws of the heart: *Starling's* and *Laplace's*.

The first one is physiological in nature, born around 1914, after a series of experiments carried out in London by Ernest Henry Starling (1866–1927) and his collaborators. The second is physical, to be found in a monumental opera, *La Mecanique Celeste*, authored by the French mathematician and astronomer Pierre Simon Laplace (1749–1827) and published in several volumes between 1790 and 1825.

However, there is more about its origins, different contributions, and its derivation (see below).

3.3 Starling's Law

The force of contraction increases as the initial fiber tension goes up (the nowadays so called *preload*). Although first described for the heart, it is also valid for skeletal muscle. Figure 3.2 is a record obtained from a naturally contracting turtle ventricle while, after each beat, the initial tension (it could be called "mechanical bias") was manually increased by a small amount using a specially designed spring mechanism attached to a force transducer.

The inset (right side of the same figure) displays the relationship between the force of contraction (measured as the difference between the maximum and the minimum per beat) and the corresponding stretch applied to the whole ventricle, both in grams (force), measured from the base line to the minima. Both, the temporal record and the plot, clearly show the important increase in the contraction

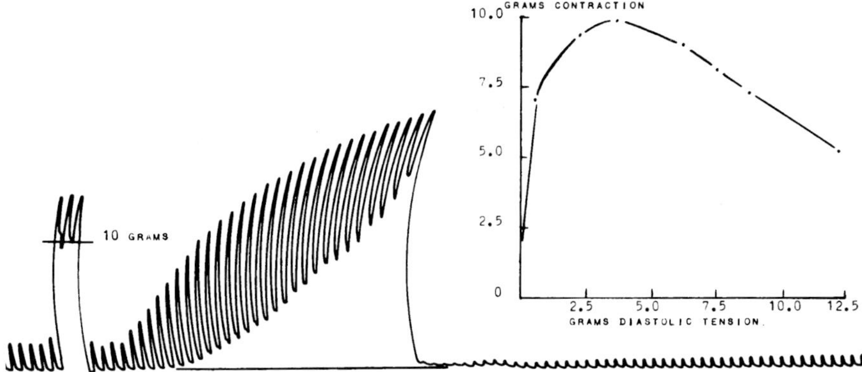

Figure 3.2: Starling's law of the heart. Ventricular spontaneous contractions of a turtle's heart as the organ were gradually stretched by means of a specially designed mechanical arrangement, which included a myographic transducer. Notice the increase in contraction with stretch up to a maximum value. Thereafter, because of overstretch, there was a decrease in contraction. The graph to the right shows clearly such maximum. Experimental data obtained from a turtle at the Physiology Department of Baylor College of Medicine, Houston, TX, 1966.

force. However, as the stretch proceeds, after reaching a maximum, the force goes down, indicating entrance into a dangerous operating region where probably, the overstretch may cause damage to the fibers. This is an easy and highly demonstrative experiment. Its foundation was presented by Starling himself in the Linacre Lecture, at Cambridge, England, in 1915, and published a few years later in 1918. His latter phrasing was "The law of the heart is thus the same as the law of muscular tissue generally, the energy of contraction, however measured, is a function of the length of the muscle fibre" (Patterson and Starling, 1914; Starling, 1918). This law is also called the Frank-Starling Mechanism, after the German physiologist Otto Frank (1865–1944), and often it is stated as, *the stroke volume of the heart increases in response to an increase in the volume of blood filling the heart (the end-diastolic volume) when all other factors remain constant.* However, there were several much earlier reports indicating that the phenomenon was known (Katz, 2002). Even some controversy aroused with an Italian physiologist and physician, Dario Maestrini, for which reason in Italy the law is often named as Starling-Maestrini's (Maestrini, 1915, 1925; Mazzoni,

2005; Farnetani, 2007). Physiologically, it should be added that stroke volume might increase, too, because of greater ventricular contractility during exercise, independently of the end-diastolic volume, as a direct result caused by the release of cathecholamines. Thus, the Frank-Starling mechanism appears to make its greatest contribution to increasing stroke volume at lower work rates, while contractility has its largest influence at higher work rates.

3.4 Laplace's Law

It refers to a film under tension, the so-called *surface tension* (T), which is defined as the *tangential force* (*tangent to the surface*) *applied perpendicularly to a unit length lying on the surface*. The theorem of Laplace applies to a membrane separating two spaces of any shape with tension in it. Any surface can always be decomposed into many "parachute-like caps", each defined by two mutually perpendicular circumferences of respective radii, R_1 and R_2, which are called the *principal radii of curvature*. Within the balloon, the pressure P (actually, it is the difference between the inside and the outside pressure, which most frequently is the atmospheric pressure) is given by

$$P = T \left[\frac{1}{R_1} + \frac{1}{R_2} \right] \tag{3.5}$$

where T is the surface tension, for example, in dynes/cm, and the radii, R's, say, in cm. It is easily seen that pressure comes out in dynes/cm^2. When radii are equal, we have a sphere (such as a bubble), and the equation simplifies into $P = 2T/R$ or $PR = 2T$. Adapted to the heart, this law states that the product of the intracavitary pressure (P_{ic}) × equivalent radius (R_{eq}) is directly proportional to the wall stress (W_s) × wall thickness h product, that is

$$P_{ic} R_{eq} = khW_S \tag{3.6}$$

where k is a constant and h stands for the wall thickness, and recalling that surface tension over wall thickness defines wall stress, as an expression of surface tension of each muscle layer moving from the epicardium to the endocardium. The relationships (3.5) and (3.6) are often found in physiology textbooks when referring to hollow organs such as the cardiac chambers, blood vessels, bladder, stomach, uterus, lungs, or the like. Its historical development was presented in detail in a series of three articles by Valentinuzzi and Kohen (2011a, 2011b, Valentinuzzi *et al.*, 2011), from where only the best mathematical derivation is reproduced herein plus a few other necessary parts.

3.4.1 *Some historical comments and a few patho-physiological applications of the law*

Sure enough, it can be said that *Le Marquis* Pierre Simon Laplace (1749–1827) was not the first who cared about the heart, vessels, and bubbles, in fact, he never thought about such anatomical organs or physical objects. Somehow, someone in the middle credited him with the law, which, incidentally, is not usually taught in regular physics courses. *The subject Le Marquis did study was surface tension of liquids contained in capillaries*, and it must be added, with due respect, in a rather cryptic way (Valentinuzzi and Kohen, 2011a). Laplace's law has been referred, too, frequently but poorly in the regular physiology textbooks, and even errors can be detected. No good and convincing mathematical derivation is given, and no clue seems to be found regarding its first inclusion in physiology — indeed disappointing.

Consider first a balloon of any shape and negligible wall thickness, when air is blown in, a certain volume needs to build up before a pressure gauge detects a small pressure value, in turn developing some counteracting surface tension. As more air gets in, the pressure inside increases and the surface tension also increases. When the balloon becomes too big, simple finger-touching indicates a large surface tension and, therefore, a possible rupture risk. These phenomena are static in nature. In a blood vessel, using Eq. (3.5), it is seen that the smaller the vessel's radius, the smaller the wall tension to balance the distending pressure, leaving pressure constant. Moreover, when pressure in a small vessel goes down, a point is reached where blood flow becomes zero, although the pressure is not zero. The latter is called *critical closing pressure* (see previous paragraph, when a balloon is inflated, and compare both statements, as the latter phenomenon

Table 3.1: Laplace's law applied to cylindrical vessels.

Vessel	P	R	$PR = T$	Wall thickness h
Type	(mmHg) (dynes/cm^2)	(cm)	(dynes/cm)	(mm)
Aorta	120			
	160,000	1.25	**200,000 C**	2.00
			170,000 G	
Vena cava	8	1.50		
	11,000	1.50	**16,500 C**	1.50
			21,000 G	

Notes: 1 mmHg = 1333 dynes/cm^2. For the aorta, we have taken the systolic pressure; for the vena cava, 8 mmHg were accepted. Check the figures of this table with others reported in the literature. C means computed with $PR = T$; G means data from Ganong's textbook of physiology (any edition would do it). Fourth column gives the surface tension.

describes its counterpart). Also, notice that surface tension (in a thin film) appears as loosely and confusedly mixed with the concept of wall stress (where thickness plays a role). A few numerical values illustrate this (Table 3.1).

An aneurysm, literally meaning widening, from Greek, is a sac formed by the dilatation of the walls of an artery or vein. The elastic properties change dramatically so that the walls become more compliant and, thus, mechanically weaker. Since surface tension in a cylinder appears only tangentially and along the circumferential cross section, as the vessel weakens, a dissecting aneurysm (wall rupture) only occurs longitudinally, because the pull is perpendicular to the vessel direction. The law predicts such behavior, and clinical practice demonstrates it. It is pertinent to make a comment: Traditionally, open surgery was the only treatment of an aneurysm. Greenhalgh *et al.* (2004) addressed the uncertainties about how endovascular repair compares with conventional open surgery. Between 1999 and 2003, slightly more than 1000 elective patients were randomized to receive either endoscopic or open surgery repair. Patients of both sexes aged at least 60 years were recruited for the study at 41 British hospitals. They concluded that the endoscopic procedure reduced the 30-day operative mortality by two-thirds compared with open repair. The use of stents for the aorta is nowadays widespread, and has almost fully replaced open surgery.

Back to Laplace's law; consider now any of the cardiac chambers, say, the left ventricle. A dilated heart shows a larger equivalent radius, while its wall becomes thinner than normal. Thus, the myocardium must develop a greater tension during contraction to sustain the required pressure, and in the end, poor ejection of blood may occur. Conversely, if hypertrophy takes place (say, because of hypertension or excess physical training), the ventricular wall gets thicker, partially compensating for the necessary increase in wall stress. We do not enter into the etiology of these cardiac pathologies; we underline only the law that explains the behavior.

Another example: Anatomical observations in normal ventricles showed that the radius at the basal region is larger than at the apical zone, thus approximating the left ventricle with two superimposed spheres, an upper and a lower one, as Valentinuzzi *et al.* (1987) did. Since the pressure inside is the same, Laplace's law clearly explains why the wall is thicker in the former than in the latter. Besides, the right ventricle develops a much lower pressure than the left, and, consequently, its wall is thinner (Martin and Haines, 1970).

The law also finds a place in respiratory physiology. A substance called *surfactant* controls alveolar surface tension, and T is inversely proportional to its concentration. Surfactant molecules are spread apart as alveolar size increases during inspiration (concentration gets down) and move closer together at expiration (concentration goes up), thus adjusting T during breathing. Where this is not the case, the alveoli would collapse at expiration because of the too high surface tension. Another interesting and significant phenomenon is that a small alveolus connected to a larger one empties its air content in the latter, simply because its surface tension surpasses that of the bigger alveolus. Atelectasis and hyaline membrane are the two classical examples to illustrate the importance of alveolar surface tension in the respiratory act, both directly linked to Laplace's law.

Micturition is a physiological function usually studied by cystometry, which relates intravesical volume with intravesical pressure;

hence, a nice situation to look at through the law. A normal cystometry shows a middle long flat portion indicating volume increase due to vesical filling but constant pressure, because the radius of the cavity increases with surface tension, thus compensating for the enlargement. The micturition reflex is triggered at a critical high volume and surface tension.

The pregnant uterus appears as another example where the law is very handy. It shows a behavior similar to that of the bladder because surface, or better, wall stress accommodates to the slowly increasing volume (de Snoo, 1936; Valentinuzzi Sr, 1939, 1950).

3.4.2 *Laplace's law: Mathematical foundation*

Soap bubbles, rubber balloons and distended bellies have been seen and experienced by men over centuries, giving them the idea of tension and eventually the feeling of blowing out, until someone found a physico-mathematical explanation to close the loop from practical to theoretical knowledge and so making a few guys happy, which does not mean at all that the inflated sensation will not show up again and that cattle in the country will not be treated with a stab in the tummy by a would-be-veterinary-doctor to relief bloat.

The following equation is what *Le Marquis* published in his already mentioned *Mecanique Celeste,*

$$gD(h + Z) = \left(\frac{1}{2}\right) H \left(\frac{1}{R} + \frac{1}{R'}\right) \tag{3.7}$$

Here, g is the Earth gravity, D is the the fluid density, and $(h + z)$ represents the height of the point above level, confusedly split into two terms; for the right-hand side of the equation, H is obscurely presented in another page as a "positive quantity decreasing very rapidly". To complicate things further, that H is also defined as a function of f, the latter standing for the distance between a solid differential element and point O (we assume that it is the point located above level), where an osculating tangential force is exerted. All these definitions must be extracted from different parts in the text. Equation (3.7) is the closest form in the Celestial Mechanics resembling what now is called Laplace's law (Valentinuzzi and Kohen, 2011b). Obviously, not at all a clear piece, especially if no adequate figure accompanies it to better explain how the overall system was thought. Our conclusion forces us to state that Pierre Simon, in a quite wavy and meandrous mathematical fashion, came out with a physical relationship that requires adaptation to the nowadays-accepted form. Thus, after comparing (3.7) with Eq. (3.5), respectively, equating both left-hand and right-hand members, we obtain

$$gD(h + z) = P \qquad (3.8)$$

$$\left(\frac{1}{2}\right) H \left(\frac{1}{R} + \frac{1}{R'}\right) = T \left[\frac{1}{R_1} + \frac{1}{R_2}\right] \qquad (3.9)$$

which dimensionally should be compatible if they are to express the same quantities or magnitudes, that is, the dimensions in Eq. (3.9) are length $\times$ mass/time^2 or force/length^2, obviously describing pressure. In the second equation, while the radii $R = R_1$ and $R' = R_2$, the only possibility left over is surface tension $T = (1/2)H$, measured, say, in dynes/cm or, more general, *force per unit length*. Thus, in the end, Eq. (3.9) is expressed in units of pressure, too.

This is not the place to go into details of the law's development as they were well told by Valentinuzzi and Kohen (2011a, 2011b). Briefly, it can be stated that, from a historical viewpoint, it would be Laplace who, based on the capillary effect, was, indeed, the first to obtain an equation resembling what is now called Laplace's law, and we stress the word "resembles" because his manifested equation needs the manipulations shown above to find an analogy to the current statement of the law. Other researchers were concerned with the same subject (Valentinuzzi and Kohen, 2011a, 2011b), but none of them reached or overtly printed in their reports the equation we are familiar with. Gauss seems to have been the first to mention Laplace.

3.4.3 *Laplace's law: Now a new and good demonstration*

In 2009, a small seminar was organized at the Institute of Biomedical Engineering of the University of Buenos Aires with the idea of discussing this old, ubiquitous, and slithering law. A few young electrical engineering students showed up, and one of them — Federico Armesto — now already graduated — gave us what, with his explicit

     M. E. Valentinuzzi & A. J. Kohen

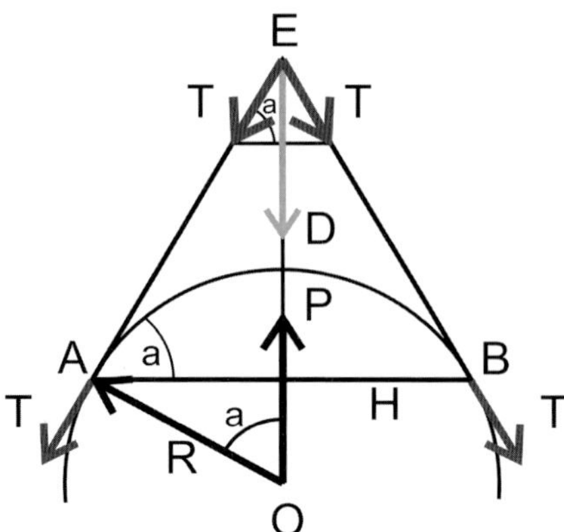

Figure 3.3: One of the principal planes. Force D: obtained after the added projections on the axis of tensions T, that is, $D = 2T \sin a$. Pressure P is constant along segment $AB = 2H$. Its associated force $F = P(2H)$. Forces D and F must balance each other to keep equilibrium: $D = F = 2T \sin(a) = P(2H)$. Angle a is equal to angles EAB, AEO, and EOA. Besides, $\sin a = H/R$, which, when replaced in the expression above, produces $T \sin a = TH/R = HP$, from which $T = PR$. In space, there is another plane perpendicular to the surface of this page. Notice that the central z-axis is the same (see also Fig. 3.4).

permission, here is translated and reproduced, including some minor changes to make it more readable (Valentinuzzi and Kohen, 2011b). To the best of our knowledge, this is one of the best demonstrations ever produced.

A volumetric body of any shape encompassing a cavity (as a Bunny-Rabbit birthday party balloon), when undergoing internal and external pressures, can be decomposed in semispherical shell elements defined by two main radii (Fig. 3.3). It is assumed with a wall of negligible thickness. The pressure difference between the external and internal medium is called P while T stands for surface tension, the latter in units of force over length. In space, the pressure is exerted in the perpendicular direction (upward, in the figure) exactly at the intersection of the two principal radii $R_1 = R_x$ and $R_2 = R_y$, which characterize the shell. Each tangential force projects to the vertical z-axis along a curved line. As pressure is force per unit area, we would get force per unit length by calculating its flux as a linear integral along line AB. Up to here, everything is

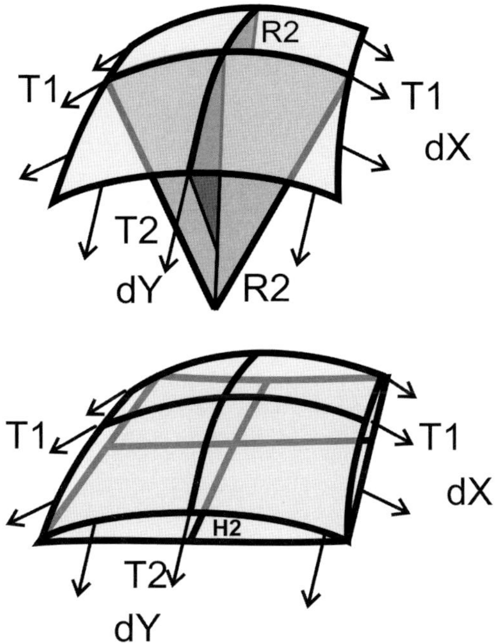

Figure 3.4: Shell element in space (above). The rectangle $H1--H2$ can be defined as a differential approximation $dX-dY$ of the curved surface (below). A flux F due to P would perpendicularly cross such rectangular area, from below upward on the z-axis. Tensions $T1$ and $T2$ are shown on both perpendicular planes.

practically a repetition of De Snoo's demonstration (1936). In three dimensions, the force F in equilibrium with D becomes the flux of pressure P crossing the rectangle formed by H_x and H_y, after differential approximation of the shell, that is,

$$F = \iint_H pdA = D \qquad (3.10)$$

or, decomposing into two linear integrals, F is obtained by linear integration over one axis with the tension on the opposite axis. It must be taken into account that the integral results in either dx or dy because sines are constant within them, so that, recalling De Snoo's geometrical equations (Fig. 3.3), we can write

$$F = \int_{dx} 2T_y \sin \alpha_y dl_x + \int_{dy} 2T_x \sin \alpha_x \, dl_x \qquad (3.11)$$

or

$$F = 2T_y \sin \alpha_y \int_{dx} dl_x + 2T_x \sin \alpha_x \int_{dy} dl_y$$

$$= 2T_y \sin \alpha_y \, dx + 2T_x \sin \alpha_x \, dy \tag{3.12}$$

where T_x and α_x and T_y and α_y correspond to the tensions and angles on the x and y axes, respectively. However, the pressure P is constant all over, and hence, it can be taken out of the integral, while T_x and T_y remain constant within the limits of the shell, meaning that they too can be shifted out. As it was done in the bidimensional demonstration, sines are replaced by each corresponding H/R. The area $dA = dx \cdot dy$, and moreover, H_x and H_y in their differential form, can be replaced by $dx/2$ and $dy/2$, while the integrals, under these conditions, can also be deleted. Thus, Eq. (3.13) becomes

$$P \cdot dx \cdot dy = F = T_y \frac{dy}{R_y} dx + T_x \frac{dy}{R_x} dy = D \tag{3.13}$$

From the latter, it easily follows Eq. (3.14) by solving for P, that is,

$$P = \frac{T_x}{R_x} + \frac{T_y}{R_y} \tag{3.14}$$

This last demonstration appears simple and rigorous, only needing the introduction of the wall thickness concept, as done earlier in the text, that is, $W_s = T/h$, and T can be, if needed, taken over the x and y directions, respectively, so giving more generality to the expression. We must add that a Russian engineer, Feodosiev (1972), came also up with a correct and general derivation, but based on a rigid (or highly inelastic) wall thickness (Valentinuzzi and Kohen, 2011b).

Summarizing this section:

Historically, there is no doubt that Le Marquis was the first to deal with this subject, although on the basis of the surface phenomena presented by the capillary effect. Armesto, a young student, produced a clever and excellent approach. We must admit and underline that the reduction of any hollow cavity to a sphere after calculating an equivalent radius (as carried out in many physiology textbooks) leads to good conceptual results and is useful in the understanding of pathologies. Even epistemological considerations have been taken into account regarding this law (Valentinuzzi *et al.*, 2011).

3.5 Fick Principle and the Continuity Equation

Blood flow, F_b, is an essential variable of the cardiovascular system (CVS), in any bed, and the total flow or cardiac output, $F_t = \text{CO}$, is obviously the main source to the rest of the circulatory network. This amount exits the left ventricle at high pressure, enters the right heart via the vena cava at very low pressure, is expelled by the right ventricle at moderately low pressure, and finally returns to the right atrium at very low pressure. If the lungs are considered as a node (Fig. 3.5), the *continuity principle* applied to blood (as carrier, in mLblood/min) and oxygen (as transported substance, in mLO_2/mLblood) establishes in the steady-state condition that,

$$F_t[V] + F_{ox} = F_t[A] \qquad \text{in } (\text{mLO}_2/\text{min}) \qquad (3.15)$$

where $[V]$ and $[A]$, respectively, stand for the concentration of oxygen in venous and in arterial blood, while F_{ox} represents the net oxygen uptake in $(\text{mLO}_2/\text{min})$ via the respiratory system. Solving for F_t results in

$$F_t = \frac{F_{ox}}{\{[A] - [V]\}} \qquad (\text{mLblood/min}) \qquad (3.16)$$

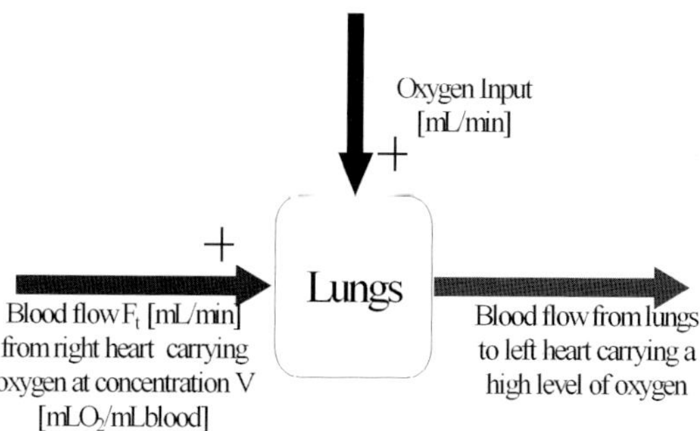

Figure 3.5: Fick's principle. In the steady state, everything that goes in (per unit time) must also go out. It is a concept similar to Kirchhoff's Second Law in the nodes of an electric circuit.

which is the famous and well-known Fick's formula, conceptually similar to the second Kirchhoff's law we derived above (section 2.6.1, Eq. (2.iv)) and is commonly applied to nodes in electric networks, that is, the total current converging to and diverging from a node. After all, recall that current is nothing but the amount of electric charge per unit time (1 amp = 1 coulomb/s) and in Eq. (3.15) we have amount of oxygen also per unit time. The numerator in Eq. (3.16) is usually obtained from a metabolimeter (an easy measurement). A normal adult at rest may take about 250 mLO$_2$/min. A sample of blood from any artery (the method, thus, requires arterial puncture) and, subsequent determination in the biochemistry lab, gives the arterial concentration of oxygen. The venous concentration of oxygen is not easy. Samples from a peripheral vein are not acceptable because the oxygen consumption varies from tissue to tissue. *A representative sample has to be a mixture coming from all tissues.* Only the right atrium, or better, the right ventricle, or the best, the pulmonary artery, carries venous blood meeting such requirement. Hence, a probing catheter must be introduced to any of these vascular places in order to withdraw a few milliliters of blood to be tested in the lab for oxygen content. Typical expected normal values are 20 mLO$_2$/100 mLblood, for the oxygenated blood, and about 15 mLO$_2$/100 mLblood, for the mixture of venous blood. Thus, the arteriovenous difference is about 5. These units are often referred to by physiologists as 15 volume percent. When the above figures are replaced in Eq. (3.16), the result is a steady-state value of 5 L/min. The historical development of cardiac output was slowly and painfully obtained, it is highly attractive, and shows many human facets (Valentinuzzi and Leder, 2013).

3.6 The Pressure–Volume Diagrams

3.6.1 *Origins*

Nicolas Léonard Sadi Carnot (1796–1832) was a French physicist and engineer who, in 1824, published a little and yet seminal contribution to science and technology: *Reflexions sur la puissance motrice du feu et sur les machines propres a developer cette puissance* (Thoughts about the power of fire and about the proper engines to develop such power). It was the foundational hallmark of *thermodynamics*, indeed, essential for the understanding and design of the *internal combustion machines* (such as the steam, gasoline, and diesel engines). In them, the pressure and volume handled by the cylinders with movable pistons, which, in the end, transmit the force to produce the rotation of a shaft, give rise to the so-called *pressure–volume diagrams* or *loops*.

Theoretical and practical thermodynamics developed rapidly all through the XIXth century. Cardiac physiology, in turn, made significant progress. It was Otto Frank (1865–1944), in Germany in 1899, the first to bring the concept of pressure–volume loop into physiology, when with very rudimentary means he produced such a diagram from a frog's ventricle. However, probably because of technical difficulties, the idea did not advance much for considerable time. In the 1970s and 1980s, the group of the late Kiichi Sagawa (1926–1989), at Johns Hopkins University, after a careful and long series of studies, brought the concept back and putting this engineering concept in its adequate place. Moreover, they interpreted it within the overall frame of cardiac physiology, opening with it a complementary new and fresh vision (Sagawa *et al.*, 1988).

3.6.2 *Generation, description of a PV-loop and the FSSS line*

Electronics Engineering students are familiar with the Lissajous figures. They are predictable nice looking composite patterns obtained by combining two sinusoidal signals injected, simultaneously and respectively, to the x-horizontal and y-vertical plates of an oscilloscope after disconnecting the horizontal internal sweep. The pressure–volume diagrams are some sort of Lissajous patterns except that the component signals, while still periodic, are not sinusoidal. One is the intraventricular pressure, P_{iv}, and the other, the intraventricular volume, V_{iv}, both as time course events experimentally obtained. In fact, any of the four cardiac chambers can produce such a loop.

Figure 3.6 displays 10 beats recorded from an experimental dog. The upper channel is the left intraventricular volume, calibrated in

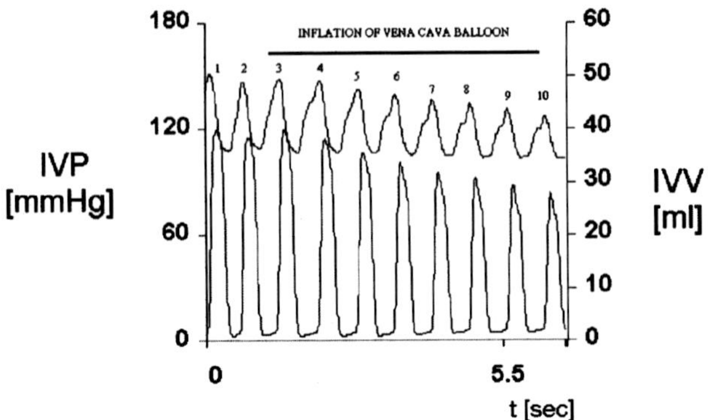

Figure 3.6: Intraventricular volume and pressure records. They are displayed as time course events during a preload maneuver. Volume above (calibration on the right side) and pressure below (calibration on the left). Obtained at the Department of Bioengineering, UNT, 1995, from an experimental dog.

milliliters and detected by a set of electrodes which measure blood conductance within the chamber (conductance is proportional to volume), while the lower channel is the left intraventricular pressure, calibrated in mmHg and detected by a miniature transducer, also placed within the chamber. For each beat, volume, as time proceeds, shows a downstroke from a maximum value (*end-diastolic volume*) to a minimum level (*end-systolic volume*), thus, describing *ejection*. Thereafter, it follows a rising limb back to a peak value and, hence, marking the *filling phase* of the left ventricle. Observe that the ventricle never empties completely. Pressure, instead, always starts from zero (or almost zero) abruptly climbing to the maximum value and displaying a rectangular-like shape.

Figure 3.7 combines beats 2 and 3 shown in Fig. 3.6 to yield the Lissajous-like shape of two typical PV loops, both almost coincident. Roughly, the corners of this quasi-quadrangular figure identify the four characteristic points of the *cardiac cycle*. They signal the openings and closures of the cardiac valves:

$A(V_{ed}, P_{ed})$, defined by the *end-diastolic volume* and *pressure*, respectively; $B(V_{ed}, P_{op})$, defined by the same *end-diastolic volume* and a much higher pressure than the preceding one, which can be called the *opening pressure* because it is larger than the aortic

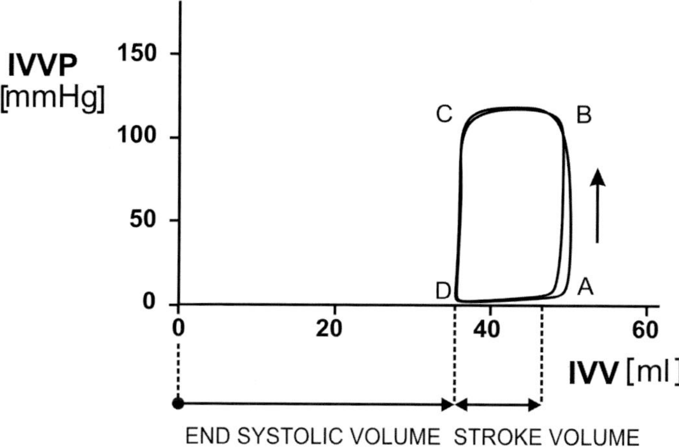

Figure 3.7: Intraventricular volume-pressure diagrams. Beats 2 and 3 of the previous figure were combined as orthogonal signals producing a Lissajous-like pattern, that is, two almost coincident PV-loops. The two isovolumetric phases (relaxation, on the left, and contraction, on the right) take place along the levels, respectively, of the end-systolic volume, and of the end-systolic volume plus stroke volume = end-diastolic volume. Points A and C mark the beginning and end of the mechanical systole (subdivided in the isometric contraction and ejection phases). The vertical arrow on the right shows the counter-clockwise rotation of each beat. On its way back to A, the cycle enters into its diastolic portion (subdivided, in turn, into relaxation and mainly passive ventricular filling, the latter from D to A). The four corners correspond to the openings and closures of both left cardiac valves. Obtained at the Department of Bioengineering, UNT, in 1995, from an experimental dog.

pressure and, thus, it opens the aortic valve, so starting the *ejection phase*. Ventricular contraction started at A, which coincides with the closure of the *mitral valve* (for the intraventricular pressure becomes higher than left intra-atrial pressure), intraventricular pressure rises to P_{op} and that contraction keeps on going sustaining pressure during all ejection until the pressure inside the chamber falls below the aortic level triggering the *closure of the aortic valve*. That is point $C(V_{es}, P_{es})$ defined by the *end-systolic volume* and *pressure*, respectively. The same point marks the end of ventricular mechanical systole, as point A characterizes its beginning. The canine experimental records of Figs. 3.6 and 3.7 were obtained with a laboratory custom-made admittance meter (Herrera *et al.*, 1986).

Thereafter, the myocardium relaxes *isovolumetrically* (at constant volume) at a fast decreasing pressure, until it reaches zero or almost zero, when the low but higher left intra-atrial pressure opens the *mitral atrioventricular valve*. It is point $D(V_{es}, P_o)$, defined by the previous *end-systolic volume* and an almost zero intraventricular pressure, and signaling the beginning of ventricular *filling phase*, which ends at A, to start all over again in the next beat. Hence, the whole cardiac cycle is divided in two halves: systole, between A and C (closure of the mitral valve and of the aortic valve), and diastole, between C and A (closure of the aortic valve and of the mitral valve). Each semi-cycle is, in turn, subdivided in two phases: First, the *isovolumetric contraction* of the ventricle, between A and B, that is, when both valves are closed and, thus, the chamber cannot change its volume (blood is essentially incompressible). Most of the energy consumption takes place during that short phase. Second, the already introduced *ejection phase*, between B and C, when the aortic valve is fully opened and blood is propelled into the elastic aortic reservoir. Finally, at diastole, we find the *isovolumetric relaxation*, between C and D, and the last *filling phase*.

The section title above anticipates an *FSSS line*. What is it? Figure 3.8 displays eight loops after composing eight pressure and volume beats, both as time course events, during inflation of an intra-caval balloon, as indicated in Fig. 3.6 (upper continuous bar). This is called in the hemodynamics jargon a *preload maneuver*. Transient occlusion of that great vessel (10–30 s, at the most) significantly reduces the venous return to the left atrium, in turn reducing the amount of blood to the left ventricle. According to Starling's law, a reduction both in pressure and in stroke volume must be expected, as clearly shown in the temporal records. The manifestation of such a phenomenon in the PV-plane is a sliding of the loops to the left, with the end-systolic point (C point) describing a straight line which is called the *Frank–Starling–Suga–Sagawa* (or *FSSS*) *line*. It has been demonstrated that that line (essentially its slope) can be used as an estimator of the contractility of the myocardium, or, in other words, as a measure of the myocardium inotropic condition. A linear regression of the end-systolic points obtained by a preload maneuver

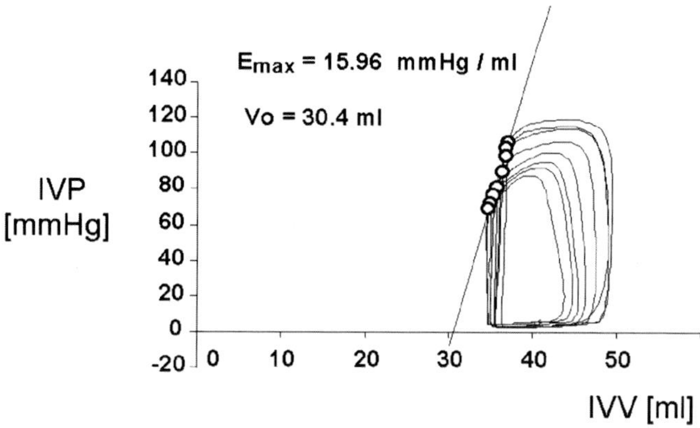

Figure 3.8: End-systolic line (or Frank–Starling–Suga–Sagawa Line or FSSS line). It was obtained with a preload maneuver, that is, a partial occlusion of the venous return to the right heart, producing a shift to the left and downward of the PV loops. The latter behaved as "if hanging from the line and sliding down". Obtained from a patient at the Institute of Cardiology of Tucumán with the technical assistance of the Department of Bioengineering, UNT, 1996.

produces an equation of the type,

$$P_{\mathrm{es}} = E_{\mathrm{max}}(V_o - V_{\mathrm{es}}) \tag{3.17}$$

where E_{max} stands for its slope, and V_o, for the horizontal axis intercept. There has been considerable discussion over the detailed definition of the C points on the loops and over the meaning of the volume for zero pressure. One relatively well-accepted stance is that the end-systolic point occurs where the absolute slope of the PV-loop is maximum, that is, when P/V reaches a maximum. However, the E_{max} is a statistical value obtained from the collection of perhaps 8–10 such maximum slope data points in a single maneuver. When pressure is zero, there is a residual volume which Sagawa and collaborators named the *dead volume*. Much has been written regarding its interpretation. It shows wide spread and its true meaning still remains controversial, especially when now and then some negative values come up.

The filling phase of the cardiac cycle corresponds to the passive increase in pressure as volume is steadily increased in a balloon, an experiment easily carried out in any lab. If one measures the increments ΔV, with a calibrated syringe, and simultaneously one reads the resulting increments ΔP, with a manometer, the overall passive PV-relationship can be plotted, so graphically describing the elastic properties of the material. Under ideal conditions, it is linear, following the well-known Hooke's law, that is,

$$P = a + kV \qquad (3.18)$$

from which, by differentiation, one gets $dP/dV = k$, the *elastance* of the material, obviously the inverse of the compliance.

Let us underline that during a preload maneuver, the PV-loops shift progressively to the left, with the end-systolic points sliding down the FSSS line, as if "hanging" from a cable. The base, that is, the filling phase DA, also slides to the left and slightly downwards, following the passive elastic PV-relationship. Such relationship, however, is not linear, being usually approximated by an exponential function; thus, the latter acts as some sort of lower rail. An inflatable balloon can also be inserted in the thoracic aorta via, say, one of the femoral arteries. During inflation for a few seconds, it tends to hinder the outflow of blood from the left ventricle leading to an increase in intraventricular pressure and a transient accumulation of blood in it. This is called an *afterload maneuver*. The PV-loops shift upward and to the right, but always "hanging" from the FSSS line and sliding over the passive non-linear PV-relationship. In normal physiological conditions, the heart moves constantly in this fashion, adjusting itself almost beat by beat to the demands. Even more, certain substances (as, e.g., epinephrine, in the case of exercising) can modify the position of the FSSS line (steeper slope) and, hence, offering more room for the left ventricle to operate. Everything that has been said so far for the left ventricle is also applicable to the right, keeping in mind the lower pressure it works with. An ischemic ventricle, instead, will contract with less force and the line would drop (lower slope). Records shown in Figs 3.6–3.8 were obtained with a conductance catheter and an admittance meter (see next section).

3.6.3 *Conductance catheter and the admittance technique*

Volume has been an elusive variable to measure, especially in irregular containers and even more in elastic cavities like those found in living organisms, which, on top of all that, show contracting characteristics including compliance changes. Baan *et al.* (1981), in Holland, after considerable previous trials, came up with the impedance catheter to continually recording intraventricular volume. Independently, a group led by M. E. Valentinuzzi, in Argentina, developed a similar system running many animal tests and also in a few patients (Herrera *et al.*, 1986; Valentinuzzi and Spinelli, 1989; Spinelli *et al.*, 1989; Herrera *et al.*, 1993; Herrera *et al.*, 1997).

Ever since those initial years (roughly, between 1980 and 2000), several advances took place, especially applied to transgenic rats, using much better and sophisticated technology while demonstrating significant improvements in the understanding of cardiac mechanics and the genesis of myocardial disease. The literature abounds and this is not the place to even trying a review; thus, we just mention briefly a few significant contributions.

Uemura *et al.* (2004), using *Bluetooth* technology, developed an implantable telemetry system for measurement of the left ventricular pressure–volume relation in conscious, freely moving rats. The telemetry system consisted of a pressure–conductance catheter connected to a small implantable transmitter. Calibrations were also conducted telemetrically. The telemetric SV closely well correlated with the flowmetric SV during inferior vena cava occlusions ($y = 0.96x + 7.5, r^2 = 0.96, n = 4$). These authors conclude that the telemetry system enables the estimation of the pressure–volume relation with reasonable accuracy and reproducibility, at least in conscious untethered rats.

Note: *Bluetooth* is a wireless way of transmitting signals from one device to another. It may be also described as a protocol for exchanging data over short distances (using frequencies in the range 2400–2500 MHz) from fixed and mobile devices, creating Personal Area Networks (PAN's). *Bluetooth* signals travel through walls, doors and usually have a good reach. It was named after Harald "Bluetooth" Gormsson (born ca 935),

King of Denmark and Norway. Authors' guess: He probably had a visible stained tooth!

Technology was of concern, so that the same year, Raghavan *et al.* (2004), from the University of Texas at Austin, USA, used a tetrapolar catheter inserted in the left ventricle of transgenic mice and a dual-frequency technique involving 1 and 100 kHz. It had been established that the imaginary part (capacitive reactance) of the complex admittance of the cardiac muscle is much smaller in the lower frequency than at the higher frequency. The design involved generation of an accurate frequency source for both frequencies, careful selection of operational amplifiers for the current conversion stage so that the current is not too large to kill the animal and that it is capable of performing at high frequencies. The overall circuit was designed with minimal shift in the phase due to the circuit elements alone. The authors applied LabVIEW and a digital oscilloscope for effective data acquisition even at high frequencies.

In a very nice experimental and theoretical paper, Wei *et al.* (2005) explored the effects of the non-uniform electric field on the conductance–volume relationship. The classic equation formerly introduced by Baan *et al.*, (1981) is linear and cannot predict the conductance–volume relationship very well, especially in larger volume ranges. Even in the mouse small LV volume range, a volume offset is unavoidable by use of the classic equation. If the empirical equation proposed by these authors is used instead, volume error can be reduced to only about 10% of the original value. They concluded that the nonlinear empirical conductance-to-volume equation showed the best overall performance.

The same group of researchers (Wei *et al.*, 2007) proposed that both the capacitive and the resistive components of the myocardium are substantial, and neither should be ignored. Hence, the measured result should be labeled admittance rather than conductance. They measured the admittance magnitude and its phase angle of the left ventricle in the mouse, showing that they increase with frequency. Further, this more accurate technique suggests that the myocardial contribution to measured admittance varies between end-systole and end-diastole, contrary to previous literature. Moreover, these authors tested these hypotheses both with numerical finite-element models for a mouse left ventricle constructed from magnetic resonance images and with *in vivo* admittance measurements in the murine left ventricle. Finally, they also propose a new method to determine the instantaneous myocardial contribution to the left ventricular admittance that does not require saline injection or other intervention to calibrate. Two other papers, from the same group led by Feldman (Porterfield *et al.*, 2009; Raghavan *et al.*, 2009), deal with corrections in murine ventricular volume determinations, the second one in particular using the permittivity concept.

In a more recent paper, Raghavan *et al.* (2011) described the design, construction, and testing of a device to measure pressure–volume loops in the left ventricle of conscious ambulatory rats. Pressure was measured with a standard sensor, but volume was derived from a tetrapolar catheter using a novel admittance technique, which has, according to these authors, two main advantages, that is,

(1) The contribution from the adjacent muscle can be instantaneously removed.
(2) The admittance technique incorporates the nonlinear relationship between the electric field generated by the catheter and the blood volume.

They used a low-power instrument weighing 27 g to take pressure–volume loops every 2 minutes and running for 24 hours. Pressure–volume data were transmitted wirelessly to a base station. The device was first validated in 13 rats with an acute preparation with 2D echocardiography to measure true volume. From an accuracy standpoint, the admittance technique is superior to both the conductance technique calibrated with hypertonic saline injections, and calibrated with cuvettes. The device was then tested in six rats with a 24-hour chronic preparation. The paper also states that small mammals are a good choice for many experimental protocols because of their low cost and ease of handling while the ability to measure left ventricular pressure and volume in real time during chronic preparations is valuable to the physiologist and for drug discovery. The paper presents a prototype instrument to be used in conscious, ambulatory rats. There are three challenges in designing such a device. The first challenge is to minimize power, weight and size. The second challenge aims at minimizing measurement error using a realistic understanding of electric field theory together with accurate calibrations. The third challenge refers to designing an apparatus that is comfortable for the animal.

The application of the admittancimetric technology, still considered as emerging, was studied in the context of myocardial infarction and LV remodeling by Clark and Marber (2013), from King's College, in London. Using a combination of high-resolution ultrasound and LV conductance catheters, they compared measures of LV function using an admittance system and a traditional conductance-derived pressure–volume (PV) system. A set of mice were subjected to focal myocardial ischaemia reperfusion by transient ligation of the left anterior descending (LAD) coronary artery and assessed cardiac function with different systems to determine the reliability and accuracy of these methods to distinguish between normal and dysfunctional ventricles. The admittance PV system provided a straightforward solution for assessing LV function in mice and, using this technique in combination with other established methods, they measured LV dysfunction following coronary artery occlusion and reperfusion, which can be ameliorated using a known preconditioning agent, and found that functional readouts are representative of other methods. It was found that, especially in diseased tissue, LV pressure–volume loops derived from complex admittance provided a reproducible and reliable method of determining LV function without the need for technically challenging calibration. It is suggested that admittance records accurately LV cavity volumes when compared to other invasive methods in the same animal. This emerging technology is both effective and reproducible in measuring LV function and dysfunction in the mouse, without the need for complicated interventions to calibrate the measurements or training in a new technology.

van Hout *et al.* (2014) developed an admittance-based pressure–volume system (AS). The technique has been validated predominantly in small animals. In large animals, it has only been compared to three-dimensional echocardiography (3DE) where the AS showed to overestimate left ventricular (LV) volumes. To fully determine the accuracy of this device, these Dutch investigators compared the AS with gold standard cardiac magnetic resonance imaging (CMRI) in a porcine model of chronic myocardial infarction (MI). Fourteen pigs were subjected to 90 min closed chest balloon occlusion of the LAD artery. After 8 weeks of follow up, pigs were consecutively subjected to LV volume measurements by the AS, CMRI, and 3DE under general anesthesia. The AS overestimated end-diastolic volume and end-systolic volume but not ejection fraction compared to CMRI. Good correlations of EDV and EF between the AS and CMRI were observed. EF measured by the AS and 3DE also correlated significantly. After subjection of pigs to MI, the AS very moderately overestimated LV volumes and showed accurate measurements for EF compared to CMRI. Thus, the AS appears as useful to determine cardiac function and dynamic changes in large animal models of cardiac disease.

Although somewhat older contributions, we should mention White *et al.* (1997) and Mazzei (1999). The former dealt with an interesting application of the conductance catheter to evaluating the outcome after repair of tetralogy of Fallot in 13 patients, 3–35 years after surgery. Pulmonary regurgitation was assessed from right ventricular pressure–volume loops generated by conductance and microtip pressure catheters. The latter article is a didactic and well-organized presentation for those readers not familiar with the subject.

Briefly summarizing the section:

Either conductance or admittance-based techniques to measure intra-cavitary volumes belong to the impedance concept, and, after a preliminary development period (1980–2000), have entered into a rapidly emerging, more advanced and solid stage, although for the time being, still as a rather research tool. Medical and veterinary applications are foreseen, so encouraging young scientists to pick up the challenge (see http://www.transonic.com/resources/research/

hemodynamic-measurements-of-left-ventricle-lv-function-post-myoc-
ardial-infarct-mi-using-pressure-volume-pv-loops-in-rodents/).

3.7 Vascular Impedance

The concept of vascular impedance has been borrowed from Electrical Engineering (EE), where clearly and unequivocally, signals are mostly pure sinusoids. Thus, only the ratio of two sinusoidal waves functions of frequency, as voltage and current usually are, can define the impediment opposed by a given circuit to the passage of current flow. Vascular impedance instead starts with two complex temporal waves (blood pressure and flow), from which the individual simple harmonic components are separated out to produce a number of impedance moduli and angles. It appears as some kind of inverse case for it goes from the complex to the simple. Such a number depends on the sample size, which in practical terms lies between 10 and 20. No wonder, then, if electrical engineers react either with doubt or as though they faced a conceptual error. Most of the material herein given has been drawn from Kohen *et al.* (2011), where a quick review of the historically most significant contributions is also given.

In simple words, in any vessel (say, an artery, the aorta, the pulmonary artery, or even a vein), the ratio of a pressure harmonic to the flow harmonic at the same frequency is called the input impedance of the vessel at that frequency. The instantaneous ratios of pressure to flow *in vivo* in their temporal courses cannot be used because the waveforms are composed of different frequencies. Thus, using Fourier expansion or discrete Fourier transform (DFT) or fast Fourier transform (FFT), the experimental pressure and flow signals, which are obviously functions of time, are decomposed in their respective simple sinusoidal components taken from them the first 10–20 terms, at the most, as representative pieces. From the Fourier coefficients, the modulus and phase of any harmonic can be easily calculated. Precisely, vascular impedance spectra, from the fundamental frequency (heart rate) to the, say, 10th harmonic, are obtained after dividing each harmonic pressure modulus through its corresponding flow modulus, and by making the differences of the respective phases. All this has been well developed and explained in the literature (Milnor, 1982; O'Rourke, 1982; Valentinuzzi, 2004). This section intends to clarify and mathematically sustain the validity of the vascular impedance concept, especially for newcomers from the EE area. For that matter, two methods will be used:

(1) CFS (Classical Fourier Series) followed by the classical Fourier transform (CFT)
(2) Time signals (TS) discretely multiplied by a sequence of shifted delta functions (SDF) followed by convolution of their DFT.

As a word of warning, in these two methods, even though it may appear as well known, we must stress that the complex hemodynamic signals we deal with reside within the time domain while the signals used for impedance calculation are simple harmonic ones within the frequency domain. Thus, it should always be clearly kept in mind what domain we are in with whatever mathematical expression there is in hand. Besides, in all this, just one cardiac cycle may be enough or, perhaps better, averaging of several beats is required, in which case a reference initial point must be determined to properly align all beats. A word about discretization or digitalization is also needed, for sampling in both domains must be carried out, meaning that the respective running indices should not be mixed up. Another point that should be mentioned is that these physiological signals are not strictly periodical for the cardiac rhythm shows variations around a mean frequency; however, if adequate timepieces are analyzed this potential drawback does not appear as significant or insurmountable.

3.7.1 *CFS followed by the CFT*

Assuming blood pressure $p(t)$ and blood flow $q(t)$ as periodic experimental functions, which are never known explicitly, both with the same fundamental heart frequency f_0, their respective Fourier expansions or series are described by the well-known expressions,

$$p(t) = P_0 + \sum_{n=1}^{n=\infty} P_n \cos(n\omega_1 t + \theta_n) \qquad (3.19)$$

which by simple trigonometry is transformed into

$$p(t) = P_0 + \sum_{n=1}^{n=\infty} P_n \cos(n\omega_1 t) \cos\theta_n - \sum_{n=1}^{n=\infty} P_n \sin(n\omega_1 t) \sin\theta_n \quad (3.20)$$

and

$$q(t) = Q_0 + \sum_{n=1}^{n=\infty} Q_n \cos(n\omega_1 t + \psi_n) \quad (3.21)$$

which similarly becomes

$$q(t) = Q_0 + \sum_{n=1}^{n=\infty} Q_n \cos(n\omega_1 t) \cos\psi_n - \sum_{n=1}^{n=\infty} Q_n \sin(n\omega_1 t) \sin\psi_n$$
$$(3.22)$$

so producing the set of simple harmonic components we talked about above. In all of them, the angular frequency $\omega_1 = 2\pi f_1$. Now, Eq. (3.20) needs to be Fourier transformed to get into the frequency domain, hence becoming the following expression, that is,

$$P(\omega) = 2\pi P_0 + \pi \sum_{n=1}^{n=\infty} P_n \left[\delta(\omega - n\omega_1) + \delta(\omega + n\omega_1)\right] \cos\theta_n$$

$$+ j\pi \sum_{n=1}^{n=\infty} P_n \left[\delta(\omega - n\omega_1) + \delta(\omega + n\omega_1)\right] \sin\theta_n \quad (3.23)$$

where the terms $\delta(\omega - n\omega_1)$ are zero for $n > 0$, so that the previous equation (3.23) can be rewritten as,

$$P(\omega) = 2\pi P_0 \delta(\omega) + \pi \sum_{n=1}^{n=\infty} P_n \delta(\omega - n\omega_1) \cos\theta_n$$

$$+ j\pi \sum_{n=1}^{n=\infty} P_n \delta(\omega - n\omega_1) \sin\theta_n \quad (3.24)$$

$$P(\omega) = 2\pi P_0 \delta(\omega) + \pi \sum_{n=1}^{n=\infty} P_n \delta(\omega - n\omega_1)(\cos\theta_n + j\sin\theta_n) \quad (3.25)$$

$$P(\omega) = 2\pi P_0 \delta(\omega) + \pi \sum_{n=1}^{n=\infty} P_n \delta(\omega - n\omega_1) e^{j\theta_n} \quad (3.26)$$

By the same token, the flow's Fourier Transform can be obtained as,

$$Q(\omega) = 2\pi Q_0 \delta(\omega) + \pi \sum_{n=1}^{n=\infty} Q_n \delta(\omega - n\omega_1) e^{j\psi_n} \tag{3.27}$$

In all the derivation above, δ stands for the well-known Dirac function and the Fourier transforms can be found in any table or adequate mathematical text. The impedance $Z(\omega)$ is given by the ratio formed by (3.26) over (3.27), that is,

$$Z(\omega) = \frac{P(\omega)}{Q(\omega)} \tag{3.28}$$

Or

$$Z_{(\omega)} = \frac{2\pi P_0 \delta(\omega) + \pi \sum_{n=1}^{n=\infty} P_n \cdot \delta(\omega - n\omega_1) \cdot e^{j\theta_n}}{2\pi Q_0 \delta(\omega) + \pi \sum_{n=1}^{n=\infty} Q_n \cdot \delta(\omega - n\omega_1) \cdot e^{j\psi_n}} \tag{3.29}$$

leading to

$$Z_{(0)} = \frac{P_0}{Q_0} \tag{3.30}$$

and

$$Z(n\omega_1) = \frac{P_n}{Q_n} e^{j(\theta_n - \psi_n)} \quad \forall\, n > 0 \tag{3.31}$$

Vascular impedance (VI), in this way, is defined only for integer multiple numbers of the fundamental cardiac frequency f_0, which do not allow a general expression of $Z(\omega)$; however, such situation does not hinder its usefulness in hemodynamics and gives good theoretical support to the concept. Based on a similar rationale, we could start with the discrete Fourier series (DFS) of the experimental signals followed by, either the DFT or FFT and, thereafter, make the division sample by sample of each corresponding coefficients.

3.7.2 *Time Signals (TS) discretely multiplied by a sequence of SDF followed by convolution of their DFT*

Let us assume a signal $f(t)$ along with its frequency spectrum $F(\omega)$, where $f(t) \in \Re$ and $F(\omega)$ is defined for $\omega \in [-\omega_1, \omega_1]$. The signal

$f(t)$ is sampled with a sampling frequency $\omega_s \geq 2\omega_1$ producing $g(t)$ as a result, that is,

$$g(t) = f(t) \sum_{k=-\infty}^{\infty} \delta(t - k \cdot T_s) \tag{3.32}$$

where $T_s = 2\pi/\omega_s$. In other words, $g(t)$ appears as the discrete product of $f(t)$ and the sequence of SDF.

In the frequency domain, we will have the following convolution product:

$$G(\omega) = F(\omega) * \left[\frac{2\pi}{T_{\mathrm{s}}} \sum_{k=-\infty}^{\infty} \delta\left(\omega - \frac{k2\pi}{T_{\mathrm{s}}}\right) \right] \tag{3.33}$$

where the bracketed factor is the Fourier transform of the temporal sequence of delta functions. We normalize in time and frequency making $u = t/T_{\mathrm{s}}$ and $\Omega = \omega \cdot T_{\mathrm{s}}$ Observe that the DFT samples the $G(\Omega)$ spectrum between 0 and 2π, that is, it explores one full period. Once the DFTs for pressure and flow are obtained, and considering Eqs. (3.34) and (3.35), both DFTs are divided sample by sample to get Eq. (3.36). This is identical to dividing the Fourier coefficients, since the factor $1/M$ cancels out (Proakis and Manolakis, 1996). That is,

$$a_k = \frac{1}{M} \sum_{n=0}^{M} x[n]\mathrm{e}^{-j \cdot k \cdot (2 \cdot \pi/M) \cdot n} \tag{3.34}$$

$$X(k) = \sum_{n=0}^{M} x[n]\mathrm{e}^{-j \cdot k \cdot (2 \cdot \pi/M) \cdot n} \quad \text{for } 0 \leq k \leq M - 1 \tag{3.35}$$

$$Z_k = \frac{P_k}{Q_k} \tag{3.36}$$

or, in plain words, the concept of vascular impedance is well validated. Thus, we can say as the old times mathematicians used to say, QED (*quod erat demonstrandum*).

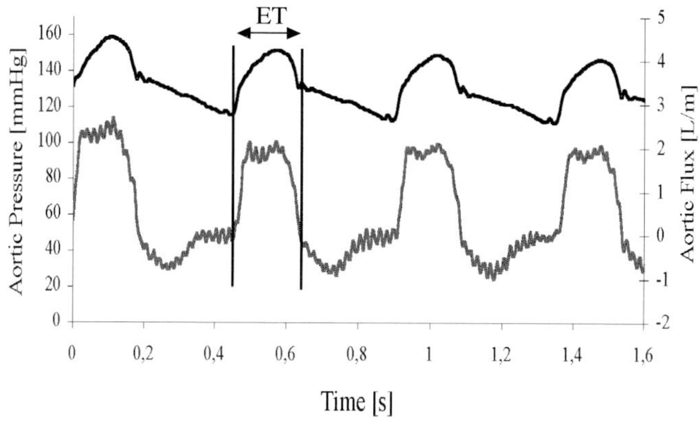

Figure 3.9: Aortic pressure (upper curve) and aortic flow (lower curve). Experimental time course records from which the aortic input impedance can be calculated. The dicrotic notch in the upper channel marks the end of ejection. ET is the ejection time, between the opening and the closure of the aortic valve. Obtained from an experimental dog at the Department of Bioengineering, UNT, 1994.

3.7.3 *Vascular impedance and its graphics*

Pressure and flow, respectively analogous to voltage and current in electric circuits, constitute two basic variables of the CVS. Both are periodic and mathematically unknown functions of time t. However, *they are not sinusoidal*. Figure 3.9 displays aortic pressure (upper channel), picked up by a miniature transducer placed almost at the root of the vessel with a catheter inserted via a femoral artery, and aortic flow (lower channel), detected by means of an electromagnetic flowmeter embracing the artery after the arch. The animal was an anesthetized dog. The foot of each beat in the pressure record (diastolic pressure) marks the opening of the aortic valve, hence, ejection starts, and the record below shows a rapid upstroke. The indentation after the maximum value (systolic pressure) marks the closure of the same valve. It is called the *dicrotic notch*, DN, a handy flag present in any good quality arterial pressure record. As a consequence, flow drops to zero after having shown a maximum. The two vertical bars clearly bound *ejection time*, ET, which in this particular case is in the order of 300 ms.

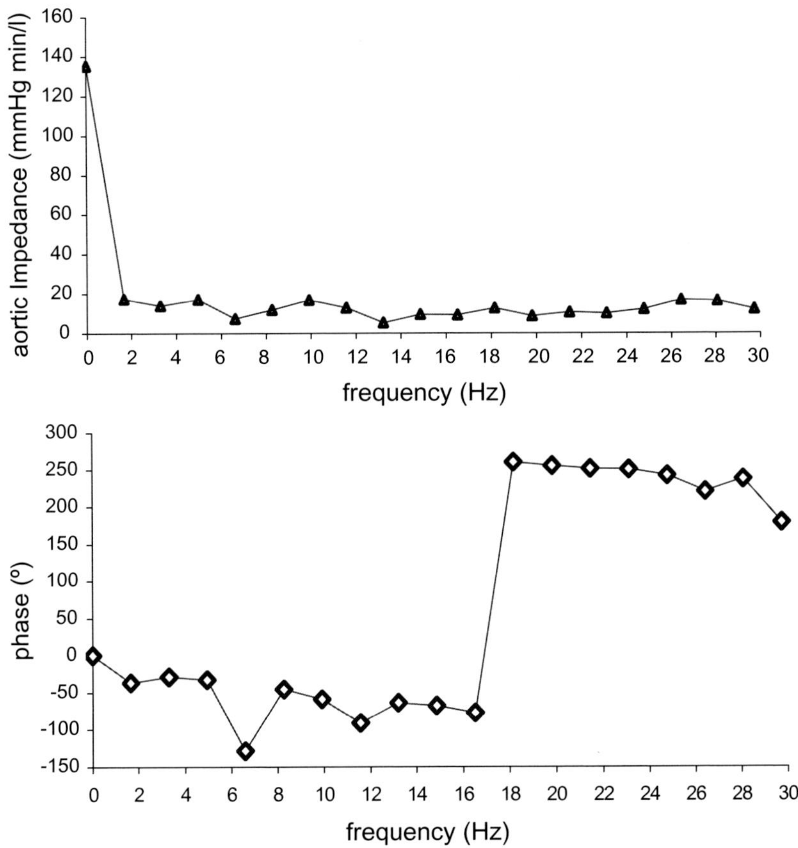

Figure 3.10: Aortic impedance. The calculations made with the data obtained from Fig. 3.9 led to these two graphs: Impedance modulus above and phase below, both as functions of the frequency expressed in Hertz.

After having obtained experimental digital records of pressure, $P = p(t)$, and of flow, $F = q(t)$, they are subjected to spectral analysis, usually by means of the FFT. In other words, each signal is decomposed in its dc (P_o, Q_o), and sinusoidal components (p_n, q_n), in which the fundamental frequency, f_1, coincides with the heart rate, HR. Figure 3.10 displays the usual way to graphically describe the vascular impedance concept, in this particular case, the so-called aortic impedance, both in modulus and in phase.

3.7.4 *Discussing vascular impedance*

The concept of vascular impedance, of relevance and significance in cardiovascular pathophysiology and hemodynamics, may strike the mind of electrical engineers and physicists, but standard and well-known mathematical techniques clearly validate it. The two methods herein used (the first one is traditional, while the latter is relatively recent), in fact, represent only spectral samples, undoubtedly valid, but the information they provide comes just from the sampled points and not from the overall continuous signals. Nonetheless, even under such limitation, the numerical approximation is acceptable and applicable in practical situations. The assumption of linearity behind all this development represents a touchy aspect that someone might also argue against. It is perhaps more philosophical than real, in particular when one finds the concept useful. In spite of the longtime the concept was introduced in hemodynamics, its use in medical practice did not find a place, one reason might be the mathematical background it requires, which the physician is obviously not prepared for; however, and considering the current computer technologies it is not needed. To drive a car, the driver needs not be a mechanic nor does the computer user have to be a computer expert. The only extra knowledge the physician requires involves what impedance embraces: its components and their influence on the vascular system.

Over 30 years ago, Merillon *et al.* (1984) aimed at the evaluation of aortic impedance in patients with congestive heart failure. Their impedance curves displayed increased values of the impedance modulus at 0 Hz (peripheral resistance) and at low frequencies. The characteristic impedance and phase were not different from normal subjects. Studies during nitroprusside infusion (a vasodilator) produced a decrease in impedance modulus at 0 Hz and at low frequencies. They concluded that the changes were due to greater peripheral resistance and wave reflection.

Aortic input impedance represents a major determinant in left ventricular afterload and consequent hypertrophy. Later on, Kobayashi *et al.* (1996), after a series of experiments in Wistar rats, concluded that sustained early systolic loading due to an increase in characteristic impedance was accompanied by less concentric, reduced hypertrophy, whereas the sustained late systolic loading due to the augmented arterial wave reflection was accompanied by concentric hypertrophy.

There seems to be a renewed interest in the medical applications of this beautiful clear-cut concept as evidenced in some papers. Sharp *et al.* (2000), for example, performed flow and pressure measurements in the ascending aortas of six pediatric patients (1–4 years old, 7.2–16.4 kg) to calculate their input impedance. Their results were interesting, indeed. Total vascular resistance decreased with increasing patient weight and was approximately one to three times higher than those of adults were. Strong inertial character was observed in the impedance of four of the six patients. The results suggested that the peripheral resistance component developed more rapidly with patient weight than did the characteristic impedance. Compliance values increased with increasing patient weight and were 3–16 times lower than adult values.

Closer in time, Sweitzer *et al.* (2007) investigated changes in the large and medium arteries of young people with Type 1 diabetes and demonstrated an increase in characteristic impedance. Such an increase was associated with a higher level of pressure pulsatility in the diabetic aorta for a given level of flow. These changes may be related to lower aortic diameter.

To close the section:

The mathematical background of the vascular impedance concept has been updated, and its predictive value in pathological hemodynamics has been briefly mentioned, so encouraging biomedical engineers and medical practitioners to come to terms with the concept. The subject was reviewed by Gow (2011), in the APS Handbook of Physiology. The angular component remains an elusive part that requires further and careful research.

3.8 Blood Pressure Regulation (Silvano Zanutto contributed to this section)

Long-term blood pressure regulation has been a traditional problem of concern in the medical and research environment. Experiments have shown that the mean arterial blood pressure (MAP) cannot be regulated after chemo and cardiopulmonary receptor denervation. Neuro-physiological information suggests that the nucleus tractus solitarius (NTS) is the only structure that receives information from its rostral neural nuclei and from the cardiovascular receptors and projects to nuclei that regulate the circulatory variables. From a control theory perspective, to answer if the cardiovascular regulation has a set point, it should be found out whether in the cardiovascular control there is something equivalent to a comparator evaluating the error signal (between the rostral projections to the NTS and the feedback inputs). The NTS would act as a comparator if,

(a) its lesion suppresses cardiovascular regulation;
(b) the negative feedback loop still responds normally to perturbations (such as mechanical or electrical) after cutting the rostral afferent fibers to the NTS;
(c) perturbation of rostral neural structures (RNS) to the NTS modifies the set point without changing the dynamics of the elicited response; and
(d) cardiovascular responses to perturbations in neural structures within the negative feedback loop compensate for much faster than perturbations in the NTS rostral structures.

The above-mentioned conditions seem to have enough support, that is, the NTS functions, indeed, as a comparator between its RNS and the cardiovascular afferents and projects to nuclei that regulate the circulatory variables. In turn, mean arterial pressure (MAP) is regulated by the feedback of chemo and cardiopulmonary receptors while the baroreflex would stabilize the short-term pressure value to the prevailing carotid MAP. The discharge rates of rostral neural projections to the NTS would function as the set point of the closed and open loops of cardiovascular control.

From this perspective, stress can cause hypertension via set point changes, so offering an answer to an old question. Even though the local blood flow to tissues is influenced by circulating vasoactive hormones and also by local factors, there is yet significant sympathetic control. It is well established that the state of maturation of sympathetic innervations of blood vessels at birth varies across animal species and it takes place mostly during the postnatal period. During ontogeny, chemoreceptors are functional; they discharge when the partial pressures of oxygen and carbon dioxide in the arterial blood are not normal. A model is a simple biological plausible adaptative neural network to simulate the development of the sympathetic nervous control. It can be hypothesized that during the embryonic development, from the RNS afferents to the NTS, the optimal level of each sympathetic efferent discharge is learned through the chemoreceptors' feedback. Its mean discharge leads to normal oxygen and carbon dioxide levels in each tissue. Thus, the sympathetic efferent discharge sets at the optimal level if, despite maximal drift, the local blood flow is compensated for by autoregulation. Such an optimal level produces minimum chemoreceptor output, which must be maintained by the nervous system. Since blood flow is controlled by arterial blood pressure, the long-term mean level is stabilized to regulate oxygen and carbon dioxide levels. After development, the cardiopulmonary reflexes play an important role in controlling efferent sympathetic nerve activity to the kidneys and modulating sodium and water excretion. Starting from fixed RNS afferents to the NTS and random synaptic weight values, the sympathetic efferents converge to the optimal values. When learning is completed, the output from the chemoreceptors becomes zero because the sympathetic efferents led to normal partial pressures of oxygen and carbon dioxide. A computer model can simulate (Fig. 3.11) how the NTS, as *emergent property* (see note below), acts as comparator and how its rostral afferents behave as set point. Say, for each tissue (k), the sympathetic efferent discharge (E_k), the blood flow (F_k), the venous partial pressure of oxygen and carbon dioxide (pO_2v_k) and (pCO_2v_k), and the arterial partial pressure after the lung gases diffuse (pO_2a) and (pCO_2a) are shown along with their respective trends. Finally, chemoreceptors discharge (C) is depicted. R_k represents the rostral neural nuclei inputs from the NTS, and Wr_k and Wc_k represent the synaptic weights. The equations and criteria employed in the model were described elsewhere (Zanutto *et al.*, 2010, 2011).

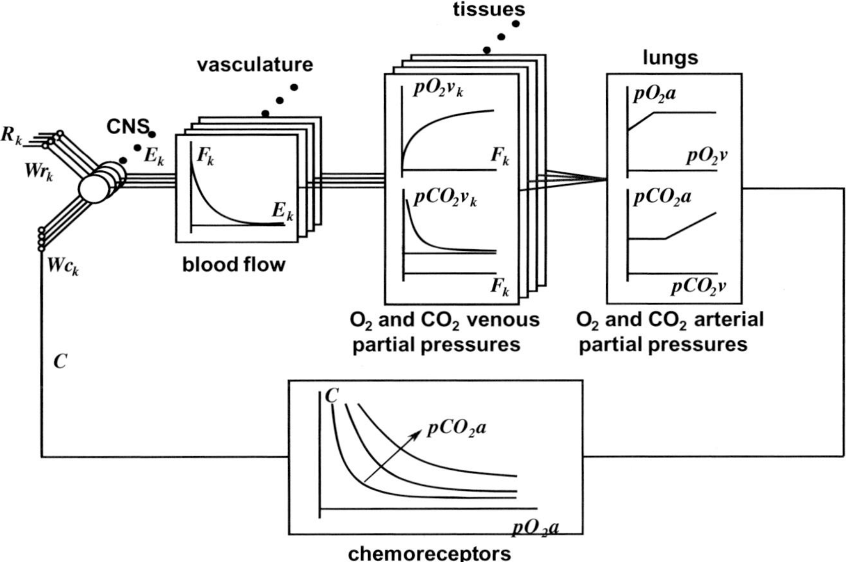

Figure 3.11: Neural network model of the sympathetic BP regulation. For each tissue (k), the sympathetic efferent discharge (E_k), the blood flow (F_k), the venous partial pressure of oxygen and carbon dioxide (pO_{2vk}) and (pCO_{2vk}), and the arterial partial pressure after the lung gases diffuse (pO_2a) and (pCO_2a) are shown. Finally, chemoreceptors discharge (C) is depicted. R_k represents the rostral neural nuclei inputs from the NTS, and W_{rk} and W_{ck} represent the synaptic weights.

An *emergent property* appears in a complex system, not in its individual parts. For example, the taste of saltiness *only* belongs to sodium chloride, the salt, but it is not a property either of sodium or chlorine or both constituents; thus, saltiness is an *emergent* or a *supervenient property* of salt. In our case above, the comparator represents the complex system.

3.9 Summarizing

This chapter covered mechanical concepts of the CVS, its basic laws and derivations, and briefly outlines the possible neural regulation of blood pressure, one of its essential variables.

References

Baan J, Jong TT, Kerkhof PL, Moene RJ, van Dijk AD, van der Velde ET, Koops J. Continuous stroke volume and cardiac output from intraventricular dimensions obtained with impedance catheter. *Cardiovasc Res* 15:328–334, 1981.

Banerjee MK, Ganguly R, Amitava Datta A. Effect of pulsatile flow waveform and Womersley Number on the flow in stenosed arterial geometry. *ISRN* (*International Scholarly Research Network*) *Biomathematics*, ID 853056, 17pp, Hindawi Publishing Corp; http://dx.doi.org/10.5402/2012/853056; www.hindawi.com/isrn/biomathematics/2012/853056/.

Clark JE, Marber MS. Advancements in pressure–volume catheter technology–stress remodelling after infarction, *Exp Physiol* 98(3):614–621, 2013. doi: 10.1113/expphysiol.2012.064733. Epub 2012 Oct 12.

de Snoo K. Die Bedeutung des Spannungsgesetzes für die Periode der Ausdehnung [in German, Meaning of the tension law during the mechanism of the dilatation period]. *Z Gynäkol* (Sept):162, 1936.

Farnetani I. Biographical Note about Dario Maestrini, In Treccani.it — L'Enciclopedia Italiana — Dizionario Biografico degli Italiani Vol. 67, 2007. http://www.treccani.it/enciclopedia/dario-maestrini_%28Dizionario-Biografico%29/.

Feodosiev VI. Materials' Resistance (in Spanish, Resistencia de Materiales), Edigtorial MIR, Buenos Aires; translated from Russian and originally published in Moscow, 1972.

Fung YC. Biomechanics: Circulation. Springer, New York, 571 pp., 1997.

Gourlay T, Taylor KM. Pulsatile cardiopulmonary by-pass. In: Gravlee GP, Davis RF, Kursz M, Utley JR, eds. Cardiopulmonary Bypass: Principles and Practice. 2nd ed., Chap. 10. Lippincott, Williams & Wilkins, Philadelphia, 2000. http://tele.med.ru/book/cardiac_anesthesia/text/gr/gr010.htm.

Gow BS. Circulatory correlates: Vascular impedance, resistance, and capacity. In: Handbook of Physiology: The Cardiovascular System. Vascular Smooth Muscle, American Physiological Society; online: 1 Jan 2011. doi: 10.1002/cphy.cp020214.

Greenhalgh RM, Brown LC, Kwong GP, Powell JT, Thompson SG, EVAR trial participants. Comparison of endovascular aneurysm repair with open repair in patients with abdominal aortic aneurysm (EVAR trial 1). 30-day operative mortality results: Randomised controlled trial. *Lancet* 364(9437):843–848, 2004 (4 Sept Issue). EVAR trial participants listed at end of report.

Herrera MC, Clavin OE, Spinelli JC, Valentinuzzi ME, Cabrera Fischer EI, Pichel RH. Multichannel tetrapolar admittance meter (MY) for intracardiac volume measurements in animals. *Med Prog Technol* 11:43–49, 1986.

Herrera MC, Valentinuzzi ME, Olivera JM. Volume profiles obtained by a conductimetric method. *J Biomed Eng* (London) 15:267–273, 1993.

Herrera MC, Valentinuzzi ME, Poliche AV, Crottogini, A, de la Serna F (h), Olivera JM, Martínez RJ, Negroni J, Lascano E, Luciardi HL, Berman SG,

Osatinski M. Evaluación técnica de la volumetría por conductancia en tres casos de severa isquemia: Número Especial Ingeniería en Cardiología. *Rev Fed Arg Cardiol* 26(4):537–547 and *Rev Arg Bioing* 3(3):96–108, 1997. published in co-edition.

Katz AM. Ernest Henry Starling, his predecessors, and the "Law of the Heart". *Circulation* 106:2986–2992, 2002. http://circ.ahajournals.org/content/106/23/2986.1.full.

Kobayashi S, Yano M, Kohno M, Obayashi M, Hisamatsu Y, Ryoke T, Ohkusa T, Yamakawa K. Matsuzaki M. Influence of aortic impedance on the development of pressure-overload of left ventricular hypertrophy in rats. *Circulation* 94:3362–3368, 1996. doi: 10.1161/01.CIR.94.12.3362; http://circ.ahajournals.org/content/94/12/3362.full.

Kohen AJ, Krouchov N, Valentinuzzi ME. Vascular impedance revisited. *Int J Biomed Eng Technol* 6(2):208–215, 2011. http://www.inderscience.com/browse/index.php?journalID=226&year=2011&vol=6&issue=2.

Maestrini D. L'influenza del peso sulla corrente d'azione e sul lavoro meccanico del muscolo cardiaco. *Arch Farmacol Sci Affini* 20:114, 1915.

Maestrini D. La legge del cuore in clinica, communication to the society of medicine and surgery, *Ospedale Abruzzesi*, August 30 session, in *Cuore e Circolazione* 9:506–513, 1925.

Martin RR, Haines H. Application of Laplace's law to mammalian hearts. *Comp Biochem Phys* 34(4, June 15):959–962, 1970. doi:10.1016/0010-406X(70)91019-4; Available online 17 March 2003.

Mazzoni M. La legge del cuore. *Notizie dalla Delfico* XIX(1–2):21–22, 2005; quarterly journal of the "Melchiorre Delfico" Public Provincial Library of Teramo, Abruzzo Region, Central Italy. http://www.provincia.teramo.it/biblioteca/pubblicazioni/notizie-dalla-delfico-5.

Mazzei WJ. The Theory and Significance of PV Loop Measurements, Canada, didactic Power Point Presentation in Cardiac Physiology, July 18, 1999. http://www.powershow.com/view/f330c-MTYyZ/Cardiac_Pressure_Volume_Relationship_powerpoint_ppt_presentation.

Merillon JP, Fontenier G, Lerallut JF, Jaffrin MY, Chastre J, Assayag P, Motte G, Gourgon R. Aortic input impedance in heart failure: Comparison with normal subjects and its changes during vasodilator therapy. *Eur Heart J* 5(6):447–455, 1984.

Milnor WR. Hemodynamics. Williams & Wilkins: Baltimore/London, 390, 1982 (see Chapters 7 and 12).

O'Rourke MF. Vascular impedance in studies of arterial and cardiac function. *Physiol Rev* 62(2):570–623, 1982.

Patterson SW, Starling EH. (1914) On the mechanical factors which determine the output of the ventricles. *J Physiol* 48:357–379, 1914.

Porterfield J, Kottam A, Raghavan K, Escobedo D, Jenkins J, Trevino R, Valvano JW, Pearce JA, Feldman MD. Dynamic correction for parallel conductance, Gp, and gain factor, α, in invasive murine left ventricular volume measurements. *J Appl Physiol* 107:1693–1703, 2009.

Proakis JG, Manolakis DG. Digital Signal Processing: Principles, Algorithms and Applications. 3rd ed. Prentice-Hall International, Inc., 968 pp., 1996, 4 appendices.

Raghavan K, Feldman MD, Porterfield JE, Larson ER, Jenkins JT, Escobedo D, Pearce JA, Valvano JW. Bio-telemetric device for measurement of left ventricular pressure–volume loops using the admittance technique in conscious, ambulatory rats. *Physiol Meas* 32(6):701–715, 2011. http://www.ncbi.nlm.nih.gov/pmc/articles/PMC3176664/.

Raghavan K, Porterfield J, Kottam A, Feldman MD, Escobedo D, Valvano JW, Pearce JA (2009) Electrical conductivity and permittivity of murine myocardium. *IEEE Trans Biomed Eng* 56(8):2044–2053, 2009.

Raghavan K, Wei CL, Kottam A, Altman DG, Fernandez DJ, Reyes M, Valvano JW, Feldman MD, Pearce JA. Design of instrumentation and data-acquisition system for complex admittance measurement. *Biomed Sci Instru* 40:453–457, 2004.

Sagawa K, Maugham L, Suga H, Sunagawa K. Cardiac contraction and the pressure–volume relationship. Oxford University Press, New York, Oxford, 480 pp., 1988.

Sharp MK, Pantalos GM, Minich L, Tani Y, McGough EC, Hawkins JA. Aortic input impedance in infants and children. *J Appl Physiol* 88(6):2227–2239, 2000.

Son HS, Sun K, Fang YH, Park SY, Hwang CM, Park SM, Lee SH, Kim KT, Lee IS. The effects of pulsatile versus non-pulsatile extracorporeal circulation on the pattern of coronary artery blood flow during cardiac arrest. *Int J Artif Organs* 28(6):609–616, 2005.

Spinelli JC, Valentinuzzi ME, Clavin OE, Cabrera Fischer EI, Pichel RH. Stability and repeatability of the multielectrode admittancimetric intraventricular volume measurements. *Automedica* 11(4):343–356, 1989.

Starling, EH. The Linacre Lecture on the Law of the Heart. Longmans, Green and Co, London, England, 1918.

Sweitzer NK, Shenoy M, Stein JH, Keles S, Palta M, LeCaire T, Mitchell GF. Increases in central aortic impedance precede alterations in arterial stiffness measures in type 1 diabetes. *Diabetes Care* 30(11):2886–2891, 2007. http://care.diabetesjournals.org/content/30/11/2886.full.

Uemura K, Kawada T, Sugimachi M, Zheng C, Kashihara K, Sato T, Sunagawa K. A self-calibrating telemetry system for measurement of ventricular pressure–volume relations in conscious, freely moving rats. *Am. J. Physiol Heart Circ Physiol* 287:H2906–H2913, 2004.

Valentinuzzi M Sr. Sobre algunas nociones de física del útero gravido: presión, tensión, tono, contracción y trabajo [in Spanish, About some physics notions of the gravid uterus: pressure, tension, tone, contraction and work]. *Bol Soc Obstetr Ginecol Buenos Aires* [*Bull Buenos Aires Soc Obstetr Gynecol*] 18(3, June 13):83–124, 1939.

Valentinuzzi M Sr. Contribución al estudio físico de la contracción uterina [in Spanish, Contribution to the physics of the uterine contraction]. Doctoral

Dissertation, Medical School, University of Buenos Aires (UBA), Argentina, 328 pp., 1950.

Valentinuzzi ME. Understanding the human machine: A primer for bioengineering. World Scientific Publishers: Singapore, London, 396 pp., 2004.

Valentinuzzi ME, Kohen AJ. Laplace's Law (part 1/3): What it is about, where it comes from, and how it is often applied in physiology. *IEEE Pulse* 2(4):74–81,84, 2011a. http://www.magazine.embe.org.

Valentinuzzi ME, Kohen AJ. Laplace's law: Its mathematical foundation (part2/3). *IEEE Pulse* 2(5):72–81, 2011b. http://www.magazine.embe.org.

Valentinuzzi ME, Kohen AJ, Zanutto BS. Laplace's law: Epistemological aspects (part 3/3). *IEEE Pulse* 2(6):71–76, 2011. http://www.magazine.embe.org.

Valentinuzzi ME, Leder RS. Cardiac output: Since when, who and how? *IEEE Pulse* 4(6, Nov–Dec):50–64, 2013.

Valentinuzzi ME, Niveiro MH, Spinelli JC, Guadix EM, Pratt G, Puglisi JL. Laplace's law of the heart and left ventricular wall thickness. *Rev Bras Eng Cad Eng Biomed* 4:5–17, 1987.

Valentinuzzi ME, Spinelli JC. Intracardiac measurements with the impedance technique. *IEEE Eng Med Biol Mag* 8(1):27–34, 1989.

van Hout GPJ, Jansen of Lorkeers SJ, Gho JMIH, Doevendans PA, van Solinge WW, Pasterkamp G, Chamuleau SAJ, Hoefer IE. Admittance-based pressure–volume loops versus gold standard cardiac magnetic resonance imaging in a porcine model of myocardial infarction. *Physiol Rep*, 2(4):e00287; online 2014 Apr 23. doi: 10.14814/phy2.287; http://www.ncbi.nlm.nih.gov/pmc/articles/PMC4001878/.

Wei CL, Valvano JW, Feldman MD, Pearce JA. Nonlinear conductance–volume relationship for murine conductance catheter measurement system. *IEEE Trans Biomed Eng* 52(10):1654–1661, 2005.

Wei CL, Valvano JW, Feldman MD, Nahrendorf M, Peshock R, Pearce JA. Volume catheter parallel conductance varies between end-systole and end-diastole. *IEEE Trans Biomed Eng* 54(8):1480–1489, 2007.

White PA, Chaturvedi RR, Shore D, Lincoln C, Szwarc RS, Bishop AJ, Oldershaw PJ, Redington AN. (1997) Left ventricular parallel conductance during cardiac cycle in children with congenital heart disease. *Am J Physiol Heart Circu Physiol* 42:H295–H303, 1997.

Zanutto BS, Cernuschi Frías B, Valentinuzzi ME. Blood pressure long term regulation: A neural network model of the set point development. *BioMed Eng OnLine* 10:54, 2011. http://www.biomedical-engineering-online.com/content/10/1/54.

Zanutto BS, Valentinuzzi ME, Segura ET. Neural set point for the control of arterial pressure: Role of the nucleus tractus solitarius. *BioMed Eng OnLine* 2010, 9:4, 2010. http://www.biomedical-engineering-online.com/content/9/1/4.

CHAPTER 4

WHAT ABOUT
THE RESPIRATORY SYSTEM

Max E. Valentinuzzi and Chi-Sang Poon

Is Earth so unique as to be alone in the Universe, coated with its feeble oxygen film? Can you imagine the challenge for a human being to survive, say, in Mars? And man is getting closer and closer to it. . .

Now being practical . . . do not forget that in case of a subject in medical emergency the first question is, does he/she breath???

Abstract

A quick review of the physiology of respiration is made assuming that the reader is acquainted with the subject to proceed, thereafter, to more challenging subareas of this essential function. After clearly defining what normal respiration is, high altitudes and deep sea responses are visited trying to update as much as possible. The theoretical side of the problem is also reviewed. This approach may help in a better understanding of respiratory distress and also in the design of ventilators.

4.1 Introductory Words: What is Normal Respiration?

In simple words, respiration is the optimal adjustment of the depth and rate of breathing to meet the metabolic demands of the body (input of O_2 and output of CO_2) with minimal work of the respiratory muscles. This is a fundamental and long-time advocated hypothesis (Poon, 1987), a paper that incorporated in a model both chemical and respiratory neuromechanical feedbacks to describe the steady-state ventilatory responses to CO_2 inhalation and exercise. All this is valid for any mammal, including humans. Probably,

111

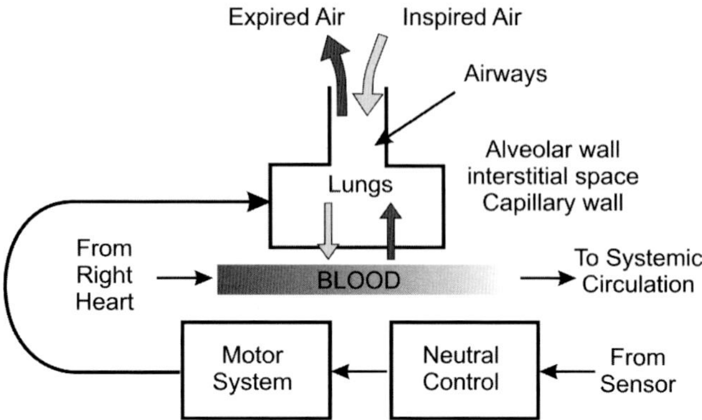

Figure 4.1: Respiratory system. The lungs, mechanically operated by the motor system (respiratory muscles), in turn driven by a nervous control system, which also receives signals from specific sensors, on its external side moves air in and out, and on its internal interface exchanges oxygen and carbon dioxide with the pulmonary circulation through an exchanger network formed by the alveolar membrane, the interstitial space, and the capillary wall. Blood low in oxygen and high in carbon dioxide enters the pulmonary circulation (left side of figure, shaded area) and leaves it with these concentrations reversed (right side of figure, white area).

the same statement is valid for other biological phyla or classes. The respiratory system relates physically to the cardiovascular system through the pulmonary capillaries, which act as an exchanger (Fig. 4.1). On one side, the airways via the mouth and nose connect the organism to the external atmosphere. On the inner side, there is the pulmonary circulation and its blood stream. The inspired air contains O_2 at 158 mmHg and CO_2 at 0.3 mmHg while partial pressures in the expired mixture are, respectively, 116 and 32 mmHg. The normal concentration of these two gases in alveolar air is 100 and 40, respectively, always expressed in mmHg, clearly marking an inward gradient of O_2 and an outward one for CO_2. Blood from the right heart carries oxygen at 40 mmHg and carbon dioxide at 46 mmHg while, after equilibration at the outflow heading to the left heart, the respective levels are 100 and 40 mmHg.

However, the respiratory system also takes care of other secondary, but still significant functions:

(1) It cooperates in the maintenance of the *acid–base equilibrium.*
(2) It helps in the *regulation of water loss.* (3) It helps in the

heat loss process (especially when the environmental temperature goes up). (4) It collaborates in the *circulation of blood* with its mechanical pumping action. (5) It plays a role in *emotional manifestations*. (6) It is essential in *protective reflexes*. (7) It is central to *other reflexes* of imprecise and yet not well understood nature. (8) It supports *other behavioral acts* such as vocalization, breathholding, Valsalva maneuver, blowing, and suckling. Let us enter into more details:

(1) Respiration calls for a well-kept *acid–base equilibrium* in body fluids, where H^+ concentrations significantly affect their pH. Remember that hydrogen ions react with many compounds, often disrupting their function. Hydrogen ion concentrations must guarantee a blood pH of about 7.4, that is, slightly alkaline. *Acidosis*, caused by accumulation of CO_2 in the blood, means a drop in blood pH. Shallow breathing may cause acidosis. The opposite condition, *alkalosis*, describes a rise in pH due to excessive CO_2 excretion. Hyperventilation commonly leads to alkalosis, but is much less serious than respiratory acidosis. Heavy exercise and some diseases may also lead to metabolic acidosis/alkalosis. Values of pH above 7.7 or below 7.0 are incompatible with life, that is, the range is rather small.

(2) Evaporative *water loss* from the respiratory tract is usually enclosed in the term *insensible water loss* (*insensible* because the individual is not aware of it) along with water that passes through the skin and is evaporated. This is solute-free water (no salt content). Via the skin, it can be in the order of 400 mL/day, while through the respiratory tract it is about the same value in an unstressed adult, but it varies according to the condition (e.g., in a ventilated patient that value increases). Measurement of these volumes is not an easy task (Brandis, 2013). Transepidermal water loss (TEWL) describes a constant loss by passive diffusion through the epidermis. If too high, the skin can dehydrate, potentially leading to infection or transepidermal passage of deleterious agents. Watson *et al.* (2002) validated the use of an evaporimeter for the accurate assessment of TEWL in the canine, provided the subject was completely still during

the measurements, the latter requirement obtained by training. The mean TEWL dropped, on average, 47% compared to that of untrained animals. Besides, TEWL tended to be higher in adult (2–7 yrs) than in older dogs (8–11 yrs), hence suggesting that an aging process could have influenced the skin.

(3) It is of interest bringing up the following example, different from the human case even though in the latter there is some heat loss through the respiratory tract. The sweating mechanism for *body temperature regulation* is very poor in dogs, mostly located along the base of their feet, and functionally not significant. Panting, instead, is their way of fighting the heat. Dogs eliminate excess body heat by panting, as they do not tolerate high temperatures (their normal body temperature lies between 101 and 102°F or 38 and 39°C). The tongue plays an important role in this process, as its surface area facilitates evaporation, so helping dogs to cool off. High temperature may cause a dry mouth and nose, abnormal breathing and heart rate, seizures, pale or dark gums, all symptoms of heat stroke. In such cases, immediate medical attention is needed (Nair, 2011).

(4) As regards *circulatory activity*, respiration may enhance venous return (VR) to the heart and hence, cardiac output through a pumping effect. Conversely, non-respiratory activities such as being on positive pressure ventilation or doing a forced expiration against a closed glottis (Valsalva maneuver) reduce VR and cardiac output (Valentinuzzi *et al.*, 1973). Respiratory movements also affect VR through changes in right atrial pressure, which is an important component of the pressure gradient for VR. Increasing right atrial pressure impedes VR, while lowering this pressure facilitates it. Respiratory activity can affect the diameter of the thoracic vena cava and cardiac chambers, which either directly (e.g., vena cava compression) or indirectly (by changing cardiac preload) influence VR (McArdle *et al.*, 1991; Klabunde, 2011).

It is of interest to comment on some particular situations, as the case is of total cavopulmonary connection (Hjortdal *et al.*, 2003). Not much is known about blood flow and its relationship to respiration during exercise in children with this vascular connection originated in a birth defect. Superior vena cava (SVC), inferior vena cava (IVC), and ascending aorta blood flow were measured under inspiration and expiration during supine lower limb exercise. These authors concluded that, in the IVC and aortic but not SVC, flows increase with supine leg exercise. Inspiration facilitates IVC flow at rest but not as much during exercise, when the peripheral pump seems to be more important.

(5) *Emotional manifestations* (crying, sobbing, weeping, sighing, wailing, moaning, and laughing) are related to emotional states somehow involving the respiratory system (Wikipedia, 2013a, 2013b; Miceli and Castelfranchi, 2003). The paper by Hjortal *et al.* (2003) offers a good review including a rather updated list of contributions.

(6) The so-called *protective reflexes*, say, coughing, sneezing, and vomiting, are frequent when bread crumbles, air dust or an irritating odor reaches our throat or nose. Under some conditions, as, for example, when sick with the flu, the subject may become more sensitive to such stimuli.

(7) There are still a couple of *other reflexes* (yawning and hiccupping) that carry unanswered questions and doubts. A *yawn* manifests itself as a simultaneous inhalation of air and stretching of the extremities, followed by exhalation of breath, which can also be called *pandiculation* (strange and funny word, from Latin, *pandiculatus*, meaning "stretching oneself"), that is, the act of yawning and stretching simultaneously (Fig. 4.2). Doesn't it sound like a word game? Anyhow, yawning in adults occurs most often immediately before, after sleep, and during tedious activities. It is often associated with tiredness and boredom. Due to its intrinsic contagiousness, it is an example of positive feedback, also seen in chimpanzees, dogs, and other species. It constitutes a subject of active ongoing research (Provine *et al.*, 2005; Schürmann *et al.*, 2005; Gallup, 2007; Senju *et al.*, 2007; Provine, 2010; Helt and Eigsti, 2010; Thompson, 2010; Norscia and Palagi, 2011). The latter two researchers underline that

Figure 4.2: Yawning and stretching. Left: Joseph Ducreux's pandiculating; self-portrait *ca.* 1783. Ducreux (26 June 1735–24 July 1802) experimented with self-portraiture by creating expressive, humorous, and rather unorthodox images of himself when stretching and yawning (This image is available for download, without charge, under the Getty's Open Content Program. See http://www.getty.edu/art/gettyguide/artObject Details?artobj =663). Right: One of the authors of this chapter (MEV), also pandiculating, as seen by our artist Gustavo Idemi.

humans are the primates with the most complex social networks, relying on the ability to share others' emotions to engage in successful social interactions. Such a phenomenon, known as empathy, has as basis a perception–action mechanism. Empathy has also been demonstrated during breath holding in a group (see, http://www.sciencedirect.com/science/article/pii/S156990 4811003995).

The involuntary re-enactment of an observed behavior may arise in the observer by recruiting neural mechanisms that, during the perception of an action or of a facial expression, activate shared representations. Contagious yawning, widely demonstrated in human and non-human primates, also involves a similar action–perception mechanism.

A spasm of the diaphragm (myoclonic jerk) resulting in a rapid, involuntary inhalation that is stopped by the sudden closure of the glottis and accompanied by a sharp, distinctive sound, is called *hiccupping*, or just a *hiccup* (obviously, an onomatopoeia or sound imitative word). It may repeat several times per minute and, occasionally, can become rather annoying; it involves a reflex arc. The time between hiccups tends to be relatively constant. Coughing, rapid eating, intense emotions, carbonated soft drinks, spicy foods, and/or laughing are common causes (Straus *et al.*, 2003; Wikipedia, 2013b).

There is a *mechanics in any of these respiratory movements.* Air inflow to the pulmonary chambers requires their expansion and return to the original volume means air outflow. Two basic mechanisms are responsible for such events:

(1) A downward movement of the diaphragm (during inspiration) and an upward movement of the same muscle (during expiration). Thus, this upwardly convex or domed muscle behaves as some kind of piston increasing and decreasing the thoracic cage volume. Approximately 70–75% of the normal respiration takes place by this mechanism. The phrenic nerves (one on each side) command, respectively, the two halves of the diaphragm (which is a skeletal muscle).

(2) Rising and lowering of the frontal face of the ribs, which pivot on the vertebral column. Hence, the net effect is an increase and decrease of the anteroposterior diameter of the thorax and almost no change on the lateral diameter (Fig. 4.3). The external intercostal muscles act as reins or straps to lift up the ribs. This mechanism accounts for the other 25–30% of the respiratory act. Figure 4.4 illustrates the action of the intercostal muscles after sectioning of the two phrenic nerves in an experimental animal. Respiration was detected with two electrodes connected to an impedance meter. During inspiration, there was an increase in impedance and expiration corresponds to a decrease (fourth channel). Breaths X and Y were produced only by the action of the intercostal muscles. Electrical stimulation of the two phrenic nerves at BPNS (during approximately 21 s) produced a good diaphragmatic contraction and consequent long inspiratory phase, as manifested by the plateau. Hence, even though the nerves had been cut, they were still responsive. Breaths Z and W were also due to the intercostal muscle contractions. In forced respiration, either voluntary or called for by need (as in exercise or distress), other chest and abdominal muscles enter in action.

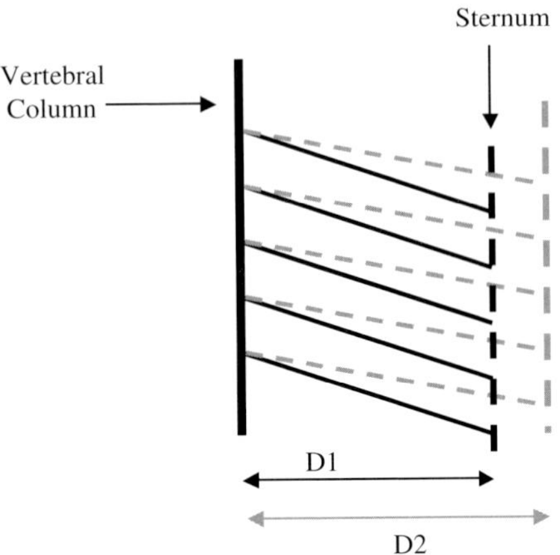

Figure 4.3: Anteroposterior diameter change. Lateral view. The arrow indicates pulling up of the external intercostal muscles. As the ribs rotate, the diameter increases from D1 to D2. No change occurs in the perpendicular direction (lateral). Only during forced expiration the internal intercostal muscles pull down the ribs to decrease the diameter; otherwise, this occurs passively.

Vocalization should be perhaps separated out, as part of an eighth very special respiratory act (see list above, third paragraph of section 4.1), in many respects unique to the human being. It is highly important for singers, especially lyric ones, who usually receive special training to better control their respiratory mechanics in order to improve phonation, intensity and duration of a given sound. Today, there are two main schools but with different approaches that involve, either

(A) Supporting the breath by compressing the abdomen during phonation (i.e., on exhalation), or its opposite,

(B) Relaxing the abdominals as much as possible during inhalation and phonation, allowing the diaphragm to work on the inhalation, and riding its relaxation on the outgoing breath (i.e., phonation).

In fact, all this calls for contracting the abdominal muscles, creating higher pressure in the abdomen and thorax, so allowing the diaphragm's relaxation (and upward rise) to be more carefully controlled. Remember that it is more difficult to relax a muscle than

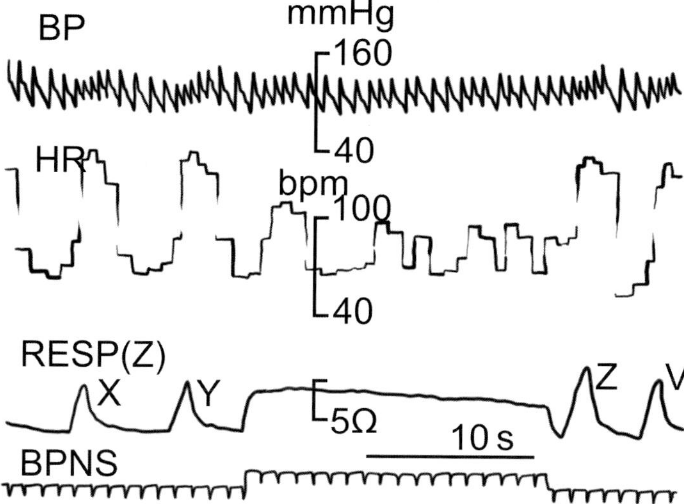

Figure 4.4: Intercostal and diaphragmatic components of the respiratory movements. Records obtained from an experimental dog whose two phrenic nerves had been sectioned. Thus, the animal was breathing only with its intercostal muscles. First channel: Carotid artery blood pressure. Second channel: Heart rate in beats/min. Third channel: Respiration detected with two transthoracic lateral electrodes connected to an impedance meter. Inspiration is shown by an increase in impedance and expiration by a decrease. Calibration of the third channel was 22.5 ohms per 500 mL of respired air. Records obtained by the author (MEV) at the Department of Physiology, Baylor College of Medicine, Houston, TX, 1970.

to contract it; performers have such a means of controlling their phonation. Most teachers recommend that the upper torso, especially the shoulder girdle, be as relaxed as possible even during the most extreme vocal demands. Ultimately, one wants access to all the breath resources available without jeopardizing the ability to phonate freely, that is, without unnecessary tension (Newham, 1994). The latter author makes an interesting addition that deserves the quotation here:

The Greek word *psyche,* meaning *soul,* has the same root as *psychein,* meaning *to breathe,* and the Greek word *pneuma,* meaning *spirit,* also means *wind.* Furthermore, the Latin words *animus,* meaning *spirit,* and *anima,* meaning *soul,* come from the Greek *anemos,* which is another word for *wind.* Similar connections also exist in

Arabic and German and they remind us that in many cultures the notions of psyche, spirit and soul have been related to the idea of the movement of air. Such a connection between air and soul is also contained in the fact that the human voice as the audible expression of the psyche can only be created through the emission of air from the body. Quite a concept, indeed!

The mechanics of breathing acquires relevance in certain professions, such as the mentioned singers, teachers, TV, and radio speakers and in some pathological conditions, but should not be forgotten by the general population, which frequently makes overt abuse of its phonation system, unnecessarily shouting or always speaking uncontrollably too loud. Significant information has been reported in recent years. During phonation, the vocal folds act as energy transducers that convert the aerodynamic power produced by the respiratory system into the acoustic power heard as voice. Phonation requires the coordination of breathing with glottic partial closure/vibration under the control of specific neuronal populations in the pons and medulla of the brainstem that are active during the post-inspiratory phase of the respiratory rhythm. A novel population of post-inspiratory driver neurons in the pons has been shown to play an important part in the control of glottis closure (Song *et al.*, 2014).

Several parameters are amenable to measurement (say, fundamental frequency, intensity, jitter, shimmer, and glottal waveform). Attention has been placed on the aerodynamic inputs of vocal function, such as sub-glottal pressure or phonation threshold pressure (PTP), which is defined as the minimum sub-glottal pressure starting vocal fold oscillations and producing voice. The relationships between PTP and the physiological parameters of the vocal folds have been studied using excised larynx experiments. Dehydration, for example, may increase the PTP. Hence, measurement of PTP in conjunction with other diagnostic tools could improve the diagnosis of vocal pathologies. A new parameter, PTF, defined as the minimum glottal airflow to produce phonation has been proposed. This is accomplished by assessing the reliability and range of PTF measurements using excised canine larynges under controlled elongation and comparing these measurements to their corresponding PTP measurements. The hypothesis states that PTF is directly related to PTP in the ranges of pressure and flow where phonation occurs. The vocal folds were elongated to adjust their biomechanical properties, and PTF and PTP measurements were taken. PTF was positively correlated with vocal fold elongation and PTP for small magnitudes of elongation. In other words, PTF may be indicative of the biomechanical properties of the vocal folds, thus providing a possible tool for the clinical evaluation of laryngeal function (Jiang *et al.*, 2008).

4.2 Pulmonary Capacities and Volumes

A physiological air volume detecting apparatus (a spirometer) shows changes with time that shift from one level to another, either upward or downward, as the subject breathes. As illustrated in Fig. 4.5, there are four important levels to define:

Maximal inspiratory level (MIL), reached when the subject is asked to take air in all the way up; normal inspiratory level (NIL), which is a maximum value during normal respiration and, thus, it may be either higher or lower; the resting normal expiratory level (NEL), a minimum obtained also during a normal respiratory act; and the maximum expiratory level (MEL), only reached when the subject voluntarily expels all the air he/she is able to.

The difference between NIL is the *tidal volume* TV (about 400–500 mL of air/breath in the normal adult). The gas remaining in the lung at the end tidal position NEL is called functional reserve capacity (FRC). If one keeps in mind that "capacities" are composed of one or more "volumes", the lung volume subdivisions are easy to deduce. Say, a maximal expiration measured from NEL is the expiratory reserve volume (or NEL – MEL). The air remaining in the lungs at the end of this maneuver is termed residual volume (RV, about 1.2 L).

Therefore, the sum of the two volumes, RV and ERV, results in FRC (about 2.2 L). Similarly, the sum of tidal volume (TV) and inspiratory reserve volume (IRV) equals inspiratory capacity (IC, in the order of 3.8 L), remembering that IRV (about 3.3 L) is obtained after (MIL – NIL). In turn, the sum of IC and FRC equals total

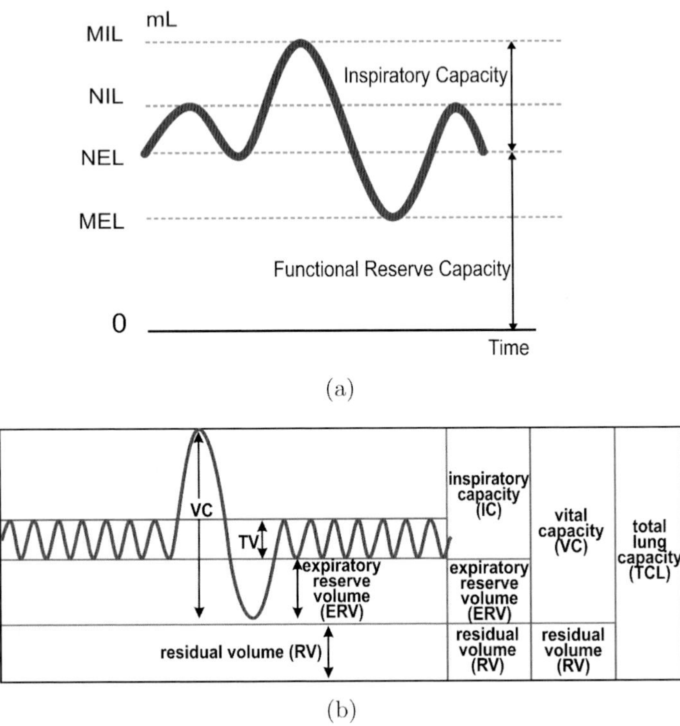

Figure 4.5: Pulmonary capacities and volumes. A spirometer may show changes with time that shift from one level to another as the subject breathes following instructions. Drawn by Gustavo Idemi. MIL: maximal inspiratory level; NIL: normal inspiratory level; NEL: normal expiratory level; MEL: maximal expiratory level, which define the following volumes and capacities, NIL – NEL = TV = tidal volume; MIL – MEL = VC = vital capacity; NEL – MEL = ERV = expiratory reserve volume; MEL = RV = residual volume; MIL – NEL = IC = inspiratory capacity; NEL – MEL = ERV = expiratory reserve volume; MIL – MEL = VC = vital capacity; MIL = TLC = total lung capacity, as can be checked in any physiology textbook. See http://humanphysiology.hubpages.com/hub/Lung-Volumes-and-Capacities#slide8222292.

lung capacity (TLC, 5–6 L), that is, from zero level up to MIL. The subject can never expel the RV, even after full expiratory exertion. The development of these measurements is long and large, with many contributions ranging from breakthrough pieces to even useless units only aimed at making money (Valentinuzzi and Johnston, 2014). This is a good point to suggest a thinking little problem, significant in legal medicine to help in the investigation of a possible crime: If a newborn is found dead, how can it be determined whether it died before or after birth? If a corpse is found in a lake, how can it be known whether the subject drowned or was killed before?

4.3 Respiratory Variables

We briefly remember these variables: Respiratory rate RR or respiratory frequency, perhaps the most obvious, with typical values in a normal adult between 12 and 16 breaths/min. When tidal volume is multiplied by rate, *ventilation* is obtained, that is, it is flow of air. Accepting 15 breaths/min and 500 mL of air/breath as average normal resting values, ventilation turns out to be 7.5 L of air/min. During exercise, these values change considerably depending on the physical fitness of the individual. However, an apparent normal ventilation value does not necessarily mean that the subject is well ventilated. It is required that air reaches the alveolar space in order to actually accomplish the gaseous exchange. A fraction of the air in the respiratory compartments (airways such as mouth, nose, throat, trachea, bronchi, and bronchioles) does not get in contact with the diffusion alveolar surface and, as a consequence, does not participate in the respiratory function. All this non-operative volume is called the *dead space*, V_D. As a rule of thumb and for a normal adult, either the body weight directly in pounds, or the same body weight in kilograms multiplied by two, both yield a number approximately equal to the dead space in mL. As an example, a person weighing 83 kgs ($\approx$166 lbs) would have a $V_D = 166$ mL. Thus, assuming a normal tidal volume of 500 mL, about 30% of this air is useless in the sense that it does not participate in gas exchange. Physiologic dead space includes all the previously mentioned non-respiratory parts of the anatomic dead space plus alveoli which are well ventilated but poorly perfused (or vice versa) and are, therefore, less efficient in the gas exchange process. In healthy individuals, the anatomic and physiologic dead spaces are roughly equivalent, as all areas of the lung are well perfused. However, in disease states where portions of the lung are poorly perfused, the physiologic dead space may be considerably larger than the anatomic dead space. Hence, physiologic dead space is a more clinically useful concept than is anatomic dead space. The dead space is significantly augmented in emphysema and silicosis.

From the functional point of view, *alveolar ventilation Q_A* is the important variable, for it removes the undesirable effect of the dead space. It is defined as

$$Q_A = \text{RR} \times (\text{TV} - \text{V}_D) \tag{4.1}$$

and so clearly establishing that only a fraction of the tidal volume (about 70%) is actually involved in the gas exchange. For 15 breaths/min, TV = 500 mL/breath and V_D = 150 mL/breath, the alveolar ventilation is only 5.25 L/min and not the 7.5 predicted by the overall ventilation.

Although this formula has been used to describe anatomic DS and alveolar DS interchangeably in the literature, the effects of anatomic DS (or external DS) are different from those of alveolar DS in terms of neural control of breathing. This is because the respiratory controller in the brain may perceive a series DS (anatomical or external) partly as a parallel dead space (like alveolar DS) and partly as CO_2 breathing because of the rebreathing of CO_2. It is therefore erroneous to treat anatomical DS the same as alveolar DS in modeling (Poon and Tin, 2013).

Pressures play obviously a role, say, *alveolar pressure P_A, pressure in the pleural cavity P_{PL}*, and the *atmospheric or barometric pressure P_B*. Hence, three differences can be defined,

$$P_A - P_B = P_{tw} \quad \text{or } transairway\ pressure \tag{4.2}$$

$$P_{PL} - P_A = P_{tp} \quad \text{or } transpleural\ pressure \tag{4.3}$$

$$P_{PL} - P_B = P_{tt} \quad \text{or } transthoracic\ pressure \tag{4.4}$$

From the three latter equations, it is seen that $P_{tt} = P_{tp} + P_{tw}$ or, in words, the *transthoracic pressure is the sum of the transpleural and the transairway pressures.* These pressures vary over the respiratory cycle and the most important and interesting is P_{tp}. The pressure in the pleural cavity is always lower than the atmospheric pressure;

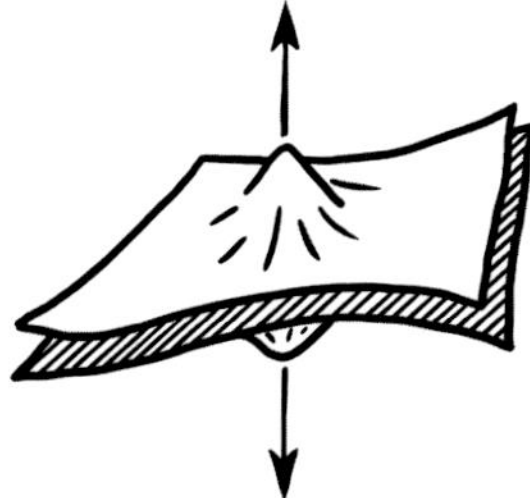

Figure 4.6: Generation of a subatmospheric pressure. Two wet rubber sheets tend to stick together when one is set over the other. A *virtual space* or *cavity* is generated.

thus, P_{tp} is always negative (slightly subatmospheric). Let us describe its mechanism of generation.

Two wet rubber sheets tend to stick together when one is set over the other (Fig. 4.6). If we try to separate them off by pulling out perpendicularly in opposite directions, a partial vacuum volume will be created by simple forced expansion. The inner thoracic wall and the outer pulmonary wall are in apposition, filled with a small amount of fluid so that they can easily slide over each other. Due to their intrinsic elastic properties, the former tends to recoil outwardly and the latter tends to do it inwardly. The net result is a phenomenon similar to the subatmospheric pressure effect illustrated in the picture.

In fact, there is no space between the two sheets or between the lungs and the thorax; thus, it is called a *virtual space* or *cavity*. Under resting expiratory conditions, intrapleural pressure is about 2 mmHg lower than atmospheric pressure because of the basic passive mechanism depicted in Fig. 4.6. Due to the action of the inspiratory muscles, the thorax expands and intrapulmonary pressure decreases. Meanwhile, the added pulling out of the chest wall decreases intrapleural pressure even more. The net result is air flowing in with a concomitant increase in intrapulmonary pressure ending up inspiration when lung recoil counteracts chest recoil. Thereafter, intrapleural pressure goes up, but always below atmospheric, because of passive exhalation (not forced expiration) due to temporary predominance of lung recoil and everything returns to the initial resting

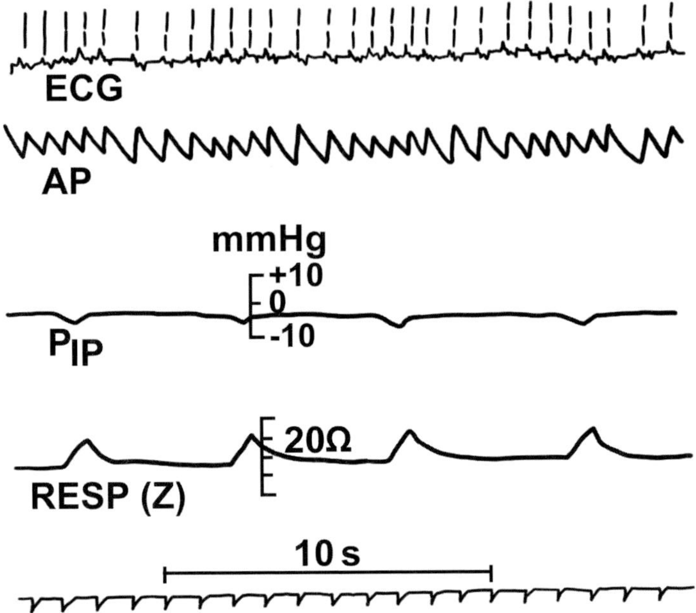

Figure 4.7: Intrapleural pressure. These records were obtained from an experimental dog. Intrapleural pressure (third channel) was recorded with a cannula inserted through the ribs in the pleural cavity and connected to a sensitive transducer. Its base line lies below the zero pressure level (which coincides with the atmospheric pressure). Records obtained by the author (MEV) in the Department of Physiology, Baylor College of Medicine, Houston, TX, 1970.

situation. During maximal normal inspiration, intrapleural pressure can be 6–7 mmHg below atmospheric.

The records of Fig. 4.7 were obtained from an experimental dog. Intrapleural pressure (third channel) was recorded with a cannula inserted through the ribs into the pleural cavity and connected to a sensitive transducer. Its base line lies below the zero pressure level (which coincides with the atmospheric pressure). The dog has only one pleural space because it has no mediastinum, as man has. The decrease during inspiration is clearly associated with the increase in impedance (inspiration) shown by the fourth channel, recorded with two lateral electrodes and an impedance meter. Besides, there is also a good respiratory heart rate response manifested by the varying rate in the ECG and in the blood pressure channels. The anesthesia given to this animal favored such response: inspiration associated with higher RR. If air gets into the pleural cavity, as the case is in chest stabbing or in some car accidents, it is called a pneumothorax. Due to the intrinsic elastic properties of the lung walls, they recoil and collapse while the thoracic wall pulls in the opposite direction. The person is unable to breathe, it is painful and, obviously, it means an emergency. If it is a dog, since it has a single pleural cavity (no mediastinum dividing the thorax in two), artificial respiration must be instituted and the air accumulated in the cavity has to be removed using well-known and not too difficult techniques. In man, instead, with a mediastinum and two pleural cavities (left and right), the situation is somewhat easier because frequently only one side gets air in it (it depends on the accident), and the person, although painfully, can continue breathing with one lung. However, it is also an emergency that requires immediate attention. We leave out the important and traditional subjects of compliance, work of respiration, pulmonary circulation, gas exchange and respiratory control, as they were dealt with acceptably in the preceding edition of this text and we rather go into the perhaps more challenging and attractive areas that follow.

4.4 Respiration in High Altitudes

Adaptation to high altitudes means to put into action a set of biological mechanisms that permit species at large to survive in extreme mountainous environments. It is opposed to short-term adaptation, or more properly acclimatization, which is an immediate physiological response to a simpler changing of environment (as when moving from a hot to a cold climate or vice versa); *high-altitude adaptation* refers instead to long-term physiological responses. The phenomenon has been well documented in human populations, such as Tibetans, South Americans, and Ethiopians, living in the Himalayas, Andes, and Ethiopia, respectively. The Peruvian physiologist Alberto Hurtado (1901–1983), no doubt, is considered one of the leading pioneers in the studies on the adaptations of human beings to high altitude. He excelled in his research work, carried out at times under very hard and primitive conditions. Famous is his first paper published in 1932 (Hurtado, 1932).

Humans have occupied High Mountain areas for more than 10,000 years and very many, indigenous and otherwise currently live there, at altitudes exceeding 3000 m above sea level (asl). While, to some extent, acclimatization can accommodate the one-third decrease in oxygen availability, having been born and raised at altitude appears to confer substantial advantages over sea level people. A number of characteristics have been postulated to a high-altitude phenotype; however, developmental and genetic adaptation to the acquisition of this phenotype has not yet been resolved. A complex trait is influenced by multiple genetic and environmental factors and, in humans, it is inherently very difficult to determine what proportion of the trait is dictated by an individual's genetic heritage and what proportion develops in response to the environment in which the person was born. Evidence for a genetic contribution to high-altitude adaptation in humans has been the subject of several reviews (Rupert and Hochachka, 2001), including rather recent contributions (Huerta-Sanchez, 2014).

The human body performs best at sea level, where the atmospheric pressure is about 101 kPa (i.e., 760 mmHg, or 1 atm, by definition). The concentration of atmospheric oxygen (O_2) at sea-level air is almost 21%, so that the partial pressure of O_2 (pO_2) is, say, 20 kPa. In healthy individuals, this saturates hemoglobin, the oxygen-binding red blood cells. The figures given above vary with the geographical place, day, and climatic conditions. Atmospheric pressure decreases exponentially with altitude, while the O_2 fraction remains constant to about 100 km, so pO_2 decreases exponentially with altitude as well. It is about half of its sea-level value at 5000 m, and only a third at 8850 m. When pO_2 drops, the body tries to compensate for by using a series of physiological mechanisms (Table 4.1).

The Pascal (Pa, or the kPa $= 1000$ Pa) are perhaps not very fortunate or practical units because numerically cannot be easily handled. Remember that 1000 mmHg $= 133.32$ kPa and 760 mmHg $= 101.33$ kPa. From a medical viewpoint, three regions have been recognized

Table 4.1: Pressure and oxygen versus altitude.

Altitude	Barometric pressure		O$_2$
(m)	(kPa)(mmHg)	(mmHg)	(%)
0	101	760	100
1000	90	679	89
2000	81	604	80
3000	72	537	71
4000	63	475	63

as hazardous due to the lowered amount of available oxygen in the atmosphere

— High altitude = 1500–3500 m
— Very high altitude = 3500–5500 m
— Extreme altitude = above 5500 m

Travel to these levels can lead to health problems and also risk of periodic breathing (PB), from mild symptoms of *mountain sickness* to the potentially fatal *high-altitude pulmonary edema* and *high-altitude cerebral edema*; the higher the altitude, the greater the risk.

Almost 150 million people live above 2500 m. These populations have ways of compensating for the lower oxygen levels. Compared with acclimatized newcomers, these natives have better oxygenation at birth, enlarged lung volumes, and a higher capacity for exercise. The mortality rate is significantly lower for permanent residents of high altitudes, where frequently residents older than 100 years are found. At high altitude, in the short term, the lack of oxygen is sensed by the carotid bodies, which causes an increase in the breathing rate. However, hyperventilation also causes respiratory alkalosis, so inhibiting the respiratory center from enhancing the respiratory rate as much as would be required. Besides, the heart beats faster, stroke volume is slightly decreased, and non-essential body functions are suppressed (as, e.g., there is a diminished gastrointestinal activity) in an effort to favor respiration. Full acclimatization, however, requires days or even weeks. Gradually, the body compensates for the respiratory alkalosis by renal excretion of bicarbonate. The body has lower lactate production, decreased plasma volume, increased hematocrit leading to polycythemia (up to 70–75%), increased red blood cell mass, a higher capillarization in skeletal muscle, increased myoglobin, increased mitochondria, and right ventricular hypertrophy, due to pulmonary artery pressure increase in an effort to oxygenate more blood. Full hematological adaptation to high altitude is achieved when the increase of red blood cells reaches a plateau. A rule of thumb states that by multiplying the altitude in kilometers by 11.4 days produces an estimate of the adaptation time (e.g., to adapt to 4000 m would require 45.6 days). Current research indicates that,

(a) The pattern of growth at high altitude, due to limited nutritional resources, physical growth in body size is delayed, but growth in lung volumes is accelerated because of hypoxic stress).

(b) Low-altitude male and female urban natives can attain functional adaptation to high altitude by exposure to high-altitude hypoxia during growth and development.

(c) Experimental studies indicate that exposure to high altitude during the period of growth and development results in the attainment of a large residual lung volume.

(d) This developmentally acquired enlarged residual lung volume and its associated increase in alveolar area when combined with increased capillarization and increase in red blood cells and hemoglobin concentration contributes to the successful functional adaptation of the high-altitude native to hypoxia (Frisancho, 2013). Figure 4.8 shows the changes in ventilation as partial pressure of oxygen decreased from the normal sea level value to a very low one of only 42 mmHg. A decrease in heart rate and an increase in respiratory frequency are also shown (to the left of the horizontal axis).

The paper by Reeves and Grover (2005) is a moving and vivid account of the heroic, even sacrificed periods spent in Andean altitudes. It had been suspected that pulmonary hypertension existed in high-altitude natives. A team of Peruvian scientists, led by Dante Peñaloza, provided not only the first clear evidence that it was so, but they demonstrated that it was chronic and occasionally severe.

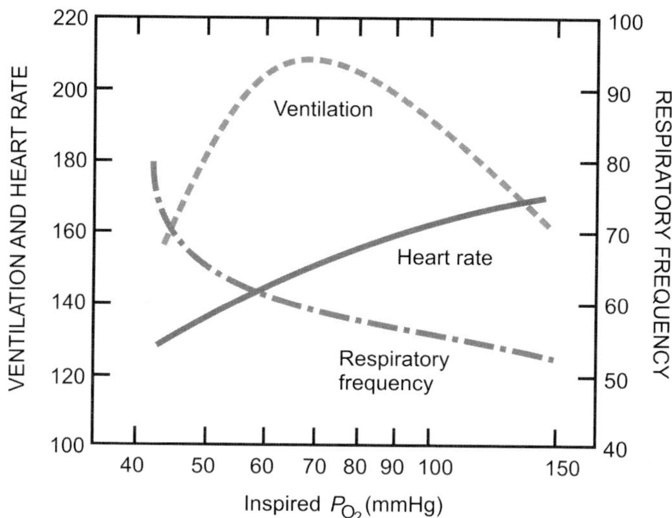

Figure 4.8: Ventilation and partial pressure of oxygen. See Fig. 1 in Schoene's paper. Limits of human lung function at high altitude. *J Exp Biol* 204: 3121–3127, 2001. Data from the 1981 American Medical Research Expedition to Everest, showing that maximal exercise ventilation, in liters per minute (BTPS) (dashed line), increased as the inspired partial pressure of oxygen decreased from sea-level values (150 mmHg) to approximately 60 mmHg (at an altitude of approximately 6300 m), but decreased as climbers approach the extreme altitude of the summit of Mount Everest, where the inspired partial pressure of oxygen was 42 mmHg. Redrawn by Gustavo Idemi.

More importantly, they showed that this was a consequence of structural changes in the pulmonary vascular bed. Histological findings indicated that hypoxia-induced thickening of the pulmonary arteriolar walls was the primary cause of the elevated pressure. Because the hypertension was not promptly reversed by vasodilators (oxygen inhalation or acetylcholine infusion), they found it differed from acute hypoxic pulmonary vasoconstriction. The team's other novel findings included a delay in the normal fall in pulmonary vascular resistance after birth and, in adults, a lack of vasodilation with muscular exercise. Furthermore, the altitude-related pulmonary hypertension resolved over time at sea level. The paper by Reeves and Grover (2005) shows a nice picture of the Morococha Laboratory at 4540 m asl, in the Peruvian Andes, taken in those early research days.

Early European colonists of the Andes had difficulties in having children, a fact that brought up the hypothesis of some kind of deleterious effects on reproduction for humans at high altitudes. Yet a 16th-century missionary wrote, "... the Indians are healthiest and where they multiply the most prolifically is in these same cold air-tempers, ... [yet most children of the Spaniards] when born in such regions do not survive." In opposition to the first hypothesis, it was suggested that humans at high altitudes are subjected to strong natural selection from hypoxia, cold and limited food sources and, furthermore, that human populations can and have adapted, and continue to adapt, to these conditions. Analyses of the determinants of natural fertility suggest that behaviors (breast/infant feeding practices in the Andes, and marriage practices and religious celibacy in the Himalaya) are major factors of fertility in high altitude populations. Furthermore, other data demonstrate that fecundity is not impaired in these indigenous *altiplano* populations, and that the risk for early pregnancy loss is not elevated by environmental hypoxia but does vary seasonally with the agricultural cycle (Vitzhum, 2013). The bibliography is huge, indeed, and we leave here a few references for the interested reader. The subject area calls for further research because still there are several gaps to fill (Moore, 2001; Beall, 2006; 2007; Peñaloza and Arias-Stella, 2007; Simonson *et al.*, 2010; Scheinfeldt *et al.*, 2012; Bigham *et al.*, 2013).

4.5 Respiration in Deep-Sea Diving or Diving Reflex

The mammalian diving response is a remarkable behavior that overrides basic homeostatic reflexes. Studied in large aquatic mammals, vertebrates show it, too. Pelagic mammals have developed several physiological adaptations to conserve intrinsic oxygen stores, but the apnea, bradycardia, and vasoconstriction is shared with those terrestrial and is neurally mediated (Panneton, 2013; Panneton *et al.*, 2014). The so-called *pelagic fish* live in the zone of ocean waters, that is, neither close to the bottom nor near the shore, in contrast with *demersal fish*, which do live on or near the bottom. The marine pelagic environment is the largest aquatic habitat on Earth. The word *pelagic* comes from Greek, πέλαγος or *pelagos*, that is, *open sea*.

SCUBA diving has become a rather widespread sport, especially in tropical seaside regions, in which divers use an underwater breathing unit (the so-called SCUBA, or Self Contained Underwater Breathing Apparatus). It is also a professional activity applied to deep-sea exploration or recovery. Scuba divers carry their own source of breathing gas (usually compressed air) plus a number of accessories that form the overall gear. The diver swims underwater at different depths by using fins attached to the feet (Cousteau and Dumas, 1954; Wikipedia, 2014a,b).

Diving has been practiced for centuries to find food, pearls, or corals, even for military purposes. The helmet, or aqualung, with air supplied from the surface was started during the second half of the 18th century being continually perfected up to these days; however, such piece carries with it several limitations. The breakthrough came in 1943, when Émile Gagnan (1900–1979) and Jacques-Yves Cousteau (1910–1997), after several drawbacks, successfully tested in deep waters an autonomous breathing system, free of an umbilical feeding hose. All this took place during the German occupation of France. Surprisingly, the current technology has not changed much. Depths between 20 and 40 m are commonly reached, while under special conditions the diver swims as far down as 100 m or even somewhat deeper.

4.5.1 *Pressure effects*

Water pressure impacts our bodies in several ways, but seasickness can happen above water, while on a boat getting ready to dive. Some people are susceptible to it, others much less or not at all. Simple facts may cause such body response, as the size of the boat, previous food or drinks, or how rested the individual is. Once underwater means being under hydrostatic pressure (a column of water over the swimmer), and that pressure increases rapidly with depth, say,

meters H_2O	mmHg
1	73.6
10	735.6 (almost 1 atm)

Observe how fast pressure grows as the subject goes down. Even in swimming pools, where at most they may have depths of 5 m, when a child is asked, as a challenge, to show how he/she does by doing a full lap touching the bottom, he/she will be subjected at one point to about 300 mmHg or slightly more and that may be risky.

Gas laws that get into play:

Boyle's law, as gas pressure increases, gas volume decrease. *Charles' law*, with constant pressure, gas volume increases and decreases with its temperature. *Dalton* adds that in a mixture of gases, the total gas pressure is equal to the sum of their individual pressures. Finally, *Henry's law* states that the amount of gas that dissolves in a liquid is a function of its partial pressure; it also says how easily the liquid absorbs the gas.

These fundamental physics laws impact the human physiology. In accordance with Henry's law, the increasing pressure correlates to increased absorption of nitrogen (remember there is slightly more than 79% in air). As pressure decreases when the body goes up from deeper waters towards the surface, absorbed nitrogen needs time to be released from the body. Ascending too quickly permits nitrogen bubbles to become too large to be eliminated through breathing, and that can have dire consequences. Other effects of underwater pressure include (1) *shallow water blackout*, (2) *oxygen toxicity*, (3) *carbon monoxide toxicity*, (4) *nitrogen narcosis*, and (5) *decompression sickness* (Blickenstorfer, 2014).

(1) Water Blackout

Except in hypoxic environments such as at high altitude, it is *not* the lack of oxygen that creates the urge to breathe, but mainly the build-up of carbon dioxide. Normally, as the oxygen level in blood falls, carbon dioxide raises leading to the need to breathe. When diving, instead, a breath-hold diver may rapidly take before a number of deep breaths. That hyperventilation removes carbon dioxide and thus increases the period of time before the urge to breathe returns. Once underwater, the diver may use up oxygen without feeling a need to breathe and eventually he may black out from insufficient oxygen in the blood (*hypoxia*). A similar thing may happen when a breath-hold diver goes deeper. Upon ascent, the oxygen pressure drops below the level to sustain consciousness. This is *deep-water blackout.*

(2) Oxygen Toxicity

It literally means having too much oxygen in one's body. When diving, the rising water pressure increases, according to Dalton's law, the partial pressure of oxygen. Oxygen represents about 20% of air at sea level. However, while that percentage stays the same at depth, we inhale many more oxygen molecules. At the 40 m dive limit for recreational divers, the number of inhaled oxygen molecules would represent 100% oxygen at the surface making oxygen atoms roam as dangerous free radicals. Despite their beneficial activities, these reactive oxygen species clearly can be toxic to cells. By definition, radicals possess an unpaired electron, which makes them highly reactive and thereby able to damage all macromolecules, including lipids, proteins and nucleic acids. One of the best-known toxic effects of oxygen radicals is damage to cellular membranes, which begins with a process known as *lipid peroxidation.* A common target for peroxidation is unsaturated fatty acids present in membrane phospholipids becoming the cause of *central nervous systems (CNS) oxygen toxicity,* the kind most often experienced by divers (called the *Paul Bert effect*), showing *tremors and seizures, ringing ears, nausea, tunnel vision, and a dry cough.* This can lead to drowning or decompression damage when a diver ascends too quickly in response to those symptoms. Recreational divers do not have to worry about CNS as it is not much of an issue for depths not more than 50 m. Another type, *pulmonary, or whole-body, oxygen toxicity,* also called the *Lorraine Smith effect,* can cause irreversible lung damage, but generally occurs from very long exposure.

(3) Carbon Monoxide Toxicity

Carbon monoxide is odorless and tasteless. It combines with blood, hemoglobin really, much more easily than oxygen, some 200 times more readily. Carbon monoxide poisoning can happen if a faulty or poorly maintained air compressor adds carbon monoxide into the tank or if it sucks in already contaminated air. At depth, while the partial pressure of carbon monoxide remains the same, the diver inhales many more carbon monoxide molecules, enough for poisoning symptoms such as headaches, confusion, and tunnel vision. Divers may pass out because there is no longer enough oxygen. Signs of carbon monoxide poisoning are flushed lips and cheeks.

(4) Nitrogen Narcosis

Nitrogen also becomes an anaesthetic under pressure. So, most divers experience what has been called *Rapture of the Deep*, at around 30–40 m or so. Signs of nitrogen narcosis are euphoric feelings, fixations, lapses in concentration, a loss of good judgment, a bit similar to the effects of alcohol. The onset of nitrogen narcosis varies from person to person, and it goes away instantly as the diver ascends.

It can be easily realized how complex the subject is and how little prepared the human being is to undergo these effects, as adaptation is not possible.

(5) Decompression Sickness (DCS)

Again, Henry's and Dalton's laws are at work: More gas dissolves in body tissues under pressure (Henry), and there is more of each gas under pressure, though the percentages stay the same (Dalton). During descend, nitrogen gets absorbed into the body tissues. Those tissues high in fat absorb a lot, other tissues less. Tissues with a large blood flow absorb and release gas more quickly than tissues with less blood flow. The net result is a lot of nitrogen absorption that will have to be released again when coming back up. If the pressure drops too quickly, nitrogen bubbles form in the blood and that can be very harmful because large bubbles can get stuck and create blockages. They may block circulation and compress nerves because those are surrounded by fatty tissue that easily absorbs nitrogen. This is called *bends* or having been *bent*, since nitrogen bubbles blocks circulation in the small veins of joints, causing pain that the diver seek to relief by bending those joints. The literature abounds and here we offer a few more references to proceed further in case of interest (Luria and Kinney, 1970; Adolfson and Berghage, 1974; Hesser *et al.*, 1978; Passmore and Rickers, 2002; Butler, 2004; Longphre *et al.*, 2007; Lang and Sayer, 2013). One important conclusion: A subject can be trained to dive, but there is no adaptation to deep waters pressure, and the latter concept must be repeated and underlined.

4.5.2 *The pearl fishers*

Just a brief paragraph for this terrible activity, which well qualifies as such, at least looking at it with the eyes of the 21st century. The lives of these young fishermen or anglers were constantly at risk and their span was significantly decreased. There are several publications telling its history (Bari and Lam, 2010; Wikipedia, 2014a, 2014b). Although it is no longer among the mysteries of the world, there is still a lot of unknown about pearls. It has been discovered that all shells can produce them. As a consequence, a whole range of pearls may exist reflecting an immense variety of mollusks and shells. The reference to Wikipedia given above recounts the history of the pearl saga; it takes into account new scientific discoveries. Before 1900, the only means of obtaining pearls was by manually gathering very large numbers of pearl oysters (or mussels) from the ocean floor. More than a ton of oysters were searched to find 3–4 quality pearls. Divers were often forced to descend to over 35 m on a single breath, exposing them to the dangers of hostile creatures, waves, and drowning, frequently because of water blackout on resurfacing. The natural pearls were a rare bonus for the divers, even though superb specimens were found over the years. Edwin William Streeter was one of the leading and most influential English jewelers in the 19th century. He outfitted a ship, which he sailed to go himself pearl fishing in 1880. Streeter furthermore led a consortium to compete with Baron Rothschild to lease ruby mines in Burma (Streeter, 1886).

Fortunately, things have changed in this respect. Today, pearls are grown; they are *cultured pearls*, created under controlled conditions. They can be farmed using two very different groups of bivalve mollusk, the freshwater river mussels, and the saltwater pearl oysters (see https://en.wikipedia.org/wiki/Cultured_pearl, for more information). A good way of satisfying human vanity.

4.6 Mathematical Models of Respiration: Grodins, Milhorn, and the New Era

The respiratory system shows properties similar to those of a control system of the regulatory type. A pioneering contribution by Fred S. Grodins and collaborators marked the starting line in 1954 regarding this kind of view (Grodins *et al.*, 1954). Carbon dioxide in central neurons regulates respiration, while oxygen concentration plays a secondary role. A sudden entrance of CO_2 via inhaled air causes arterial and tissue carbon dioxide to rise. In response, alveolar ventilation increases trying to restore the normal level. It acts as a negative feedback loop. Howard T. Milhorn and collaborators, first in 1965 and later on in his well-known textbook, developed and improved the model in an excellent set of equations describing normal and disease situations (Milhorn *et al.*, 1965; Milhorn, 1966).

By and large, each model has tried to simulate (periodic breathing) PB under established conditions, but often they contain 15 physiological parameters or even more. Some parameters show a wide range of values in a population, so that the simulations are not able enough to test breathing patterns of individuals. Furthermore, it is impractical to run direct experimental validations. Carley and Shannon, from the Massachusetts General Hospital, in 1988, presented what they called a minimal model suitable for direct validation. For that matter, they used (1) CO_2 sensitivity, (2) cardiac output, (3) mixed venous CO_2, (4) circulation time, and (5) mean lung volume for CO_2. This model showed to be consistent with previous models and experimental data regarding the degree of hypoxia or congestive heart failure required to produce PB (Carley and Shannon, 1988). A few years before, Michael Khoo and his group had also carried out extensive research on PB (Khoo *et al.*, 1982). These researchers developed a general model that accounts for all kinds of PB resulting from instability in respiratory control, say, in normal people during sleep and on acute exposure to high altitude, in sleeping infants, and in patients with cardiovascular or neurologic lesions. It was found that in almost every case, the ventilatory oscillation is mediated predominantly by the peripheral controller. System stability decreased due to hypoxia, hypercapnia, increased lung washout times, prolonged lung-chemoreceptor delays, and high controller sensitivity. Stability was enhanced by large lung CO_2 and O_2 storage volumes but little affected by body tissue stores. The model predicted that the mean cycle time of PB decreases from about 30 s at sea level to 20 s at 14,000 ft (4267 m), in agreement with data from other studies. Allometric scaling of the relevant parameters also showed close agreement between model predictions and data obtained on infants.

In a later report (Khoo *et al.*, 1991), the authors tried to elucidate the mechanisms that lead to sleep-disordered breathing. They proposed a mathematical model that allows for dynamic interactions among the chemical control of respiration, changes in sleep–waking state, and changes in upper airway patency. The increase in steady-state arterial PCO_2 accompanying sleep is shown to be inversely related to the ventilatory response to CO_2. Chemical control of respiration becomes less stable during the light stage of sleep, despite a reduction in chemoresponsiveness, due to a concomitant increase in responsiveness of blood gases to ventilatory changes. The withdrawal of the *wakefulness drive* during sleep onset represents a strong perturbation to respiratory control: Higher magnitudes and rates of withdrawal of this drive favor instability. These results may account for the higher incidence of PB observed during light sleep and sleep onset. Periodic ventilation can also result from repetitive alternations between sleep onset and arousal.

The potential for instability is further compounded if the possibility of upper airway occlusion is also included. In systems with high controller gains, instability is mediated primarily through chemoreflex overcompensation. However, in systems with depressed chemoresponsiveness, rapid sleep onset and large blood gas fluctuations trigger repetitive episodes of arousal and hyperpnea alternating with apneas that may or may not be obstructive. Between these extremes, patterns that are more complex can arise from the interaction between chemoreflex-mediated oscillations of shorter cycle-duration (approximately 36 s) and longer wavelength (approximately 60–80 s) state-driven oscillations.

In another and relatively more recent study, Poon *et al.* (2007) stated in their abstract: "Homeostasis is a basic tenet of biomedicine and an open problem for many physiological control systems". Among them, none has been more extensively studied and intensely debated than the dilemma of exercise hyperpnea — a paradoxical homeostatic increase of respiratory ventilation, that is, geared to metabolic demands instead of the normal chemoreflex mechanism. Classical control theory has led to a plethora of "feedback/feedforward control" or "set point" hypotheses for homeostatic regulation, yet so far, none of them has proved satisfactory in explaining exercise hyperpnea and its interactions with other respiratory inputs. Instead, the available evidence points to a far more sophisticated respiratory controller capable of integrating multiple afferent and efferent signals in adapting the ventilatory pattern toward optimality relative to conflicting homeostatic, energetic and other objectives. This optimality principle parsimoniously mimics exercise hyperpnea, chemoreflex, and a host of characteristic respiratory responses to abnormal gas exchange or mechanical loading/unloading in health and in cardiopulmonary diseases — all without resorting to a feedforward "exercise stimulus". Rather, an emergent controller signal encoding the projected metabolic level is predicted by the principle as an exercise-induced "mental percept" or "internal model", presumably engendered by associative learning (operant conditioning or classical conditioning) which achieves optimality through continuous identification of, and adaptation to, the causal relationship between respiratory motor output and resultant chemical–mechanical afferent feedbacks. This internal model self-tuning adaptive control paradigm opens a new challenge and exciting opportunity for experimental and theoretical elucidations of the mechanisms of respiratory control — and of homeostatic regulation and sensorimotor integration in general."

Katiyar, from the Indian Institute of Technology, reviewed various mathematical models for respiratory mechanics, which are the basis for most clinically applied methods for the analysis of the mechanics of breathing (Katiyar, 2009). The same year, 2009, a group of Polish researchers reported a hybrid model of the human respiratory system that enables connecting respirators with computerized virtual lungs. A simulation of the artificial ventilation of lungs, with the use of the hybrid model and a Siemens Servo was made. Waveforms of pressure inside the lungs, flow in the respiratory tract, and the lung volume during the simulated artificial ventilation were recorded. Compliance and resistance of the hybrid model were calculated based on the inspiratory pause algorithms and compared to the values set in the model. The initial tests showed that the calculated values of the parameters differ by 20% (worst result) from the values set in the model (Michnikowski *et al.*, 2009).

It is worth mentioning a doctoral dissertation with rather attractive concepts. The author recalls that in patients where mechanical ventilation is required, the aim is to ensure adequate gas exchange, while avoiding ventilator-induced lung injury (VILI). Uncertainties exist regarding the causes and prevention of VILI. The effect of mechanical ventilation on gas distribution, perfusion, and gas-exchange is not yet fully understood, especially the effect of gravity. Furthermore, it has not been determined how alveoli behave during mechanical ventilation. Mogensen aimed first at understanding how healthy lungs respond to mechanical ventilation. Thereafter, a mathematical model was developed subdivided in four submodels to describe (1) pulmonary ventilation, (2) perfusion, (3) blood chemistry, and (4) gas-exchange during mechanical ventilation. The ventilation model simulates pressure–volume relationships and the perfusion model describes the pulmonary perfusion during mechanical ventilation, both models stratified with respect to the effects of gravity. The third one, blood model, describes blood acid–base chemistry, and the last one simulates oxygen and carbon dioxide distributions. The models are validated against experimentally obtained data and simulate well a wide range of physiological parameters during breathing. The model has indicated that alveoli in the healthy subjects do not collapse. Furthermore, simulation results show that gravity affects the gas exchange more than what is experimentally observed. This leaves room to speculate that other effects such as anatomical gradients, hypoxic vasoconstriction, and bronchodilation may compensate for the effects of gravity on the regional ventilation and perfusion of the lungs. As general conclusion, this author states,

(1) A comprehensive stratified physiological model of the total respiratory system has been developed. The model is composed of four submodels, that is, a ventilation model, a perfusion model, a blood model, and a gas exchange model.
(2) The model is able to reproduce physiological data observed in the literature, of, for example, PV or pressure-volume curves, distribution of lung density, ventilation and perfusion, total capillary perfusion, capillary transition time of the red blood cells, arterial, mixed venous and end tidal partial pressures of O_2 and CO_2.
(3) Simulated and experimentally measured lung densities indicate that alveoli do not collapse in the healthy lungs and that the hysteresis of the PV curves is mainly due to the hysteresis of surfactant and not to recruitment of alveoli.
(4) The simulated lungs were unable to sustain sufficient oxygenation of the arterial blood in upright position, it is, therefore, concluded that other mechanisms counteract the effects of gravity (Mogensen, 2011).

4.7 Summary

We have widely reviewed the respiratory system of mammalian animals and man, from its basics to rather elaborated concepts playing a role in it. The act of respiration appears clearly as a complex result of neural, humoral, and perceptual factors, which, in turn, are influenced by changes in the respiratory mechanics. Besides, there is also a close relationship with the cardiovascular system. The lungs are particularly prone to malignant tumors, especially and unfortunately because smoking is still rather widespread among people. Air contaminants of different origin — smog in daily language — appear also as biasing factors. Prevention is always the best therapy but, even so, there are other oncogenic factors. Molecular biology is one of the big hopes of humanities in this respect. Gene-based therapies for cancer find their background on the augmentation of the host's antitumor immunity or the augmentation of sensitivity to antincoplatic drugs. The available information goes beyond any possible intended full review while it keeps growing steadily, as with other subject areas clearly showing a time compression effect (Arini *et al.*, 2014).

References

Adolfson J, Berghage T. Perception and performance under water. John Wiley & Sons, 1974. ISBN 0-471-00900-8.

Arini PD, Bianchi J, Valentinuzzi ME. Scientific discoveries and technological inventions: Their relativistic history effect. *IEEE Pulse* 4(3):64–74, 2014.

Bari H, Lam D. Pearls: The time of the great fisheries (1850–1940), see especially Chapter 4, 189–238, 2010; SKIRA/Qatar Museums Authority.

Beall CM. Andean, Tibetan, and Ethiopian patterns of adaptation to high-altitude hypoxia. *Integr Comp Biol* 46(1):18–24, 2006. doi:10.1093/icb/icj004.

Beall CM. Detecting natural selection in high-altitude human populations. *Respir Physiol Neurobiol* 158(2–3):161–171, 2007. doi:10.1016/j.resp.2007.05.013.

Bigham AW, Wilson MJ, Julian CG, Kiyamu M, Vargas E, Leon-Velarde F, Rivera-Chira M, Rodriquez C, Browne VA, Parra E, Brutsaert TD, Moore LG, Shriver MD. Andean and Tibetan patterns of adaptation to high altitude. *Am J Hum Biol* 25(2):190–197, 2013. doi:10.1002/ajhb.22358.

Blickenstorfer CH. Diving Physiology; by ScubaDiverInfo.com, a magazine publisher and a SCUBA instructor, 2014. http://www.scubadiverinfo.com/2_physiology.html.

Brandis K. On-line physiology tutorials: Fluid physiology — An on-line text, 2013. http://www.anaesthesiamcq.com/FluidBook/fl3_2.php; under Creative Commons License; downloaded on Dec 29.

Butler FK. Closed-circuit oxygen diving in the U.S. Navy. *Undersea Hyperb Med* 31(1):3–20, 2004.

Carley DW, Shannon DC. A minimal mathematical model of human periodic breathing. *J Appl Physiol* 65(3):1400–1409, 1988. http://www.ncbi.nlm.nih.gov/pubmed/3141356.

Cousteau, J-Y, Dumas F. Le Monde du silence. Éditions de Paris, Paris, France, 237, 1954. There are several editions, still available in the market.

Frisancho AR. Developmental functional adaptation to high altitude: A review. *Am J Hum Biol* 25(2):151–168, 2013.

Gallup AC. Yawning as a brain cooling mechanism: Nasal breathing and forehead cooling diminish the incidence of contagious yawning. *Evol Psychol* 5(1):92–101, 2007.

Grodins FS, Gray JS, Schroeder KR, Norins AL, Jones RW. Respiratory responses to CO_2 inhalation. A theoretical study of a nonlinear biological regulator. *J Appl Physiol* 7:283–308, 1954.

Helt M, Eigsti I-M. Contagious yawning in autistic and typical development. *Child Dev* 81(5):1620–1631, 2010. doi:10.1111/j.1467-8624.2010.01495.x.

Hesser CM, Fagraeus L, Adolfson J. Roles of nitrogen, oxygen, and carbon dioxide in compressed-air narcosis. *Undersea Biomed Res* 5(4):391–400, 1978.

Hjortdal VE, Emmertsen K, Stenbøg E, Fründ T, Rahbek Schmidt M, Kromann O, Sørensen K, Pedersen EM. Effects of exercise and respiration on blood flow in total cavopulmonary connection: A real-time magnetic resonance flow study. *Circulation* 108:1227–1231, 2003. http://circ.ahajournals.org/content/108/10/1227.full.pdf.

Huerta-Sanchez, E. Altitude adaptation in Tibetans caused by introgression of denisovan-like DNA. *Nature* 512:194–197, 2014. doi:10.1038/nature13408.

Hurtado A. Respiratory adaptation in the Indian natives of the Peruvian Andes. Studies at high altitude. *Am J Phys Anthropol* 17(2):137–165, 1932.

Jiang J, Regner MF, Tao CH, Pauls S. Phonation threshold flow in elongated excised larynges. *Ann Oto-Rhino-Laryngol* 117(7):548–553, 2008. http://www.ncbi.nlm.nih.gov/pmc/articles/PMC2922005/.

Katiyar VK (2009) Mathematical modeling of the respiratory system: A review, *Indian J Biomech* (Special Issue), NCBM 7–8, March, 56–60, 2009. http://www.iitr.ac.in/ISB/uploads/File/ISB/pdf/devdatta.pdf.

Klabunde RE. Factors promoting venous return. In Cardiovascular physiology concepts. 2nd ed. Lippincott Williams & Wilkins, 2011. http://www.cvphysiology.com/Cardiac%20Function/CF018.htm, 1998–2013.

Khoo MC, Gottschalk A, Pack AI. Sleep-induced periodic breathing and apnea: A theoretical study. *J Appl Physiol* 70(5):2014–2024, 1991.

Khoo MC, Kronauer RE, Strohl KP, Slutsky AS. Factors inducing periodic breathing in humans: A general model. *J Appl Physiol* 53(3):644–659, 1982.

Lang MA, Sayer MDJ. Proceedings of the Joint International Scientific Diving Symposium, American Academy of Underwater Sciences and European Scientific Diving Panel, Curaçao, October 24–27, 2013. http://cima.uprm.edu/~n_schizas/reprints/Sherman_etal_2013_AAUS.pdf.

Longphre JM, DeNoble PJ, Moon RE, Vann RD, Freiberger JJ. First aid normobaric oxygen for the treatment of recreational diving injuries. *Undersea Hyperb Med* 34(1):43–49, 2007. ISSN 1066-2936.

Luria SM, Kinney JA. Underwater vision. *Science* 167(3924):1454–1461, 1970. doi: 10.1126/science.167.3924.1454.

McArdle WD, Katch FI, Katch VL. Exercise physiology. Lea & Febiger, Philapelphia and London, 853 pp., 1991.

Miceli M, Castelfranchi C. Crying: Discussing its basic reasons and uses. *New Ideas Psychol* 21(3):247–273, 2003. doi:10.1016/j.newideapsych.2003.09.001.

Michnikowski M, Glapiński J, Guć M, Gólczewski T, Darowski M. A hybrid model of the respiratory system. *Biocybern Biomed Eng* 29(1):71–80, 2009.

Milhorn HT, Benton R, Ross R, Guyton AG. A mathematical model of the human respiratory control system. *Biophysical J* 5:27–46, 1965.

Milhorn HT. The application of control theory to physiological systems. W.B. Saunders Co, Philadelphia, 386 pp., 1966.

Mogensen ML. A physiological math model of the respiratory system. Doctoral Diss, Center Model-Based Medical Decision Support, Dept Health Science &

Technol Aalborg Univ, Denmark. 2011. http://vbn.aau.dk/files/56706346/ PhDThesis_MadsLauseMogensen.pdf.

Moore, LG. Human genetic adaptation to high altitude. *High Alt Med Biol* 2(2):257–279, 2001. doi:10.1089/152702901750265341. Lorna G. Moore is with the Department of Anthropology, University of Colorado, Denver, USA.

Nair S (2011) Why do dogs pant?, 2011. http://www.buzzle.com/articles/why-do-dogs-pant.html; September 23.

Newham P (1994) the singing cure: An introduction to voice movement therapy. Shambhala Publisher, Boston, 256, 1994. ISBN-10 0877739978 and ISBN-13 978-0877739975; http://www.yorku.ca/earmstro/journey/respiration.html.

Norscia I, Palagi E. Yawn contagion and empathy in Homo sapiens. *PLoS ONE* 6(12):e28472, 2011. doi:10.1371/journal.pone.0028472.

Panneton WM. The mammalian diving response: An enigmatic reflex to preserve life? *Physiology* (Bethesda), 28(5):284–297, 2013. http://www.ncbi.nlm.nih.gov/pubmed/23997188.

Panneton WM, Anch AM, Panneton WM, Gan Q. Parasympathetic preganglionic cardiac motoneurons labeled after voluntary diving. *Front Physiol* 5:8, 2014. doi: 10.3389/fphys.2014.00008. eCollection 2014; http://www.ncbi.nlm.nih.gov/pubmed/24478721.

Passmore MA, Rickers G. Drag levels and energy requirements on a SCUBA diver. *Sports Eng* 5:173–182, 2002, Blackwell Science Ltd.

Peñaloza D, Arias-Stella J. The heart and pulmonary circulation at high altitudes: Healthy highlanders and chronic mountain sickness. *Circulation* 115(9):1132–1146, 2007. doi:10.1161/CIRCULATIONAHA.106.624544.

Poon, CS. Ventilatory control in hypercapnia and exercise: Optimization hypothesis. *J Appl Physiol* 62(6):2447–2459, 1987.

Poon CS, Tin C, Yu Y. Homeostasis of exercise hyperpnea and optimal sensorimotor integration: The internal model paradigm. *Respir Physiol Neurobiol* 159(1): 1–20, 2007; online March 7; http://www.ncbi.nlm.nih.gov/pmc/articles/PMC2225386/.

Poon CS, Tin C. Mechanism of augmented exercise hyperpnea in chronic heart failure and dead space loading. *Respiratory Physiol Neurobiol* 186(1):114–130, 2013. Epub 2012 Dec 27; http://www.ncbi.nlm.nih.gov/pubmed/23274121.

Provine, RR. Yawning as a stereotyped action pattern and releasing stimulus. *Ethology* 72(2):109–122, 2010. doi:10.1111/j.1439-0310.1986.tb00611.x.

Provine R, Tate BC, Geldmacher LL. Yawning. *Am Sci* 93(6):382–93, 2005. doi:10.1511/2005.6.532.

Reeves JT, Grover RF. Insights by Peruvian scientists into the pathogenesis of human chronic hypoxic pulmonary hypertension. *J Appl Physiol* 98:384–389, 2005. doi:10.1152/japplphysiol.00677.2004.

Rupert JL, Hochachka PW. Genetic approaches to understanding human adaptation to altitude in the Andes. *J Exp Biol* 204:3151–3160, 2001.

Scheinfeldt LB, Soi S, Thompson S, Ranciaro A, Woldemeskel D, Beggs W, Lambert C, Jarvis JP, Abate D, Belay G, Tishkoff SA. Genetic adaptation to high altitude in the Ethiopian highlands. *Genome Biol* 13(1):R1, 2012. doi:10.1186/gb-2012-13-1-r1.

Schürmann M, Hesse MD, Stephan KE, Saarela M, Zilles K, Hari R, Fink GR. Yearning to yawn: The neural basis of contagious yawning. *NeuroImage* 24(4):1260–1264, 2005. doi:10.1016/j.neuroimage.2004.10.022.

Senju A, Maeda M, Kikuchi Y, Hasegawa T, Tojo Y, Osanai H. Absence of contagious yawning in children with autism spectrum disorder. *Biol Lett* 3(6):706–708, 2007. doi:10.1098/rsbl.2007.0337.

Simonson TS, Yang Y, Huff CD, Yun H, Qin G, Witherspoon DJ, Bai Z, Lorenzo FR, Xing J, Jorde LB, Prchal JT, Ge R. Genetic evidence for high-altitude adaptation in Tibet. *Science* 329(5987):72–75, 2010. doi:10.1126/science.1189406.

Song G, Tin C, Poon CS. (2014) Multiscale fingerprinting of neuronal functional connectivity. *Brain Struct Funct* 220(4):2967–2982, 2014. http://www.ncbi.nlm.nih.gov/pubmed/25056933.

Straus C, Vasilakos K, Wilson RJ, Oshima T, Zelter M, Derenne JP, Similowski T, Whitelaw WA. A phylogenetic hypothesis for the origin of hiccough. *BioEssays* 25(2):182–188, 2003. doi:10.1002/bies.10224.

Streeter E. Pearls and pearling life. George Bell & Sons, London, 1886. https://archive.org/details/pearlspearlingli00stre. Available in print at the Smithosian Library.

Thompson S. The dawn of the yawn: Is yawning a warning? Linking neurological disorders. *Med Hypotheses* 75:630–633, 2010. doi:10.1016/j.mehy.2010.08.002.x.

Valentinuzzi ME, Baker LE, Powell T. Heart rate response to the Valsalva maneuver. *Proc X ICMBE* (Dresden), paper 17.7, p 249. Full paper in *Med Biol Eng* 1974, 12(6):817–822, 1973.

Valentinuzzi ME, Johnston R. Spirometry: A historical gallery up to 1905. *IEEE Pulse* 5(1):73–76, 2014.

Vitzhum VJ. Fifty fertile years: Anthropologists' studies of reproduction in high altitude natives. *Am J Hum Biol* 25(2):179–189, 2013. doi: 10.1002/ajhb.22357.

Watson A, Fray T, Clarke S, Yates D, Markwell PJ. Reliable use of the ServoMed Evaporimeter EP-2 to assess transepidermal water loss in the canine. *J Nutr* 132(Suppl 2):1661S–1664S, 2002.

Wikipedia (2013a). http://en.wikipedia.org/wiki/Crying, as for Dec.

Wikipedia (2013b). http://en.wikipedia.org/wiki/Hiccup, as for Dec.

Wikipedia (2014a). http://en.wikipedia.org/wiki/Pearl_hunting, as for Sept.

Wikipedia (2014b). http://en.wikipedia.org/wiki/Scuba, as for June.

CHAPTER 5

THE RENAL SYSTEM: ALSO A SOURCE FROM GENERAL PHYSIOLOGY

Max E. Valentinuzzi and Alfredo Coviello

From the salty waters we come, slowly aiming at the infinity of the mind.

Abstract

First, we recall our aquatic origin showing how, steadily and very slowly, evolution brought us to the present state, where the intellect has become the central core — the kernel — governing all our steps immersed in the internal sea. The renal functions take place within it, giving us freedom to move about and beyond. Convolution and deconvolution are thereafter introduced, as mathematical tools, to find the renal retention function, essential physiological concept derived from renography. Finally, the renin–angiotensin–cardionatrine system is dealt with, even though some reader may complain it had better belong to the endocrine area.

5.1 From Fish to Philosopher

The objectives of the renal function are essential for life as all have a fundamental regulatory action over the different concentrations of metabolites in blood, osmotic pressure, fluid volumes, electrolytes levels, and blood pressure (BP). The exchanger associating the kidneys to the cardiovascular system is a highly complex arrangement, delicate in its adjustment mechanisms and of the utmost importance for a free life, as expressed by Claude Bernard (1813–1878), French physician and scientist who, in the 19th century, introduced the concept of *internal environment* (the extracellular fluid [ECF] or plasma plus interstitial fluid). Smith (1959), in his profound book, says,

> *The story of how the kidneys operate and how they came to function in the way they do is the vertebrate story, from which man is the most notable and clever actor and, besides, is the only philosopher.*

Briefly stated, the renal system centers its *homeostatic* activity on the total body water and, more specifically, on the ECF compartment, where all the cells are immersed, as in a small and personalized sea. *Homeostasis* refers to the dynamic regulation and readjustment processes of the physiological variables sustaining life; Walter Bradford Cannon (1871–1945) devised the word. His momentous book, *The Wisdom of the Body*, published in 1932, described how the human body maintains steady levels of temperature and other vital conditions such as the water, salt, sugar, protein, fat, calcium, and oxygen contents of the blood. His own words went more or less like the following, as he shaped up the concept over the years (Cannon, 1929, 1932, 1941),

> *The highly developed living being is an open system having many relations to its surroundings, in the respiratory and alimentary tracts and through surface receptors, neuromuscular organs and bony levers. Changes in the surroundings trigger reactions in this system, so that internal disturbances of the system are elicited. Such disturbances are normally kept within narrow limits, because automatic adjustments within the system are brought into action, and thereby wide oscillations are prevented and the internal conditions are held fairly constant.*

The kidneys play an essential role in such a complex task, in regulating the chemical composition of body fluids by removing metabolic wastes and retaining the proper amounts of water, salts, and nutrients. Waste is removed from the body by the kidneys in the form of urine. They produce approximately 1 mL of urine per min ($\approx$1.5 L/day), and maintain an average ECF at an osmolarity level of 300 mOsm/L.

5.2 Basic Renal Anatomy and Functions

All the blood volume must pass through the kidneys and it does so approximately 30 times per day or about 160 L/day (assuming a blood volume of 5–6 L). It enters via the two renal arteries (one per kidney) getting out through the renal veins to return to the general circulatory stream. Between inflow and outflow, there is a highly complex structure of vessels and tubules. Besides, each kidney has an exit duct, the *ureters*, both connected to the bladder (a temporary urine reservoir) ending in the *urethra*, the final path for urinary excretion.

Each renal artery branches off until it reaches the level of very many and small caliber *afferent arterioles* that get into a minute capillary network called *glomerulus*, in turn contained into the capsule of Bowman, some kind of basket-like structure, from which the *efferent arteriole* comes out (Fig. 5.1). The *proximal convoluted tubule* originates in this capsule continuing, thereafter, with the *loop of Henle* (descending and ascending legs), the *distal convoluted tubule* and the *collecting duct*. The glomerulus and its associated tubular-loop system constitute the renal unit called *nephron*. Each human kidney has approximately 1,000,000 nephrons while in the dog, instead, the

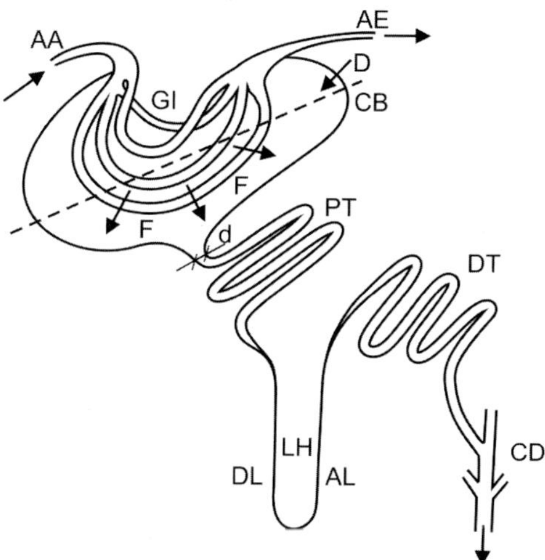

Figure 5.1: Functional renal unit: the nephron. AA: afferent arteriole. AE: efferent arteriole. GI: glomerulus with its glomerular capillaries. The total area of these capillary walls, in man, is in the order of 1.5 m^2. CB: capsule of Bowman, with a diameter D $\approx 200\mu$m. TP: proximal tubule. Its output from the capsule has a diameter d $\approx 55\mu$m. LH represents the loop of Henle, with its descending and ascending limbs DL and AL, respectively. As a continuation, the distal convoluted tubule DT is depicted ending at the colleting duct CD. Arrows F within the capsule show the direction of filtration. Redrawn by Gustavo Idemi, from the previous edition of this book.

number of nephrons is about 400,000, always per kidney. Up to a point, the kidney size depends on the number of nephrons. All collecting ducts converge to the minor, major calyces and the renal pelvis that, finally, end up in the ureter. Calyces act as some sort of funnels to direct the urine to its destination (the bladder). Blood, after traversing the glomerular capillaries, exits this minute system via the efferent arteriole, which, in turn, branches off to form the *peritubular capillaries*. They surround and wrap the tubular system in such a way as to establish the *renal exchanger*, which permits the back and forth shift of substances between blood and the intratubular fluid. These capillaries successively converge (or fan in) into venules and larger diameter vessels until reaching the renal vein to get back to the general circulation. The *renal parenchyma* is the soft and well-wet tissue found between the peritubular capillaries and the tubular system.

5.2.1 *Renal processes*

All blood getting into the glomeruli is *ultrafiltrated*, that is, only particles of molecular weight smaller than 75,000 are able to pass through the glomerular capillary thin walls and appear in the capsule's volume in order to continue their trip along the tubules. Thus, ultrafiltration is a special very fine filtration. Cellular somas and proteins, which are larger than the size stated above, cannot pass the capillary walls and are never supposed to appear in normal urine. If they do, the fact would signal pathology. Filtration does not involve local expenditure of metabolic energy and depends only on the hydrostatic pressure imparted to the blood by the heartbeat.

The glomerular filtrate, loaded with substances, undergoes highly significant and physiologically important modifications as it moves within the tubular system until it reaches the ureter. There are processes of *secretion* and *reabsorption*, that is, by way of passive and mostly active mechanisms, substances pass from the blood in the peritubular capillaries to the renal parenchyma to the intratubular fluid and, vice versa, substances traverse an inverse pathway, that is, from the intratubular fluid to the blood.

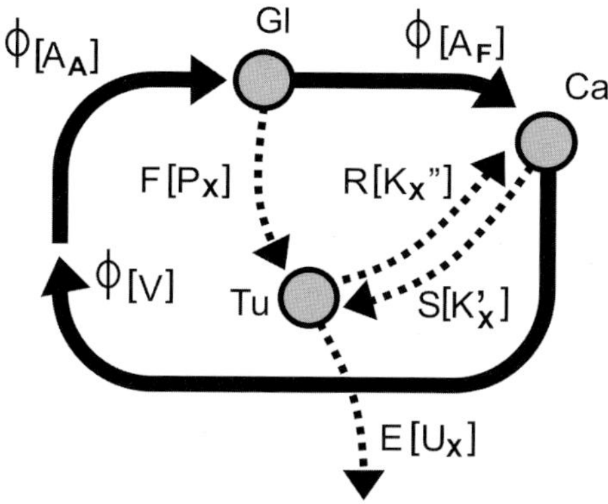

Figure 5.2: Flow diagram of the renal system. The renal equations for the three basic renal processes are obtained after application of the continuity principle to the nodes GI, Tu, and Ca (see text). Drawn by Gustavo Idemi.

A simple linear old model describes filtration, reabsorption, and secretion (Valentinuzzi *et al.*, 1968), where all the glomeruli become a single node GI, all the peritubular capillaries another node Ca, and all the tubules a third node Tu (Fig. 5.2). If considered as a hydraulic system, let us apply the continuity principle to node Tu, which is equivalent to one of Kirchhoff's law in electric networks. In other words, in steady-state conditions, what gets into a node must be equal to what gets out, or the algebraic sum of the flows to a node is zero. Flow is interpreted as the amount of fluid per unit time (say, mL/s), or the amount of a dissolved substance carried by the fluid (say, mg/s), or charge per unit time (say, coulombs/s, which mean current in amperes in an electric circuit). In the case of Fig. 5.2, there are four branches converging to node T_u, thus,

$$F[P_x] + S[K'_x] - R[K''_x] - E[U_x] = 0 \qquad (5.1)$$

where the subindex x represents the substance dissolved in the fluid, F, S, R, and E are the rates of filtration, secretion, reabsorption, and excretion, respectively, expressed in mL/min, and the variables between brackets stand for the respective concentrations of the substance x within the indicated branches of the system. Obviously, the net units of the equation are mg of substance/min. Equation (5.1) describes a dynamic equilibrium of the flow of substance with respect to node Tu.

5.2.1.1 *Filtration*

Let us assume a substance that is only filtrated (as, e.g., *inulin*, frequently used in the determination of glomerular filtration). In such cases, secretion and reabsorption are absent and Eq. (5.1) reduces to

$$F[P_x] - E[U_x] = 0 \qquad (5.2)$$

which corresponds to a straight line when the *excreted load* $E[U_x]$ is represented as a function of the plasmatic concentration of substance P_x (Fig. 5.3, left upper panel, line a). The slope F is precisely a measure of the *glomerular filtration rate* (also called GFR). By definition of ultrafiltrated fluid, the concentration of x in plasma is equal to the concentration in the capsule fluid. Experimentally, it has been found that the inulin filtration rate is about 120–130 mL/min, or in the order of 180 L/day (which is about 30 times the blood volume, as already mentioned previously).

If Eq. (5.2) is divided through by the plasmatic concentration $[P_x]$, we obtain

$$C_x = F = \text{GFR} = \frac{E[U_x]}{[P_x]} \qquad (5.3)$$

which is the GFR and, by definition, it is *the excreted load over the plasma concentration*, called also the *clearance* of substance x. In other words, clearance of x is the excreted load of that substance per unit concentration of the same substance in plasma. The numerator is measured in mg/min and the denominator in mg/mL, meaning that clearance is expressed in mL/min (i.e., it is a flow). Another common definition in medical practice and in renal physiology states that clearance is the volume of plasma that in the unit time (say, one minute) is completely cleared of the substance x. For inulin or for any substance that only is being filtered by the kidneys, C_x is constant with respect to P_x (Fig. 5.3, right upper panel, horizontal line a). Creatinine, a breakdown product from creatine phosphate, which is

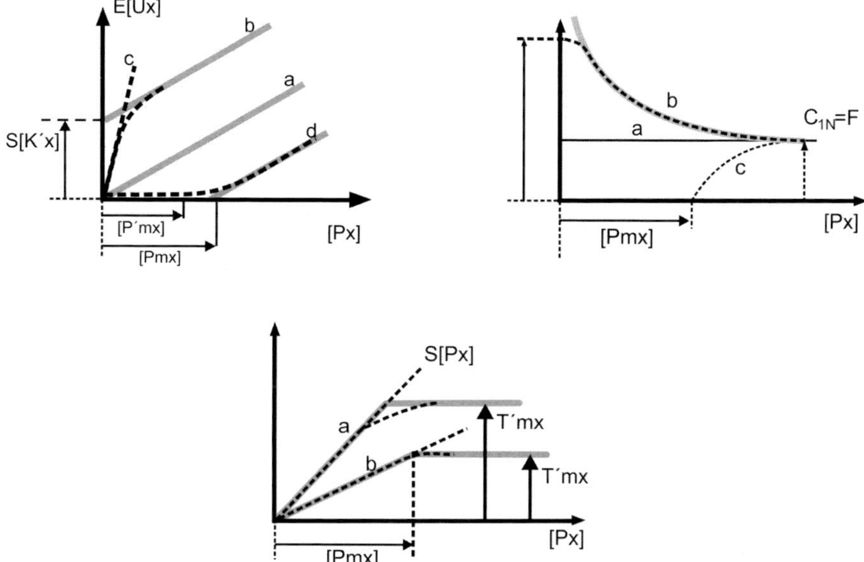

Figure 5.3: The three basic renal processes. Left upper panel: Excreted load as a function of the plasmatic concentration. For pure filtration (a), for secretion first (c), and filtration predominance thereafter (b), and reabsorption followed by filtration predominance (d). Right upper panel: Clearance as a function of the plasmatic concentration. Filtration (a), secretion (b) and reabsorption (c). At high plasmatic concentrations, all three processes tend to the same constant GFR, which equals the inulin filtration value. Lower panel: Secreted and reabsorbed loads as functions of the plasmatic concentration. Curve (a), secretion, and curve (b), reabsorption. Notice that, in the case of reabsorption, the maximum plasmatic concentration [P$_{mx}$] determines the breaking points of (d), excreted load, (c), clearance, and (b), reabsorbed load. Redrawn by Gustavo Idemi.

naturally found in blood, is another substance that is mainly filtered by the kidneys, although not quite as accurate as inulin because about 10% is reabsorbed, but often used in medicine since no injection is required.

Other frequently used concepts are *extraction ratio* (ER) and *filtration fraction* (FF); the former is the amount of compound entering the kidney and excreted in the final urine. The following equation describes it, that is,

$$ER = (P_a - P_v)/P_a \tag{5.4}$$

where P_a stands for the concentration in the renal artery and P_v represents the concentration in the renal vein. For instance, *para-aminohippuric acid* (PAH) is almost completely excreted in the final urine, and thus almost none is found in the venous return ($P_v \sim 0$); therefore, the ER of PAH is ~ 1. This is why PAH is used in PAH clearance to estimate renal plasma flow (see below). FF is defined as the ratio of the GFR to the renal plasma flow (RPF), that is,

$$FF = GFR/RPF \tag{5.5}$$

The FF measures the proportion of fluid reaching the kidneys that passes into the renal tubules. It is normally about 20%. The GFR on its own is the most common and important measure of renal function. In a condition such as renal artery stenosis, the blood flow to the kidneys is reduced, so that filtration must increase in order to perform the normal balancing of fluid and electrolytes in the body. This would be reflected in a higher FF and the kidneys have to over-work. Diuretics tend to decrease the FF. Conversely, *catecholamines* increase the FF by vasoconstriction of afferent and efferent arterioles. Severe hemorrhage will also result in an increased FF.

5.2.1.2 *Secretion*

Essentially, the kidneys filtrate all substances, and some are also secreted into the tubular fluid, but not reabsorbed. Tubular secretion is the transfer of materials from peritubular capillaries to renal tubular lumen and occurs mainly by active transport. It is the tubular secretion of H^+ and NH_4^+, from the blood into the tubular fluid, that helps keeping blood pH at its normal level. A good place to check is the so-called Boundless (2014). The general equation (5.1) given above becomes

$$F[P_x] + S[K'_x] - E[U_x] = 0 \tag{5.6}$$

because $R = 0$. If it is accepted that $[K'_x] = [P_x]$, we can define

$$T'_x = S[P_x] \tag{5.7}$$

as the *secreted load* (Fig. 5.3, lower panel, line *a*). That straight line, as the plasma concentration increases, reaches a plateau or saturation value T'_{mx}. Experimental curves show that the breaking point is not abrupt but rather there is a smooth bending to reach the maximum level (dashed curved lines in Fig. 5.3). This phenomenon is due to the spread out or splays of the individual nephron behaviors.

Clearance is also defined from Eq. (5.6) dividing it through by the plasma concentration and solving for the quotient of the excreted load to the same concentration, that is,

$$C_x = \frac{E[U_x]}{[P_x]} = F + \frac{T'_x}{[P_x]} \tag{5.8}$$

which is represented in Fig. 5.3 (right upper panel, curve b). When the plasmatic concentration is low, the secretory process dominates and the clearance curve starts at a high value C_o (not infinite, as incorrectly predicted by the equation, meaning that the model is far from being perfect). The curve falls rather sharply with increasing plasmatic concentrations tending to the inulin clearance value at high concentrations, when there is net filtration predominance. This is clearly seen in Eq. (5.8), as its mathematical limit is taken for $[P_x] \to \infty$. In turn, the excreted load displays a steep slope first $(F + S)$, as shown in Fig. 5.3 (upper left panel, line c) to become equal to the GFR (same figure, line b) as the plasmatic concentration goes up. Thus, line b runs parallel to line a, and with an upward shift. The passage from the first slope to the second one in actual experimental curves is smooth and the breaking point represents only a theoretical limiting value. A typical substance frequently used in renal secretion studies is PAH, with a maximum secreted load of 80 mg/min and a clearance of 650 mL/min at low plasmatic concentrations (Boundless, 2014).

5.2.1.3 *Reabsorption*

The third process does not have secretion and the general equation, with $S = 0$, becomes now,

$$F[P_x] - R[K''_x] = E[U_x] \tag{5.9}$$

since the excreted load was moved to the right-hand side. This is represented by the straight line d in Fig. 5.3 (left side panel), shifted to the right with respect to line a, and crossing the horizontal axis at $[P_{mx}]$. If $[K_x''] = [P_x]$, it is possible to define the *reabsorbed load* as

$$T_x'' = R[P_x] \qquad (5.10)$$

which is represented by line b in Fig. 5.3 (lower panel). Breaking points, as mentioned before, are really defined by the respective projections of the straight lines, and experimental curves display a smooth transition from one to the other, as shown in the figure. The reabsorbed load reaches a maximum value (saturation) when the tubules are no longer able to take more substance and that happens beyond a given plasmatic concentration specific for each substance. Research is still going on regarding this particular aspect of renal physiology. At lower concentrations, instead, there is full reabsorption. As we did before, clearance is obtained by dividing Eq. (5.9) by $[P_x]$ leading to

$$C_x = \frac{E[U_x]}{[P_x]} = F - \frac{T_x''}{[P_x]} \qquad (5.11)$$

The latter describes a hyperbola (Fig. 5.3, upper right) that tends to the constant value given by inulin as the plasmatic concentration increases. For low concentrations, the reabsorptive process dominates and the substance in full returns to the circulation. This is the case for substances of high physiological value, as glucose is. The maximum reabsorbed load lies in the order of 380 mg/min, at a plasmatic threshold level of about 300 mg/100 mL. In practice, however, because of the splay phenomenon, it spreads between 180 and 200 mg/100 mL (Fig. 5.3; A*d*, B*c*, and C*b*, where A, B, and C stand, respectively, for the upper left, upper right, and lower panels). A diabetic person will have a high level of glucose in blood (*hyperglycemia*), usually beyond the renal plasma threshold. The reabsorptive capacity is saturated and glucose appears in urine (*glucosuria*). In the old days, the physician used to taste the urine for its sweetness in order to determine whether a patient was diabetic. Do not smile, it is true! Fortunately, techniques are more sophisticated nowadays, with devices that permit a quick and easy determination.

All substances go through the filtration process in the kidneys, including exogenous substances as drugs are. Almost common knowledge is the fact of a person taking vitamin B who, a few hours later, produces reddish urine, typical of that substance. Many times the attending physician will warn the patient not to be scared in such a case. Some substances are reabsorbed, others are secreted and there is a group that is both absorbed and secreted (as urea and creatinine, with a predominance of secretion, the latter much more than the former). Sodium, instead, is also secreted and reabsorbed, but approximately in equal amounts. In a quantitative sense, the filtration and reabsorption of ions and water are by far the most significant operations of the mammalian kidney. The body cannot afford, for example, to lose potassium and, thus, it reabsorbs and retains about 93%. The interplay of these processes, in the end, is responsible for the proper electrolyte balance.

Foreign substances, such as marijuana, cocaine, heroin, or others, usually lead to a number of renal pathologies, and often they seriously affect the functions described before, especially when their use has become chronic. The literature abounds and exceeds the aims of the present text, but a few comments are pertinent. Renal disease in cocaine and heroin users is associated with the nephritic syndrome, acute glomerulonephritis, and interstitial nephritis. The pathophysiologic basis of cocaine-related renal injury involves renal hemodynamic changes, glomerular matrix synthesis and degradation, and oxidative stress and induction of renal atherogenesis. Heroin is the most commonly abused opiate in the United States, and a spectrum of renal diseases in heroin users has been identified. Administration of cocaine in animal models resulted in nonspecific glomerular, interstitial, and tubular cell lesions. Apparently, there is no animal model of heroin-associated renal disease. The heterogeneity of responses that are associated with heroin is not consistent with a single or simple notion of nephropathogenesis. Although there is a paucity of evidence to support a heroin-associated nephropathy, the evidence from *in vitro* cellular and animal studies to support the existence of cocaine-induced renal changes tends to be convincing (Crowe *et al.*, 2000; Jaffe and Kimmel, 2006).

5.2.2 *Other renal mechanisms*

5.2.2.1 *Osmosis*

It is important to review first the concept of osmosis as an essential phenomenon in renal physiology. When a membrane, permeable to a solvent but not to a solute, separates a solution and pure solvent, the solvent passes into the solution by *osmosis*. The *osmotic pressure* is that particular hydrostatic pressure which must be applied to the solution to prevent the entry of solvent. In other more general words, water passes from a solution with lower concentration to a solution with higher concentration so that both concentrations tend to equilibrate. Osmotic pressure is one of the properties of solutions and depends on the number of particles per unit volume of solvent, not on their chemical characteristics. Similar to the law of gases relating pressure, volume, and temperature, van't Hoff's equation, — due to the Dutch chemist Jacobus Henricus van't Hoff (1852–1911), in 1884, and Nobel Prize for chemistry in 1901 — states that

$$\Pi = C_{\mathrm{m}} \times R \times T \tag{5.12}$$

where Π represents the osmotic pressure or "water attraction" generated by the solution, C_{m} stands for the molar concentration of the solution, R is the gas constant ($= 0.08$ atm $\times$ L/mole $\times$ K), and T is the absolute temperature. Since the molar concentration $C_m = n/V$, with n being the number of moles and V the volume in liters, thus expressing it in moles/L, Eq. (5.12) can be also written as

$$\Pi \times V = n \times R \times T \tag{5.13}$$

which well reminds the law of gases because pressure P in one is replaced by osmotic pressure Π in the other. Eq. (5.13) defines 1 osmol — a unit — as the osmotic strength generated by a concentration of 1 mole per liter at the measured temperature T in Kelvin, or

$$1 \text{osmol} = 1\frac{\text{mole}}{\text{liter}} \times 0.08 \frac{\text{atm} \times \text{liter}}{\text{mole} \times \text{K}} \times 310\,\text{K} = 25.4\text{atm} = 19,304\text{mmHg} \tag{5.14}$$

Above, we take the temperature as equal to the body temperature, $37°$C, and adding 273 to obtain the Kelvin. If the temperature instead were $0°C$, that is, $T = 273$ K, 1 osmol becomes equivalent to a pull of 22.4 atm. Summing up the molar concentrations of all the ions and non-dissociating molecules gives the osmotic strength of a solution. Some numerical examples will help in the understanding of the subject. For any substance, one gram-molecular weight (or one mole $= 1$ M) contains $N = 6.02 \times 10^{23}$ molecules (*Avogadro's number*). If the substance is glucose, sucrose, or any non-dissociating compound, the osmotic strength is 1 osm. Now, let us consider a substance that dissociates when in solution. One gram-molecular weight of NaCl consisting of 6.02×10^{23} molecules dissociate into twice this number of ions in solution. Thus, 1 M of NaCl exerts an osmotic effect of nearly 2 osmols (we say "nearly" because the degree of dissociation depends on the concentration; the lower the concentration, the higher the dissociation). If it were a substance dissociating in three ions (such as Na_2SO_4), the effect would approach 3 osmols. The osmotic concentration of plasma, interstitial fluid, and intracellular fluid are all kept in man within the band (283 ± 11) milliosmoles/L (mOsm/L), basically, by the ingestion and excretion of water. Gain of water induces prompt water diuresis, whereas loss of water induces thirst and antidiuresis. Such a value, roughly equal to 300 mOsm/L, produces at body temperature an osmotic effect or pull equivalent to 7.62 atmospheres or 5791 mmHg, which is quite an impressive value (check the calculation with the relationship given above). In other words, this is the pressure needed to counteract the tendency of plasma (or to any solution similar to plasma) to draw water.

The osmotically active solutes are, largely electrolytes (around 90% or more), such as sodium, chloride, and bicarbonate. Glucose, amino acids, and urea contribute not more than 10%. Hence, the osmolar concentration is a way of measuring the "total concentration" of a fluid. Besides, and as practical information, the weight concentration of a substance, in g/L, is given by the molecular weight (MW, a number) multiplied by the molar concentration, in moles/L, or $C_w = MW \times C_m$. A 5.4% solution of glucose means a weight concentration of 54 g in 1 L which, with a molecular weight of 180 yields the molar concentration $C_m = 300$ mM/L at $37°C$. By the same token, a 0.9% NaCl solution means a weight concentration

$C_w = 9\,\text{g/L}$ which, with MW $= 58$, produces a molar concentration $C_m = 155$ mM/L and the same osmolar strength of plasma.

Another piece of useful information refers to the freezing technique: 1 mole/L of ideal solute will depress the freezing point by $1.86°\text{C}$, from which the freezing point depression (in $°\text{C}$) experimentally obtained from a given sample divided by $1.86 \times 10^{-3}°\text{C}$ yields directly the number of mOsm/L of the sample solution. "Osmolarity" refers to the number of osmols per liter of solution (the solution contains the solute), while "osmolality" is the number of osmols per kilogram of solvent (the solvent does not contain the solute). These terms are common in the texts as the terms "molarity" and "molality" are:

> Amadeo Avogadro (1776–1856), Italian chemist, stated in 1811 that "equal volumes of all gases under the same conditions of temperature and pressure contain the same number of molecules" (Avogadro's Hypothesis), but he never tested it. The name "Avogadro's Number" is just an honorary name attached to the calculated value of the number of atoms or molecules in a gram-mole of any chemical substance. If we used some other mass unit for the mole, such as "pound-mole", the "number" would be different from 6.022×10^{23}. The first person to have actually calculated the number of molecules in any mass of substance was Josef Loschmidt (1821–1895), an Austrian high school teacher, who in 1865 obtained the number of molecules in one cubic centimeter of gaseous substance under ordinary conditions of temperature and pressure to be around $\mathbf{2.6 \times 10^{19}}$ **molecules**. This is usually known as "**Loschmidt's Constant.**"

Maintenance of osmolarity is essential for life, and it is kept by water ingestion and excretion. An individual can survive many days without food but not too long without water. If a large amount of water is drunk, a diuresis of diluted, uncolored, and almost odorless urine reestablishes normal osmolarity. Conversely, if there is dehydration (as in diarrhea, or heavy sweating during a sunny day), oliguria restricts urine outflow and the reduced amount excreted is characterized by high concentration (high osmolarity), amber color (yellowish to brownish), and penetrating offensive odor. Besides, the mechanisms of thirst are activated so that the subject is driven to drink water. The renal system is much more efficient in defending the organism against dilution (water excess) than against dehydration.

Children and old persons are particularly sensitive to the latter and sports people must be always warned of the risks faced when heavy exercise is practiced under high-temperature conditions, especially during summer time. In the end, the thirst mechanism leads to compensating for the water deficit. The kidneys cannot do this by themselves for they only are able to either remove or conserve water. There are special sensors, the osmoreceptors, located in the central nervous system (CNS), that constantly check osmolarity to effect thirst and the proper renal actions.

The kidneys maintain a strong osmolar concentration gradient from their deep medullar region, where osmolarity is 1200 or even 1400 mOsm/L, to gradually decreasing down to 300 mOsm/L at the cortical region. There is a group of nephrons with their glomeruli placed at the level of the renal cortex and the loop of Henle and collecting ducts penetrating deep into the renal medulla. Besides, their peritubular capillaries run almost parallel to Henle's limbs, which show a U-like shape (Fig. 5.1). When a subject is dehydrated, central osmoreceptors located in the hypothalamus order the secretion of antidiuretic hormone (ADH), also called *vasopressin*, from the *posterior hypofisis*. ADH acts on the distal convoluted tubules and the collecting ducts increasing their permeability to water which, due to the concentration gradient across the tubular wall, as fluid moves along the tubules and duct, water goes easily from the intratubular fluid to the interstitium by simple passive osmotic gradient and from there to the blood in the peritubular capillaries. Thus, retention of water occurs. An opposite situation takes place when the subject drinks too much water. The osmoreceptors suppress the secretion of ADH and the permeability to water of the distal tubules and collecting ducts decreases. Hence, in spite of the concentration gradient, water cannot traverse the walls and there is diuresis. In these two extreme situations, the total urine excretion in 24 hours may be of only 0.5 L, in dehydration, at a concentration of 1200 or even 1400 mOsm/L. The latter means an excreted load of 600–700 mOsm. On the other extreme, urine excretion may reach values up to about 20 or 24 L in dilution, because of excess water at a concentration of only 30 mOsm/L, meaning an excreted load also of 600 or 720 mOsm. In other words, the excreted load remains essentially constant in both situations, a remarkable stable feature, indeed. In this respect, desert animals should be mentioned for their outstanding ability to concentrate urine. The kangaroo rat rarely ever drinks any water as it gains water from the dry food it eats. This rodent can produce urine that is twice as concentrated as seawater. In a similar way, dehydrated camels have the ability to conserve water by concentrating the urine. They often reach levels above 2000 mOsm/L, quite amazing (Etzion and Yagil, 1986; Schmidt-Nielsen, 1997).

5.2.2.2 *Countercurrent mechanisms*

We have seen in the previous paragraphs that the kidneys require a medullo-cortical osmolar gradient to regulate the osmolar-excreted load and, with it, to keep the ECF osmolarity. Let us explain now how this gradient builds up and is maintained. Each operation makes use of a countercurrent system, a principle well known by chemical engineers in many types of industrial exchangers to improve the exchanging efficiency.

5.2.2.3 *Countercurrent multiplication*

There is a basic mechanism between the ascending and the descending limb of the loop of Henle, at any of its levels, that by active transport of sodium ions from the ascending branch into the descending one generates a constant transversal difference of 200 mOsm/L. This is a physiological renal property. The difference includes the interstitial fluid, too (Fig. 5.4). The ascending limb walls, we underline, are impermeable to water, while the descending branch as the collecting duct are not, implying that water cannot get into that portion of the tubules and dilute its content. Because of the U-like shape of the loop, the fluid in both limbs flow in a countercurrent way (opposite directions) so that there is a multiplying effect that, in the end, becomes a longitudinal gradient.

Figure 5.4 (left panel, drawn horizontally) explains the build-up process step by step, that is,

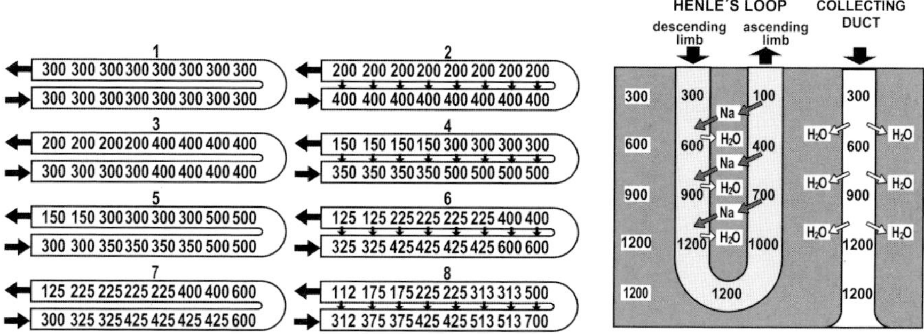

Figure 5.4: Mechanism of countercurrent multiplication. Left panel: Stage 1 depicts the loop of Henle full of fluid at 300 mOsm/L. At stage 2, due to a basic active process, a transverse osmolar difference of 200 mOsm/L is established between the descending and ascending limbs. Right panel: It shows the countercurrent exchange relationship between Henle's loop and collecting duct. See text for further details. Redrawn after Valentinuzzi (2004) by Gustavo Idemi.

At stage 1, fluid gets into the loop filling it fully at the same concentration, entering via the descending limb and exiting along the ascending portion. Thereafter, at stage 2, due to the basic active transport of sodium, a 200 mOsm/L is transversely generated all along the loop (say that the descending side rises its concentration to 400 mOsm/L while the other side lowers it to 200 mOsm/L). However, flow continues and a moment later, stage 3 depicts the situation when fresh fluid, at 300 mOsm/L gets in shifting the whole column by a certain length. Again the active mechanism restores the osmolar difference (stage 4), but observe that the region of the bent (right side in the figure, right upper panel) begins to increase its concentration creating a difference with respect to the entrance. Successive fluid shifts followed always by the basic transversal osmolar gradient build-up plus the countercurrent flow give rise to a longitudinal much larger gradient between the entrance (at 300 mOsm/L) and the medullar region (reaching there 1200–1400 mOsm/L). The mechanism is most interesting and ingenious, so much that it can be qualified as outstanding. A neat result of evolution or perhaps a design of the Great Engineer!

5.2.2.4 *Countercurrent exchanger*

The medullo-cortical concentration gradient, however, tends to equilibrate, as a ball placed on a ramp tends to roll down unless something is done, either to prevent it or at least to partially brake it. Another mechanism is required to keep such gradients. The *vasa recta* of the peritubular capillaries that run parallel to Henle's limbs take care of that function (Fig. 5.4, or just check any anatomical sketch of this part, for example, in Pitts, 1966). Notice the specialization of the medullary nephrons. Blood gets into the *vasa recta* at 300 mOsm/L and, as it moves down in their descending branches, water traverses the capillary walls because of the osmotic pull from outside while active osmotic particles get into the blood, also due to a small transverse gradient. As blood goes up following the ascending limbs of the *vasa recta*, the opposite shifts take place, that is, water gets into blood and solutes go out because at any level the ascending side has a slightly higher concentration than the descending one. Thus, the countercurrent exchanger reduces excessive loss of osmotically active solutes from the inner medulla. Blood in the *vasa recta* remove sodium and water. Loss of the medullocortical osmotic gradient would be disastrous for animal or human life.

This mechanism is similar to the countercurrent heater exchangers widely used in industrial plants. It has been amply studied, both theoretically and experimentally, and its better efficiency has been fully demonstrated as compared to heater exchangers with parallel streams in the same direction. Penguins have in their legs a circulatory arrangement of the same kind that helps them in keeping a temperature gradient from the trunk to the feet. Does it make the feet warmer? No, but the upper legs, the abdomen, and the upper body better conserve body temperature. The extremities of the sloth also have the same circulatory arrangement; however, its function is unknown, and especially if we recall that they live in tropical regions, such as NE Brasil. Another stimulating subject for the inquisitive mind. Zobel *et al.* have reported an industrial application in biotechnology not long ago, from Germany (2014).

5.2.2.5 *Osmotic exchanger*

The distal convoluted tubules and the collecting ducts (more the latter than the former) regulate the final osmotic urine adjustment by the action or not of the ADH, which, as mentioned before, controls their wall permeability to water. Most of the water is recovered and only a minor amount is excreted, either as hypotonic or as hypertonic urine (measured with respect to the plasmatic 300 mOsm/L), and depending on the hydration degree of the subject. Sodium is also recovered along this final pathway together with other ions (such as phosphate and bicarbonate). Thus, the final urine equilibrates with the hypertonic interstitium of the renal medulla and papillae. The major osmotically active constituents of urine are sodium and chloride ions and urea. The osmolar clearance may be calculated from a formula derived from the clearance definition given above, that is,

$$C_{\text{osm}} = \frac{[U_{\text{osm}}]V}{[P_{\text{osm}}]} \tag{5.15}$$

In this expression (5.15), $[U_{\text{osm}}]$ represents the collected urine osmotic concentration, V denotes the collected volume in a given period, and the denominator stands for the plasma osmolarity. In words, the equation above describes the volume of plasma per unit time completely cleared of osmotically active solutes or, also, gives the osmolar excreted load per unit of plasmatic osmotic concentration.

5.2.3 *Renal blood flow*

Let us go back to Fig. 5.2 writing now the continuity equations for nodes Gl and Ca, that is,

$$\Phi[A_{\text{A}}] - F[P_x] - \Phi'[A_E] = 0 \tag{5.16}$$

$$\Phi'[A_E] - S[K'_x] + R[K''_x] - \Phi[V] = 0 \tag{5.17}$$

By solving the two previous expressions for $\Phi'[A_E]$, equating, and considering Eq. (5.1), the following is easily obtained:

$$\Phi[A_{\text{A}}] - \Phi[V] = E[U_x] \tag{5.18}$$

where ϕ stands for the renal plasma flow, so that,

$$\Phi = \frac{E[U_x]}{[A_{\mathrm{A}} - V]} \tag{5.19}$$

or in words: The excreted load of a given substance divided by the renal arteriovenous concentration difference yields the renal plasma flow. If the test substance is fully extracted from blood in its passage through the kidneys, then the venous concentration becomes zero. Besides, if the other tissues do not extract that substance from blood, the renal artery concentration will be equal to the concentration in any systemic artery or vein, that is, it will be equal to $[P_x]$. With all this in mind, Eq. (5.19) simplifies to

$$\Phi = \frac{E[U_x]}{[P_x]} \tag{5.20}$$

the latter coincident with the plasmatic clearance of the substance given before. *para*-Aminohippurate is one substance that approximately meets the described conditions. Consequently, the measurement of the plasmatic renal flow is relatively simple: After administration of PAH, a volume of urine is collected for a specific time span to have E. Immediately, thereafter, the PAH concentration in urine is determined. From a venous blood sample, its PAH concentration is also determined. Finally, Eq. (5.20) produces the renal plasma flow. Normal adult human kidneys yield a value of about 650 mL/min. Taking the hematocrit into account (around 0.45), the renal blood flow becomes in the order of 1200 mL/min, which is 20% of cardiac output.

As recommended general and didactic reference for some kidney aspects, we should mention a chapter written by Guyton and Hall (2006).

5.3 Partial Summary

The renal system is a complex exchanger connected, on one side, with the cardiovascular system, and on the other, with the external world to excrete waste products. The metabolic needs of the kidneys do not justify the large perfusion they receive; rather, this comes about because several times per day all the blood volume must pass through them in order to guarantee the adequate homeostasis of the internal environment. Most remarkable is the urine concentration–dilution process, by which water and other substances are conserved. Mainly, the collecting ducts use a large medullo-cortical osmolar gradient. At every level of Henle's loops ascending branches, a basic transversal difference of 200 mOsm/L is established by active sodium transport as compared to the descending limb. The ascending branch is impermeable to water, so preventing the immediate loss of that gradient. The basic gradient is longitudinally multiplied by the countercurrent flow of the intratubular fluid. The peritubular *vasa recta*, also working in a countercurrent arrangement, reduce losses and tend to conserve the medullo-cortical osmotic difference. Finally, the collecting ducts act as true osmolar exchangers controlled by the ADH. In this way, only a small volume of fluid is lost per day. Such fluid carries in solution different substances and electrolytes. A person can survive and have a normal life with one kidney but life is incompatible without renal function. Thus, when we think in terms of renal failure, immediately we think of transplantation, dializers, and eventually total kidney replacement by an artificial one. From the perspective of Biomedical Engineering, there is a lot to offer to the problems posed by this area, from theoretical, physiological to technological aspects.

5.4 Renography

Injection into a vein of a radioisotope is the first step to obtaining a nuclear renogram. The isotope flows through the blood vessels of the kidneys and is filtered by the glomeruli and/or secreted by the renal tubules. As the isotope flows into the collecting system, a nuclear camera placed behind the kidneys detects its passage. The amount of isotope filtered and drained by the kidneys, in counts per sec, produces a collection of points as time proceeds, thus giving graphic information regarding drainage from the kidneys (the so-called *retention function*, because it tells how long the kidneys *retain* the foreign substance, also related to the *residence time* of the substance). The renogram produces data useful for detecting obstructions, compromised blood flow or relative function of one kidney to the other or to a normal organ. Hence, *renography* is a dynamic study where time appears as important dimension for kidney function evaluation. The upslope of the curves demonstrates kidney uptake while their downslopes refer to elimination. The two most common radiolabeled pharmaceutical agents used are Tc99m-MAG3 (mercaptoacetyltriglycine) and Tc99m-DTPA (diethylene triamine pentacetic acid). MAG3 is by far a better diagnostic agent than Tc-99m-DTPA, particularly in neonates, patients with impaired function, and patients with suspected obstruction.

Dynamic renal studies lend themselves to analysis by mathematical discrete deconvolution. The response of a kidney to a bolus injection of radioisotope into the renal artery can be derived from the renogram, which results from the time-varying input of marker from the blood into the kidneys. Figure 5.5 shows a schematic of the system and Fig. 5.6 displays the experimental curves, including the

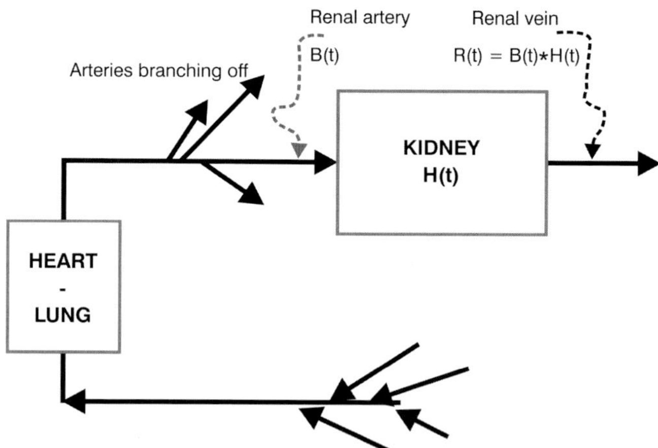

Figure 5.5: Schematic of the system to experimentally obtain the renal response $R(t)$. The radioisotope (the marker) is injected in a peripheral vein. The bolus reaches the heart–lung block coming out from the aorta to branch off into the overall circulation. One of these arteries is the renal artery (one per kidney), where activity is detected as the signal $B(t)$, input to the kidneys. Thus, the observed renogram — curves $B(t)$ and $R(t)$ — will be a convolution of the input function from the blood to the kidney with the impulse response function $H(t)$, which is the *retention function*, to be determined by deconvolution. Drawn by Gustavo Idemi.

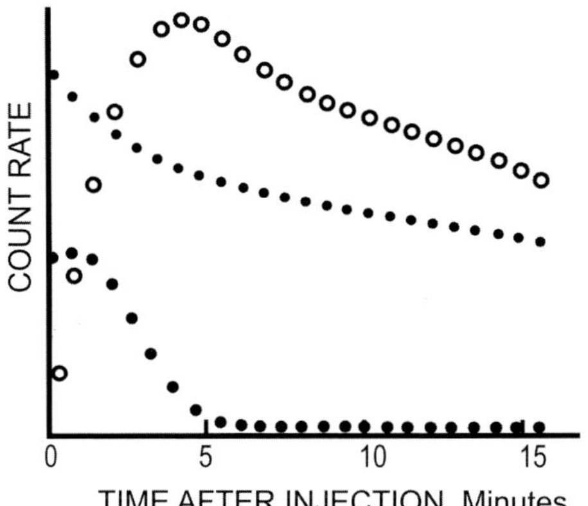

Figure 5.6: Renal retention function [RRF = $H(t)$]. The upper curve (open circles) is the kidney activity or the response of the kidneys to the input BA (blood activity, middle steadily decaying curve). The lower bumped curve (flattening at zero to the right) represents the impulse response of the kidneys, the *retention function*, obtained by mathematical deconvolution.

result of numerically deconvolving $R(t)$ by $B(t)$. Symbolically, the operation in its two directions can be written as

$$B(t)*H(t) = R(t) \quad \text{convolution product} \qquad (5.21)$$

and

$$H(t) = R(t)*/*B(t) \quad \text{convolution division or deconvolution} \tag{5.22}$$

for the latter operation (5.22) is the inverse of the former, in a way similar to the basic arithmetical algorithms (addition–subtraction, multiplication–division, taking a number to a power n — taking the nth root of a number). By and large, inverse operations give ambiguous results that must be resolved using practical criteria. The sign $*/*$ is suggested here to indicate "convolution division", or deconvolution, in a way similar to $*$, which means "convolution product" (Gonzalez *et al.*, 2016). The historical development of convolution is long and rather complex. A good account was given by Domínguez (2015).

The literature on deconvolution applied to renography and discussing its pitfalls is large ever since the technique was introduced (Valentinuzzi and Montaldo Volachec, 1975; Diffey *et al.*, 1976). We will mention just a few other contributions. Over two decades ago, in Barcelona, Spain, transit time and relative kidney function were studied by González *et al.* (1994) using two different tracers; transit times were in the order of 300 s ($\pm100\,$s). Other contributions are those by Lawson (1999), Stevens *et al.* (2006), and Thomas *et al.* (2006). A report carried out at the University of Jordan, in 2010, compared matrix inversion deconvolution with the Rutland–Patlak (R–P) plot, which is a very specific method (not to be described in this text). The values of the renal parenchymal mean transit time obtained by applying matrix inversion were significantly higher than that obtained by the (R–P) plot. However, a strong positive cor- relation was found between the values obtained by applying both methods. These authors believe that (R–P) analysis is expected to be more reproducible than the matrix inversion method because the latter relies heavily on the accuracy of the first point (Al-Shakhrah, 2010). No doubt, whatever the variants in the approach may be, deconvolution appears as a good complementary mathematical tool in renal function analysis (Gonzalez *et al.*, 2016).

5.5 The Renin–Angiotensin–Cardionatrine System

In a simplified manner, Fig. 5.7 outlines an accessory system to reg- ulate BP and body fluids. Many concepts are relatively recent, still with questions highly attractive to the physiologist, clinician and bioengineer. Let us consider BP as the central variable assuming a momentary decrease in its value. As a consequence, there will be less stretch of the juxtaglomerular apparatuses in the kidneys, located at the afferent arterioles. Thus, the hormone called *renin* will be secreted which, as it enters into the circulation, triggers a complex chain of events that end up with the formation of the polypeptide angiotensin II (Ang II), in itself a potent vasopressor. This simple negative feedback loop tends to compensate for the initial decrease in BP; however, that is not the main pathway, for Ang II stimulates the adrenal cortex to secrete aldosterone. The latter acts on the kidneys enhancing sodium reabsorption, water retention, and its consequent increase in extracellular volume, which has also a compensatory effect

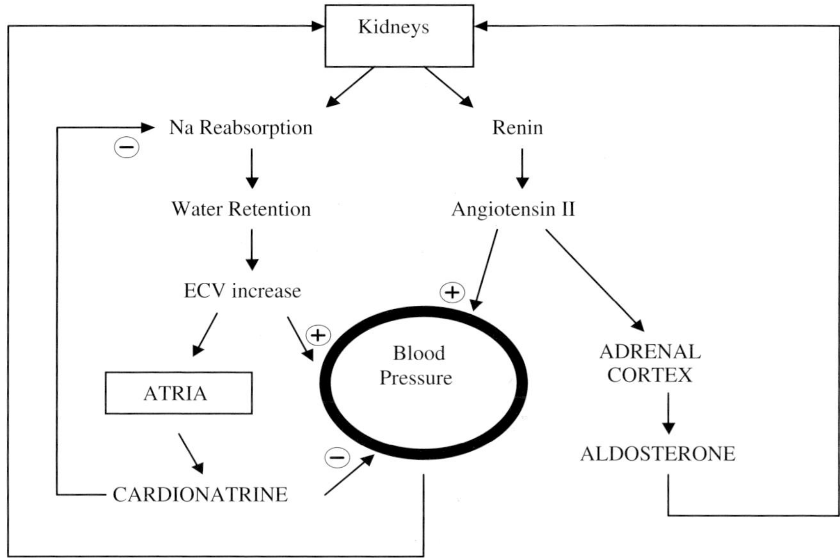

Figure 5.7: Renin–angiotensin–cardionatrine system. Angiotensin and cardionatrine have opposite effects. Negative and positive signs indicate activation and inhibition, respectively. It is a slow acting system. Arterial blood pressure is presented as the regulated variable that acts on the juxtaglomerulus apparatuses of the kidneys: when it goes down, they secrete renin. Production of angiotensin II occurs in the blood stream, after a complex chain of biochemical reactions.

on BP. This is a second negative feedback loop. The atria, in turn, have stretch volume receptors that activate the secretion of the atrial natriuretic peptide or cardionatrine. Surprisingly, the heart appears also as an endocrine organ. This relatively new substance stimulates the kidneys to excrete sodium (natriuresis), opposite to sodium reabsorption. Besides, it has a vasodilating effect, thus, tending to lower BP.

Natriuretic peptides are a group of naturally occurring substances that act in the body to oppose the activity of the renin–angiotensin system (RAS). Heart failure is a leading cause of morbidity and mortality. In the USA, there are more than 5 million patients with heart failure and over 500,000 newly diagnosed cases each year. Sodium and water retention play a significant role in this disease. There are three major natriuretic peptides: atrial natriuretic peptide (ANP), synthesized in the atria; brain natriuretic peptide (BNP), synthesized in the cardiac ventricles; and C-type natriuretic peptide, synthesized in the endothelium. Both ANP and BNP are released in response to atrial and ventricular stretches, respectively, and cause vasorelaxation, inhibition of aldosterone secretion in the adrenal cortex, and inhibition of renin secretion in the kidneys. Both ANP and BNP cause natriuresis and a reduction in intravascular volume, effects amplified by antagonism of the ADH. Increased blood levels of natriuretic peptides have been found in certain disease states, suggesting a role in the pathophysiology of them, including congestive heart failure (CHF), systemic hypertension, and acute myocardial infarction (MI). De Bold *et al.* (1981) first demonstrated that atrial extracts contain a substance that produced natriuresis and diuresis. Soon after that, the ANP molecule was purified and sequenced (Flynn *et al.*, 1983), and not a long time later, other data were added (de Bold, 1985).

Compression of the renal artery and its consequent generalized hypertension to compensate for the kidney ischemia was the experiment of the 1930s decade that brought light to the BP regulatory subject (Fig. 5.8). The first observation linking the kidneys to the circulatory system goes back to 1836, when the British physician Richard Bright (1789–1858) found a high incidence of ventricular hypertrophy in patients who had died of renal disease. In 1898, the studies made by Tigerstedt and Bergman reported the pressor effect of renal extracts; they named the renal substance *renin* (Tigerstedt and Bergman, 1898; Marks and Maxwell, 1979).

Obviously, all this previous knowledge was a significant antecedent to the Harris Goldblatt (1891–1977) renal artery clamp technique introduced in 1934, when he induced experimental hypertension in dogs (Goldblatt *et al.*, 1934; Van Epps, 2005). The latter reference offers an excellent historical account. About 1936, simultaneously in the Medical School of the University of Buenos Aires, Argentina, and in the Eli-Lilly Laboratories in Indianapolis, two independent groups, using the Goldblatt technique, demonstrated renal secretion of a pressor agent similar to renin. In the following years, both teams described the presence of a new compound in the

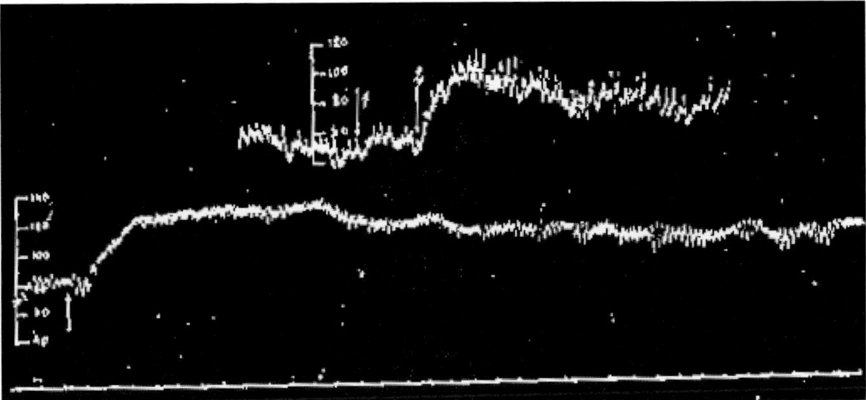

Figure 5.8: Hypertensive response of an experimental ischemic kidney. Upper channel: Carotid artery BP; lower channel: renal artery BP. Observe the delay (about 1 min 20 sec) between the renal clamp (upward arrow, left) and the systemic BP increase due to renin release. Time marks (5 seconds apart) on the lowest horizontal line. This record must be qualified as historic because, even though it was obtained during the 1950's at the Physiology Institute of the *Universidad Nacional de Tucumán* (UNT, Argentina), it still made use of a famous an obviously obsolete technology, clearly underlining the difficulties the then young investigator Coviello was facing. Kymographic record courtesy of Prof. Dr. Alfredo Coviello.

renal vein blood of ischemic kidneys. This agent had a short pressor effect. The final conclusion was that renin acted enzymatically on a plasma protein to produce the new substance. In Buenos Aires, it was called *hypertensin*, and in the United States, *angiotonin*. In 1958, Eduardo Braun Menéndez, from Argentina, and Irving H. Page, from the USA, agreed to name it *angiotensin*. Fyhrquist and Saijonmaa (2008) gave another good review, where these authors mention that new components and functions of the RAS are still being unraveled. The classical RAS, as it looked in the middle 1970s, consisted of circulating renin acting on angiotensinogen to produce angiotensin I, which, in turn, was converted into angiotensin II (Ang II) by the angiotensin-converting enzyme (ACE). Ang II, still considered the main effector of RAS, was believed to act only as a circulating hormone via angiotensin receptors, AT1 and AT2. Since then, an expanded view of RAS has gradually emerged. Most organs show local tissue RAS systems. Recently, evidence for an intracellular RAS has been reported. The new wider view of RAS, therefore, covers endocrine, paracrine, and intracrine functions.[1] Other peptides of RAS have biological actions, too, such as angiotensin III, with actions similar to those of Ang II. Furthermore, the angiotensin 3–8 hexapeptide (Ang IV) exerts its actions via insulin-regulated amino peptidase receptors. Finally, there is angiotensin 1–7, commonly named Ang 1–7, for short. The discovery of another angiotensin-converting enzyme-2 (ACE2) was an important complement to this picture, while the news also of renin receptors has made the RAS picture unexpectedly complex. Besides, clinical benefits of ACE inhibitors and AT1 receptor blockers, plus the appearance of renin inhibitors demonstrated the relevance of RAS in cardiovascular disease. Indeed, the RAS regulates more and diverse physiological functions than previously believed, not the least, if cardionatrine is also included as part of the system.

To add complexity, new components have been recently incorporated, Ang-(1–9), Ang-A and Alamandine, the last two formed by replacement of asparagine by alanine, a process involving decarboxylation of the aspartate residue. In a multiauthor paper (just 32, really not frequent), Lautner *et al.* (2013) characterized *alamandine* as a novel component of the RAS. For that matter, they reported identification of this substance, a new heptapeptide generated by catalytic action of ACE2, angiotensin A or directly from angiotensin-(1–7). Using mass spectrometry, they observed that alamandine circulates in human blood and derives from angiotensin-(1–7) in the heart. Alamandine produces several physiological actions that resemble those produced by angiotensin-(1–7), including vasodilation, antifibrosis, antihypertensive, and central effects. In addition, oral administration of an inclusion compound of alamandine/β-hydroxypropyl cyclodextrin produced a long-term antihypertensive effect in spontaneously

---

[1] *Paracrine signaling* is a form of cell-to-cell communication in which a cell produces a signal to induce changes in nearby cells, altering their behavior. Signaling molecules known as *paracrine factors* diffuse over a relatively short distance, as opposed to endocrine hormones, which travel longer distances via the circulatory system. *Intracrine*, instead, refers to a hormone that acts inside a cell, regulating intracellular events. Steroid hormones, for example, are of the intracrine type.

hypertensive rats and antifibrotic effects in isoproterenol-treated animals.

As conclusion, they state that the identification of these two novel components of the RAS, alamandine and its receptor, provides new insights for the understanding of the physiological and pathophysiological role of the RAS and may help to develop new therapeutic strategies for treating human cardiovascular diseases and other related disorders. Another recent paper uncovering angiotensins news is that by Passos-Silva *et al.* (2015).

5.5.1 *Biological effects of the renin–angiotensin system*

The effects of angiotensin II are four, say, *production of oxidative stress, production of inflammation, tendency to tissue remodeling,* and *eventual endothelial dysfunction.* Let us briefly explain them.

5.5.1.1 *Oxidative stress*

The so-called oxidative stress takes place via the action of oxidases (enzymes that catalyze an oxidation–reduction reaction, especially when involving molecular oxygen as the electron acceptor). In reactions involving donation of a hydrogen atom, oxygen is reduced to water (H_2O) or hydrogen peroxide (H_2O_2). At first, research pointed toward the renin–angiotensin–aldosterone system as BP regulator and with hydrosaline balancing action, that is, for its vasoconstrictor, antinatriuretic, and antidiuretic effects with rapid responses over renal sodium reabsorption and the adrenergic system, both central and peripheral. In the brain, renin and angiotensin appear as neurotransmitters effecting memory and influencing depressive conditions, too. The distributions of angiotensin AT1 and AT2 receptors have been mapped by *in vitro* autoradiography throughout most tissues of many mammals, including humans. In addition to confirming that AT1 receptors occur in sites known to be targets for the physiologic actions of Ang II, such as the adrenal cortex and medulla, renal glomeruli and proximal tubules, vascular and cardiac muscle, and brain circumventricular organs, many new sites of action have been demonstrated. In particular, AT1 localizes in the cerebral cortex, hippocampus, limbic system, and several hypothalamic nuclei (Allen *et al.*, 1999).

5.5.1.2 *Inflammation*

Studies in the last few years have documented new roles for Ang II as a pro-inflammatory molecule and, more recently, as a possible pro-fibrotic agent that contributes to progressive deterioration of organ function in disease. Binding of Ang II to its receptors, in particular AT1, mediates intracellular free radical generation that contributes to tissue damage by promoting mitochondrial dysfunction. Blocking Ang II signaling protects against neurodegenerative processes and promotes longevity in rodents. Altogether, these findings open the unanticipated perspective for exploring Ang II signaling in therapeutic interventions in inflammatory diseases and aging-related tissue injury. Ang II facilitates free radical liberation and their consequent inflammatory cytokines, which lead to arteriosclerosis development (Benigni *et al.*, 2010). The latter reference is a review that extends from the discovery of Ang II and its implications in renal and cardiovascular physiology up to its roles in the inflammation process, tissue injury, autoimmunity, oxidative stress, and aging.

During such processes, *upregulation* and *downregulation* may become significant mechanisms, the former making cells more responsive to stimuli, like hormones by increasing the number of receptors on the surface of the cell, and the latter, which is opposite, where cells become less sensitive to stimuli. These *downregulated cells* are said to *desensitize*, so reflecting the fact that more hormones are needed to stimulate the cells (Sampson *et al.*, 2015).

5.5.1.3 *Tissue remodeling*

The observation about undesired trophic actions of Ang II due to growth factors stimulation in smooth muscle vessels, and also over the prostate is quite attractive. Conversely, clinical experience showed that blocking of the RAS with enalapril or losartan tends to prevent sclerotic processes in heart, kidneys, and vascular beds. Grobe *et al.* (2007), for example, stated that RAS has previously been established to play an important role in the progression of cardiac remodeling, and inhibition of a hyperactive RAS provides protection from cardiac remodeling and subsequent heart failure. They demonstrated that overexpression of ACE2 prevents cardiac remodeling and hypertrophy during chronic infusion of Ang II. Altogether, their findings indicate an antiremodeling role for ANG-(1–7) in cardiac tissue, which is not mediated through modulation of BP or altered cardiac angiotensin receptor populations and may be at least partially mediated through an ANG-(1–7) receptor.

Xu *et al.* (2008) mention the hypothesis that angiotensin II receptor blockade combined with exercise training after MI could attenuate post-MI left ventricular remodeling and preserve cardiac function. Sprague Dawley rats underwent ligation of the left descending coronary artery, resulting in MI. Losartan treatment and exercise training were initiated one week after infarction, and continued for eight weeks, either as a single intervention or combined. Collagen volume fraction in the sedentary MI group was significantly higher than other MI groups treated with exercise training and/or losartan. Compared with the sedentary group, hearts of rats receiving exercise and/or losartan treatment had lower tissue inhibitor of matrix metalloproteinase. Matrix metalloproteinase, 2 or 9, did not differ among all groups. Additionally, the level of Ang II receptor type 1 (AT1) protein significantly decreased in response to exercise training. Furthermore, ACE binding was markedly lower in hearts receiving exercise training than in the infarcted sedentary hearts. Cardiac function was preserved in rats receiving exercise training, and the beneficial effect was further improved by exercise combined with losartan treatment in comparison to the sedentary group.

In conclusion:

The suggestion is that post-MI exercise training and/or Ang II receptor blockade reduces TIMP-1 (a tissue inhibitor of metalloproteinases) expression and mitigates the expressions of ACE and AT1 receptor. These improvements, in turn, attenuate myocardial fibrosis and preserve post-MI cardiac function. Furthermore, the same group of researchers (Xu *et al.*, 2010), after recalling that increased oxidative stress and decrease in antioxidant enzymes, were suggested to be involved in the pathophysiology of MI, reported the use of treadmill exercise training and losartan treatment in rats one week post-MI during eight weeks. Their experiments indicated that both exercise and losartan may reduce lipid oxidative damage plus other beneficial effects, all potentiated when combining exercise with Ang II receptor blockade.

In turn, Pushpakumar *et al.* (2013) reported that sustained hypertension induces renovascular remodeling by altering extracellular matrix components. Matrix metalloproteinases are enzymes that regulate these components turnover in concert with their inhibitors. The purpose of this study was to determine the effect of metalloproteinases on Ang-II induced renal remodeling. For that matter, C57BL/6J (wild type) and TIMP2 knockout mice were infused with Ang-II for four weeks. Blood pressure was measured weekly and end-point laser Doppler flowmetry was done to assess cortical blood flow. Besides, immunohistochemical staining was performed for collagen and elastin analyses while metalloproteinases activity was also determined. Ang-II induced similar elevation in mean BP in TIMP2 and wild-type mice. In TIMP2 mice, Ang-II treatment was associated with a greater reduction in renal cortical blood flow and barium angiography demonstrated decreased vascular density compared with Ang-II treated wild-type mice. Periglomerular and vascular collagen deposition was increased and elastin content was decreased causing increased wall-to-lumen ratio in TIMP2 mice compared with wild-type mice receiving Ang-II. The latter increased the expression and activity of metalloproteinases predominantly in TIMP2 mice than in wild-type mice. It is suggested that TIMP2 deficiency exacerbates renovascular remodeling in agonist-induced hypertension by a mechanism that may, in part, be attributed to increased activity of those proteinases. As reminder: A *knockout mouse* is a genetically modified animal in which an existing gene was inactivated by replacing or disrupting it with an artificial piece of DNA.

5.5.1.4 *Endothelial dysfunction*

Atherosclerosis begins in childhood, progresses silently through a long preclinical stage, and eventually manifests clinically, usually from middle age. Over the last 40 years, it has become clear that the initiation and progression of disease, and its later activation to increase the risk of morbid events, depends on profound dynamic changes in vascular biology. The endothelium has emerged as the key regulator of vascular homeostasis, in that it has not merely a barrier function but also acts as an active signal transducer for circulating influences that modify the vessel wall phenotype. Alteration in endothelial function precedes the development of morphological atherosclerotic changes and eventually contributes to lesion development and later clinical complications (Ross, 1993, 1999; Vita and Keaney, 2002; Schechter and Gladwin, 2003).

5.5.2 *RAS components and other aspects worth calling the attention*

In recent years, the RAS components have incorporated new elements, not only regarding its hormones, corresponding agonist, and non-peptidic antagonists, but also in its receptors and coding genes, including their tissue localization, in particular in the CNS (Almeida *et al.*, 2000). Radioimmunoassay and immunochemical luminescence have undoubtedly played an important role in such advances.

The formation of Ang II in plasma is the result of a series of enzymatic reactions that start with renin appearance, from pro-renin, due to a convertase activity. The renin substrate angiotensinogen (of hepatic origin) releases a decapeptide angiotensin I, with no biological action, from which finally, due to a converting enzyme, the octapeptide Ang II is formed. The latter substance is responsible of about 60 different tissue effects. Any biochemistry textbook explains in detail this chain of complex and highly significant physiological events.

de Kloet *et al.* (2010) have addressed the metabolic syndrome from the point of view of the RAS. These authors recall that the RAS, mostly known for its critical roles in the regulation of cardiovascular function and hydromineral balance, has regained the spotlight for its potential roles in various aspects of the metabolic syndrome. It may serve as a causal link among obesity and several co-morbidities. Drugs that reduce the synthesis or action of angiotensin II (the primary effector peptide of the RAS) have been the means to treat hypertension for decades, while clinical trials have determined the utility of these pharmacological agents to prevent insulin resistance. Moreover, there is evidence that the RAS contributes to body weight regulation by acting in various tissues. This reference reviews a relatively recent knowledge of the actions of the RAS in the brain and throughout the body to influence various metabolic disorders. Special emphasis is given to the role of the RAS in body weight regulation (see Fig. 2 of this paper by de Kloet for a quick overall block diagram).

In turn, CHF is often associated with excitation of the sympathetic nervous system. This event very likely appears as a negative predictor of survival in CHF. Sympatho-excitation and central angiotensin II (Ang II) have been causally linked. Recent studies have shown that NAD(P)H oxidase-derived reactive oxidant species (ROS) are important mediators of Ang II signaling. In the study carried out by Gao *et al.* (2004), the following hypothesis was tested:

That central Ang II activates sympathetic outflow by stimulation of NAD(P)H oxidase and ROS in the CHF state. Their results demonstrated intense stress in autonomic areas of the brain in experimental CHF and provided evidence for a tight relationship between Ang II and ROS (reactive oxygen species) as contributors to sympatho-excitation in CHF. As pertinent comment: Nicotinamide adenine dinucleotide (NAD) is a coenzyme found in all living cells. Since it consists of two nucleotides joined by phosphate groups, a P is added; instead, when acts as reducing agent, an H is added because it donates an electron.

5.5.3 *The toad: Important character in the RAS saga*

Argentine toad, formerly *Bufo arenarum*,
later reclassified as *Rhinella arenarum*.
Picture by MEV.

The Danish Nobel Prize August Krogh (1874–1949) used to say that for each physiological problem there is always an animal of choice, and the frog is a good example in this case because amphibian experimentation (frog and toads) have greatly contributed to the knowledge of physiological mechanisms. Houssay and Biasotti (1930) first demonstrated the diabetogenic action of growth hormone, as anticipated above, in toads and, later on, in dogs and man. The effect of Ang II in sodium and water transport was first demonstrated in isolated toad kidney, and also in its ventral skin and bladder, both membranes showing functional similarity with the mammalian distal nephron.

The Ussing chamber is an apparatus for measuring epithelial membrane properties. It can detect and quantify transport and barrier functions of living tissues. The Danish zoologist and physiologist Hans Henriksen Ussing (1911–2000) introduced the technique, in 1946. It is used to measure the short-circuit current (SCC) as an indicator of net ion transport across an epithelium. Besides, this technique and that of the German physiologist Moritz Nussbaum (1854–1915) were early used to study the effects of Ang II in Na^+ transport. Nussbaum, in order to elucidate if the mechanism of formation of urine was a vital or a mechanical process (at that time renal physiology was poorly developed), used a perfusion technique in the isolated amphibian kidney, as shown in Fig. 5.9. The actual problem was to know whether the Ang II effect on urinary sodium excretion was a vascular or a tubular process, the latter depending on active Na^+ transport. The experiment allowed perfusion of the proximal convoluted tubule via the renal vein (Coviello, 1969). It must be remarked that in bigger specimens, like the toad *Bufo paracnemis*, catheterization of this vein is easier. This experiment demonstrated for the first time that Ang II increased sodium transport in the proximal convoluted tubule (Fig. 5.10). The effect was later on confirmed in the rat by Harris and Young (1977), in the rabbit by Schuster *et al.* (1984), and in man by Seidelin *et al.* (1989).

They used physiological doses of Ang II, and lithium clearance techniques to measure proximal tubular reabsorption. It is important to point out the magnitude of proximal sodium reabsorption by considering that, if glomerular filtration is, say, 180 L/day, nearly 1 kg of sodium chloride is reabsorbed to excrete only few grams in urine.

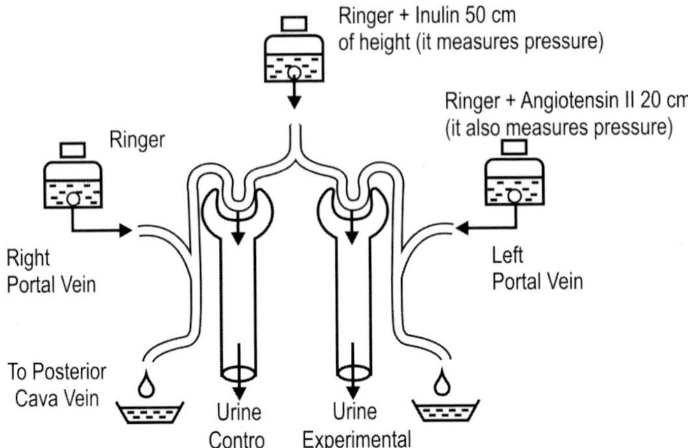

Figure 5.9: The Nussbaum perfusion technique in isolated frog kidney (proposed in 1878). To avoid vascular effects of Ang II, the hormone was infused through the portal renal vein. Ringer solution with inuline was infused through the aorta to measure glomerular filtration. Urine was collected to, thereafter, calculate reabsorption. Drawn by Gustavo Idemi.

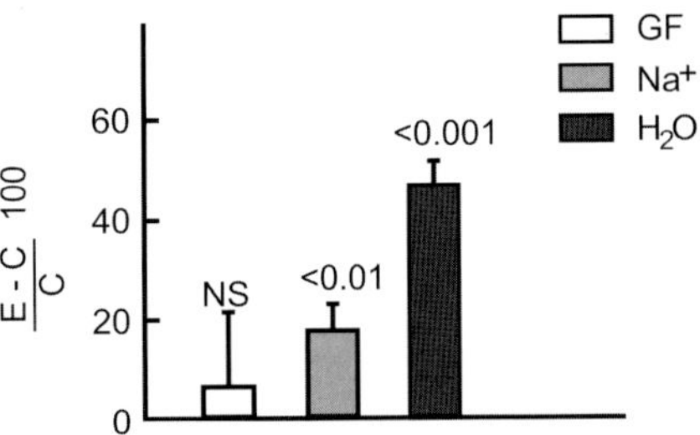

Figure 5.10: Angiotensin II increased sodium and water reabsorption in the proximal tubule. Toad *Bufo paracnemis* isolated kidney. Ang II infused into the renal vein at $(120 \pm 1.2\,\mathrm{ng/min})/100$ g to avoid the glomerular circulation (E, experiment). Control kidney (C) infused with Ringer. Results shown as % increase over C. GF: glomerular filtration, NS: not significant increase. Significant increase ($p < .01$) in Na reabsorption. Significant increase ($p < .001$) in water reabsorption. Number of measurements 14.

Weder, in 1986, with lithium obtained evidence that proximal tubule reabsorption is increased not only in hypertensive people but also in their children, a fact confirmed by Simsolo *et al.* (1999). Angiotensin increases the short circuit current, which is a marker of active sodium transport across the ventral toad's skin. It is necessary to underline that the effects of Ang II on sodium transport have a great variability depending on the amphibian species (Norris *et al.*, 1988). Moreover, there is firm impression that changing the type of angiotensin affects the reabsorption response (Proto *et al.*, 1983). Besides, the effect of Ang II in smooth arterial muscle is also inhibited by the tissue levels of NO (Marañón *et al.*, 2009).

The ventral skin transports Na^+ and absorbs water by osmosis (frogs do not drink water), and the renin–angiotensin–aldosterone system is present and active as it also is the ADH. During the measurements, and to prevent fluid loss, the animal's cloaca was ligated. The subcutaneous injection of Ang II led to water and sodium uptake, as shown in the experimental curves (Fig. 5.11). If the surrounding bath contains sodium, absorption is larger than when the bath is distilled water (dotted lines). The weight of the toad increases

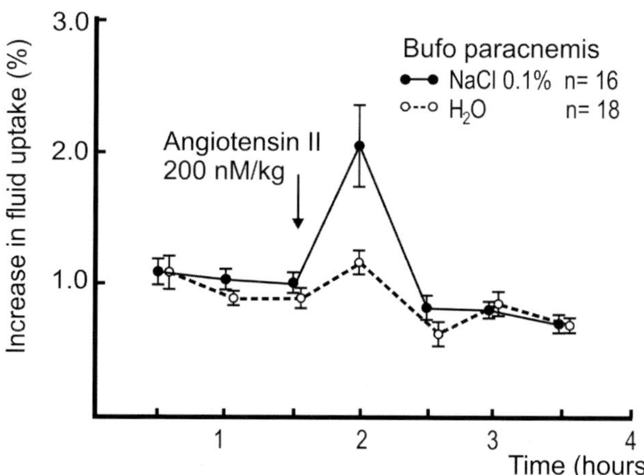

Figure 5.11: Fluid uptake (in %) as a function of time (in hours). The subcutaneous injection of Ang II leads to water and sodium uptake, as shown in the experimental curves shown here. If the surrounding bath contains sodium, absorption is larger than when the bath is distilled water (dotted lines). After Coviello (1969).

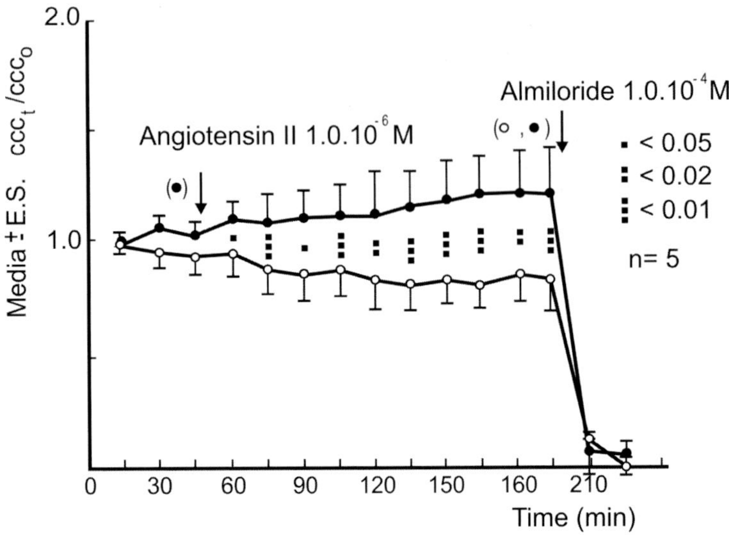

Figure 5.12: The abdominal skin of *Bufo arenarum* was divided in two halves and mounted in Ussing chambers to measure the SCC. One-half (black dots) stand for the experimental set receiving angiotensin II (left arrow). The other half remained as control (white dots). The ordinate axis represents SCC and the abscissa scales time in minutes. Whereas the control showed a progressive decline, the experimental curve under the effect of angiotensin II increased SCC significantly. At the right arrow, amiloride (a blocker of sodium transport) was added to both halves leading to a reduction in the SCC to 0.

due to fluid uptake. After several control periods, subcutaneous injection of Ang II induced a peak of fluid uptake (potentiated by NaCl in the bathing fluid). Besides, there was also BP increase (Coviello 1970; Coviello and Brauckman, 1973, Coviello *et al.*, 1974). When added to the dermal side of the frog skin, Ang II produced a significant increase in sodium transport that was abolished by amiloride (Fig. 5.12). It is of interest to underline that the frog skin and bladder constitute epithelia with functional properties similar to those of the mammalian renal collecting ducts. All these tests were demonstrated at different Ang II concentrations (Norris *et al.*, 1988). Besides, tissue NO content inhibits Ang II response. Experiments in the isolated toad *Bufo arenarum* aorta showed smooth muscle contraction.

5.6 Summary

The last section dealt with the RAS, including cardionatrine, pointing out to several effects and aspects, and underlining the complexity of the mechanisms involved along with many questions that still are under research. Certainly, theoretical models, new technologies and further experimentation open the field in a broad fan highly attractive for the young bioengineer (Randall *et al.*, 2006). Lastly, and to close the summary, it should be mentioned that Pérez-Rosas and Rodríguez González (2011), in Mexico, developed a mathematical model of the system to emulate the response of the RAS in humans. The model consists of a set of differential equations paying special attention to the estimation of all the model parameters from reported experimental data. These equations allow to modeling hypertensive and normotensive patients and pharmaco-therapeutic approaches to treatment. Besides, dose-response curves of BP and biochemical components of the RAS can be predicted and so reproducing clinical outcomes, thus giving way for the exact sciences in these decades preoccupying subject.

References

Allen AM, Zhuo J, Mendelsohn FA. Localization of angiotensin AT1 and AT2 receptors. *J Am Soc Nephrology* 10(Suppl 11):S23–S29, 1999.

Almeida RS, Ferrari MFR, Fior-Chadi DR. Quantitative autoradiography of adrenergic, neuropeptide Y and angiotensin II receptors in the nucleus tractus solitarii and hypothalamus of rats with experimental hypertension. *Gen Pharmacol* 34(5):343–348, 2000.

Al–Shakhrah I. A comparison between the values of renal parenchymal mean transit time by applying two methods, matrix inversion deconvolution and Rutland–Patlak Plot. *World Appl Sci J* 8(10):1211–1219; ISSN 1818–4952, 2010.

Benigni A, Cassis P, Remuzzi G. Angiotensin II revisited: New roles in inflammation, immunology and aging. *EMBO Mol Med* 2(7):247–257, 2010.

Boundless. Tubular Secretion. Boundless Anatomy and Physiology, 2014. Retrieved 28 November; see https://www.boundless.com/physiology/textbooks/boundless-anatomy-and-physiology-textbook/the-urinary-system-25/kidney-physiology-240/tubular-secretion-1175-412/. Boundless Learning, Inc., 207 South St., Boston, MA 02111, USA. *Boundless* is an enterprise that provides ready-to-use educational online material by means of Cloud Technology. http://en.wikipedia.org/wiki/Cloud_computing.

Cannon WB. Organization for physiological homeostasis. *Physiol Rev* 9:399–431, 1929.

Cannon WB. The wisdom of the body. WW Norton & Co., New York, 312 pp., 1932.

Cannon WB. The body physiologic and the body politic. *Science* 93:1–10, 1941.

Coviello A. Tubular effect of angiotensin II on the toad kidney. *Acta Physiol Latinoam* 19:73–82, 1969.

Coviello A. Natriuretic and hydrosmotic effect of angiotensin II in the toad *Bufo paracnemis*. *Acta Physiol Latinoam* 20:349–358, 1970.

Coviello A, Brauckmann ES. Hydrosmotic effect of angiotensin II: Isolated toad skin. *Acta Physiol Latinoam* 23:18–25, 1973.

Coviello A, Orce G, Causarano J. Effect of a competitive antagonist (8-Leu-angiotensin II) of angiotensin II on sodium and water transport in toad skin. *Acta Physiol Latinoam* 24:409–413, 1974.

Crowe AV, Howse M, Bell GM, Henry JA. Substance abuse and the kidney: Review. *QJM* 93:147–152, 2000.

de Bold AJ, Borenstein HB, Veress AT, Sonnenberg H. A rapid and potent natriuretic response to intravenous injection of atrial myocardial extract in rats. *Life Sci* 28:89–94, 1981.

de Bold AJ. Atrial natriuretic factor: A hormone produced by the heart. *Science* 230:767–770, 1985.

de Kloet AD, Krause EG, Woods SC. The renin angiotensin system and the metabolic syndrome. *Physiol Behav* 100(5):525–534, 2010.

Diffey BL, Hall FM, Corfield JR. The ^{99m}Tc-DTPA dynamic renal scan with deconvolution analysis. *J Nucl Med* 17(5):352–355, 1976.

Domínguez A. A history of the convolution operation. *IEEE Pulse Mag* 6(1):38–49, 2015.

Etzion Z, Yagil R. Renal function in camels (*Camelus dromedarius*) following rapid rehydration. *Physiol Zool* 59(5):558–562, 1986.

Flynn TG, de Bold ML, de Bold AJ. The amino acid sequence of an atrial peptide with potent diuretic and natriuretic properties. *Biochem Biophys Res Commun* 117:859–865, 1983.

Fyhrquist F, Saijonmaa O. Renin–angiotensin system revisited. *J Intern Med* 264(3):224–236, 2008.

Gao L, Wang W, Li Y, Schultz HD, Liu D, Cornish KG, Zucker IH. Superoxide mediates sympathoexcitation in heart failure: Roles of Angiotensin II and NAD(P)H oxidase. *Circ Res* 95:937–944, 2004.

Goldblatt H, Lynch J, Hanzal RF, Summerville WW. Studies on experimental hypertension. The production of persistent elevation of systolic blood pressure by means of renal ischemia. *J Exp Med* 59(3):347–379, 1934. doi: 10.1084/jem.59.3.347.

González A, Ros D, Pavia J. Estimate of relative function and transit time in renographic studies. *J Nucl Biol Med* 38(3):502–507, 1994.

Gonzalez S, Valentinuzzi ME, Arini PD. Deconvolution: It fans back, out and ahead. *IEEE Pulse Mag* 6, 7(4):54–61, 2016.

Grobe JL, Mecca AP, Lingis M, Shenoy V, Bolton TA, Machado JM, Speth RC, Raizada MK, Katovich MJ. Prevention of angiotensin II-induced cardiac remodeling by angiotensin-(1–7). *Am J Physiol Heart Circ Physiol* 292(2):H736–H742, 2007.

Guyton A, Hall J. Chapter 26: Urine formation by the kidneys. Glomerular filtration, renal blood flow and their control. In: Gruliow R, ed. Textbook of Medical Physiology. 11th ed. Elsevier, Philadelphia, PA, pp. 308–325, 2006.

Harris PJ, Young JA. Dose dependent stimulation and inhibition of proximal tubular reabsorption by Angiotensin II. *Pflügers Archiv* 367:295–297, 1977.

Houssay BA, Biasotti AB. Hypophysectomie et diabète pancréatique chez le crapaud. *C R Soc Biol* 104(18):407, 1930.

Jaffe JA, Kimmel PL. Chronic nephropathies of cocaine and heroin abuse: A critical review. *Clin J Am Soc Nephrol* 1:655–667, 2006. doi: 10.2215/CJN.00300106.

Lawson R. Quantitative methods in renography. *Semin Nucl Med* 29:146–159, 1999.

Lautner RQ, Villela DC, Fraga-Silva RA, Silva N, Verano-Braga T, Costa-Fraga F, Jankowski J, Jankowski V, Sousa F, Alzamora A, Soares E, Barbosa C, Kjeldsen F, Oliveira A, Braga J, Savergnini S, Maia G, Peluso AB, Passos-Silva D, Ferreira A, Alves F, Martins A, Raizada M, Paula R, Motta-Santos D, Klempin F, Pimenta A, Alenina N, Sinisterra R, Bader M, Campagnole-Santos MJ, Santos RA. Discovery and characterization of alamandine: A novel component of the renin–angiotensin system. *Circ Res* 112(8):1104–11011, 2013. doi: 10.1161/CIRCRESAHA.113.301077; Epub 2013 Feb 27.

Marks LS, Maxwell MH. Tigerstedt and the discovery of renin: An historical note. *Hypertension* 1:384–388, 1979.

Marañón RO, Joo Turoni CM, Coviello A, de Bruno MP. Reactivity of isolated toad aortic rings to angiotensin II: The role of nitric oxide. *J Comp Physiol B* 179 (4):401–409, 2009.

Norris B, Concha J, Contreras G, González C. Stimulatory effect of angiotensin II on the electrical properties of the isolated toad skin. *Biochem Pharmacol* 37:3005–3009, 1988.

Nussbaum M. Über die Secretion der Niere (in German, On the secretion of the kidneys). *Arch Gesamte Physiol Menschen Tiere* (now named *Eur J Physiol*), 16(1):139–143, 1878.

Passos-Silva DG, Brandan E, Santos R. Angiotensins as therapeutics targets beyond heart disease. *Trends Pharmacol Sci* 36 (5):310–320, 2015.

Pérez-Rosas N, Rodríguez-González J. Pharmacological modulation of the renin–angiotensin system by mathematical modeling. *Proc Western Pharmacol Soc* (Mx) 54:24–26, 2011.

Pitts RF. Physiology of the kidney and body fluids. Year Book Medical Publishers, Chicago, 243 pp., 1966.

Proto MC, Coviello A, Khosla MC, Bumpus M. Effects of frog-skin angiotensin II in amphibians. *Hypertension* 5(Suppl V):V22–V28, 1983.

Pushpakumar S, Kundu S, Pryor T, Givvimani S, Lederer E, Tyagi SC, Sen U. Angiotensin-II induced hypertension and renovascular remodelling in tissue inhibitor of metalloproteinase 2 knockout mice. *J Hypertens* 31(11):2270–2281, 2013.

Randall TS, Layton AT, Layton HE, Moore LC. Kidney modelling: Status and perspectives. *Proc IEEE* 94(4):740–752, 2006.

Ross R. The pathogenesis of atherosclerosis: A perspective for the 1990s. *Nature* 362:801–809, 1993.

Ross R. Atherosclerosis: An inflammatory disease. *N Engl J Med* 340:1928–1929, 1999.

Sampson AK, Irvine JC, Shihata WA, Dragoljevic D, Lumsden N, Huet O, Barnes T, Unger T, Steckelings UM, Jennings GL, Widdop RE, Chin-Dusting JP. Compound 21 prevents endothelial inflammation and leukocyte adhesion *in vitro* and *in vivo*. *Br J Pharmacol*, 2015; ISSN 0007-1188; doi:10.1111/bph.13063.

Schechter AN, Gladwin MT. Hemoglobin and the paracrine and endocrine functions of nitric oxide. *N Engl J Med* 348:1483–1485, 2003.

Schmidt-Nielsen, K. Animal physiology: Adaptation and environment. 5th ed. Cambridge University Press, Cambridge, 613 pp., 1997.

Schuster VL, Kokko JP, Jacobson HR. Angiotensin II directly stimulates sodium transport in rabbit proximal convoluted tubules. *J Clin Invest* 73:507–515, 1984.

Seidelin PH, McMurray JJ, Struthers A. Mechanisms of the antinatriuretic action of physiological doses of angiotensin in man. *Clin Sci* 76:653–658, 1989.

Simsolo RB, Romo MM, Rabinovich L, Bonanno M, Grunfeld B. Family history of essential hypertension versus obesity as risk factors for hypertension in adolescents. *Am J Hypertens* 12(3):260–263, 1999.

Smith HW. From fish to philosopher. CIBA ed., Boston, MA, 304 pp., 1959.

Stevens LA, Coresh J, Greene T and Levey AS. Assessing kidney function: Measured and estimated glomerular filtration rate. *N Engl J Med* 354(23):2473–2483, 2006.

Thomas SR, Layton AT, Layton HE, Moore LC. Kidney modelling: Status and perspectives. *Proc IEEE*, 94(4):740–752, 2006.

Tigerstedt R, Bergman PG. Niere und Kreislauf (in German, Kidneys and Circulation). *Scand Arch Physiol* 4:223–271, 1898.

Valentinuzzi ME. Understanding the human Machine: A primer to bioengineering. World Scientific Publisher, New Jersey & Singapore (xii + 396) pp, 2004, 8 chapters. Series on Bioengineering & Biomedical Engineering, vol. 4, ISBN 981-238-930-X & ISBN 981-256-043-2.

Valentinuzzi ME, Geddes LA, Baker LE, Hoff HE. The node equation and its applications in physiology. *Med Biol Eng* 6(4):387–397, 1968.

Valentinuzzi ME, Montaldo Volachec EM. Discrete deconvolution. *Med Biol Eng* 13(1):123–125, 1975.

Van Epps HL. Harry Goldblatt and the discovery of renin. *J Exp Med* 201(9):1351, 2005. See http://jem.rupress.org/content/201/9/1351.full.

Vita JA, Keaney JF. Endothelial function: A barometer for cardiovascular risk? *Circulation* 106:640–642, 2002.

Weder AB. Red-cell lithium-sodium countertransport and renal lithium clearance in hypertension. *N Engl J Med* 314:198–201, 1986.

Xu X, Wan W, Ji LL, Lao S, Powers AS, Zhao W, Erikson JM, Zhang JQ. Exercise training combined with angiotensin II receptor blockade limits post-infarct ventricular remodelling in rats. *Cardiovasc Res* 78(3):523–532, 2008.

Xu X, Zhao W, Wan W, Ji LL, Powers AS, Erikson JM, Zhang JQ. Exercise training combined with angiotensin II receptor blockade reduces oxidative stress after myocardial infarction in rats. *Exp Physiol* 95(10):1008–1015, 2010.

Zobel S, Helling C and Ditz R, Strube J (2014) Design and operation of continuous countercurrent chromatography in biotechnological production. *Ind Eng Chem Res*; see http://pubs.acs.org/doi/pdf/10.1021/ie403103c.

CHAPTER 6

STILL WITH THE SOURCES: NOW, THE GASTROINTESTINAL SYSTEM

Max E. Valentinuzzi

Man is the only species often using the GI system for pleasure ... and sometimes ... so it goes!

Abstract

After food ingestion and the digestion process in the stomach, the intestinal contents move from the duodenum through the rest of the small intestine, while more and more micronutrients are given off by the food. They must be absorbed into the surrounding cells and distributed to all the body tissues. Most absorption takes place by diffusion, so the available surface area and the concentration gradient determine the rate of absorption across the cell membrane. The intestine is specialized to maximize both. Absorption is an essential function for life maintenance. Malabsorption implies defective entrance of a dietary constituent resulting from interference with its digestion or absorption. In small animals, it is typically due to exocrine pancreatic insufficiency, whereas most cases of absorption failure are caused by small intestinal disease. This chapter will try to go deeper in it while devoting also space to the hepatic exchanger.

6.1 Overview of the Whole System

The gastrointestinal system (GIS) is as an input–output physiological unit designed to supply "combustible" to the organism. Its objective is to transform (digest) the ingested food into simple nutritive substances that are to be transferred (absorbed) to the cardiovascular system, which, as already described, distributes and delivers

191

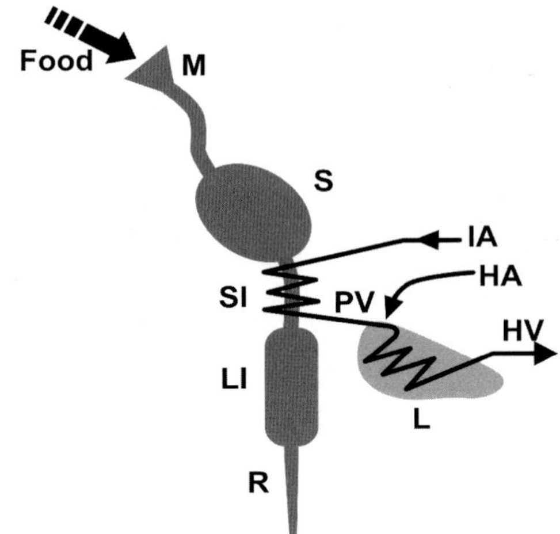

Figure 6.1: Schematic of the alimentary canal. From top to bottom, left side: M, mouth; S, stomach; SI, small intestine; LI, large intestine; R, rectum. The intestinal artery IA vascularizes the SI in a complex network of capillaries that constitutes the first section of the GIS exchanger. Blood loaded with nutritive substances comes out of the intestinal capillaries via the portal vein PV, which enters the liver L after a short pathway. The liver receives also blood from the hepatic arteries HA. The hepatic capillaries or sinusoids form the second section of the GIS exchanger. Blood returns to the general circulation via the hepatic veins. Redrawn by Gustavo Idemi.

them to the tissues. Chewing and swallowing, digestion, absorption, and defecation are all processes controlled by autonomic, nervous, and hormonal mechanisms. Digestive glands act to provide moisture, lubrication, emulsification, and enzymes for digestion of proteins, carbohydrates, and lipids. In a simplified manner, the gastrointestinal tract (GIT, Fig. 6.1) is a canal that,

(1) receives food through the mouth;

(2) receives secretions from the salivary glands, liver, pancreas, stomach, and intestines to digest food;

(3) absorbs the digested nutritive substances, especially at the level of the small intestine;

(4) supplies the absorbed nutritive substances to the liver via the portal vein;

(5) elaborates and stores other substances in the liver; and

(6) excretes the residues.

The entire tract has an associated muscular mechanics, a vasculature, and a complex regulatory system. Some important endocrine activities take also place in it.

6.2 Mechanics

Any individual, daily feels and easily detects movements all along the alimentary canal. The experience of the hunger pangs just before dinnertime or the bowel noises, especially after having eaten certain gas-producing foods (such as milk or soya beans), is well known by everybody. They are GI mechanical manifestations. Such movements are complex in nature and rather irregular and they originate in the smooth muscles that cover the different GI structures. This musculature is controlled by the autonomic nervous system, by hormones, by local mechanisms (which, by and large, involve stretching due to the entrance of material into the gastric cavity or the intestinal lumen) and, also, originates in intrinsic automatism (i.e., there are temporary pacemakers).

Basically, motility takes the form of two types of movements, that is, as *segmental contractions* and as *peristaltic waves* (peri and *stalsis* = contraction, from *Greek*). The former mix and churn the intestinal content (*chyme*), whereas the latter essentially propel the *chyme* along the intestine. Segmental contractions are ring-like contractions that appear at fairly regular intervals along the gut, then disappear and are replaced by another set of contractions in the segments between the previous contractions. When the intestinal wall is stretched, a deep circular contraction or peristaltic wave forms behind the point of stimulation and passes along the intestine toward the rectum at rates varying from 2 to 25 cm/s. Such response to stretch is the *myenteric reflex*. Movement normally takes place in the aboral (oral–caudal) direction; occasionally, and in pathological conditions, antiperistalsis (antidromic propulsion) is possible. Figure 6.2 illustrates the spontaneous mechanical activity of a piece of rabbit small intestine, placed in a thermic physiological bath, and attached to a sensitive isotonic transducer. These contractions, as shown in the figure, are usually regular in amplitude, frequency, and

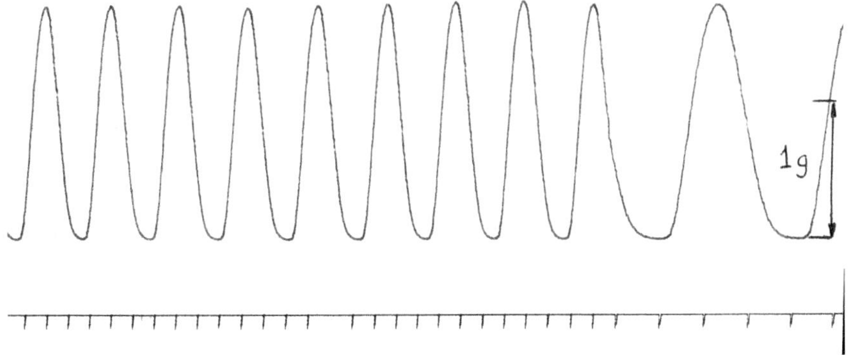

Figure 6.2: Rabbit ileum activity. Spontaneous contractions of a small intestine sample (ileum) immersed in an adequate physiological solution, temperature controlled, and with gentle air bubbling. The maximum force of contraction was about 1.75 g and very regular. The pattern was also regular and quasi-sinusoidal. Lower marks are 1 s apart. To the right of the record, paper speed was increased to better visualize the waveform. An isotonic transducer attached to one end of the sample while the other end was anchored to the holder longitudinally detected these contractions. Thus, they cannot be identified either as segmental or as peristaltic. Records obtained by the author at the Department of Bioengineering, UNT, 1980.

pattern, allowing the recording of rather long strips. In this particular case, there was a pacemaker activity of about 20 periods/min. Such intrinsic automaticity may vanish or may be enhanced (e.g., by pharmacological stimulation) decreasing in the oral–caudal direction (*Alvarez' law*). The latter means that always a sample from either the duodenum or the jejunum or the ileum will show a higher spontaneous frequency than a sample from lower regions, as say, the large intestine (ascending, transverse, or descending colon). Observe the sinusoidal-like waveform of the record shown in the latter (Fig. 6.2), a rather interesting feature that can be explored a little further. When the sample is too short (say, less than 1 cm), no activity is detected because no pacemaker site is in it; conversely, if the sample is longer than 1 cm and about 2 cm, a waveform resembling the beating effect of two near sinusoidal frequencies can be produced (Fig. 6.3). This phenomenon is typical of coupled systems, sometimes modeled with a second-order differential equation. If the sample is even longer, then more than two pacemakers may be picked up and a complex waveform will be the result.

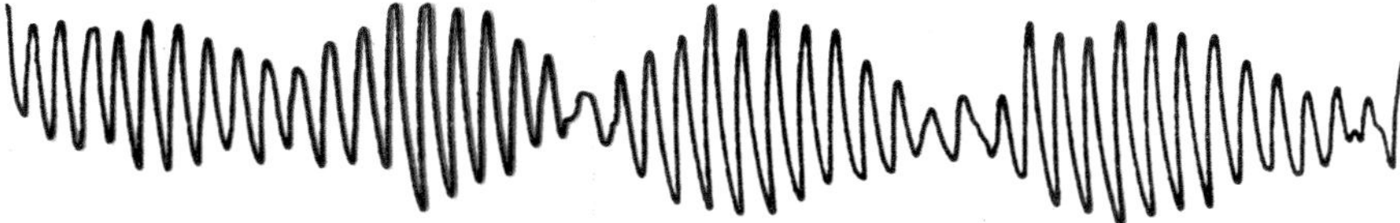

Figure 6.3: Beating effect from a rabbit ileum sample: The sample was about 2 cm in length; thus, it probably caught two pacemakers with slightly different intrinsic frequencies (recall Alvarez' law: frequency decreases towards the tail, hence, the caudal portion of the segment should show a slightly lower activity than the upper one). The net result was the beating effect as manifested by the slow modulation. Time marks are 5 s apart. The carrier had a frequency of about 15 cycles/min, while the beat had a very low 1.5 cycles/min. Records obtained by the author at the Department of Bioengineering, UNT, 1980.

Bayliss and Starling, in 1899–1901, formulated the *Law of the Intestine* (not to be confused with *Alvarez' law*, which refers to an oral-caudal rhythmicity gradient) to provide an explanation for peristalsis. They found that the response of the small intestine to a local stimulus consisted of contraction of the *muscularis externa* layer immediately above, and relaxation immediately below the point of stimulation. It was attributed to a reflex that involved the myenteric plexus and was independent of the external innervations of the intestine. Cannon, in 1911, called it a *myenteric reflex*. Later workers thereafter, as Alvarez (1915, 1950), Bozler (1949), and Connell (1961), improved this knowledge reporting their respective results in several papers over the years. Alvarez, in particular, devoted a large part of his research and during a long time to the intestinal system. Bozler, for example, found that the introduction of a bolus in the duodenum of a dog strongly increases the strength of the rhythmic contractions on the oral side, with little or no effect caudally. Besides, such activity differs from spontaneous contractions only by their greater strength and participation of the circular muscles. In strong peristalsis, a single powerful and prolonged contraction sweeps over the intestine. Connell, and that was 1961, at the Middlesex Hospital, London, recorded ileum motility from an awake human volunteer by means of a radio pill, obviously an advanced and early biomedical instrumentation contribution.

There are longitudinal and circular smooth muscles all along the intestinal walls. Contractions of the former and of the latter musculature during peristalsis are out of phase by 90°. On distension of the lumen, the longitudinal muscle contracts, followed by progressive contraction of the circular layer. The circular layer begins contraction when contraction of the longitudinal layer is half-complete; contraction of the circular layer is complete when relaxation of the longitudinal is half-complete. The definition of peristalsis appears rather superficially in many scientific and medical dictionaries. The consensus appears to be that it is a vermiform or a progressive, wavelike movement in tubular organs, consisting of alternating waves of relaxation and contraction in the muscular coat so propelling the content. Contraction of a smooth muscle cell is associated with change in potential of the cell membrane; the potential depends on the distribution of electrolytes between the cell and the extracellular space, much like in skeletal muscle. Two main types of spontaneously arising changes in the membrane potential may be detected, that is, slow potential variations and spike potentials. The latter are superimposed on the former and are associated with mechanical activity. Thus, the electrophysiology of the GIT appears as another research field still with many unresolved problems and unknowns. An overall account is to be found in a text by Keet (1993, 1998), especially directed to the pyloric sphincter.

Age is one factor with possible effects on the GIT mechanics. *In vivo* and *in vitro* rates of intestinal rhythmic contractions were measured in rats varying in age from very young to the senescent. Contraction rates, in both situations, were similar. Rhythmic contractions, at all ages, were fastest in the duodenum, slower in the jejunum, and slowest in the ileum; hence, showing an aboral gradient, as Alvarez' law states. Contractions at 10 days of age were significantly slower than at all other ages. Old age does not seem to influence the rate at which rhythmic contractions occur in the small intestine of the rat, and very possibly, these results are applicable to other species including humans (Caruolo, 1990).

6.3 Kinetics of Intestinal Absorption

Glucose, perhaps the most relevant and ubiquitous nutrient, is absorbed through the intestine by a transepithelial transport system. Stümpel *et al.* (2001), a group from Göttingen, Germany, and Lausanne, Switzerland, evaluated this rather complex process by using an isolated intestine perfusion system and found that glucose may be released out of the cells by a membrane traffic-based pathway. However, the absorption of other substances (as drugs) via the oral route is a subject of continuous investigation, especially in the pharmaceutical industry, because of the importance for a given drug to reach the circulation by mouth, route obviously preferred by patients. Oral dry absorption is affected by both drug properties and the physiology of the GIT, also by the patient's specifics, including drug dissolution of the dosage form, the manner in which the drug interacts with the aqueous environment and the membrane, permeation of the membrane, and irreversible removal by first-pass organs such as the intestine, liver, and lung. Pang (2003), from the University of Toronto, Canada, highlighted in a minireview the processes governing drug bioavailability when the drug is already in solution, emphasizing the roles of intestinal transporters and metabolism on oral bioavailability. This author underlined that, besides the liver, the intestine represents a section of the GIT that regulates the absorption extent of orally administered substances, since the intestine and the liver are involved in first-pass removal. For the most part, drug absorption occurs at the small intestine because of the large effective surface area due to the presence of villi and microvilli, which greatly increase the absorptive area. The duodenum and jejunum clearly excel in such capacity while the ileum does not as much. Intestinal blood circulation is unique because of its portal system to the liver, which accounts for 75% of the total liver blood flow (see below). For drugs that are highly cleared by the intestine, the contribution of the liver or lung to drug metabolism becomes reduced, whereas for drugs that are poorly extracted by the intestine, the substrate is able to reach the liver and lungs for removal. Substance concentration entering the guts and the intestinal flow rate certainly alter the rate of drug delivery and affect the degree of saturability of the intestinal enzymes. Moreover, the presence of other drugs or food also modifies transit times within the GIT and somehow modulates the absorption of substrates by the intestine. In the paper by Pang (2003), the concept of effective permeability coefficient (EPC) is defined as,

$$\text{EPC} = P_{\text{eff}} = (Q_{\text{lum}}/2\pi RL)\ln(C_{\text{o,lum}}/C_{i,\text{lum}}) \tag{6.1}$$

where Q_{lum} is the luminal chime flow rate, R is the lumen radius, L is the lumen length, $C_{\text{o,lum}}$ is the concentration of substance leaving the intestinal lumen, $C_{i,\text{lum}}$ is the concentration of substance getting into the lumen. The EPC depends on the physicochemical properties of the substrate, including lipophilicity, molecular size, hydrogen bonding capacity, and polar surface area.

Gruzdkov *et al.* (2012), from the Pavlov Institute of Physiology, in Saint Petersburg, Russia, investigated the kinetics of maltose hydrolysis and glucose absorption in the isolated loop of the small intestine in a wide range of concentrations. The processes were also simulated using mathematical models taking into account the mechanisms of hydrolysis and transport of nutrients and geometric characteristics of the intestinal surface. Their results, obtained from chronic experiments and mathematical simulation, showed that, under conditions close to physiological ones, the glucose transport mediated by sodium glucose transporters (SGLT) is the main mechanism of absorption in comparison with the unsaturated component. This was demonstrated not only at low, but also at high substrate concentrations. They concluded that correct evaluation of the relative contribution of different mechanisms in glucose transport through the intestinal epithelium requires taking into account the geometric specificities of its surface. By way of clarification, it must be added that glucose transporters are a wide group of membrane proteins that facilitate the transference of glucose through the plasma membrane. Since glucose is a vital source of energy for all life, these transporters are present in all phyla, most significantly in mammalian cells located in the intestinal mucosa (enterocytes) and also in the proximal nephron tubules, where they contribute to renal glucose reabsorption. In case of too high plasma glucose concentration

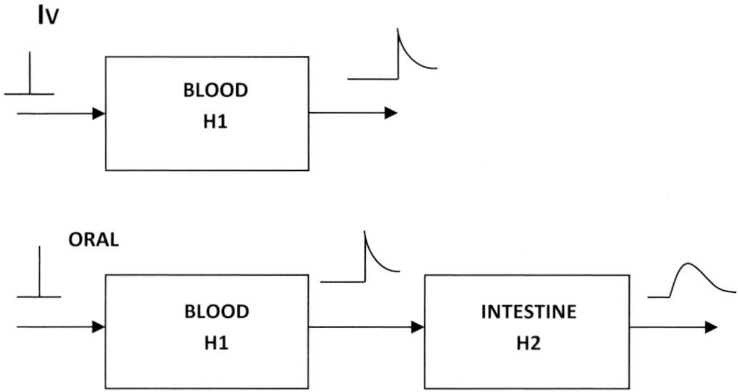

Figure 6.4: The intestinal tract as a transfer system. Top: A sudden substance shot, say, an impulse $e_1(t)$ similar to the delta function $\delta(t)$, is injected into the blood stream intravenously. The response is $r_1(t) = h_1(t)$, or impulse response, with H_1 standing for the blood transfer function in the s-domain. Bottom: An oral amount $e_2(t)$, in a way also similar to the delta function, of the same substance is ingested by the subject producing a response $r_2(t)$ between the first and the second block and obviously impossible to record that acts as input to the intestines. The final response is $r_3(t)$, detectable at the lower right-hand end of the figure.

(hyperglycemia), when there is glucosuria, the SGLT become saturated with the filtered monosaccharide.

6.3.1 *GI absorption kinetics analyzed by deconvolution*

And we come back to this numerical mathematical procedure already described in the previous chapter. Segre (1967), in Italy, reported intestinal absorption results obtained from experimental measurements and by application of discrete deconvolution. Figure 6.4 briefly outlines the basic concept, where the intestinal tract is viewed as a transfer system. In the upper part, a sudden shot of a substance is given intravenously. Thus, it can be regarded as an impulse, $e(t)$, introduced in a very short time. This reminds the Dirac delta function $\delta(t)$.

The response is $r_1(t) = h_1(t)$, or the impulse response, with H_1 standing for the blood compartment transfer function. In the lower section of the figure, an oral amount, $e(t)$, in a way also similar to the delta function, of the same substance is ingested by the subject producing a response $r_2(t)$, which should be detected between the first and the second block and obviously impossible to record. It acts as input to the intestines. The final response is $r_3(t)$, detectable at the lower right-hand end of the figure. From all this, we can write down the following general convolution relationships, but in the s-domain, that is,

$$R_1 = E * H_1 \tag{6.2}$$

$$R_2 = E * H_2 \tag{6.3}$$

$$R_3 = R_2 * H_1 = E * H_2 * H_1 = R_1 * H_2 \tag{6.4}$$

In the three equations shown above, capital letters represent, respectively, the Laplace transforms of the time-dependent small letter variables, while the symbol $(*)$ stands for *convolution product*. It is clear that deconvolving R_3 with R_1 would lead to an estimate of the intestinal transfer function, $H_2(s)$, whose impulse response is $h_2(t)$. Either of these two functions, the first one in the s-domain and the second in the t-domain, quantitatively describes the intestinal kinetics. As stated above, Segre (1967) introduced the method, which was further discussed by Valentinuzzi *et al.* (1975). Many other researchers have applied it, including some modifications (such as the use of mathematical filters) to improve the results and other tests for assessing its validity (Gonzalez *et al.*, 2016).

6.3.2 *GI absorption by tracer methods*

In 1961, Aldo Rescigno and Giorgio Segre, respectively, from New South Wales, Australia, and the University of Camerino, Italy, produced an outstanding textbook devoted to pharmacokinetics, later on translated into English (Rescigno and Segre, 1966). It is a book that no doubt has a bioengineering approach, using mathematical concepts and tools, and still valid in many respects of its overall philosophy. It deals with absorption, elimination, and distribution of substances while considering the fruitful idea of physiological compartment. They also introduced the so-called precursor–successor relationship, as a substance at one point may be the precursor of the same substance at another point, when diffusion or transport forces intervene. In essence, the relationship searches for a link between two points, 1 and 2, within the physiological structure, where diffusion or chemical processes take place.

In particular, intestinal absorption appears as well amenable to the application of tracers (Turco *et al.*, 1966). Several substances can be used to assess how efficient the transference process is. For example, by determining the proportion given by mouth, which is not absorbed, say, of ^{131}I-labeled oleic acid, or by measuring the percent of urinary excretion of tritium-labeled pyridoxine. Other tests determine the level in blood of substances given by mouth, as the folic acid test. Radiohippuran can produce good results to calculate the kinetics of passage from the GI lumen into the circulating blood. It is a substance passively absorbed, does not undergo appreciable biotransformation, and the kidneys excrete it rapidly. The model these authors used is the same as that depicted in Fig. 6.4 (bottom), where the first box on the left represents the transfer from the guts to blood and the second box stands to the blood kinetics when the tracer is injected intravenously. Numerical deconvolution, as explained above, was applied to obtain the intestinal describing function.

Birge *et al.* (1969) developed a physical model of calcium absorption. They analyzed data obtained from 23 subjects, including 13 patients having a variety of abnormalities of calcium metabolism. This technique provides a quantitative description of the rate of entry of an oral dose of calcium into the circulation as a function of time by analysis of serum or forearm radioactivity in response to intravenous and oral administration. The kinetics of the absorption process was characterized by an initial delay of 15–20 min, a maximal rate of absorption at 40–60 min after ingestion, and 95% completion thereafter. The isotope was also introduced at various levels of the gut. Although the region of the duodenum was found to have the greatest rate of absorption per unit length in normal subjects, it was least responsive to stimulation by parathyroid hormone and suppression by calcium loading. Furthermore, the response of the gut to parathyroid hormone was delayed, whereas the suppression of absorption by intravenous or oral calcium loading was rapid and dramatic.

In the last 30 years, many have been the number of advances in tracer methods applied to the GIT absorptive processes. We will mention just a few. Ostlund *et al.* (2002), which is a paper available in the WEB, studied phytosterols and phytostanols absorption in humans (stanols are a reduced form of sterols). The oral-to-intravenous tracer ratio in plasma, a reflection of absorption, was measured by a sensitive negative ion mass spectroscopic technique and became constant after 2 days. Percent absorption of the oral tracer was determined as the quotient of intravenous/oral tracers in plasma averaged over 2–4 days divided by the administered ratio times 100. It is pertinent to mention that phytosterols are cholesterol-like molecules found in all plant foods, especially in vegetable oils. They are absorbed in trace amounts, but inhibit the absorption of intestinal cholesterol including recirculating endogenous biliary cholesterol, a key step in cholesterol elimination. Reduction of cholesterol absorption can be measured at a dose of only 150 mg, suggesting that natural food phytosterols may be clinically important. Increasing the aggregate amount of phytosterols consumed in a variety of foods may be an important way of reducing population cholesterol levels and preventing coronary heart disease (Ostund, 2002).

Another contribution deals with the absorption of strontium (Höllriegl *et al.*, 2006). Intestinal absorption of strontium from an oral test dose was studied in healthy human volunteers using double tracer techniques with two stable strontium isotopes as tracers. Defined amounts of one isotope were administered orally, while tracer amounts of the second isotope were injected intravenously. The authors used two different methods to assess the total fraction absorbed. Fractional intestinal strontium absorption came from the ratio of the two isotopes in plasma or urine samples or the convolution integral technique. The latter additionally provides information on the absorption kinetics in the GIT. Why the interest in strontium? It has a promoting action on calcium uptake into bone at moderate strontium supplementation, while it shows a rachitogenic action at higher dietary levels (*rachitogenic* refers to favoring the development of *rickets*, a softening of bones in mammals due to deficiency or impaired metabolism of vitamin D, phosphorus, or calcium).

Biotin, also known as vitamin H or coenzyme R, is a water-soluble B-vitamin (vitamin B7). It is involved in the synthesis of fatty acids, isoleucine, valine, and in gluconeogenesis, and is essential for normal cellular functions; its deficiency leads to a variety of clinical abnormalities. Mammals obtain biotin from exogenous sources via intestinal absorption, a process mediated by a sodium-dependent transporter. Chronic alcohol use in humans is associated with a significant reduction in plasma biotin levels, and animal studies have shown inhibition in intestinal biotin absorption by chronic alcohol feeding. Subramanya, Subramanian *et al.* (2009) reported that in rats and transgenic mice, chronic alcohol feeding led to a significant inhibition in carrier-mediated biotin transport events across jejunal areas. Besides, chronic alcohol feeding also inhibited carrier-mediated biotin uptake in rat colon. A good review covers our current understanding of the mechanisms involved in intestinal absorption. It underlines especially water-soluble vitamins at large, including their regulation, the cell biology of the carriers involved and the factors that negatively affect these absorptive events (Said, 2011).

Prediction of human intestinal absorption is an important piece of information in the design and selection of drugs for oral delivery. Various techniques are available to evaluate how well protein absorption takes place, including screening protocols to assess the process. All this considers also a range of preclinical methodologies, such as *in silico*, *in vitro*, *in situ*, *ex vivo*, and *in vivo* tests. The truth is that human intestinal permeability cannot be accurately predicted based on a single pre-clinical method. A review by Antunes *et al.* (2013) compared the above-mentioned methods to predict human intestinal absorption, giving special attention to the absorption of therapeutic peptides and proteins.

It is appropriate to recall the meaning of the terminology: *In silico* (coined in 1989) refers to simulation by computer. *In vitro* (Latin, within glass) are experiments carried out within a test or essay tube; fertilization *in vitro* is a typical example. *In situ*, from Latin, translates to "on site", meaning, "in the very place the phenomenon occurs without isolating it from other systems". *Ex vivo* experiments are those performed on biological tissues within an environment under minimal changes with respect to the normal conditions. *In vivo* tests are undertaken under real and controlled conditions. Clinical trials appear as good examples.

To finish this quick overview of GI absorption, let us now mention small bowel transplantation, practice that has become a recognized treatment of irreversible, permanent, and subtotal intestinal failure. Ordóñez *et al.* (2013) tried to assess intestinal absorption at the time of weaning from parenteral nutrition in a series of children after intestinal transplantation. They treated 24 children (age range: 14–115 months) with intestinal transplantation, together with the liver in 6 children and the colon in 16 children. Parenteral nutrition was slowly tapered while increasing enteral tube feeding. The absorption rate was measured from a stool balance analysis performed a few days after the child had weaned from parenteral nutrition to exclusive enteral tube feeding. Results followed the resting energy expenditure (REE) concept, as proposed by Schofield (1985), that is, it represents the amount of calories required for a 24-hour period by the body during a non-active period (see also http://www.vacumed.com/293.html for more details). All children were weaned from parenteral nutrition between 31 and 85 days post-transplantation. As conclusion, there had been a suboptimal intestinal graft absorption capacity with fat malabsorption, which requires energy intakes of at least twice the REE.

6.4 The GI and the Hepatic Exchangers

Mesenteric circulation and splanchnic circulation are terms often used as synonyms; however, the former actually refers specifically to the intestinal vasculature, whereas the latter encompasses the blood flow to all the viscera within the abdominal cavity. Figure 6.5 summarizes the main avenues. It can be observed that the liver receives blood from two important inputs: from the hepatic artery, to take care of its tissue needs, and from the portal vein, to supply via its exchanger the hepatic parenchyma with the substances to be metabolized and stored in it. The first one, supplying in the order of 700 mL of blood/min, feeds the hepatic artery, HA, which carries into the liver 500 mL/min to satisfy the tissue needs of this organ, and distributes also blood to the stomach, St, the spleen, Sp, and the pancreas, Pa. The SMA carries another 700 mL/min with ramifications to the pancreas, the small intestine, SI, and the colon, Co. Finally, the IMA, with 400 mL/min, completes the blood supply to the colon. The portal vein, PV, high in nutritive contents loaded mainly through the small intestinal capillaries, returns blood to the liver at a rate of 1300 mL/min, by far much more than its tissue needs because this blood goes into the delicate hepatic sinusoids. The liver parenchyma and its hepatocytes is the place of essential metabolic and storing processes. The hepatic veins, HV, return blood to the vena cava and, from there, on to the right heart.

The liver is an essential station for the sustenance of life. Claude Bernard, around 1860, already recognized its importance. Any ingested substance, good or bad, is metabolized in the liver. Nutrients, such as glucose, are processed, elaborated, and stored in the liver to be delivered later on to satisfy body needs. The liver acts, for example, as a glucostat to regulate the blood sugar level. Many times, say at mid-afternoon, we feel some dizziness and headache, especially when already tired after some demanding work. However, by sheer will we keep on going and soon the headache and weakness

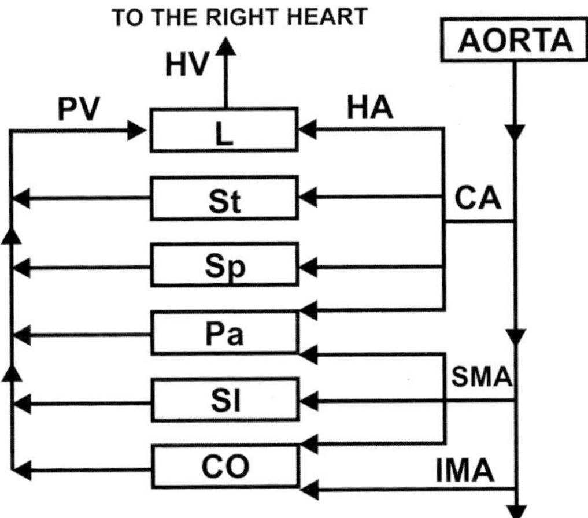

Figure 6.5: Simplified diagram of the splanchnic circulation. The descending abdominal aorta gives off three main branches: the celiac (CA), the superior mesenteric (SMA), and the inferior mesenteric (IMA) arteries.

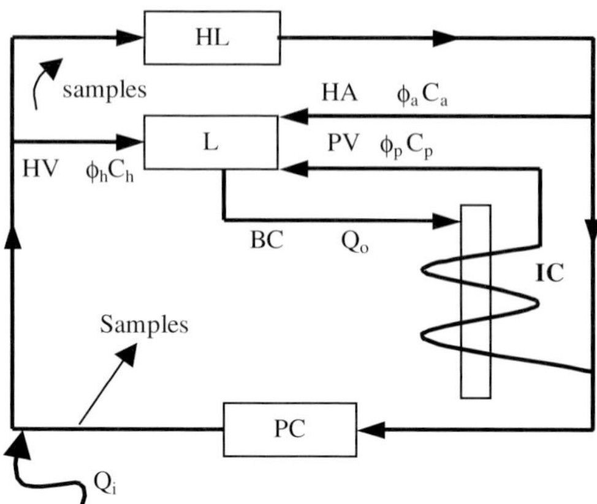

Figure 6.6: Hepatic blood flow determination. The heart–lung HL supplies blood through the aorta (above). One of its branches is the hepatic artery HA going into the liver. It carries a flow ϕ_a, in mL/min, which has a concentration C_a of an indicator substance, in mg/mL. The indicator is constantly infused via a peripheral vein at a rate of Q_i mg/min. The intestinal capillaries IC are represented by the loop on the lower right, receiving blood from the arterial supply and converging into the portal vein PV, with flow ϕ_p carrying a concentration C_p of indicator. From the liver exits also the biliary canal BC which dumps bile into the intestine at a rate of Q_o mL/min.

vanish. It was hypoglycemia, that is, our sugar level went down. We could have quickly solved the discomfort by taking a cup of tea with a couple of cookies to restore the normal glycemia, but we did not, and the liver sensing the deviation released glucose into the blood stream from its glycogen reserves. Thus, it was an "internal cup of tea" served by the liver. No wonder then that knowledge of the hepatic blood supply is of paramount significance. Based on a simple linear model, similar to that developed previously for the kidneys, a method for its determination will be outlined (Valentinuzzi, 1971). Let us remind the definition of a portal system, schematically represented in Fig. 6.6, as one that connects two capillary exchangers, in this case the intestinal network with the hepatic sinusoids.

Figure 6.6 depicts succinctly the procedure: The liver is considered a node where the continuity principle can be applied under steady-state conditions, that is,

$$\phi_a C_a + \phi_p C_p = \phi_h C_h + Q_o \tag{6.5}$$

where ϕ_a, ϕ_p, and ϕ_h are, respectively, the average blood flows of the hepatic artery, portal vein, and hepatic veins, while C_a, C_p, and C_h stand for the corresponding concentrations of the indicator. A fraction Q_o, expressed in mg/min, is the amount of indicator coming out of the biliary duct. The indicator substance is constantly infused via a peripheral vein, at a rate of Q_i mg/min, until the stationary condition is reached when $Q_i = Q_o$. The latter condition takes place if a substance, such as sulfobromophthalein sodium (a triphenylmethane derivative), is not taken up by any tissue and only is excreted through the liver to the intestine through the biliary duct (Goodman and Kingsley, 1953). Besides, and for the same reason, the arterial and portal concentrations are constant and equal, that is, $C_a = C_p = C$. As a consequence, Eq. 6.5 simplifies to

$$(\phi_a + \phi_p)C = \phi_h C_h + Q_i \tag{6.6}$$

from which the hepatic blood flow ϕ_h can be solved for as

$$\phi_h = \frac{Q_i}{[C - C_h]} \tag{6.7}$$

because $\phi_a + \phi_p = \phi_h$. Notice that the latter equation (6.7) is nothing but the dilution principle so many times used in cardiovascular physiology. The numerical value of Q_i is the rate of indicator administration, C is obtained by analyzing blood samples removed from a peripheral vein and C_h requires catheterization of one of the hepatic veins or a surgical procedure to reach that site. In the dog, a typical value is 35 mL/min per kg of body weight, that is, a 30 kg animal is expected to have a hepatic flow of about 1 L/min. Since cardiac output is in the order of 8–9% of body weight (or $\cong 2.5$ L/min, at rest), hepatic flow would represent 40% of the total cardiac outflow. In man, hepatic flow is about 25–35% of cardiac output in resting adults.

We must underline that the gastrointestinal exchanger is highly complex and, actually, it has two sections: the intestinal exchanger, with its capillary network, and the hepatic sinusoids or hepatic capillaries, both connected by the portal vein, as already mentioned before. The portal blood, especially after a meal, is well loaded with nutritive substances (carbohydrates, amino acids, and lipids). Transport from the intestinal lumen to the capillary blood is essential for life. This is the absorptive function of the intestine, as described above. Such a function takes place mainly at the level of the jejunum and ileum, with a total length of about 6.5 m in the adult. The caliber goes from 3 cm in the duodenal–jejunal angle, decreasing gradually to 2 cm at the beginning of the large intestine. The ileocecal valve marks the limit between both intestinal sections. After that, the large intestine becomes much larger in diameter. Several hundred grams of carbohydrates are being absorbed per day plus 100 or more grams of fatty acids, including monoglycerides and cholesterol, 50–100 g of amino acids, 50–100 g of different ions (such as Na, K, Mg, and the like), and 8–10 liters of water. However, the absorptive capacity of the small intestine is much greater than these values. The intestinal mucosa acts as an amplifier. Its surface area is approximately 100 times the skin body surface, that is, it is in the order of 200 m^2 (an adult's body surface area ranges from 1.5–2 m^2, typically 1.75). Such an enormous contact area is attained by successive convolutions: intestinal loops, mucosal convolutions, intestinal cilia or villi, and epithelial microcilia. Each villus has a capillary network with a small arterial input; a small output venula and a central exit lymphatic vessel, everything in a countercurrent arrangement to improve the exchange efficiency (see the Renal System). Intestinal lymphatics play a significant role in the absorption of fatty acids. Absorptive mechanisms are passive and active, and many are still not well understood.

6.4.1 *The hepatocytes*

The liver parenchyma is the functional component of the liver, made up of the hepatocytes that filter blood to remove toxins. The stroma, instead, is the connective tissue that supports the liver and creates a framework for the hepatocytes to grow on. Hepatocytes make up 70–85% of the liver's cytoplasmic mass. These cells are involved in

- Protein synthesis
- Protein storage

- Transformation of carbohydrates
- Synthesis of cholesterol, bile salts and phospholipids
- Detoxification, modification, and excretion of exogenous and endogenous substances
- Secretion of bile.

Patients with liver disorders develop damage of the liver parenchyma leading to a number of malfunctions that may compromise life itself. Individual hepatocytes grow in hexagonal units called lobules. Each lobule is arranged around a central vein, with a framework of cells around it. At the points where hexagons meet, arteries, veins, and bile ducts transport materials to and from the liver. Much of this organ's blood supply is venous, consisting of blood that needs to be filtered before it can be oxygenated and returned to circulation. A number of individual lobes make up the liver; these should not be confused with the much smaller lobules, which perform the day-to-day functions of this organ. Damage to the lobes can result in internal bleeding because of the liver's substantial blood supply. It can also decrease the efficiency of the liver, making it hard to process blood to remove compounds that might be hazardous.

6.4.2 *Derangements of the hepatic system*

Liver diseases constitute a big and complex subject of clinical medicine; medical literature describes many kinds in a long list. Viruses cause some of them, such as hepatitis A, hepatitis B, and hepatitis C, others can be the result of drugs or poisons. If the liver forms scar tissue because of an illness, it is called cirrhosis. Jaundice, or yellowing of the skin, can be one sign of liver disease, too. Cancer can affect the liver, while other liver diseases can be inherited, such as hemochromatosis (Cohen, 2013).

Most common is *hepatomegaly*, defined as the condition of having an enlarged liver, in itself not a disease but better described as a non-specific medical sign recognizing several causes, which can broadly be broken down into infection, direct toxicity (often due to alcohol), hepatic tumors, or metabolic disorder. Frequently, it presents a prominent abdominal mass. Depending on the cause, it may sometimes develop along with jaundice.

6.5 Malabsorption Conditions

The GIT possesses a huge epithelial surface area. Disorders of intestinal absorption and secretion comprise a variety of different diseases. In principle, impaired small intestinal function can occur with or without morphological alterations of the intestinal mucosa. Therefore, in the work up of a malabsorptive syndrome an early small intestinal biopsy is recommended in conjunction with breath tests and stool analysis to guide further management. In addition, there is an array of functional tests. Early diagnosis of the underlying pathophysiology is most important in order to initiate proper therapy. Specific and relatively common pathologies are coeliac disease, Whipple's disease, giardiasis, and short bowel syndrome. Furthermore, bacterial overgrowth, carbohydrate malabsorption and specific nutrient malabsorption (say, for iron or vitamins), and protein-losing enteropathy are presented with obligatory and optional tests as used in the clinical setting (Schulzke *et al.*, 2009). By way of clarifying names and terms, Whipple's disease or intestinal lipodystrophy, described by George Hoyt Whipple (1878–1976) in 1907, is a rare bacterial infection that most often affects the GIT. It interferes with normal digestion by impairing the breakdown of foods, such as fats and carbohydrates, and hampering the body's ability to absorb nutrients (Whipple, 1907).

6.6 Secretions of the GIT

Many substances are secreted into the alimentary canal. Some are *hormones* (which are internal secretions, that is, they get into the circulatory stream). The first to mention is saliva. The salivary glands in the mouth produce about 1200 mL/day of saliva (an external secretion) at a pH of 6–7, that is, from acid to neutral.

The stomach secretes a hormone, *gastrin*, whose principal effect is to stimulate the production of pepsine (an *enzyme*) and of gastric juice, both from the stomach itself. They are exocrine secretions, that is, they do not get into blood. Gastric juice contains a variety of substances, it shows a very low pH (from 1 to 3.5, which means high acidity, due to the predominance of hydrochloric acid), and is secreted by parietal cells at a rate of about 3000 mL/day. Such high acid level should damage the gastric wall, however, the surface membrane of the mucosal cells and the tight junction between cells seem to act as a protective barrier. Substances that tend to break the barrier (such as aspirin or vinegar) may lead to gastric irritation, from mild to severe. Vagal activity also stimulates acid secretion from the stomach cells. This perhaps explains the rather common and so-called *heartburns* (upper gastric burning sensation) reported by worried or overstressed persons who may show an increased tone of the vagi. As many as 33–44% of Americans experience heartburn at least once a month, and up to 13% have heartburn each day. The likelihood of having heartburn increases with age and among women who are pregnant. Having heartburn every once in a while is something almost everyone experiences, but if it occurs two or more days per week, it can be a sign of a more serious problem called *gastroesophageal reflux disease.*

Excess of gastrin represent a pathology, hence, gastrinomas are defined as gastrin-secreting tumors that are associated with Zollinger–Ellison syndrome (ZES), which is characterized by elevated fasting gastrin serum levels and clinical symptoms such as recurrent peptic ulcer, gastroesophageal reflux, and occasional diarrhea. Gastrinomas can be hereditary, but by and large, duodenal gastrinomas are small and solitary. Sporadic gastrinomas occur in the duodenum or in the pancreas, while the hereditary gastrinomas are located in the duodenum. The incidence of sporadic duodenal gastrin-producing tumors is increasing, possibly due to optimized diagnostic procedures. In contrast, pancreatic-associated gastrinomas seem to be extremely rare (Zollinger and Ellison, 1955; Anlauf *et al.*, 2006). Now a typical bioinstrumentation problema. Gastrin measurements are performed primarily for the diagnosis of these gastrin-producing tumors. This hormone circulates as several bioactive peptides, however, and the peptide pattern in gastrinoma patients often deviates from normal. Therefore, it is necessary to measure all forms of gastrin. Only immunoassays are useful for measurement of gastrin in plasma. Most clinical chemistry laboratories use commercial kits. Because of cases of kit-measured normogastrinemia in patients with ZES symptoms, the diagnostic sensitivity and the analytical specificity of the available kits have been examined. It was shown that gastrin kits frequently measure falsely low concentrations because they measure only a single gastrin form, obviously a design defect. Gastrinomas are neuroendocrine tumors, some of which become malignant. A delay in diagnosis leads to fulminant ZES, with major, even lethal complications. Consequently, it is necessary that the diagnostic sensitivity of gastrin kits be adequate (Rehfeld *et al.*, 2012).

The small intestine secretes two hormones, *secretin* and *cholecystokinin-pancreozymin* (CCK-PZ). They both act on the exocrine pancreatic function stimulating the secretion of pancreatic juice (1200 mL/day, pH 8.0–8.3, rich in enzymes to break up carbohydrates, proteins, and lipids). Besides, CCK-PZ stimulates the contraction of the gall bladder to inject bile (produced by the liver) into the duodenum. Bile is temporarily stored in the gall bladder. There is a production of about 700 mL of bile per day at a constant pH of 7.8, that is, it is on the alcaline side. Secretin was the first hormone ever found, discovered by Bayliss and Starling, in 1902. It causes the secretion of a watery, alcaline pancreatic juice. Its action on the duct cells of the pancreas is mediated by cyclic AMP (adenosine monophosphate). This hormone is produced by cells located deep in the glands of the mucosa of the upper portion of the small intestine. A stimulating factor was isolated in crude form from hog intestines by William Maddock Bayliss and Ernest Henry Starling (the latter is the same of the law of the heart), very early in the 20th century. These scientists, realizing the messenger role of a chemical substance, coined a new word — hormone (which means "I move" in Greek) — and called it *secretin*. It turned out, however, to be rather difficult to achieve its isolation from gut mucosa in pure form. Jorpes and Mutt of the Karolinska Institute, in Stockholm in 1961, finally obtained the pure hormone about 50 years later. The same group was also successful in determining the sequence of the 27 amino acids constituting the peptide chain of secretin. Figure 6.7 summarizes in a block diagram the main secretions (hormones and juices) of the GIS.

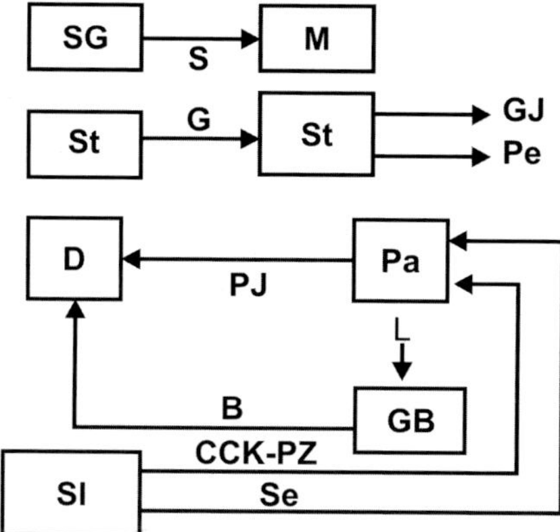

Figure 6.7: Main secretions of the GIS. SG, salivary glands; Sa, saliva; M, mouth. The stomach, St, secretes gastrin, G, which in the same stomach stimulates the production of gastric juice, GJ, and pepsin, Pe. Small intestine, SI, secretes cholecystokinin-pancreozymin, CCK-PZ, and secretin, Se, to stimulate pancreatic juice secretion and the gall bladder contraction, respectively. The liver, L, sends bile into the gall bladder, GB, and the latter into the duodenum, D. Redrawn by Gustavo Idemi.

6.7 Summary

All the GIS activity aims at the exchange of substances that in the end must sustain viable cells and tissues. The exchanger possesses two separated sections: an absorptive one — intestinal — and a metabolic exchanger — hepatic — the latter storing in its parenchymal cells a number of essential compounds. The portal vein connects both sections. It is easy then to understand the harmful and even lethal effects of the so-called malabsorption syndromes (due, e.g., to the lack of certain enzymes or to surgical partial ablation). By the same token, hepatic degenerative processes (as cirrhosis) can also lead to serious and sometimes irreversible conditions. Deconvolution was described as a numerical mathematical tool to evaluate the transfer function of the intestinal absorptive tract, either in the s-domain or in the t-domain, and so evaluating the kinetics of the transport process. Besides, tracer techniques were also mentioned to study this section of the GIS. A few paragraphs have been devoted to pathologies, especially those associated with gastrin. Finally, as we have here introduced already the concept of *secretion*, it is a good time to better clarify the current recognized types and their respective definitions. They are also referred to as types of chemical communication or mediation. The suffix *-crine* derives from the *Greek* word *krinein*, to separate or elaborate products, and is attached to a prefix that modifies its meaning. Etymologically, "secretion" is the Latin equivalent and has exactly the same meaning. These substances can also be considered as mediators. Thus, *exocrine secretion or mediation*: As *exo* means "outside", it is a secretion that goes out, usually via a duct. Saliva, pancreatic juice, and bile are typical examples. *Endocrine secretion or mediation*: As *endo* means "inside", it refers to substances secreted into the blood vessels to act on distant target cells. They are called collectively "hormones" and the paragraphs above have already introduced a few of them, such as secretin, which was the first to be discovered.

References

Alvarez, WC. The rate of rhythmic contraction in the small intestine of the rabbit. *Am J Physiol* 37:267–281, 1915.

Alvarez WC. The muscular versus the nervous origin of the rhythmic contractions of the gut, Chapter X. In: An Introduction to Gastroenterology. 4th ed. Heinemann, London, 1950.

Anlauf M, Garbrecht N, Henopp T, Schmitt A, Schlenger R, Raffel A, Krausch M, Gimm O, Eisenberger CF, Knoefel WT, Dralle H, Komminoth P, Heitz PU, Perren A, Klöppel G. Sporadic versus hereditary gastrinomas of the duodenum and pancreas: Distinct clinico-pathological and epidemiological features. *World J Gastroenterol* 12(34):5440–5446, 2006.

Antunes F, Andrade F, Ferreira D, Mørck Nielsen H, Sarmento B. Models to predict intestinal absorption of therapeutic peptides and proteins. *Curr Drug Metab* 14:4–20, 2013.

Birge SJ, Peck WA, Berman M, Whedon GD. Study of calcium absorption in man: A kinetic analysis and physiologic model. *J Clin Invest* 48:1705–1713, 1969.

Bozler E. Reflex peristalsis of the intestine. *Am J Physiol* 157(2): 338–342, 1949.

Caruolo EV (1990) The effect of age on the rhythmic contractions of the rat small intestine. *Lab Anim* 24:207–212, 1990. http://lan.sagepub.com/content/24/3/207.

Cohen DE (ed.). Seminars in Liver Disease. Vol. 33, issue 4, November. Georg Thieme Verlag KG, Stuttgart, Germany, 2013. doi: 10.1055/s-003-25902.

Connell AM. The motility of the small intestine. *Postgrad Med J* 37:703–716, 1961; doi: 10.1136/pgmj.37.434.703.

Gonzalez SA, Valentinuzzi ME, Arini PD. Deconvolution: It fans back, out and ahead. *IEEE Pulse Mag* 7(4):54–61, 2016.

Goodman RD, Kingsley GR. Sulfobromophthalein clearance test. *JAMA* 153(5):462–466, 1953. doi:10.1001/jama.1953.02940220006003.

Gruzdkov AA, Gromoval LV, Grefner NM, Komissarchik YY. Kinetics and mechanisms of glucose absorption in the rat small intestine under physiological conditions. *J Biophys Chem* 3(2):191–200, 2012. doi:10.4236/jbpc.2012.32021.

Höllriegl V, Li WB, Oeh U, Roth P. Methods for assessing gastrointestinal absorption of strontium in humans by stable tracer techniques. *Health Phys* 90(3):232–240, 2006; http://www.ncbi.nlm.nih.gov/pubmed/16505620.

Keet AD. The pyloric sphincter cylinder in health and disease. Springer-Verlag, Berlin, 39 chapters, 198 pp., 1993, 1998. Internet edition, PLiG, London, UK; http://med.plig.org/2/8.html.

Ordóñez F, Barbot-Trystram L, Lacaille F, Chardot C, Ganousse S, Petit LM, Colomb-Jung V, Dalodier E, Salomon J, Talbotec C, Campanozzi A, Ruemmele F, Révillon Y, Sauvat F, Kapel N, Goulet O. Intestinal absorption rate in children after small intestinal transplantation. *Am J Clin Nutr.* Online 6 February 2013. doi:10.3945/ajcn.112.050799; April 2013 ajcn.050799.

Ostund RE. Phytosterols in human nutrition. *Ann Rev Nutr* 22:533–549, 2002.

Ostlund RE, McGill JB, Zeng Ch-M, Covey DF, Stearns J, Stenson WF, Spilburg CA. Gastrointestinal absorption and plasma kinetics of soy Δ^5-phytosterols and phytostanols in humans. *Am J Physiol Endocrinol Metab* 282:E911–E916, 2002. http://ajpendo.physiology.org/content/282/4/E911.

Pang KS. Modeling of intestinal drug absorption: Roles of transporters and metabolic enzymes (for the Gillette Review Series). *Drug Metab Dispos* 31(12):1507–1519, 2003. http://dmd.aspetjournals.org/content/31/12/1507.full.pdf+html.

Rehfeld JF, Bardram L, Hilsted L, Poitras P, Goetze JP. Pitfalls in diagnostic gastrin measurements. *Clin Chem* 58(5):831–836, 2012.

Rescigno A, Segre G. (1966) Drug and Tracer Kinetics, Blaisdell Publishing Co., Waltham, 209 pp., 1966. Translated from Italian by Piero Ariotti, published 1961, as *La Cinetica dei Farmaci e dei Traccianti Radioattivi*, Editore Boringhieri.

Said HM. Intestinal absorption of water-soluble vitamins in health and disease. *Biochem J* 437(3):357–372, 2011. doi:10.1042/BJ20110326; http://www.ncbi.nlm.nih.gov/pubmed/21749321.

Schofield WN (1985) Predicting basal metabolic rate, new standards and review of previous work. *Hum Nutr Clin Nutr* 39(1):5–41, 1985. http://www.ncbi.nlm.nih.gov/pubmed/4044297.

Schulzke JD, Tröger H, Amasheh M (2009) Disorders of intestinal secretion and absorption. *Best Pract Res Clin Gastroenterol* 23(3):395–406, 2009.

Segre G. Compartmental models in the analysis of intestinal absorption. *Protoplasma* 63(1–3):328–335, 1967.

Stümpel F, Burcelin R, Jungermann K and Thorens B (2001) Normal kinetics of intestinal glucose absorption in the absence of GLUT2: Evidence for a

                    M. E. Valentinuzzi

transport pathway requiring glucose phosphorylation and transfer into the endoplasmic reticulum. *Proc Natl Acad Sci USA* 98(20):11330–11335, 2001. http://www.pnas.org/content/98/20/11330.full.

Subramanya SB, Subramanian VS, Kumar JS, Hoiness R, Said HM. *Am J Physiol Gastrointest Liver Physiol*, 300(3):G494–G501, 2009. Online 9 December 2010. doi: 10.1152/ajpgi.00465.2010; http://www.ncbi.nlm.nih.gov/pmc/articles/PMC3064116/.

Turco GL, de Filippi P, Segre G. The kinetics of intestinal absorption of radiohippuran in control subjects and in patients with intestinal malabsorption. *J Nucl Biol Med* 10(2):52–57, 1966.

Valentinuzzi ME. A mathematical model of the hepatic portal system. *Med Biol Eng* 9(3):213–220, 1971.

Valentinuzzi ME, Montaldo Volachec EM. Discrete deconvolution. *Med Biol Eng* 13(1):123–125, 1975.

Whipple GH. A hitherto undescribed disease characterized anatomically by deposits of fat and fatty acids in the intestinal and mesenteric lymphatic tissues. *Bull Johns Hopkins Hosp* 18:382–391, 1907.

Zollinger RM, Ellison EH. Primary peptic ulcerations of the jejunum associated with islet cell tumors of the pancreas. *Ann Surg* 142(4):709–723, 1955.

CHAPTER 7

AND NOW THE ENDOCRINE SYSTEM, ANOTHER INFORMATION SOURCE

Max E. Valentinuzzi

Perhaps, most of our acts and reactions are driven by the internal secretions, sometimes unfortunately, for the outcome may be regretted!

by Gustavo Idemi

Abstract

Who really knows, but many of the unknowns still hidden in the endocrine system (ES) may contain the answers to the flows and ebbs of human behavioral misfortunes and states of elation. It is a system with a high "engineering content". In it, control is omnipotent, showing delicate high sensitivities all over, with extremely low concentrations of hormones in many cases. Its derangements lead always to unhappy endings. Its functional subsystems include the hypothalamic–hypophyseal axis, the catecholamine system (adrenal medulla, its central gland), the thyroid–parathyroid system for calcium regulation, the insulin–glucagon system (pancreas), the renin–angiotensin–cardionatrine system, and the relatively recently revisited pineal gland and associated Biological Clock, the two latter systems dealt with in other chapters of this book. Deconvolution is once more used, this time in a quite physiological ingenious way. The ES controls, coordinates, and regulates different functions in the organism, many times in conjunction with the nervous system. Its actions are relatively slow to take place because the hormones — with their messages — are released into the blood stream by the internal secretion glands. Hormones are highly specific and only trigger effects on well-determined targets (cells or organs). The elicited response is usually proportional to the stimulating hormonal concentration.

7.1 Overview of the Whole System

7.1.1 *Hypothalamic–hypophyseal axis (HHA)*

The hypothalamus — a functionally ubiquitous portion of the diencephalon — is the principal center of this group for it skillfully pulls the strings to keep most of the whole system under control, as it is responsible for certain metabolic processes and other activities of the autonomic nervous system. It synthesizes and secretes neurohormones, which, in turn, stimulate or inhibit the secretion of pituitary hormones. Besides, it controls body temperature, hunger, aspects of parenting and attachment behaviors, thirst, fatigue, sleep, and circadian rhythms. It receives information from higher centers in the brain, it has vascular connections with the adenohypophysis (anterior pituitary) and it has, too, neural pathways linking it to the neurohypophysis (posterior pituitary). However, both hypophyses, while anatomically together, are embriologically different. Thus, the hypothalamic–hypophyseal axis (HHA) is, in fact, split into two, the hypothalamic–adenohypophyseal (HAH) and the hypothalamic–neurohypophyseal (HNH) axes; both project into the whole organism. Two excellent texts must be recommended, one by Johnson and Veldhuis (1995), and another by Martini (2010), in both books the referred to names act as editors and coauthors, too.

7.1.2 *Hypothalamic–adenohypophyseal system (HAHS)*

It is characterized by a portal vascular arrangement transporting minute quantities of hypothalamic releasing and inhibiting hormones directly to their target cells in the anterior pituitary, that is, they are not diluted out in the systemic circulation. The distance they travel is very short. Specific hypothalamic hormones bind to receptors on specific anterior pituitary cells, modulating the release of the hormone they produce. Thus, some of the neurons within the hypothalamus — neurosecretory neurons — secrete hormones that strictly control secretion of hormones from the anterior pituitary. Hans Selye and Roger Guillemin, in Montreal, put the hypothesis of hypothalamic mediators forward in the early 1950s and that can be considered the beginning of neuroendocrinology (Squire, 1998). The hypothalamic hormones are referred to as *releasing hormones*, reflecting their influence on anterior pituitary hormones. In the early years of neuroendocrinology, these hormones were baptized as *releasing factors*, thus, TRH (thyroid-releasing hormone) was TRF (thyroid-releasing factor), CRH (corticotropin-releasing hormone) was CRF (corticotropin-releasing factor), and so on. The hypophysis or pituitary gland (both sections), in turn, is often portrayed as the "master gland" of the body. Such praise is justified in the sense that the anterior and posterior pituitaries secrete a battery of hormones that collectively influence all cells and affect virtually all physiologic processes. Other secretions of the hypothalamus, still with unsettled functions, include the melanocyte-stimulating hormones (α-MSH and β-MSH) and the endorphins. In lower species, such as amphibians and reptiles, the two former seem to be related to skin coloration. In mammals, however, their function remains uncertain for the time being. The latter, which are morphine-like substances, have analgesic effects. All hypothalamic secretions are polypeptides, proteins, or glycoproteins.

7.1.3 *Thyroid*

The hypothalamus secretes the thyrotropin-releasing hormone (TRH), which binds to receptors on anterior pituitary basophilic cells called thyrotrophs (or thyrotropes), stimulating them to secrete the thyroid-stimulating hormone (TSH) or thyrotropin. It is a glycoprotein with a molecular weight of approximately 28,000 daltons. This pituitary hormone enters the systemic circulation and binds to their receptors on other target organs. In the case of TSH, the target organ is the thyroid gland (Fig. 7.1) that, in response, produces tri-iodothyronine (T3) and thyroxine (T4). These two hormones affect the metabolism, growth, and cell differentiation of practically all the tissues. Two negative feedback loops presumably control the blood concentration of these hormones, one detecting TSH with hypothalamic sensors and another checking T4 with pituitary sensors. Higher centers always influence the hypothalamic action, implicitly meaning

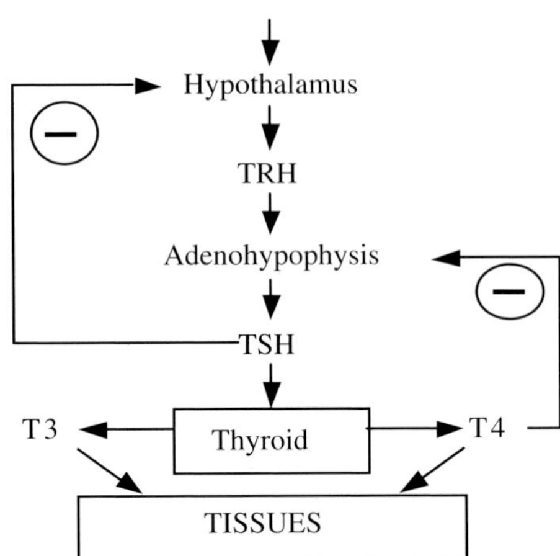

Figure 7.1: Hypothalamus–adenohypophysis–thyroid relationship. TRH is a hypotha-lamic neurosecretion carried by the portal system to the adenohypophysis where it stimulates the secretion of TSH. The latter, in turn, carried also by the circulatory stream has a specific stimulating action on the thyroid gland, which in response, produces triiodothyronine (T3) and thyroxine (T4). These two hormones affect the metabolism of practically all the tissues. Two negative feedback loops control the blood concentration of these hormones, one detecting TSH with hypothalamic sen-sors and another checking T4 with pituitary sensors. Higher brain centers have an influence, too, on the hypothalamus (indicated by the arrow on the top). In short: When T_3 and T_4 concentrations are low, the production of TSH is increased, and, conversely, when T_3 and T_4 concentrations are high, TSH production is decreased. Thus, there are negative feedback loops.

that external perturbations may also have an effect (Braverman and Cooper, 2013). The latter reference is a comprehensive textbook on the thyroid covering its anatomy, development, biochemistry, physi-ology, pathophysiology, and treatment of thyroid disorders.

Iodine is a raw element essential for thyroid hormone synthe-sis. Ingested iodine (I_2) is converted to iodide (I^-) and absorbed. The minimum daily intake that will maintain normal thyroid func-tion is 100–150 μg in the adult. Deficiency of the thyroid hormones (hypothyroidism) leads to *cretinism*, in children, and *myxedema*, in the adult. Both are serious conditions. *Goiter* is another form of hypothyroidism triggered by low-dietary content of iodine. In that case, since T3 and T4 synthesis and secretion are low, due to the feed-back loop, TSH production increases to abnormally high levels and, thus, to a permanent stimulation of the gland which responds with its excessive growth (hypertrophy) in an effort to unsuccessfully make up for the deficiency. The enlargement of the thyroid gland results in bulging of the neck that may become extremely large. Occasionally, it may cause some difficulty in breathing and swallowing. This is why it is common practice to add iodide to table salt (NaCl) as a goiter prevention measure.

Summarizing, hypothyroidism signs and symptoms may include, but not necessarily all: fatigue, increased sensitivity to cold, constipation, dry skin, weight gain, puffy face, hoarseness, muscle weakness, high blood cholesterol level, muscle aches, pain, stiffness or swelling in joints, irregular menstrual periods, thinning hair, slow heart rate, depression, and memory loss.

Conversely, excess levels in blood of T3 and T4 are generically termed hyperthyroidism. Graves' disease is its most common form. Robert Graves (Irish physician, 1796–1853) discovered it in 1835. It affects approximately 3 out of 1000 people and is more prevalent in women and in families with a history of the disorder. Graves' disease is an autoimmune disorder in which an as yet unknown immunological defect results in the production of autoantibodies to the TSH receptor located on the surface of thyroid cells. These antibodies bind the receptor and stimulate it to overproduce thyroid hormones. This activation is not subject to the normal regulatory negative feedback loop even though the blood level of TSH is lower than normal (as low as 1/100 the level of euthyroid subjects). Symptoms of Graves' disease include nervousness, irritability, weight loss, increased appetite, heat intolerance, excessive sweating, rapid pulse, diarrhea, fine tremors in fingers, and warm moist skin. About 50% of patients also develop exophtalmia (bulging eyes).

Roger Guillemin — Nobel Prize of Physiology in 1977 (see Squire, 1998) — states in his recollections:

> I consider the isolation and characterization of TRF the major event in the establishment of modern neuroendocrinology, the inflection point that separated confusion and a great deal of doubt from real knowledge. Contemporary neuroendocrinology was born of that event. Isolation of LRF (the luteinizing hormone-releasing factor, now called LRH or

LHRH), somatostatin, the endorphins, others later, were all extensions of that major event — the isolation of TRF — a novel molecule in hypothalamic extracts, with hypophysiotropic activity, the first so characterized. The event was the vindication of 14 years of hard work within the paradigm of a hypothalamic neurohumoral control of adenohypophyseal secretions. From observation of what has happened in neuroendocrinology since 1969, the isolation of TRF was also the vindication of my early decision, as a physiologist, that the most heuristic event in neuroendocrinology would be the isolation and characterization of the first one (any one) of the then-hypothetical hypothalamic hypophysiotropic factors.

And he added about the cost:

I once calculated that the first 1 mg of native, pure, ovine TRF made from 1 kg of pure, native TRF, was 2.5 times more expensive than a kilogram of moon rock brought back from the Apollo XI mission. Today the cost of synthetic TRF is a few cents per milligram.

After TRF, pioneering in neuroendocrinology ceased and became the harvesting of a new expanding science, because far more interesting and revolutionary observations were to follow. TRH (or TRF) was isolated in 1968 by Guillemin and collaborators in the Department of Physiology of Baylor College of Medicine, in Houston, TX.

7.1.4 *Adrenal cortex*

The adrenal glands, which are also called suprarenal glands, are small, triangular pinkish structures located on top of both kidneys. This gland — essential for life — is made of two parts:

The outer region (or the adrenal cortex) and the inner region (or the adrenal medulla). They are quite different from a developmental point of view as they are also functionally. We will consider now the former.

The hypothalamus produces CRH, which stimulates the adenohypophysis (or anterior pituitary gland) while the latter, in response, produces adrenocorticotropin hormone (ACTH), which, in turn, stimulates the adrenal cortex to secrete a set of hormones: aldosterone (a mineralocorticoid) and glucocorticoids (corticosterone and cortisol). Figure 7.2 briefly summarizes the relationships, including three negative feedback regulatory loops. The hypothalamus senses the blood levels of ACTH and of cortisol so that an increase in them

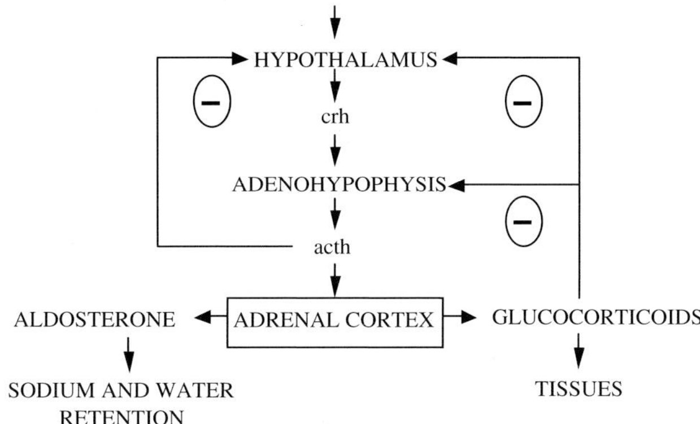

Figure 7.2: Hypothalamus–adenohypophysis–adrenal cortex system. The relationship is similar to that of the thyroid gland. Higher nervous centers always influence the hypothalamus. Corticotropin-releasing hormone (CRH) is another hypothalamic neurosecretion carried by the portal system blood to stimulate the adenohypophysis, which, in response, produces adrenocorticotropic hormone (ACTH). The latter, in turn, via the general circulation, reaches its target organ — the adrenal cortex — to elicit secretion of two types of hormones: a mineralocorticoid (aldosterone) and two glucocorticoids (corticosterone and cortisol).

elicits a decrease in CRH and in ACTH triggering in this way an opposite compensatory change. Similarly, the adenohypophysis constantly measures the concentration of glucocorticoids (mainly cortisol) to decrease or increase its ACTH production and, hence, make up for any initial increase or decrease in cortisol. Aldosterone activates the retention of sodium and water playing, as a consequence, a role in the electrolytic balance and in blood pressure regulation. Glucocorticoids have an effect on the intermediary metabolism of carbohydrates, proteins and lipids in different tissues. They also suppress inflammatory reactions in the body and affect the immune system. Besides, the adrenal cortex secretes androgenic steroids (androgen hormones). These hormones have minimal effect on the development of male characteristics.

Hans Selye (see Guillemin's Chapter in Squire, 1998), through his stress concept as related to the adrenal cortex, had a major stimulating role in orienting the early efforts in neuroendocrinology toward the study of the hypothalamus–pituitary ACTH–adrenal cortex functional relationship. According to Guillemin's comments, strangely enough, and unwittingly on Selye's part, this is probably about the worst thing that happened to nascent neuroendocrinology. The search for CRF was to prove so complex and baffling that it was not completed until 1981 through the elegant work led by Wylie Vale, one of Guillemin's students and collaborators. They should have better put the effort on some other hypothalamic secretion. Perseverance and vision are essential characteristics of an investigator.

There are important pathologies associated with adrenal cortex malfunction: *Cushing's syndrome* (excess production of glucocorticoids), *Conn's syndrome* (excess production of aldosterone), and *Addison's disease* (deficiency of corticoids). Cushing's syndrome (named after Harvey Williams Cushing, American surgeon, 1869–1939) occurs when the body's tissues are exposed to excessive levels of cortisol for long periods. Many people suffer the symptoms of Cushing's syndrome because they take glucocorticoid hormones such as prednisone for asthma, rheumatoid arthritis, lupus, and other inflammatory diseases (allergies, for example), or for immunosuppression after transplantation. Because of the negative feedback loop, the higher than normal blood concentration of these hormones inhibits the production of the physiological hormones and the gland may stop secretion even after treatment is stopped. Cushing's syndrome is relatively rare and most commonly affects adults' aged 20–50. An estimated 10–15 of every million people are affected each year.

The second condition affecting the adrenal gland is Conn's syndrome, also called primary aldosteronism (named after Jerome W. Conn, an American internist, 1907–1981). It is due to the presence of an adrenal tumor, usually benign. The excess aldosterone secreted in this condition increases sodium reabsorption and potassium loss by the kidneys and results in electrolyte maladjustments. Risk factors are being female and being between 30 and 50 years old. The incidence is 2 out of 100,000 people; fewer than 10 children have been reported in the literature with Conn's syndrome. Secondary aldosteronism originates in other causes, not directly related to the adrenal cortex.

The third significant pathology of the adrenal cortex is Addison's disease (named after Thomas Addison, English physician, 1793–1860). This primary adrenal failure may be the result of congenital or acquired lesions. In the early part of the 20th century, tuberculosis was the most common cause. Other infrequent conditions may lead to adrenal cortex damage and insufficiency. However, in recent years, it appeared that most cases represent the outcome of an autoimmune process.

7.1.5 *The gonads: sex glands*

This system follows a scheme similar to the two systems previously described; perhaps, it is somewhat more complex than they are. Figure 7.3 summarizes the hypothalamic–adenohypophysis–gonads

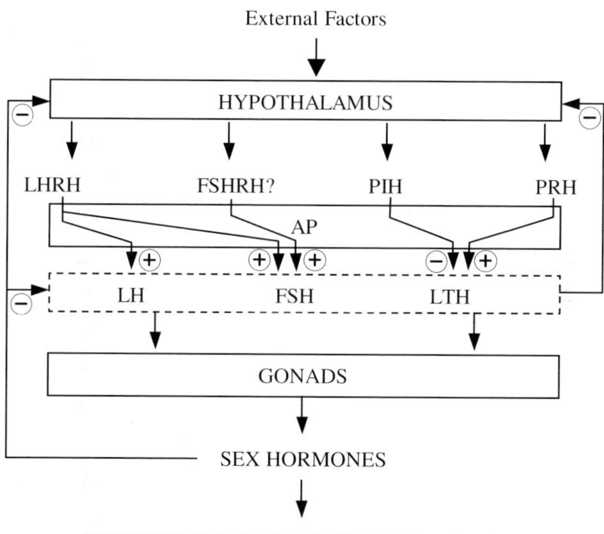

Figure 7.3: Hypothalamic–adenohypophyseal–gonads system. Four releasing hormones are produced by the hypothalamus (one of them is still uncertain, as indicated by the interrogation mark). Three are stimulating hormones (thus, a plus sign is shown) and one has an inhibiting effect (negative sign), respectively, on the adenohypophysis or anterior pituitary gland (AP). In response, the latter produces LH, FSH, and LTH, which, in turn, having the gonads as target organs, mediate the liberation of the sex hormones to act on the reproductive, developmental, and other behavioral functions. The probable negative feedback loops are also shown.

relationships. The hypothalamus produces the releasing hormones: luteinizing hormone-releasing hormone (LHRH), follicle-stimulating-hormone-releasing hormone (FSHRH) — still not fully demonstrated — prolactin-inhibiting hormone (PIH), and prolactin-releasing hormone (PRH). The first one, that is LHRH, stimulates the production of the luteinizing hormone (LH) and of the follicle-stimulating hormone (FSH) from the adenohypophysis or anterior pituitary. This peptide, first characterized and synthesized by Roger Guillemin and Andrew Schally (both shared the 1977 Nobel prize for this labor), has been also called the gonadotropic hormone-releasing hormone, or GnRH (Knobil, 1999). The second, FSHRH, if true, would stimulate the secretion of FSH. In turn, PIH and PRH have opposite actions on the anterior pituitary, the former inhibiting the secretion of luteotropic hormone, LTH, also called prolactin, and the latter stimulating its secretion. Thus, the three hormones from the adenohypophysis are LH, FSH, and LTH, with direct actions on the gonads as target organs that, in the end, release the sex hormones. There is evidence of the existence of negative feedback loops to control these hormonal blood levels.

In women, the gonadotropins (collective name of the anterior pituitary hormones aimed at the gonads) have effects on the ovaries (the female gonads, as opposed to the testicles, the male gonads): FSH stimulates follicular growth (thus, it is a trophic hormone), while LH stimulates ovulation and luteinization of the *corpus luteum*. Luteinization is the process by which a postovulatory ovarian follicle transforms into a corpus luteum through vascularization, cell hypertrophy, and lipid accumulation, the latter giving the yellow color indicated by the term (*luteus*, yellow). Both, follicle and corpus luteum, secrete estrogens (estradiol, estrone, and estriol), while the corpus luteum secretes progesterone. All these are termed the female sex hormones, responsible for uterine endometrial modifications, development of the genitalia, and secondary female characteristics. Prolactin causes the postpartum milk secretion from the breasts after estrogen and progesterone priming. Besides, prolactin inhibits the effects of gonadotropins, possibly through an action on the ovaries via negative feedback because it stimulates progesterone production.

Surprisingly, it was found that TRH also stimulates the secretion of LTH (Squire, 1998).

A fascinating phenomenon is the pulsatile nature of GnRH secretion — demonstrated in all vertebrates studied in this regard — that is, in turn, responsible for LH pulses with a frequency of approximately one per hour. Moreover, it has also been shown that each LH pulse is associated with electrical activity at the mediobasal hypothalamus (Knobil, 1999). The central event of the female cycle — ovulation — is initiated by a bolus of LH from the pituitary gland. Such surge is superimposed upon, or perhaps temporarily replaces, the pulsatile pattern of LH secretion. Interestingly enough, the pacemaker for this phenomenon is not the hypothalamus but the ovary itself, which, with a preovulatory estradiol rise, acts on the HHA to start the LH surge. It means, then, that estradiol has negative and positive feedback actions (fall in LH and FSH plasma levels and preovulatory surge), both occurring at the adenohypophysis. How this can happen still remains unknown (Knobil, 1999).

In man, LHRH is the only hypophyseal hormone with definite known stimulating action on FSH and LH, for the existence of the other three is still uncertain. FSH acts on the testicular seminiferous tubules to control spermatogenesis. These tubules would produce *inhibin*, which acting on the adenohypophysis, would inhibit FSH production and, thus, indicating a regulatory loop for sperm production. LH, in turn, has an action on the Leydig's cells in the testicles, the main source of testosterone and the essential male sex hormone with androgenic and general anabolic effects. Besides, testosterone stimulates spermatogenesis from the seminiferous tubules and feeds back to the hypothalamus to regulate LHRH release, in another negative control loop.

It must be pointed out that *inhibin* and *activin* are closely related proteins that have almost directly opposite biological effects. The latter was identified by Asashima *et al.* (1990); it enhances FSH biosynthesis and secretion, and participates in the regulation of the menstrual cycle. Many other functions have been found to be exerted by *activin*, including roles in cell proliferation, differentiation, apoptosis (Chen *et al.*, 2006), metabolism, homeostasis, immune response, wound repair (Sulyok *et al.*, 2004), and endocrine function. Conversely, *inhibin* downregulates FSH synthesis and inhibits FSH secretion (van Zonneveld *et al.*, 2003).

With the negative feedback loop in mind, the conceptual mechanism of oral contraceptives (estrogens and progesterone) in women is easily explained. Most of the women who were prisoners in concentration camps during the World War II stopped their menstrual cycles. Obviously, Hans Selye's stress idea entered no doubt in action, and the concept is also valid for the adrenal cortex. These are subjects to think about, as part of the building process of science at large, including their dramatic and often ignored effects in human responses.

7.1.6 *Growth*

The growth endocrine system (GES) does not have a single target organ for its adenohypophyseal hormones; it is a systemic gland, as the previously described thyroid or adrenal cortex or gonads. In this case, the hypothalamus controls the adenohypophysis — its main target organ — with two secretions, the growth hormone-releasing hormone (GRH) and the growth hormone-inhibiting hormone (GIH) or somatostatin, with opposing effects as their names clearly identify, for the former stimulates the secretion of growth hormone (GH) and the latter inhibits its secretion from the anterior pituitary (Fig. 7.4). GH, also known as *somatotropin*, is a protein of about 190 amino acids that is synthesized and secreted by cells called *somatotrophs*, in the anterior pituitary. It is a major participant in control of several complex physiologic processes, including growth and metabolism. It has direct and indirect effects. The former are the result of GH binding on specific cells, such as fat cells (adipocytes), stimulating them to break down triglycerides and suppressing their ability to take up and accumulate circulating lipids. The latter, instead, are mediated primarily by an insulin-like growth factor (IGF-1), a hormone that is secreted from the liver and other tissues in response to GH. IGF-1 stimulates proliferation of chondrocytes (cartilage cells), resulting in bone growth. It also appears to be the key player in muscle growth stimulating amino acid uptake and protein synthesis. GH is one of a battery of hormones that serves to maintain blood glucose within a normal range. It has anti-insulin activity because it suppresses the abilities of insulin to stimulate uptake of glucose in peripheral tissues and enhance glucose synthesis in the liver. Somewhat paradoxically, administration of GH stimulates insulin secretion, leading to hyper-insulinemia. This amazing system has demonstrated unexpected and not yet elucidated ramifications (Fig. 7.4):

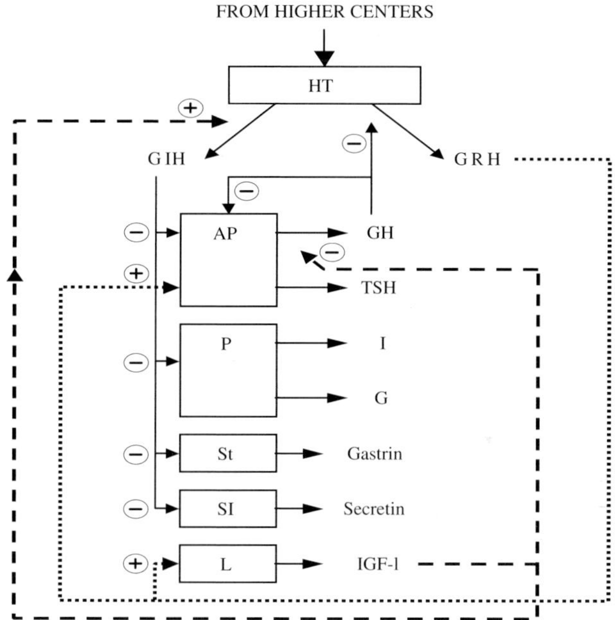

Figure 7.4: Growth endocrine system (GES). The hypothalamus (HT), always influenced from higher nervous system centers, which includes external stimulation, produces two releasing hormones with opposite actions on the adenohypophysis or anterior pituitary (AP): the growth hormone-releasing hormone (GRH) and the growth hormone-inhibiting hormone (GIH), also called somatostatin. The figure also depicts the relationships with the pancreas, stomach, small intestine, and liver. See text for more details.

Somatostatin or GIH inhibits the secretion of TSH from the ade-nohypophysis, insulin and glucagon from the pancreas, gastrin from the stomach and secretin from the small intestine. Besides, GH secretion is also part of a negative feedback loop involving IGF-1. High blood levels of IGF-1 lead to decreased secretion of GH not only by directly suppressing the somatotroph, but also by stimulating release of somatostatin from the hypothalamus. GH also feeds back to inhibit GHRH secretion (Fig. 7.4). States of either GH deficiency or excess provide very visible evidence to the role of this hormone in normal physiology. Clinically, deficiency leads to retardation or dwarfism. Conversely, excessive secretion of GH — very dependent on the age of onset — may lead to two distinctive disorders: *giantism* (very rare, usually resulting from a tumor), the result of excessive GH beginning in childhood or adolescence, and *acromegaly*, which results from excessive secretion of GH in adults.

Guillemin (see Squire, 1998), in his beautiful account of the development of neuroendocrinology, stated that the nature of the hypothalamic-releasing factor for GH, now called GRH, was not to be established until 1982. Its discovery happened in a much-unexpected way, triggered by studies made on a pancreatic peripheral tumor that was functioning as an ectopic source of the relentless searched factor. This is one example of the fascinating serendipities of science that only the alert and well-prepared mind can catch; thus, have always ready the reception antennas.

7.1.7 *Hypothalamic–neurohypophyseal system (HNHS)*

Figure 7.5 briefly synthesizes the pathways involved in this system. The neurohypophysis originates from neural tissue; it stores and secretes two hormones, oxytocin and vasopressin (antidiuretic hormone or ADH). These hormones are synthesized in the cell bodies of neurons located in the hypothalamus and transported along the axons to the terminals located in the neurohypophysis and are released in response to neural stimulation. Oxytocin acts on the uterine and vaginal smooth musculature, and also on the breasts to help sperm transport during sexual intercourse or during child delivery or milk expulsion during nursing. Mechanical receptors located in these anatomical structures send afferent neural information to the hypothalamus that, in a positive loop, enhance the effects by increasing the secretion of oxytocin. The antidiuretic hormone has a direct

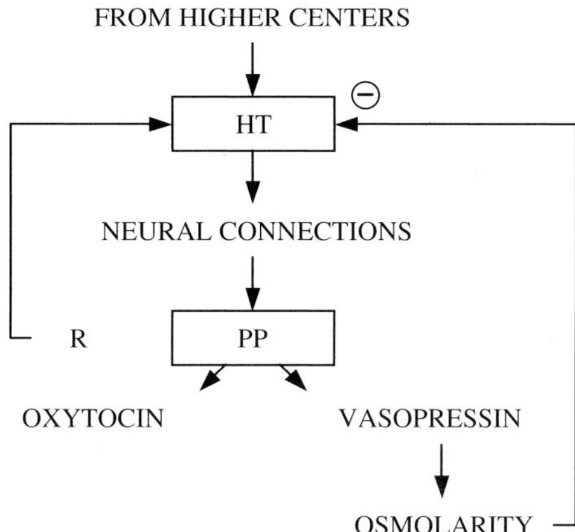

Figure 7.5: The hypothalamic–neurohypophyseal system (HNHS). The hypothalamus (HT), via neural connections, stimulates the secretion of oxytocin and vassopressin (or ADH) from the posterior pituitary (PP), also called neurohypophysis. Mechanical receptors R located in the genitalia, uterus, and in the breasts stimulate hypothalamic detectors to elicit contractions in the smooth musculature of these organs to favor sperm transport during sexual intercourse, birth delivery and milk secretion during nursing. ADH, in turn, has an effect on the extracellular fluid osmolarity that is constantly checked by other hypothalamic receptors, thus, establishing a negative feedback loop.

effect on the renal distal tubules and collecting ducts permeability to produce water retention and, thus, to regulate the extracellular fluid osmolarity. The latter is being checked by the so-called hypothalamic *osmoreceptors* so creating a regulatory feedback loop.

A typical derangement of this system is *diabetes insipidus*, which is characterized by low ADH secretion leading to *polyuria* (increased diuresis) because much less water is being retained. This urine is diluted and not sweet, as the case is with *diabetes mellitus*, that is, it does not contain glucose. There is also *polydipsia* (drinking of large amounts of fluid), when the thirst mechanism works correctly, to compensate for the renal water loss. Usually, the cause is a hypothalamic lesion. Alcohol induces diuresis because it inhibits ADH secretion.

It is of interest to call the attention to magnetic resonance imaging (MRI), as a revolutionary advance in diagnostic imaging of the hypothalamic–neurohypophyseal system (HNHS). With it, the detailed anatomy is clearly visible because it has no bony artifacts. The posterior lobe of the pituitary gland displays a characteristic bright signal on and it is distinctly separated from the anterior lobe. The bright signal is absent in patients with central diabetes insipidus, which is thought to reflect normal vasopressin storage in the posterior lobe. The signal intensity ratio of the posterior lobe to the pons is also strongly correlated with vasopressin content in the posterior lobe. In addition to the morphological evaluation, MRI provides unique information concerning the function of the HNHS. Findings such as normal condition, central diabetes insipidus, a depleted posterior lobe, an ectopic posterior lobe, or a damming-up phenomenon of the neurosecretory vesicles in the pituitary stalk have been demonstrated (Fujisawa, 2004).

Moreover, the elucidation of the genomes of a large number of mammalian species has produced a huge amount of data on which to base physiological studies. These endeavors have also produced surprises, not least of which has been the revelation that the number of protein coding genes needed to make a mammal is only about 22,333. However, this small number belies an unanticipated complexity that has only recently been revealed, thanks to genomic studies. One important and relatively recent outcome is that the HNHS elaborates many peptide hormones in addition to the classical neurohypophyseal hormones oxytocin and vasopressin.

Given its central role as a neuroendocrine organ, a comprehensive description of all of the secreted peptide products of the HNHS, and an exploration of their functions, is vital, indeed. As overall conclusion, these authors state that recent exploration of the mammalian genome and its expression has revealed a daunting complexity that is only just being clarified in terms of molecular mechanics of regulation and function. The subsequent task will be to place this complexity in physiological context (Murphy *et al.*, 2012).

7.1.8 *The catecholamine system: adrenal medulla*

The adrenal medulla, the inner part of the adrenal gland, is not essential to life, but helps a person in coping with physical and emotional stress. It consists of masses of neurons that are part of the sympathetic branch of the autonomic nervous system. Instead of releasing their neurotransmitters at a synapse, these neurons release them into the blood. Thus, although part of the nervous system, the adrenal medulla functions as an endocrine gland. It secretes epinephrine (also called adrenaline) and norepinephrine (also called noradrenaline). Both derive from the amino acid tyrosine and are collectively called *catecholamines*, a group that includes other related substances with similar properties. The former hormone increases the heart rate and force of heart contractions, blood is shunted from the skin and viscera to the skeletal muscles, coronary arteries, liver, and brain, causes relaxation of smooth muscles, and helps with conversion of glycogen to glucose in the liver. Other effects include bronchial and pupillary dilatation, hair stands on end (the so-called gooseflesh in humans), blood clotting time goes down, and ACTH secretion from the anterior lobe of the pituitary increases. The latter (norepi), instead, has little effect on smooth muscle, metabolic processes, and cardiac output, but has strong vasoconstrictive effects, thus increasing blood pressure.

This gland is innervated, via the celiac ganglion, by the splanchnic nerves, which originate in the spinal cord, at the thoracic levels 5–12. Thus, stimulation of them produces secretion of catecholamines. Any external stressful perturbation, such as fear or anger, may easily lead to the so-called alarm reaction perhaps enhanced by a positive feedback loop (Fig. 7.6). The cardiovascular effects of catecholamines converge to at least a reversible and temporary increase in blood pressure, so explaining that common mother reproach to her children when she says "don't bring me a headache with your behavior". Even getting ready for an action, as before a physical exercise (a race, a given competition, or soldiers in combat) calls for the catecholamine discharge and its physiological effects. There are also demonstrated neural tracts that descend from higher brain centers able to contribute to the phenomenon. Figure 7.7 illustrates the effects of electrical stimulation of the splanchnic nerve in an experimental animal.

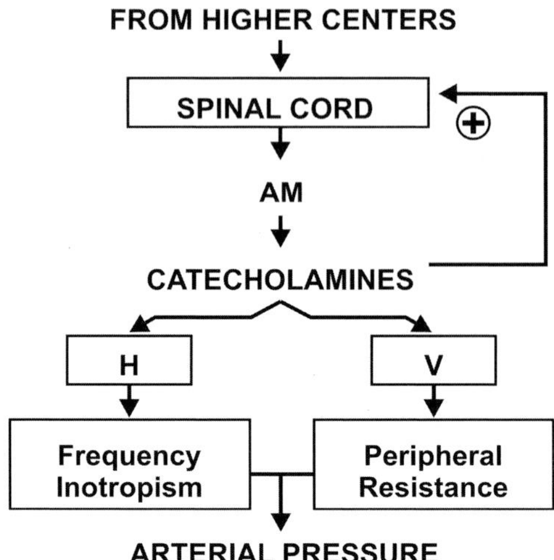

Figure 7.6: The adrenal medulla system. The splanchnic nerves via the celiac ganglion innervate the adrenal medulla. These nerves originate in the spinal cord, at the thoracic levels 5–12. Thus, stimulation of them produces secretion of catecholamines. The positive feedback loop would be part of the alarm reaction.

7.1.9 *Thyroid–parathyroid system for calcium regulation*

Calcium concentration in plasma is well kept at 10 mg/100 mL (or 5 mEq/L or 2.5 mM/L). It is an ion of paramount importance because it plays a role in several essential physiological functions, such as blood coagulation, cardiac and skeletal muscle mechanics, electrical activity of excitable tissues, and also in neuromuscular transmission. The largest calcium store is bone.

Three hormones are responsible for calcium homeostasis:

(1) 1,25-*Dihydroxycholecalciferol* (DHC), which is a steroid formed in the liver and kidneys from vitamin D;

(2) *parathyroid hormone* or *parathormone* (PTH), secreted by the parathyroid glands; and

(3) *calcitonin* secreted by the thyroid and with no relationship whatsoever with the HAS system.

Figure 7.8 summarizes the calcium regulation mechanisms. Vitamin D, supplied by the daily diet, is essential for DHC

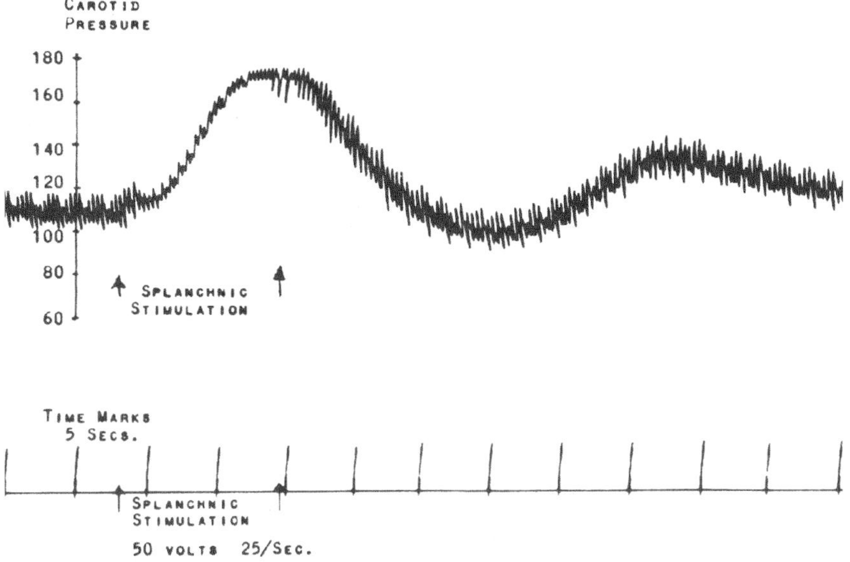

Figure 7.7: Splanchnic stimulation in an anesthetized dog. The first increase in arterial blood pressure was due to the direct action of the train of pulses applied to the nerve. Release of catecholamines was also triggered with effect manifested somewhat later, as demonstrated by the second hump, about 25 s after the stimulus was disconnected. Records obtained at the Department of Physiology, Baylor College of Medicine, Houston, TX (from a Laboratory Manual by H. E. Hoff and L. A. Geddes, 1965, which for a long time was produced by the then manufacturer of The Physiograph, Texas E&M Instruments. Also in a somewhat modified Spanish version by Max E. Valentinuzzi, 1966).

production in the liver and kidneys. In the intestine, an increase of DHC stimulates absorption of calcium (second row of blocks) and, thus, an increase in its blood level (far right block, line b). DHC and PTH cause calcium resorption (release) from bone, thus, the curves relating it to DHC and PTH increase with an increase in calcium liberation (third row, center block) and that calcium from bone contributes to an increase in blood calcium; that is represented by the input termed c in the figure. In turn, blood calcium (dashed–dotted arrows) acts as input to the thyroid and parathyroid gland, the latter decreasing its PTH production when blood calcium level increases (hence, the inverse relationship shown in the third row, left-hand block), while the former increases calcitonin secretion after the

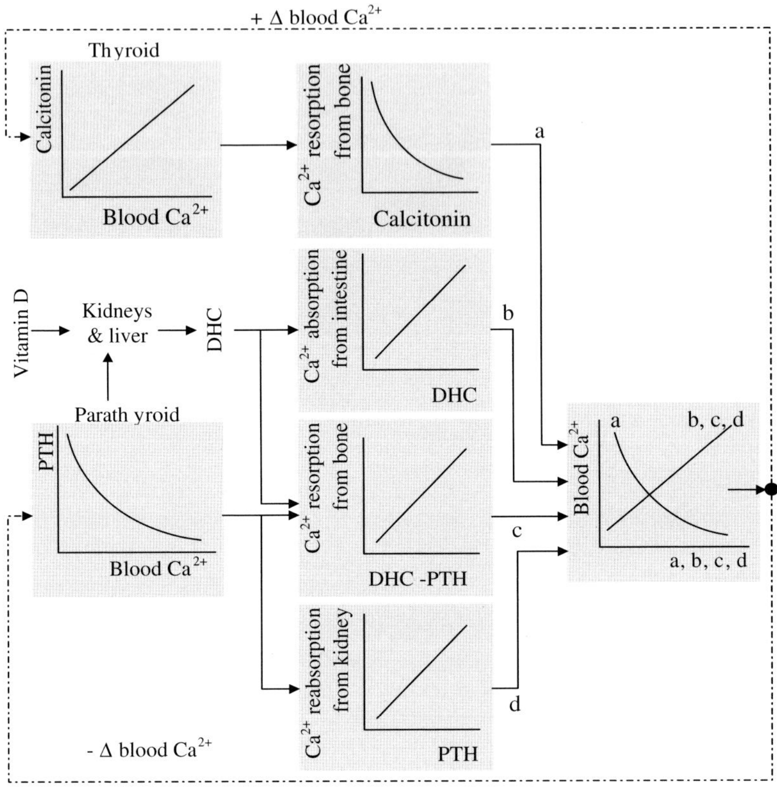

Figure 7.8: Calcium regulation system. Located on the back of the thyroid gland, the parathyroid gland consists of four pea-sized nodules. When blood calcium levels drop, this gland secretes parathyroid hormone (PTH). This hormone works to raise blood calcium levels by stimulating osteoclasts to break down bone and release calcium into the bloodstream.

same calcium increase. PTH and calcitonin levels, in a true negative feedback loop, respectively, stimulate and inhibit bone resorption. Bone resorption is the process by which osteoclasts release minerals, resulting in a transfer of calcium from bone tissue to the blood. The second effect is represented by the inverse curve (upper row, center block). PTH has also a stimulating effect in the liver and kidneys to produce DHC.

Summarizing: PTH, a polypeptide hormone produced in the parathyroid gland, along with Vitamin D, is the principal regulator of calcium and also phosphorus homeostasis. The most important actions are (1) rapid mobilization of calcium and phosphate from bone and long-term acceleration of bone resorption; (2) increase of renal tubular reabsorption of calcium; (3) increase of intestinal absorption of calcium (mediated by an action on the metabolism of vitamin D); and (4) decrease of renal tubular reabsorption of phosphate. These actions account for most of the important clinical manifestations of PTH excess or deficiency. Calcitonin, in turn, acts directly on osteoclasts (via specific receptors). Bone biopsies from patients treated with the drug show no effects on mineralization. It has a short half-life. Given as a subcutaneous injection showed significant improvements in bone density but there was a high incidence of side effects, including pain at the injection site, flushing and nausea, which limited the use of the drug. Calcitonin now is available as a nasal spray, which has made it better tolerable for patients. Calcitonin is a safe alternative to estrogen in women who cannot take estrogen. Mathematical models have also been proposed for this system (Powell and Valentinuzzi, 1974). One important surgical consequence is that in a thyroidectomy, the parathyroids must be preserved, otherwise, derangements in the calcium system are to be expected. Another interesting concept is that of the liver as an internal secretion organ.

7.1.10 *The insulin–glucagon system: pancreas*

In a previous section dealing with the gastrointestinal system, the pancreas was described as an exocrine gland collaborating in the digestive process. The pancreas, however, is also an endocrine gland secreting four hormones that originate from specific cells located in well-differentiated structures called the *islets of Langerhans*:

(a) *insulin,* produced by the *beta* or B cells (make up 60–75% of islets);

(b) *glucagon,* secreted by the *alpha* or A cells (make up 20% of islets);

(c) *somatostatin* (SS or GIH), secreted by the *delta* or D cells (10% of islets); and

(d) *pancreatic polypeptide,* secreted by the F cells (just a few percent of islets).

                M. E. Valentinuzzi

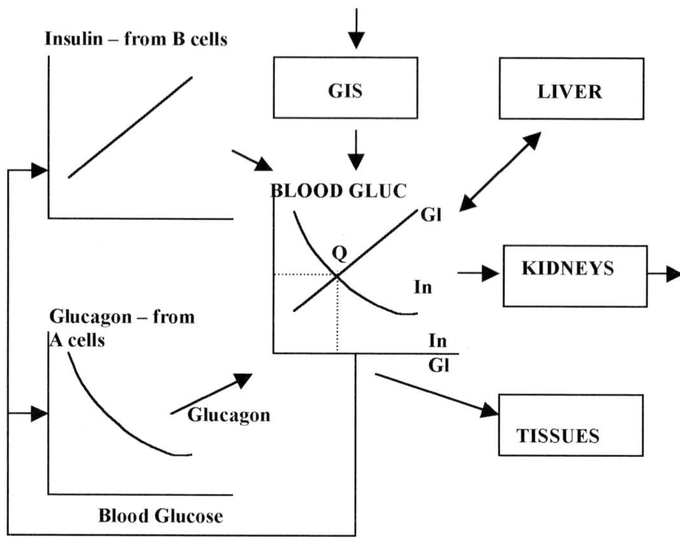

Figure 7.9: Glucose regulation. Curves shown in the central part of the figure represent blood glucose in the vertical axis and insulin and glucagon concentrations, respectively, also in blood, in the horizontal axis. The operating point is Q. Inputs act on the horizontal axes and outputs come off the vertical axes.

The two first, insulin and glucagon, are essential in the normal regulation of blood glucose, which is maintained at a level of 80 mg/100 mL (or 80 V%). The third one could have an influence in the case of diseased pancreatic cells, such as tumors leading to hyperglycemia. The fourth — sometimes called the "hunger hormone" because it might stimulate the hunger center in the hypothalamus — suppresses SS secretions from gut and pancreas and pancreatic enzyme output, too. However, its function is not known yet. All four appear in the *portal blood* and pass through the liver before getting into the general circulation.

The curves shown in the central part of Fig. 7.9 represent blood glucose (vertical axis) with insulin and glucagon concentrations, respectively, also in blood (horizontal axis). An increase in blood insulin produces a decrease in the concentration of blood glucose (curve In) while the opposite occurs when blood glucagon goes up (straight line Gl). The crossing point, Q, of both curves is the operating point. The horizontal dashed line intercepts the vertical axis at the normal blood glucose level. The *beta* or B-cells from the pancreas supply insulin to blood and increase their output as the glucose blood concentration goes up (left upper curve). By a similar token, the lower left relationship represents glucagon release from the *alpha* or A-cells as a function of blood glucose level: When the latter increases glucagon liberation goes down. The latter subsystem also feeds into blood. Glucose enters the system in the daily diet via the gastrointestinal tract and the liver constantly checks the blood glucose level; if needed, because of temporary hypoglycemia, it gives off an amount of glucose to compensate for the fall. Excretion of this sugar takes place through the kidneys when its blood concentration exceeds the threshold of about 180 mg/100 mL. Finally, the tissues in general are the users of glucose for their metabolic processes. Thus, blood acts as a dynamic reservoir.

The most important derangement of pancreatic function leads to diabetes mellitus, highly frequent, largely genetic, and caused by deficiency or lack of insulin. Juvenile diabetes is insulin dependent as opposed to diabetes of the elder, which can be taken care of by just a well-adjusted diet. It is a disorder characterized by hyperglycemia or elevated blood glucose (blood sugar). Our bodies function best at a certain level of sugar in the bloodstream (80–100 V%).

If the amount of sugar in our blood runs too high or too low, then we typically feel bad. *Diabetes is the name of the condition where the blood sugar level consistently runs too high.* Diabetes is the most common endocrine disorder. Sixteen million Americans have diabetes, yet many are not aware of it; in 2014, the global prevalence of diabetes was estimated to be 9% among adults aged 18 or more years. Diabetes has potential long-term complications that can affect the kidneys, eyes, heart, blood vessels, and nerves. The bibliography is enormous and growing because it is a field of basic and clinical research. It is a very big topic, indeed.

There are two different types of diabetes, which are similar in their elevated blood sugar, but different in many other ways: Type 1 diabetes and Type 2 diabetes. The former is due to insulin deficiency, whereas the latter stems in insulin resistance. Insulin deficiency means that the pancreas does not produce enough hormone due to a malfunction of the *beta* cells. Insulin resistance occurs when there is plenty of insulin made by the pancreas but the cells of the body (the ultimate users) are resistant to its action, which results in the blood sugar being too high. Sometimes, a third type named *gestational diabetes* is also accepted, and is defined by hyperglycemia with blood glucose values above normal but below those diagnostic of diabetes. It occurs during pregnancy. Women with gestational diabetes are at an increased risk of complications.

Glucose is absorbed from the intestines into the bloodstream after a meal. Insulin is then secreted by the pancreas in response to this detected increase in blood sugar. Most cells of the body have insulin receptors that bind the insulin, which is in the circulation. When a cell has insulin attached to its surface, the cell activates other receptors designed to absorb glucose from the blood stream into the inside of the cell. Without insulin, a person can eat lots of food and actually be in a state of starvation since many of his/her cells cannot access the calories contained in the glucose. It is as having a refrigerator full of food but closed with a padlock. This is why Type 1 diabetics who do not produce insulin can become very ill without insulin shots. Insulin is an indispensable hormone. Those who develop a deficiency of insulin must have it replaced via shots or pumps.

More commonly, people will develop insulin resistance (Type 2 diabetes) rather than a true deficiency of insulin. In this case, the levels of insulin in the blood are similar or even a little higher than in normal, non-diabetic individuals. However, many cells of Type 2 diabetics respond sluggishly to the insulin they make and, therefore, their cells cannot absorb the sugar molecules well. This leads to blood sugar levels that run higher than normal. Occasionally Type 2 diabetics will need insulin shots but most of the time other methods of treatment will work.

The first successful insulin preparations came from cows and later on from pigs. The pancreatic islets and the insulin protein contained within them were isolated from animals slaughtered for food in a similar but more complex fashion than was initially used (see below). The bovine and porcine insulin were purified, bottled, and sold. Bovine and porcine insulin worked very well for the vast majority of patients, but some could develop an allergy or other types of reactions to the foreign protein (a protein, which is not native to humans). In the 1980s, biotechnology had advanced to the point where we could make human insulin. The advantage would be that human insulin would have a much lower chance of inducing a reaction because it is not a foreign protein (all humans have the exact same insulin). The technology, which made this approach possible, was the development of recombinant DNA techniques. In simple terms, the human gene, which codes for the insulin protein was cloned and, thereafter, put inside of bacteria (*Escherichia coli*). A number of tricks were performed on this gene to make the bacteria want to use it to constantly make insulin. Big vats of bacteria now make tons of human insulin. From this, pharmaceutical companies can isolate pure human insulin.

Summarizing: The purpose of insulin is to regulate blood glucose. It forces many cells of the body to absorb and use glucose thereby decreasing blood sugar levels. Insulin is secreted in response to high blood glucose. Conversely, low blood glucose inhibits its secretion from the *beta* pancreatic cells. There is a type of tumor called *insulinoma*, which secretes large amounts of insulin, and thus, it produces severe hypoglycemia. Glucagon, instead, has actions opposite of insulin. It assists insulin in the regulation of blood glucose because it forces many cells of the body to releasing or to producing glucose. Secretion takes place in response to low blood glucose, while its secretion is inhibited by high blood glucose. Deficiency may lead to hypoglycemia, but not always. Tumors called glucagonomas cause hyperglycemia. Many mathematical models have been devised to predict responses to given stimulations and, thus, to better dose insulin. The group of Claudio Cobelli, in Padova, Italy, has extensive contributions to the field. Another area of development refers to automatic infusion pumps, either for the hospital environment or for ambulatory patients.

Behind the insulin-diabetes knowledge, there is a dramatic piece of history that must be recalled. In 1920, a Canadian surgeon, Frederick G. Banting (1891–1941) visited the University of Toronto to see the newly appointed head of the Department of Physiology, John J.

R. Macleod (1876–1935). The latter had studied glucose metabolism and diabetes, and Banting had a new idea on how to find not only the cause but a treatment for the so-called sugar disease. Late in the 19th century, scientists had realized that there was a connection between the pancreas and diabetes. The connection was further narrowed down to the islets of Langerhans. From 1910 to 1920, Oscar Minkowski and others tried unsuccessfully to find and extract the active ingredient from the islets. While reading a paper on the subject in 1920, Banting had an inspiration. He realized that the pancreas' digestive juice was destroying the islets of Langerhans hormone before it could be isolated. If he could stop the pancreas from working, but keep the islets of Langerhans going, he should be able to find the stuff. He presented this idea to Macleod, who at first scoffed at it but finally gave him lab space, 10 experimental dogs, and a medical student assistant, Charles Best (1899–1978). In May 1921, as Macleod took off for a holiday in his native Scotland, Banting and Best began their experiments. By August, they had the first conclusive results: When they gave the material extracted from the islets (called "insulin", from the Latin for "island") to diabetic dogs, their high blood sugars were lowered. Macleod, back from holiday, was still skeptical of the results and asked them to repeat the experiment several more times. They did, finding the results the same, but with problems due to the varying purity of their insulin extract. Macleod assigned a chemist, James Bertram Collip (1892–1965) to the group to help with the purification. Within six weeks, the purified insulin was tested on a 14-year-old boy dying of diabetes. Insulin lowered his blood sugar and cleared his urine. Banting and Best published the first paper on their discovery a month later, in February 1922. In 1923, the Nobel Prize was awarded to Banting and Macleod for the discovery, and each shared their portion of the prize money with the other researchers on the project, Best and Collip. Banting initially threatened to refuse the award because he felt that Charles Best's work as research assistant had been vital to the project and that he should be included in the honor. The discovery of insulin was one of the most revolutionary moments in medicine. Although it took some time to work out proper dosages and to develop manufacturing processes to make enough insulin of consistent strength and purity, the introduction of insulin seemed literally like a miracle. One year the disease was an automatic death sentence; the next, people had hopes of living full and productive lives even with the disease. Estimates show that there are more than 15 million diabetics living today who would have died at an early age without insulin. Animal experimentation was essential to obtain this knowledge.

7.2 Endocrine Impulse Functions and Deconvolution

Deconvolution, as the inverse operation of convolution, has been recalled earlier in this book in the chapters devoted to the kidneys and the intestinal tract. Gonzalez *et al.* (2016) have more recently assessed this discrete mathematical technique. The turn arrived in 1987 for neuroendocrinology by using a simple, ingenious, and clever idea. Traditional control engineering and the two previously referred to physiological applications of deconvolution analysis require the injection of a known input signal, which by and large is the impulse delta Dirac function (any other could be employed just as well, but the delta function has some advantages). Obviously, in practice, its theoretical behavior is not truly realizable and can only be approached by large and very short pulses. A salient highly significant feature is that when $\delta(t)$ is applied to a linear system, the response read at its output describes the characteristic function of the system, usually termed $h(t)$, and called the *impulse response* of the system. That is what was done when intestinal absorption and renal function were studied, which thereafter, by proper block diagram and deconvolution analysis led to the determination of the system's dynamic constants.

Several neuroendocrine glands secrete their hormones as short bursts and such bursts can be looked at as physiological imperfect delta functions. The concept, quite clever, indeed, brought forward by Veldhuis *et al.* 1987, simply said, *no need of an external generator or injection for the system has its own.* Deconvolution did the rest and a series of superb contributions greatly improved the quantification neuroendocrine process (Johnson and Veldhuis, 1995).

Five out of sixteen chapters, that is, 31% of this whole book, deal with deconvolution analysis based on the burst intrinsic glandular secretion. Such would-be-like-delta function or simil, called $S(t)$, the blood hormonal concentration, $C(t)$, as the output signal, and the system elimination function, $E(t)$, as the characteristic time equivalent to the transference function $h(t)$ in systems engineering, were the elements the former authors used (Valentinuzzi, 2010).

The research group led by Johannes D. Veldhuis, from the University of Virginia, has contributed extensively in this area. For example, it has been suggested that PTH is secreted in a pulsatile fashion. To characterize PTH secretory dynamics in healthy subjects, seven young women and six young men were studied, all of whom had hip and spine bone densities studies carried out by dual photon densitometry. PTH concentrations were measured by immunoradiometric assay in blood sampled every 2 min over 6 h. Ionized calcium levels were obtained during the second and third hours of the study. Plasma PTH profiles were subjected to deconvolution analysis, which resolves measured hormone levels into secretion and clearance components. Cross-correlation analysis was performed to assess direct or inverse correlations between serum PTH and ionized calcium concentrations at various time lags. PTH was secreted in a dual fashion, with significant basal (tonic) secretion and PTH pulses approximately every 20 min. Pulsatile PTH secretion accounted for approximately 25% of the total secreted PTH, with no differences between men and women, nor were there any significant correlations between PTH and ionized calcium concentrations. It was concluded that in normal subjects, the predominant mode of PTH secretion is tonic, with superimposed PTH pulses of small amplitude but high frequency (Samuels *et al.*, 1993).

Also, and by the same team, aiming at the study of daily cortisol production and clearance rates in a group of 18 normal unstressed pubertal males, deconvolution analysis was applied to serum cortisol concentrations. Samples were obtained every 20 min for 24 h. Subject-specific characterization of adrenocortical secretory episodes, cortisol production rate, and serum hormone half-life for nine early pubertal and nine late pubertal subjects was undertaken to assessing potential roles of sexual maturation and changing gonadal steroid hormone concentrations on glucocorticoid physiology. No differences were observed between the two pubertal groups in the secretory burst frequency and half-duration, mass of cortisol released per secretory episode, average maximal rate of hormone secretion, and serum cortisol half-life. In conclusion:

(1) Cortisol production rates as estimated by deconvolution analysis are in agreement with other independent isotopic estimates, but are lower than many previous estimates.
(2) The rise in serum gonadal steroid hormone levels is not associated with alterations in the production rate or metabolic clearance of cortisol.
(3) Increased secretory burst frequency, increased amplitude (maximal rate of cortisol secretion), and increased mass of cortisol released per adrenocortical secretory episode give rise to the normal diurnal rhythm of circulating cortisol (Kerrigan *et al.*, 1993).

Somewhat later, the same international team sought to determine whether elevated circulating GH concentrations in uremic prepubertal children were due to an increase in GH secretory activity by the pituitary gland or a decrease in the metabolic clearance of GH consequent to reduced GFR. Deconvolution analysis was applied to the night-time plasma GH profiles of 11 children with preterminal chronic renal failure, 12 children with end-stage renal disease (ESRD), and a control group of matched children with idiopathic short stature ($n = 12$). Mean half-life of endogenous GH in children with ESRD and preterminal chronic renal failure was significantly higher than in controls. GH half-life correlated inversely with GFR. The number of GH secretory bursts/10 h in ESRD was amplified compared with preterminal chronic renal failure and with controls. GH production rate varied over a broad range in the three groups: It was highest in ESRD, mainly because of an increased number of GH secretory bursts, and not statistically different in preterminal chronic renal failure and in controls. Total immunoreactive plasma IGF-I levels were indistinguishable between groups. This disruption of the normal relationship between circulating GH and total plasma

IGF-I levels in ESRD suggests a relative insensitivity to the action of GH in uremia (Tönshoff *et al.*, 1995). Insulin-like growth factor IGF-I is a hormone that functions as the major mediator of GH-stimulated somatic growth, as well as a mediator of GH-independent anabolic responses in many cells and tissues. IGF-I is a small peptide (molecular weight 7647) that circulates in serum bound to high-affinity binding proteins (Clemmons, 2014).

Hindmarsh *et al.* (2011), in turn, reported that glucocorticoid replacement therapy uses twice or thrice daily regimens of hydrocortisone (HC) with variable distribution of the dose over the day. Deconvolution analysis determines the mass of hormone that needs to be secreted to attain a particular serum concentration. These authors have used this methodology to determine the amount and distribution of cortisol over a 24-h period in 79 adults aged 60–74 years and 30 children aged 5–9 years. The subjects underwent 24-h serum cortisol profiles with samples drawn at 20 min intervals. Profiles were subjected to deconvolution analysis using a cortisol half-life of 80 min to yield the amount of cortisol released by the adrenal to generate the corresponding serum concentration. Cortisol secretion occurred in discrete bursts. Daily cortisol secretion ranged 5.6–12.4 mg/m^2 per day in adults and 7.2–16.5 mg/m^2 per day in children. Peak secretion was lower in adults than in children. It was suggested an optimal dosing and distribution regimen for HC replacement.

The same group of researchers (Peters *et al.*, 2013) reported a year after that HC therapy is based on a dosing regimen derived from estimates of cortisol secretion, but little is known of how the dose should be distributed throughout the 24 h. Again, by means of deconvolution analysis, they obtained HC dosing schedules in young children and older adults. Data were apparently obtained from the same group of patients participating in the study mentioned above. Mean daily cortisol secretion was similar between adults and children, with peak serum cortisol concentrations higher in children compared with adults. As conclusions, they stated that these observations suggest that the daily HC replacement dose should be equivalent on average to 6.3 mg/m^2 body surface area/day in adults and 8.0 mg/m^2 body surface area/day in children. Differences in distribution of the total daily dose between older adults and young children need to be taken into account when using a three or four times per day dosing regimen.

A significant paper dealt with deconvolution of serum cortisol (Faghih *et al.*, 2014), a hormone that shows pulsatile release controlled by a hierarchical system involving CRH, ACTH, and cortisol itself. Determination of the number, timing, and amplitude of these hormone secretory events represents a challenging problem. These authors modeled cortisol secretion from the adrenal glands using a second-order linear differential equation with pulsatile inputs that represent cortisol pulses released in response to pulses of ACTH. Under the assumption that the number of pulses is between 15 and 22 over 24 h, they successfully deconvolved both simulated datasets and actual 24-h serum cortisol datasets sampled every 10 min from 10 healthy women, so obtaining physiologically plausible timings and amplitudes of each cortisol secretory event with 0.96, or even slightly better, as correlation coefficient. Such information improves the understanding of pathological neuroendocrine states, and helps in designing optimal approaches for treating hormonal disorders. The model built by Faghih *et al.* (2014) is based on the stochastic differential equation model of diurnal cortisol patterns. It uses the first-order kinetics for cortisol synthesis in the adrenal glands, cortisol infusion to the blood, and cortisol clearance by the liver, while considering a doubly stochastic pulsatile input in the adrenal glands that has Gaussian amplitudes and gamma distributed inter-arrival times (Brown *et al.*, 2001). This input can be considered as an abstraction of hormone pulses and marks the timing and amplitude of the secretory events leading to cortisol secretion. It was assumed that there are between 15 and 22 secretory events over a 24-h period that result in the observed cortisol profile (Faghih *et al.*, 2011). This model is represented as follows:

$$\frac{dx_1(t)}{dt} = -\theta_1 x_1(t) + u(t) \text{ (Adrenal Glands)} \tag{7.1}$$

$$\frac{dx_2(t)}{dt} = \theta_1 x_1(t) - \theta_2 x_2(t) \text{ (Serum)} \tag{7.2}$$

where x_1 is the cortisol concentration in the adrenal glands and x_2 is the serum cortisol concentration. θ_1 and θ_2, respectively, represent the infusion rate of cortisol from the adrenal glands into the blood and the clearance rate of cortisol by the liver. The equation

$$u(t) = \sum_{i=1}^{N} q_i \delta(t - \tau_i) \tag{7.3}$$

represents an abstraction of the hormone pulses that result in cortisol secretion, where q_i stands for the amount of the hormone pulse initiated at time τ_i (q_i is zero if a hormone pulse did not occur at time τ_i), and it was accepted that impulses occur at integer minute values. N corresponds to the length of the input ($N = 1440$). Blood was collected, beginning at y_0 and then, with a sampling interval of 10 min, for M samples ($M = 144$). Faghih *et al.* (2014) display curves and a table where the numerical values can be found, as for example, amplitude and timing of hormone pulses, experimental cortisol data, and model-predicted cortisol estimates for each participant. The known circadian changes are well demonstrated. The recovered pulses are mostly small at the beginning of the scheduled sleep, with a large pulse towards the end of the sleep period or beginning of the wake period. Besides, there are multiple small and medium-sized pulses during the wake period. R^2 values close to 1.0 suggest that the model is a good estimator.

Due to the simultaneous release and clearance of hormones and the unknown timing and amplitudes of the secretory events, identifying the pulsatile input to the system and the infusion and clearance rates is challenging. Moreover, due to data collection difficulties and cost, the sampling interval is usually relatively large (10–60 min) compared to the expected inter-pulse intervals as well as secretion and clearance rates of cortisol, hence, the data resolution is low, making not easy to identify delays and potential consecutive pulses. A third obstacle is due to cortisol secretion differences in sleep and wake states and at different circadian times. Finally, there is inter-individual variation, even among healthy individuals. The approach used by Faghih *et al.* (2014) can be applied to GH, thyroid hormone, and gonadal hormones in a similar fashion. In summary, their proposed algorithm works well even when the pulses are not easily identifiable while still being applicable to cases in which pulses are identifiable by visual inspection.

7.3 Technologies to Fetch Endocrine Information

In 1977, Andrew Schally and Roger Guillemin were awarded the Nobel Prize, along with Rosalyn Sussman Yalow, for their discoveries concerning the hormone production of the brain. She was a physicist and mathematician, hence with a background not related to the biomedical sciences. Solomon Berson, her close collaborator, was a physician. Getting acquainted with their lives provides surprising and touching human insights. Radioimmunoassay (RIA), developed by them, allowed precise and minute concentration measurements of a substance, by quantitating the binding, or the inhibition of binding, of a radiolabeled substance to an antibody, a revolutionary highly quantified diagnostic process largely ignored when Yalow and Berson published it in 1960. Their primary use was intended for the study of diabetes, but today its applications are multiple. Yalow and Berson (the latter died prematurely in 1972) could have patented RIA and would have gotten enormous royalties. Instead, to their credit, they made efforts to get RIA into common use; they wanted its potential value directed to humankind ahead of its potential value to them. Quite a human lesson to underline with not too many examples to mention (one to certainly remembering is Maria Slodowska Curie). Thus, neuroendocrinology's final birth certificate, based on strong and well-behaved numerical grounds and, why not, we may add with contributions from bioengineering, can be established between the late 1950s and early 1960s (Valentinuzzi, 2010). RIA was, without doubt, its major technological password or key.

Technologies are often based on and mixed with mathematics. The deconvolution process of noisy signals, as those often found in microscopy or endocrinology, become difficult to handle, especially when the number of peak signals is limited, and the solution of the deconvolution is required to be sparse, hence the new term introduced, *sparse deconvolution*. In endocrine data, the secretion pattern appears as a *sparse series* of spikes, a problem specifically addressed by de Rooi *et al.* (2014). Selesnick (2012), from the Polytechnic Institute of New York University, delves deeply into this mathematical method. The following *Deconvolution Tutorials* can be suggested for those readers interested in the subject: http://www. cv.nrao.edu/~abridle/deconvol/deconvol. html; check https://www. youtube.com/watch?v=QIRu00hIM0s; also in Wikipedia.

7.4 Summary

The three main sections of this chapter deal, first, with the physiology of the ES, underlining the many regulatory feedback loops (that still need to be studied); second, with the application of a mathematical concept (deconvolution), to uncover kinetic information from several inner glands, and so better understand their dynamics; third, and final, a very brief mention of the enormous step forward made when RIA was introduced in endocrinology. One of the fields where endocrinological advances are going on at fast steady pace is biotechnology, as for example, in the food industry. In China, the hypothalamic hormones are applied to breed carps, a traditional fish in the Chinese diet, which used to require a significant human effort to collect it in the mountains. Now, it is conveniently harvested in well-installed breeding sites. There are systems for optimal and robust control of the automated infusion of insulin to Type I and Type II diabetic individuals. Wireless technology, insulin pumps, embedded computation, and emerging non-invasive glucose monitoring systems may be interconnected and a closed-loop control system realized. The potential for future implementation of an implanted artificial pancreas is also investigated. The young bioengineer no doubt will find in endocrinology a potential area of development either in basic or applied research, in the clinical setting or in biotechnology. All this puts well endocrinology as a physiology branch receiving very useful inputs from Bioengineering.

References

Asashima M, Nakano, H, Shimada KK, Kei I, Shibai H, Ueno N. Mesodermal induction in early amphibian embryos by activin A (erythroid differentiation factor). *Roux's Arch Deve Biol* 198(6):330–335, 1990.

Braverman LE, Cooper DS, eds. Werner & Ingbar's the thyroid: A fundamental and clinical text. 10th ed. Lippincott, Williams & Wilkins-Wolters Kluwer, Philadelphia, 912 pp., 2013. ISBN/ISSN 9781451120639.

Brown EN, Meehan PM, Dempster AP. A stochastic differential equation model of diurnal cortisol patterns. *Am J Physiol Endocrinol Metab* 280:E450–E461, 2001.

Chen YG, Wang Q, Lin SL, Chang CD, Chuang J, Chung J, Ying SY. Activin signaling and its role in regulation of cell proliferation, apoptosis, and carcinogenesis. *Exp Biol Med (Maywood)* 231(5): 534–44, 2006.

Clemmons DR. Physiology of insulin-like growth factor I, in *Up To Date Website*, March 27, 2014. http://www.uptodate.com/contents/physiology-of-insulin-like-growth-factor-i.

de Rooi **JJ**, Ruckebusch C, Eilers PHC. Sparse deconvolution in one and two dimensions: Applications in endocrinology and single-molecule fluorescence imaging. *Anal Chem* 86(13):6291–6298, 2014. doi:10.1021/ac500260h**.**

Faghih RT, Dahleh MA, Adler GK, Klerman EB, Brown EN. Deconvolution of serum cortisol levels by using compressed sensing. *PLoS ONE* 9(1):1–12; e85204, 2014. http://www.plosone.org/article/metrics/info:doi/10.1371/journal.pone.0085204#savedHeader.

Faghih RT, Savla K, Dahleh MA, Brown EN. A feedback control model for cortisol secretion. In: *2011 Annual International Conference IEEE Engineering in Medicine and Biology Society*, 716–719, 2011. doi:10.1109/IEMBS.2011.6090162.

Fujisawa I. Magnetic resonance imaging of the hypothalamic-neurohypophyseal system. *J Neuroendocrinol* 16(4):297–302, 2004.

Gonzalez SA, Valentinuzzi ME, Arini PD. Deconvolution: It fans back, out and ahead. *IEEE Pulse Mag* 7(4):54–61, 2016.

Hindmarsh P, Hill N, Dattani M, Peters C, Charmandari E, Matthews D Deconvolution analysis of 24 h serum cortisol profiles informs the amount and distribution of hydrocortisone replacement therapy. *Endocrine Abstracts* 27, 2011 OC1.5; published by bioscientifica; ISSN 1470-3947 (print); ISSN 1479-6848 (online); http://www.endocrine-abstracts.org/ea/0027/ea0027oc1.5.htm.

Johnson ML, Veldhuis JD, eds. Quantitative neuroendocrinology, vol. 28. Academic Press, New York, 433 pp., 1995.

Kerrigan JR, Veldhuis JD, Leyo SA, Iranmanesh A, Rogol AD. Estimation of daily cortisol production and clearance rates in normal pubertal males by deconvolution analysis. *J Clin Endocrinol Metab* 76(6):1505–1510, 1993. Online 1 July 2013.

Knobil E. The wisdom of the body revisited. *N Physiol Sci* 14:1–11, 1999.

Martini L ed. Neuroendocrinology: The normal neuroendocrine system. In: *Progress in Brain Research*, vol. 181. Elsevier, Amsterdam, Holland, printed in Great Britain, 2010. It is a multiauthored edited volume.

Murphy D, Konopacka A, Hindmarch Ch, Paton JFR, Sweedler JV, Gillette MU, Ueta Y, Grinevich V, Lozic M, Japundzic-Zigon N. The hypothalamo–neurohypophyseal system: From genome to physiology. *J Neuroendocrinol* 24(4): 539–553, 2012.

Peters CJ, Hill N, Dattani MT, Charmandari E, Matthews DR, Hindmarsh PC. Deconvolution analysis of 24-h serum cortisol profiles informs the amount and distribution of hydrocortisone replacement therapy. *Clin Endocrinol* 78(3): 347–351, 2013. http://onlinelibrary.wiley.com/doi/10.1111/j.1365-2265.2012. 04502.x/abstract.

Powell T, Valentinuzzi ME. Calcium homeostasis: Responses of a possible mathematical model. *Med Biol Eng* 12(3):287–294, 1974.

Samuels MH, Veldhuis J, Cawley C, Urban RJ, Luther M, Bauer R, Mundy G. Pulsatile secretion of parathyroid hormone in normal young subjects: Assessment by deconvolution analysis. *J Clin Endocrinol Metab* 77(2), 1993. http://press.endocrine.org/doi/abs/10.1210/jcem.77.2.8345044. Online: 1 July 2013.

Selesnick I. Sparse deconvolution (an MM algorithm), 2012; selesi@poly.edu; http://eeweb.poly.edu/iselesni/lecture_notes/sparse_deconv/SparseDeconv. pdf; last edit: Oct, 2014. The MM algorithm is an iterative optimization method of a function in order to find their maxima or minima. MM stands for Majorize-Minimization or Minorize-Maximization.

Squire L ed. The History of Neuroscience in Autobiography. Chapter by Roger Guillemin on the hypothalamic secretions. The Society for Neuroscience, Academic Press, San Diego, vol. 2, 96–131 pp., 1998. Rest of the book for other neuroendocrine systems.

Sulyok S, Wankell M, Alzheimer C, Werner S. Activin: An important regulator of wound repair, fibrosis, and neuroprotection. *Mol Cell Endocrinol* 225(1– 2):127–32, 2004.

Tönshoff B, Veldhuis JD, Heinrich U, Mehls O. Deconvolution analysis of spontaneous nocturnal growth hormone secretion in pre-pubertal children with pre-terminal chronic renal failure and with end-stage renal disease. *Pediatr Res* 37:86–93, 1995.

Valentinuzzi ME. Neuroendocrinology and its quantitative development: A bioengineering view (Letter to the Editor). *BioMed Eng OnLine* 9:68, 2010. doi: 10.1186/1475-925X-9-68; Nov 4. http://www.biomedical-engineering-online. com/content/9/1/68.

van Zonneveld P, Scheffer G, Broekmans F, Blankenstein M, de Jong F, Looman C, Habbema J, Velde E. Do cycle disturbances explain the age-related decline of female fertility? Cycle characteristics of women aged over 40 years compared with a reference population of young women. *Hum Reprod* 18(3):495– 501, 2003.

Veldhuis JD, Carlson ML, Johnson ML. The pituitary gland secretes in bursts: Appraising the nature of glandular secretory impulses by simultaneous multiple-parameter deconvolution of plasma hormone concentration. *Proc Natl Acad Sci USA* 84:7686–7690, 1987.

CHAPTER 8

THE NERVOUS SYSTEM: SOURCE *PAR EXCELLENCE*

B. Silvano Zanutto and Esteban Cynowiec

The very information system, still a mystery, as vast as the Universe itself... or even more.

Abstract

In a mechanistic-like oversimplification, the brain shows some similarities to a computer handling logical information, but, in fact, they are very different. In the brain, pattern recognition is an adaptive process that makes use of neural circuit changes, principally, by adjusting the strength of certain neural connections. Biological evolution "chose" the way to adjust them in order to improve individual survival. The brain has a population of hundreds of billions of neurons; hence, the interconnection of this neural circuitry is huge, so much, that it allows the brain to make each individual unique, with the ability to create art and science. In spite of such enormous and interconnected size, it has an organization that, in a way, makes easier to theorize how the brain works. Moreover, it is possible to find experimentally how neural populations encode information, and to simulate it with computational theories in an effort to find hypotheses to explain brain higher functions, such as learning and behavior.

8.1 Introduction

Our brain is able to handle logical information, but it is not a logical system like a computer, which processes programmed sentences. In the brain, there are circuits, say, as in a computer; however, they are not made of logical gates. Instead, the building blocks are neurons. Despite such apparent similarity, we underline that the brain is far from working as a computer. The processing time of a single neuron is much slower than a computer clock, nonetheless, how do we recognize a face much faster than a computer, by far a very complex action? It is because of the parallel arrangement of thousands of simple processors and few steps needed for electric signals to "travel" from the retina to achieve the face recognition.

Besides, pattern recognition is an adaptive process that puts things together always in novel ways. Such adaptation makes use of neural circuit changes, basically, by adjusting the strength of certain neural connections (called *synapses*) leading to survival improvement. Validation of this concept takes us to Lamarck's, Darwin's, and Wallace's times (Valentinuzzi and Saltor, 2016).

The brain has a population of billions of neurons, with a number of possible interconnections surpassing the number of particles in the universe. Thus, the variance of this neural circuitry is huge, so much, that allows the brain process to make each individual unique, so accounting for the ability to create poetry, music, and science. Figure 8.1 depicts a schematic representation of the brain anatomy and areas.

Evolution selected a nervous system control in a very similar way to how engineers could control a robot. Figure 8.2 intends to show diagrammatically the major components of the central nervous system (CNS) and peripheral nervous system (PNS) and their functional relationships. Stimuli from the environment convey information to processing circuits within the brain and spinal cord, which, in turn, interpret their meaning and send signals to peripheral effectors that move the body and adjust the internal organs' functions.

8.2 Cellular and Structural Organization of the Nervous System

8.2.1 *Organization*

Animals' (humans, among them) nervous system is organized as a CNS consisting of the brain and spinal cord, and a PNS consisting of nerves and ganglia on the outside of the CNS. The brain and the spinal cord have billions of specialized cells called *neurons and glia cells*, and a group of parallel nerves and ganglia, too (also

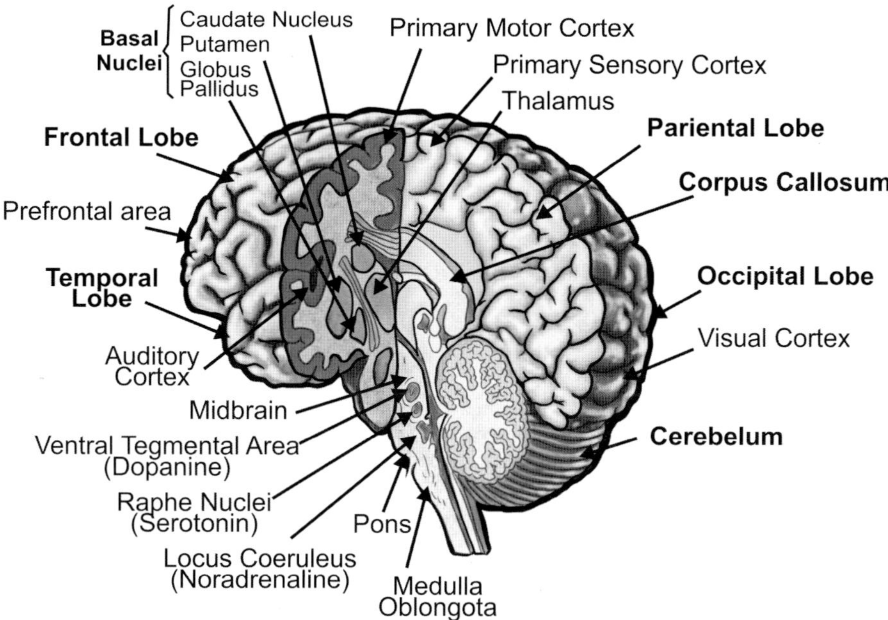

Figure 8.1: The diagram shows a brain where a quarter part is eliminated in order to show internal structures. Lobes and specific cortices are observed. In addition, nuclei with cortical modulator neurons of the midbrain, pons, and medulla oblongata are shown. A similar figure is freely available in Internet and in several textbooks. Drawn by Gustavo Idemi.

formed by neurons and glia). The PNS has two big parts, the *somatic nervous system* and the *autonomic nervous system*, the latter with two branches: the *sympathetic* nervous system and the *parasympathetic* nervous system. The sympathetic nervous system's primary task centers on the stimulation of the *survival* system, the one that activates when a dangerous situation either shows up or is subjectively perceived as dangerous (body's fight-or-flight response), while the parasympathetic nervous system is sometimes called the *resting* system ("feed and breed"). However, both are constantly active at a basic level to maintain the body homeostasis. In many cases, both systems have opposing actions, that is, while one activates a physiological response, the other system inhibits it. In a few cases, only one system is controlling the physiological responses.

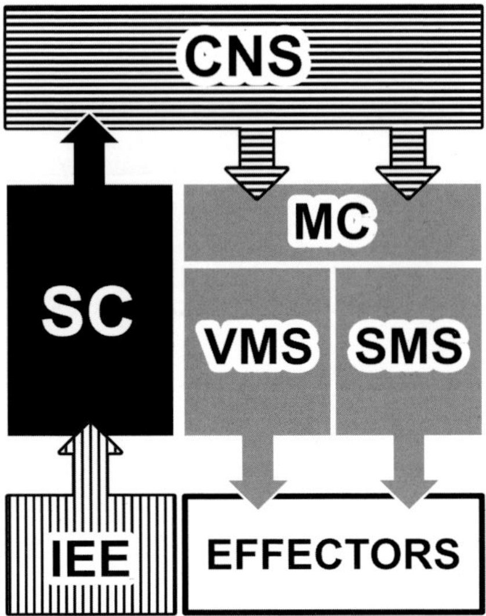

Figure 8.2: Schematic diagram of the major components of the central and peripheral nervous systems and their functional relationships Lower right: EFFECTORS, which include smooth muscles, cardiac muscle, glands (involuntary), and skeletal muscle (voluntary). Lower left block: IEE is the internal–external environment. Left central block: SC, sensory components, with sensory ganglia and nerves, and sensory receptors at body surface and within it. Other central blocks: MC, motor components, divided in visceral motor system (VMS) and somatic motor system (SMS). The former accounts for *sympathetic, parasympathetic,* and enteric divisions, including autonomic ganglia and its nerves. The SMS contains the motor nerves. Upper black: Overall and simplified block diagram of the CNS, central nervous system, with cerebral hemispheres, diencephalon, cerebellum, brainstem, and spinal cord. It takes care of analysis and integration of the sensory and motor information. Drawn by Gustavo Idemi.

8.2.2 Neurons, glia cells, synapses, brain, and cord segmental levels

8.2.2.1 Neurons, synapses, and glia cells

A neuron (Fig. 8.3) is an electrically excitable cell that processes and transmits information through electrical–chemical signals that travel along the neuron traversing one or more *synapses* to excite other cells. A typical neuron possesses a cell body or *soma,* one or more *dendrites* and an *axon.* Dendrites are prolongations that arise from the cell body, often extending for tens to thousands of micrometers

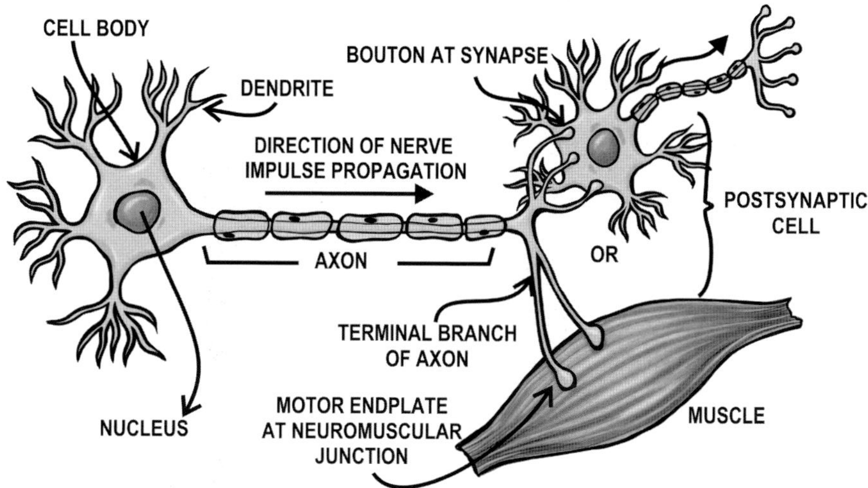

Figure 8.3: Neuron: Its parts and its synapses. It also shows a striated muscle end plate. A similar figure is freely available in Internet and in several textbooks. Redrawn by Gustavo Idemi.

(a few centimeters) and sometimes branching multiple times, giving rise to a complex *dendritic tree* (Nicholls *et al.*, 2011).

An axon arises from the cell body at a site called the *axon hillock* and travels for a distance, as far as 1 m in humans or even more in other species, until it hits one or more neurons, muscle, or organ in a *synapsis*. Thus, an axon becomes the structure that permits a neuron to pass an electrical–chemical signal through itself and to other neurons. The neuron frequently has multiple dendrites, but never more than one axon, although the axon may branch off hundreds of times before it terminates. At the majority of synapses, signals travel from the axon of one neuron to a dendrite of another. Nonetheless, there are many exceptions to these rules, such as neurons that lack dendrites, or neurons that have no axon, or synapses that connect an axon to another axon, or even a dendrite to another dendrite. Another important characteristic of the neuronal system is that all neurons mostly derive from special types of stem cells. In humans, neurogenesis (new neurons development) ceases during adulthood; however, the hippocampus and the olfactory bulb seem to be able to show it.

The neuron membrane consists of a *lipid bilayer* of molecules in which larger protein molecules are embedded as *voltage-gated ion channels*, or as other kinds of ion channels and *ionic pumps*. The lipid bilayer is highly resistant, so it functions as an insulator. The electrical properties of a cell are determined by the structure of the membrane and the inside–outside concentration ion gradient (see Chap. 2 for more details). If there is no external perturbation, the interior voltage of the neuron is negative (e.g., $-70\,\mathrm{mV}$). The electrical signal propagates along the neural path following the *Cable Theory*, only if the voltage peak value is below a *threshold*. If the peak value is above the *threshold*, the voltage-gated ion channels open for a short time, causing the interior potential to become positive. This rapid upward spike followed by a rapid fall is the *action potential*. The earliest site of action potential initiation locates just adjacent to the axon hillock, at the initial (unmyelinated) segment of the axon, where the threshold is lower.

The biological space where electrical signals travel to other neurons and to muscles or glands is the *synapsis*. There are two types of synapses: electrical and chemical. The former permits passive flow of ionic current through the *gap junction* pores from one neuron to another. At this gap, cells approach within about 3.5 nm of each other (a much shorter distance than the 20–40 nm distance that separates cells at the chemical synapses). Electrical synapses permit the transference of important intracellular metabolites between neurons. This arrangement has a number of consequences. One is that transmission in many cases can be bidirectional; that is, current can flow in either direction across the gap junction. Another important feature of the electrical synapse is that transmission is extraordinarily fast (without the delay that is a characteristic of chemical synapses), but they lack gain, that is, the signal in the postsynaptic neuron is the same or even smaller than that at the originating neuron. A more general consequence of electrical synapses is to synchronize electrical activity among populations of neurons. For example, the brainstem neurons that generate rhythmic electrical activity underlying breathing are synchronized by electrical synapses, as are populations of interneurons in the cerebral cortex, thalamus, cerebellum, and other brain

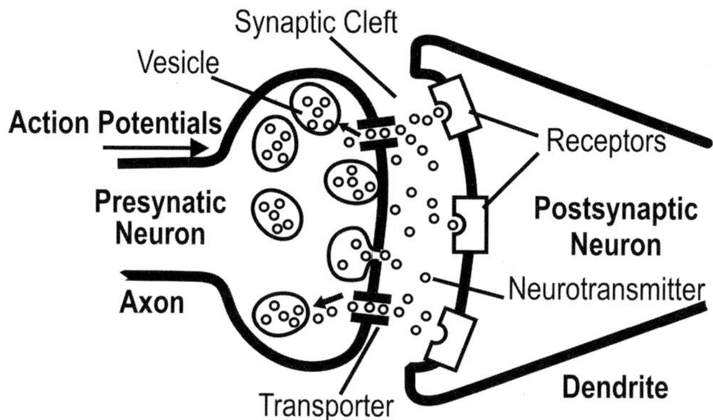

Figure 8.4: Chemical synapsis. The presynaptic axon releases neurotransmitter molecules triggered by action potentials and binds specific receptors at the postsynaptic neuron. Redrawn by Gustavo Idemi from a freely available INTERNET picture.

regions. Electrical transmission between certain neurons within the mammalian hypothalamus ensures that all cells fire action potentials approximately at the same time, facilitating a burst of hormone secretion into the circulation. The fact that gap junction pores are large enough to allow molecules, such as *ATP* and *second messengers* to diffuse intercellularly, also permits electrical synapses to coordinate the intracellular signaling and metabolism of coupled cells. This property may be particularly important for glial cells (*see below*), which form large intracellular signaling networks via their gap junctions (Purves *et al.*, 2011).

Chemical synapses (Fig. 8.4) make the brain plastic, thus allowing the possibility of learning and change their behavior. At a chemical synapse, when action potentials reach the axon terminal, molecules called *neurotransmitters* are released into a microscopic gap, adjacent to another neuron through a *synaptic cleft*. The neurotransmitters, kept within small sacs called *vesicles*, get into the synaptic cleft by *exocytosis*. These molecules bind to receptors on the *postsynaptic* neuron's side of the synaptic cleft, generating changes in the membrane potential called *postsynaptic potentials*. In postsynaptic cells, neurotransmitter receptors often trigger an electrical signal by regulating the activity of ion channels.

There are two types of neurotransmitter receptors: ligand-gated receptors or *ionotropic* receptors, and *metabotropic* receptors. The changes in the post synapsis can be positive or negative, called *excitatory postsynaptic potential* or *inhibitory postsynaptic potential*, respectively. Finally, neurotransmitters are cleared from the synapsis through different mechanisms, such as enzymatic degradation and re-uptake by specific transporters on the presynaptic cell or by glia cells (see below) to terminate the action of the transmitter.

The neuronal system has also non-neuronal cells, the *glial cells* (sometimes called neuroglia). Some glial cells function primarily as the physical support for neurons. Others are involved in the synaptic changes regulating the clearance of neurotransmitters, having an important role in learning and behavior. During embryogenesis, a type of glia cells directs the migration of neurons and produce molecules that modify the growth of axons and dendrites. Another type of cell, known as *Schwann cells*, makes the *myelin sheath*. The main purpose is to increase the speed at which impulses propagate along. Moreover, the myelin sheath promotes repair after axonal injury and is involved in regrowth of the axon.

Nerve cells that carry information toward the brain or spinal cord (or farther centrally within the spinal cord and brain) are the *afferent* neurons; nerve cells that carry information away from the brain or spinal cord are the *efferent* neurons. Interneurons or local circuit neurons only participate in the local aspects of a circuit, based on the short distances over which their axons extend. These three functional classes — afferent neurons, efferent neurons, and interneurons — are the basic constituents of all neural circuits. A simple example of the shortest neural circuit is the ensemble of cells that subserves the myotatic spinal reflex (the knee-jerk reflex).

8.2.2.2 *The CNS: Brain and spinal cord*

Human brain is incredibly compact, weighing about 3 pounds ($\approx$ 1.4 kg). Its many folds and grooves provide it with the additional surface area. The spinal cord, on the other hand, is a long bundle of nerve tissue about 18 inches long ($\approx$ 46 cm) and 3/4 inch (1.9

cm) thick. It goes from the lower part of the brain projecting down all along the spine with many nerves branching bilaterally out and off to the entire body. These make up the PNS. The brain and the spinal cord are mechanically (acting as a cushion) and immunologically protected by the *cerebrospinal fluid* within their bony confines. The bone structures protect the brain and the spinal cord: the brain by the skull and the spinal cord by a set of ring-shaped bones called vertebrae. Layers of membranes called *meninges* envelop the CNS and, in mammals, they consist of three layers (the *duramater*, the *arachnoid*, and the *piamater*).

8.2.2.3 *Cord segmental levels*

As said above, the spinal cord consists of a series of vertebral segments; it has "neurological" segmental levels, defined by the spinal roots that enter and exit the spinal column between each of the vertebral segments.

Three major neural avenues composed of long nerve fibers interconnect both sections: the pyramidal, the extrapyramidal, and the sensory tracts (Fig. 8.5). Outputs from different segmental levels of the spinal cord constitute efferent outputs. The pyramidal system originates in the anterior region of the cerebral cortex central fissure, known as the motor area. Its stimulation produces discrete movements in any skeletal muscle. Voluntary movements follow this

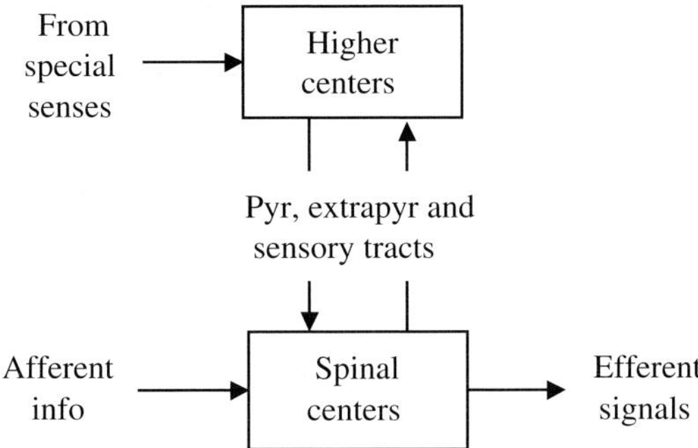

Figure 8.5: Higher and spinal centers. Three main tracts interconnect them, pyramidal, extrapyramidal and sensory). The pyramidal channel controls voluntary movements.

pathway, and injuries in this area lead to either total or partial paralysis depending on the extension of the damage. *Hemiplegia* (paralysis of one side) and *hemiparesia* (weakness of one side) represent the two typical and rather common human conditions after a cerebrovascular stroke, also called cerebrovascular accident. Since there is a crossover from one side to the other (the pyramidal *decussation*), the left hemisphere controls the right side of the body and vice versa, although the crossover is not total. We might think in terms of two domains, the body skeletal voluntary muscles and the motor cortical area, of fully different shapes and extensions, but functionally linked by the nervous tree in a predictable way.

The other pathway is called *extrapyramidal* to distinguish it from the tracts of the motor cortex that reach their targets by traveling through the *pyramids* of the medulla, which is a complex system that includes several structures. Integration of extrapyramidal activity takes place at different levels, from the very cerebral cortex to the spinal cord, and many of its actions still await elucidation. Facial expression is one important communicative behavior mediated by the extrapyramidal tract (often affected in some Parkinson patients). In contrast to the pyramidal tract, the extrapyramidal tract is an indirect, multisynaptic avenue, interchanging signals with *basal ganglia*, the *red nucleus*, the *substantia nigra*, the *reticular formation*, and the *cerebellum*.

The spinal centers receive afferent information from the different segmental levels (cervical down to coccygeal) and have efferent outputs from the same levels. The dorsal roots of the spinal cord are sensorial or afferent, and the ventral roots are efferent or motor. This anatomical rule is known as Bell–Magendie law, after Bell (1774–1842; see Bell, 1803) and François Magendie (1783–1855), English and French physicians, respectively, who described it. Thus, the spinal cord shows four different regions: the cervical, thoracic, lumbar, and sacral regions arranged in 31 segments, which give off, in turn, 31 pairs of nerves. These nerves are distributed into 8 cervical, 12 thoracic, 5 lumbar, 5 sacral, and 1 coccygeal. Dorsal and ventral roots enter and leave the vertebral column, respectively, through intervertebral foramens. The overall number of nerve fibers entering into the spinal cord is in the order of 3×10^6. The typical reflex arc exemplifies a segmental circuit: Say that while walking barefooted, we step on a sharp little stone. The painful stimulus excites pain receptors, which transmit the signal to the corresponding spinal segment, where after integration elicits a motor efferent outflow to the leg muscles in order to quickly lift the foot and avoid or minimize injury. This is a withdrawal protective involuntary reflex response. Spinal segmentation well manifests itself in lower animals, such as fish, amphibians, or reptiles. In man, the phenomenon is less marked, even though it clearly shows up in the paraplegic and in the quadriplegic subject, meaning in persons who suffered lesions (full or partial sections), low or high, in the spinal cord, respectively. Car and motorcycle accidents and wars are, unfortunately, a significant cause of this kind of patients in young people. They are candidates to participate as active members in Rehabilitation Engineering research and development groups, which have become a leading specialty of Biomedical Engineering.

8.2.2.4 *Inner brain divisions*

The higher centers located within the skull are usually divided in three basic and complex parts: *proencephalon* (or anterior cerebrum), *mesencephalon* (or medial cerebrum), and *rhomboencephalon* (or posterior cerebrum). They are well-interconnected receiving and sending off information from and to different regions. The first one, also called forebrain, contains the *telencephalon* (cerebral cortex, rinhoencephalon, striated body, and neopallium) and the *diencephalon* (thalamus, hypothalamus, metathalamus, and epithalamus). The former is responsible for the initiative actions, memory, and conditioned reflexes, while the latter handle many locomotor reflexes and emotional responses. The hypothalamus got particular attention in a previous chapter (as part of the endocrine system); however, it is also responsible for other functions: temperature regulation, thirst and appetite control, sexual behavior, defensive reactions, circadian rhythms, and the so-called motivation.

The mesencephalon or mid-brain consists of (1) a ventrolateral portion, composed of a pair of cylindrical bodies, named the *cerebral peduncles*; (2) a dorsal portion, consisting of four rounded eminences, named the *corpora quadrigemina*; and (3) an intervening passage or tunnel, the *cerebral aqueduct*, which represents the original cavity of the mid-brain and connects the third with the fourth ventricle. Several locomotor reflexes are integrated here. The posterior cerebrum or brainstem houses the cerebellum, pontine protuberance or *pons* and *medulla oblongata*. The former takes care of posture and coordination; the second structure participates in the arousal of the individual, some autonomic functions and relays sensory information between cerebrum and cerebellum, while the latter is the place of the respiratory and cardiovascular centers.

From the cranial cavity, 12 important pairs of nerves come out. They carry afferent and efferent information: the olfactory nerve (or pair I), the optic (or pair II), the common oculomotor (or pair III), the pathetic trochlear (or pair IV), the trigeminal (or V), the external oculomotor (or VI), the facial (or VII), the auditive (or VIII), the glossopharyngeal (or IX), the vagus (or X), the spinal (or XI), and the hypoglossal (or XII).

The term *basal ganglia* refers to a group of five subcortical nuclei that play an important role in the regulation and coordination of cortically originated movement. They are also involved in cognitive processes. The main components are the *dorsal striatum* (*caudate nucleus* and *putamen*), *ventral striatum* (*nucleus accumbens* and *olfactory tubercle*), *globus pallidus, ventral pallidum, subthalamic nucleus*, and the *substantia nigra pars compacta*.

Finally, the *rhombencephalon* (sometimes known as "the hindbrain") includes the medulla, pons, and cerebellum, all supporting vital bodily processes. There is a rare disease of the rhombencephalon called *rhombencephalosynapsis*, characterized by a missing *vermis* (rounded and elongated central part of the cerebellum, between the two hemispheres) resulting in a fused cerebellum that, in turn, makes patients present cerebellar ataxia. In other sense, the caudal rhombencephalon has been generally considered as the initiation site for the neural tube closure.

8.2.3 *Neuronal circuit and synaptic plasticity*

Synaptic plasticity is the ability of synapses to strengthen or weaken due to the individual–environment interaction. There are several underlying mechanisms in synaptic plasticity: by altering the release of neurotransmitters in the presynaptic neuron, by changing the function and number of its receptors in the postsynaptic cell, or both. Some patterns of synaptic activity in the CNS produce a long-lasting increase in synaptic strength known as *long-term potentiation* (LTP), whereas other patterns of activity produce a long-lasting decrease in synaptic strength, known as *long-term depression* (LTD).

Both (LTP and LTD) are broad terms that describe only the direction of change in synaptic efficacy; in fact, different cellular and molecular mechanisms can be involved in producing LTP or LTD at different synapses. Modifications by encoding of synaptic strength are widely considered one of the major cellular mechanisms involved in learning and memory.

Synaptic plasticity has been extensively characterized at chemical synapses, but it was found that the electrical synapses are also involved in LTD. Particularly, modification of electrical synapses resulting from activity in coupled neurons is likely to be a mechanism for dynamic reorganization of the neuronal networks.

Models of LTP and LTD have been found before the biochemical mechanisms. Hebb (1949) had the intuition that if two neurons are active at the same time, the synapses between them become stronger. The first mechanism supporting it was discovered in the early 1970s. The effect proposed by Hebb is known as Hebb's rule, that is, *cells that fire together, wire together*. This effect simulates the LTP; LTD is simulated by the anti-Hebbian rule, the opposite of Hebb's rule: if two neurons are active at the same time, the synapses between them are weakened.

8.3 Models in Neuroscience

8.3.1 *Model of neuronal circuits (artificial neural networks)*

Since Galileo Galilei (1564–1632) asked himself "how" natural phenomena take place, a new way of making science began, that is, discovering laws based on actual experimentation and its consequent experiment for validation. In Physics, Galileo not only described phenomena, but also experimentally found the laws that governed them. Formalized laws gave Physics an enormous explicative power. Similar methodology was later applied in other scientific and even more complex areas, such as Biology, and in particular, in systems so difficult to tackle as the brain.

Many attempts tried to model the brain, perhaps the most popular being those within the so-called artificial intelligence area, in which intelligence tends to be heuristically simulated for its application in different disciplines such as robotics or psychology. From Kohonen's viewpoint (1987), *artificial intelligence is a set of strategies found by trial and error during a long time and, afterwards, inserted within an efficient code.* Artificial *neural networks* (ANN) have a different approach; they are parallel systems where memory, learning, and behavior are *emergent properties of the interaction simple nodes* (Sejnowski *et al.*, 1988). An *emergent property* appears when two or more component objects are somewhat joined together in constraining relations to construct a higher level aggregate object; hence, it is a novel property that unpredictably comes from a combination of simpler constituents. ANN are special systems made out of simple elements (often adaptive) massively interconnected in parallel with hierarchical organization; they are intended to interact with the real-world objects in the same way the nervous system does. Two groups represent often the division made for ANN models, that is, 1-*connectionist models*, where each node corresponds to the simplest element assigned to a neuron, and 2-*semantic networks*, where each node corresponds to a concept or to some other linguistic variable.

8.3.2 *Historical overview of ANN*

The first researchers to elaborate the conceptual frame were McCulloch and Pitts (1943), who considered neurons as discrete elements, and proposed a parallel model of the eye-movement control system based on the higher centers topographic organization. Later on, Farley and Clark (1954) produced a model to study the stimulus–response relationship in random networks. Such theories were used by Rosenblatt (1958, 1961), the author of the well-known PER-CEPTRON, which was defined as a probabilistic model to store information.

Within similar lines of thought, we should mention Widrow and Hoff (1960), Cainiello (1961), and Steinbuch (1961). The latter paper was the first mathematical model of associative memory; however, despite its significance and potential applications, it was essentially ignored. Rosenblatt's PERCEPTRON generated great interest since it exhibited certain learning abilities. However, Minsky and Papert (1969) demonstrated considerable limitations regarding the one-layer type, and pointed out to its lack of learning possibilities when dealing with some logic functions, such as the *exclusive or* (*XOR*).

Minsky and Paperts (1969) also added, but without demonstration, that more layers could not overcome these limitations; hence, expectations regarding the PERCEPTRON received a good blow. Nonetheless, at that time, it was known that two layers were enough to learn the XOR function. Years later, Funahashi (1989) showed that any continuous function in multiple variables can be represented by means of a three-layer ANN called *Backpropagation*. In fact, describing this theory evolution between 1969 and 1982 is not an easy task, even though a few papers come to memory. Malsburg (1973), for example, modeled the formation of the visual cortex-oriented columns. Thereafter, Grossberg (1976), Perez *et al.* (1975), and Nass and Cooper (1975) also simulated self-organization. Marr's papers on the cerebellum (1969) and neocortex (1970) must be mentioned; in them, the architecture is highly parallel and includes some learning rules. Other parallel models, also applied to the cerebellum, were developed by means of vectors and tensors (Pellionisz and Llinas, 1979).

In 1982, again Marr (see note below) discussed the parallel approach when linked to physiological processes and dealing with the early stages of the visual system. One of the pioneers in brain function theories was Arbib (1972). Papers by Cooper (1981) and Bienenstock *et al.* (1982) analyzed learning rules for the visual cortex obtaining rather good predictions. Kohonen (1987), in turn, defined and studied the so-called self-organized processes, in particular, topological maps formation due to sensory experience. Hopfield (1982) implemented a *content-addressable memory* as an emergent property of an ANN model. After Hopfield's paper, a huge number of contributions used ANN in different areas, especially Biology, Psychology, and other disciplines. From experimental data, ANN could lead to formal theories and, thereafter, validate in newer experiments the model emergent properties, which, in turn, would permit adjustments for better theories.

8.3.3 *Models of neurons and synapses*

In any neural network, the following three basic concepts require definition, namely,

(1) The neuron model,
(2) The way to compute the synaptic weights, and
(3) The network topology.

An artificial neuron, for example, consists of a node with inputs x_j to simulate the afferent dendritic discharge, and an output y_i to stand for the axon frequency discharge. The best known model of neuronal signal transmission is the traditional produced by Hodgkin and Huxley (1952), composed of a set of nonlinear ordinary differential equations. Even though this model is the most precise, its computation becomes too slow. Neurons can be simulated as simple linear integrators of their respective inputs (Mountcastle, 1967), a way to do that is by the *integrate and fire* model. In many cases, one can use a much simpler model, like that schematized in Fig. 8.6. There, it is assumed that each synapsis is independent from the others, each neuron output can be obtained as a nonlinear function of the inputs' sum, multiplied by the synaptic weight (Cooper, 1973; Nass and Cooper, 1975).

$$y_i(t) = F[S_i(t)] \tag{8.1}$$

$$S_i(t) = \sum_{j=1}^{m} x_j(t)w_{ij}(t) - \theta_i \tag{8.2}$$

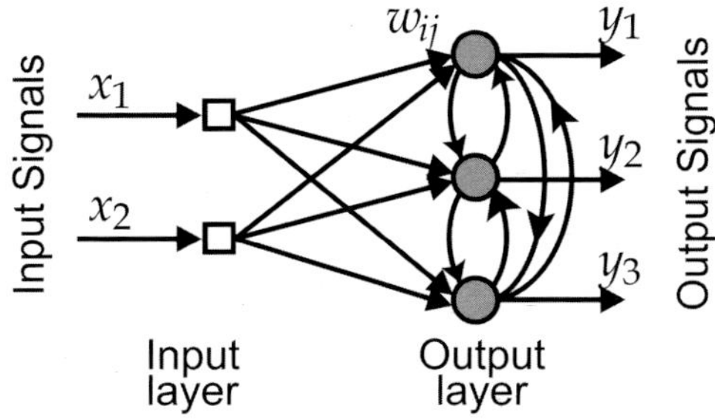

Figure 8.6: Scheme of a basic neural network. Each output (y_i) from a node (i), with m inputs (x_j), is obtained from Eqs. (8.1) and (8.2). This scheme also shows the lateral inhibition for a Kohonen neural network (see below). Drawn by Gustavo Idemi.

where m is the number of neuron inputs, x_j simulates the dendritic frequency discharge, and y_i stands for the axon's discharge rate. Both variables range lies between 0 and 1. The synaptic weight w_{ij}, from the input j to neuron i, takes values between -1 and 1. The variable t is the iteration number at the time of the simulation. Finally, F is a nonlinear function and θ_i is the threshold value of neuron i.

Some demonstrations show that learning and memory would not be viable, when the output y_i is not affected by a nonlinear function F. According to Guyton (1986), several nonlinear functions are acceptable but, largely, the most common is the sigmoid one. For each node, and for a given value of S_i, the output from the neuron y_i is determined by the function F, however, it could follow a probability distribution.

As mentioned before, the computation of the weights is an important point to solve. Variants to compute the LTP are used many times based on Hebb's postulate (1949), which says that if two neurons on either side of a connection are activated in synchrony, then the weight of that connection is increased. In this way, the synaptic weight w_{ij} at the input j of the neuron i increases if, and only if, there is simultaneous activity at the output y_i, and at the input x_j. The network weights are updated, according to,

$$w_{ij}(t+1) = w_{ij}(t) + \Delta w_{ij}(t) \tag{8.3}$$

$$\Delta w_{ij}(t) = \alpha y_i(t) x_j(t) \tag{8.4}$$

where $\Delta w_{ij}(t)$ is the incremental value of the synaptic weight at the iteration t and α is a constant with values between 0 and 1.

Thus, so formalizing Hebb's postulate, there are only unidirectional changes, since w_{ij} cannot decrease, this means that weights would reach a maximum level (hence, canceling the learning capacity). This difficulty finds different ways out (Kohonen, 1987), for example, assuming a limitation of resources leads to competition among synapses and an *active forgetfulness* that increases as the neuron y_i activity goes up. Its formalization would state,

$$\Delta w_{ij}(t) = \alpha y_i(t)[x_j(t) - w_{ij}(t)] \tag{8.5}$$

Equation (8.5) is simple and useful in many simulations to compute LTP and LTD.

8.4 Sensory and Motor Systems

8.4.1 *Somatic sensory system*

The somatic sensory system (SSS) selected by evolution is amazingly similar to robots built by engineers. It gives the body the ability to identify shapes and textures of objects, to monitor the internal and external forces acting on the body at any moment, and to detect potentially harmful circumstances. In a few words, to sense all the stimuli that can affect in one or other way its life continuity. Receptors are specialized neurons. The SSS usually follows the classification described in the next sections (Purves *et al.*, 2011).

8.4.1.1 *Mechanoreceptors and proprioceptors*

Tactile System

Tactile system: Tactile sensory processing of external stimuli starts with the activation of a diverse population of cutaneous and subcutaneous sets of receptors (mechanoreceptors) at the body surface that relay information to the CNS for interpretation and eventual ultimate action. Tactile information travels to the brain via several ascending pathways that run in parallel through the spinal cord, brainstem, and thalamus to reach the *primary somatic sensory cortex* in the *postcentral gyrus* of the parietal lobe. The primary somatic sensory cortex projects in turn to higher order association cortices in the *parietal lobe* and back to the *subcortical* structures involved in mechanosensory information processing.

Proprioceptive system: Proprioception means "sense of self". Additional receptors located in muscles, joints, and other deep structures monitor mechanical forces generated by the musculoskeletal system. In the limbs, the proprioceptors are sensors that provide information about the joint angle, the muscle length, and the muscle tension. They provide information about the position of the limb in space. The muscle spindle is one type of proprioceptor that provides information about changes in muscle length. They are found throughout the body, in parallel with muscle fibers. The Golgi tendon organ is another type of proprioceptor that provides information about changes in muscle-tendon tension.

Auditory and vestibular system: The auditory system mediates much of human communication; indeed, loss of hearing in many respects can be more socially debilitating than blindness. From a cultural perspective, the auditory system is essential not only to understand speech, but also to music, one of the most aesthetically sophisticated forms of human expression.

The auditory system — one of the bioengineering masterpieces of the human body — is an array of miniature acoustical detectors. They can faithfully transduce vibrations as small as the diameter of an atom, and they can respond a thousand times faster than visual photoreceptors. The auditory portion of the inner ear is the *cochlea*. It is a spiral-shaped cavity in the bony labyrinth (in humans making 2.5 turns), it contains the receptors (mechanoreceptors) placed at the *organ of Corti*. It allows the transduction of acoustic signals into electrical ones. The transduction occurs through vibrations of structures causing displacement of cochlear fluid and movement of *hair cells* producing action potentials. Such rapid auditory responses to acoustical cues facilitate the initial orientation of the head and body to novel stimuli, especially those that are not initially within the field of view.

Even though it is not part of the traditional senses, the peripheral portion of the *vestibular system* plays a significant role, too, as it includes inner ear structures that function as miniaturized accelerometers and inertial guidance devices, continually reporting information about self-motion perception, head position, and spatial orientation relative to gravity (Valentinuzzi, 1980). There are two types of structures, *utriculus* and *sacculus*, that are involved in the sense of gravity and acceleration, and the semicircular channel that is involved in detecting the rotation of the head in three orthogonal planes of space. The receptors are *hair cells* that transduce movement to action potentials. The information gets into brainstem and cerebellum centers, and into somatic sensory cortical levels where processing takes place. Besides, the vestibular system has important sensory functions, contributing to important motor functions, including vestibular nuclei, which also directly innervate motor neurons controlling extraocular, cervical, and postural muscles.

8.4.1.2 *Photoreceptors: Visual system*

At a single glance, either in humans or in animals, it is sufficient to describe the location, size, shape, color, and texture of objects, and if the objects are moving, their direction and speed. Today, it is not possible to do that with a computer basically because of the parallel arrangement and the way the information is processed. The retina has more than 100 million receptors (photoreceptors), and only a few steps or synapses are necessary to a conscious perception of the visual scene.

The retina is a structure with several layers of neurons interconnected by synapses. There are two types of photoreceptors: 120 million rods and 6.5 million cones. Rods provide black-and-white and low-light vision, while cones support daytime vision and the perception of color. Equally remarkable is the fact that visual information spreads over a wide range of stimulus intensities. The human visual system is extraordinary in the quantity and quality of information it supplies about the world. The primary visual pathway goes from the retina to the dorsal lateral *geniculate nucleus* in the thalamus and, from there, to the *primary visual cortex*. Parallel processing of different categories of visual signals continues in cortical pathways that extend beyond the primary visual cortex, supplying a variety of visual areas in the *occipital, parietal,* and *temporal lobes.* Visual areas in the temporal lobe are primarily involved in object recognition, whereas those in the parietal lobe are concerned with motion. Normal vision depends on the integration of information in all these cortical areas.

8.4.1.3 *Sensitive chemoreceptors: Olfactory and gustatory systems*

The chemosensory system relies on receptors (chemoreceptors) in the nasal cavity and mouth that interact with relevant molecules to generate and transmit action potentials to appropriate regions of the CNS. Odors and taste provide information about food, oneself, other animals, plants, and many other environmental aspects.

Olfactory and gustatory information can influence feeding behavior, social interactions and, in many animal species play a role, too, in the reproduction process. The taste (or gustatory sense) system detects alimentary elements that provide information about the quality, quantity, and together with olfactory system, safety of the ingested food. Both directly link to the most primitive areas of the CNS to accept or reject stimuli depending if they are appetitive or aversive.

8.4.2 *Motor system*

Skeletal muscle contraction is initiated by lower motor neurons (α motor units) in the spinal cord. Descending pathways from higher centers comprise the axons of upper motor neurons and modulate the activity of lower motor neurons by influencing this local circuitry. Upper motor neuron cell bodies are located either in cortical or in brainstem centers (such as the vestibular nucleus, the superior colliculus, and the reticular formation). The upper motor neuronal axons typically contact the local circuit neurons in the brainstem and spinal cord, which link, in turn, to the appropriate combinations of lower motor neurons. The local circuit neurons also receive direct inputs from sensory neurons, thus mediating important sensory motor reflexes that operate at the level of the brainstem and spinal cord. Therefore, lower motor neurons are the final common pathway for transmitting information from a variety of sources to the skeletal muscles (Purves *et al.*, 2011).

The *motor frontal lobe* and *premotor* areas are responsible for planning and precise control of voluntary movement's complex sequences. Most upper motor neurons, regardless of their source, influence the generation of movements by directly affecting the activity of the local circuits in the brainstem and spinal cord. Upper motor neurons in the cortex also control movement indirectly, via pathways that project to the brainstem motor control centers, which, in turn, project to the local organizing circuits in the brainstem and cord. An important function of these indirect pathways is to maintain the body's posture during cortically initiated voluntary movements.

The axons of upper motor neurons descend from higher centers to influence the local circuits in the brainstem and spinal cord that organize movements by coordinating the lower motor neurons' activity. The sources of these upper pathways include several brainstem centers, and a number of cortical areas in the frontal lobe. As in engineering, to control a process that demands much computation time, the principal processor does not spend time in routine calculation. The coprocessor does it. Natural selection chose a similar mechanism for the brain. When a football player kicks a ball, he does not think in the way to put the foot; he just concentrates on where the ball has to go. The structures involved in this learning are the *basal ganglia* and *cerebellum*. In contrast to the upper motor neurons, the efferent cells of the basal ganglia and cerebellum do not project directly, neither to the local circuits of the brainstem and spinal cord that organize movement, nor to the lower motor neurons that innervate muscles. Instead, they influence movements by modifying the activity patterns of the upper motor neurons in the cortex.

The basal ganglia neurons respond in anticipation of any movement, and their effects on upper motor neurons are required for the normal course of voluntary movements. In addition, they are involved in language and cognition. As stated before, the main components are the *dorsal striatum* (caudate nucleus and putamen), *ventral striatum* (*nucleus accumbens* and olfactory tubercle), *globus pallidus*, *ventral pallidum, subthalamic nucleus*, and the *substantia nigra pars compacta*. The basal ganglia synaptic plasticity depends on the substantia nigra pars compacta's dopamine.

The primary function of the cerebellum is to detect the difference, or motor error, between an intended movement and the actual movement, and to reduce the error through its projections to the upper motor neurons. These corrections are attained both during the course of the movement and as a form of motor learning when the correction is stored. The cerebellum sends prominent projections to virtually all upper motor neurons. Structurally, the cerebellum has two main components, that is,

(1) A laminated cerebellar cortex.
(2) A subcortical cluster of cells referred to collectively as the deep cerebellar nuclei.

Pathways that reach the cerebellum from other brain regions project to both components; thus, the afferent axons send branches to both the deep nuclei and the cerebellar cortex. The output cells of the cerebellar cortex project to the deep cerebellar nuclei, which give rise to the main efferent pathways that leave the cerebellum to regulate upper motor neurons in the cerebral cortex and brainstem. Hence, much like the basal ganglia, the cerebellum is part of a vast loop that receives projections from and sends projections back to the cerebral cortex and brainstem.

8.5 Cortical Structure Organization

8.5.1 *The visual, tactile and auditory cortical systems as examples*

Experimentally, characterization of the stimuli eliciting strong responses from cortical sensory neurons refers to the stimuli to which they exhibit a strong response. It is well known that, for example, in the visual cortex, neurons respond maximally to a moving light bar with a specific orientation. Hubel and Wiesel (1962) found that all neurons encountered in an electrode penetration perpendicular to the surface at a particular point would very likely to have the

                        B. S. Zanutto and E. Cynowiec

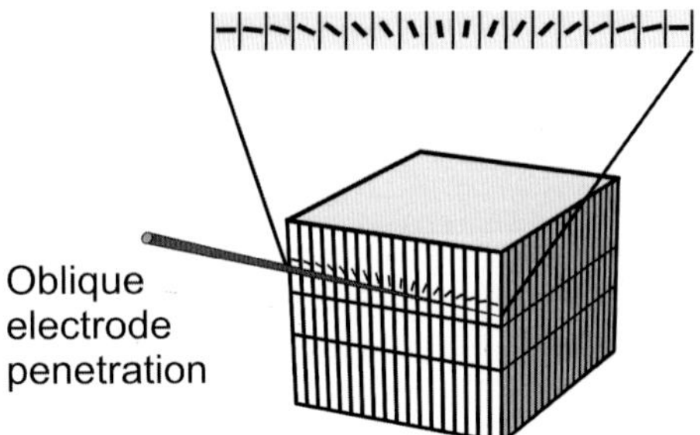

Figure 8.7: Tonotopic map. Column arrangement of neurons for receptive fields in the visual cortex. The dark sharp pointed arrow shows the oblique electrode penetration. A similar figure is freely available in Internet and in several textbooks. Drawn by Gustavo Idemi.

same orientation preference, forming a "column" of cells with similar response properties. When the stimulus orientation changes, the cell activity decreases Adjacent columns have slightly different orientation preferences; the sequence of orientation preferences along a tangential electrode penetration gradually shifts as the electrode advances forming a *tonotopic map* (Fig. 8.7).

This observation has led to the idea that cortical cells can be grouped into "columns" of neurons with similar properties. The exact size and anatomical definition of a cortical column is a matter of debate, but each is likely to contain several thousand neurons with similar receptive fields (Purves *et al.*, 2011). This population of neurons defines a *cluster of neurons*.

This arrangement of different cell response to different stimuli in cortical column can be seen in other sensory cortical areas, too. The characterization of neuronal receptive fields in the auditory system is analogous to that found in the visual cortex. For example, auditory cortical neurons can be characterized by stimulation with pure tones. Each cluster of neurons has its preferred frequency and neighboring neurons have similar preferences. In the somatosensory cortex this effect shows up again, that is, neurons that respond strongly to touch (say, fingers) are located close to each other. The concepts of receptive field and optimal stimuli are not restricted to cortex, but are also found in retinal ganglion and thalamic cells, in the olfactory bulb, or also in insect sensory systems. The properties exhibited by cortical neurons raises the question of how neurons with different receptive fields arrange themselves in a cluster of neurons.

8.5.2 *Model of neuronal cluster organization: Effect of the lateral interaction*

Lateral inhibition refers to the inhibition that neighboring neurons in brain pathways have on each other. Kohonen (1987) studied properties of this type of neural networks and made a model that can simulate a possible mechanism of how cortical cells can be grouped into columns of neurons with similar properties. He called it *self-organizing* networks. The topology is shown in Fig. 8.6. In that model, the lateral feedback is considered in the general case, with excitatory or inhibitory connections. This is achieved modulating synaptic weights neurons with a *Mexican hat* function as shown in Fig. 8.8.

This process can be simulated in a simplified manner replacing the Mexican hat function by Fig. 8.9. The output from each neuron is computed with the following function:

$$\eta_i(t) = \sigma \left[\varphi_i + \sum_{k=-L}^{+L} \gamma'_k, \eta_{i+k}(t-1) \right] \tag{8.6}$$

where L stands for the number of the last neighbor neurons involved in the lateral interaction.

When this process is simulated, a cluster of high activity (high frequency of action potential) shows up. It is centered on the neurons with maximum activity (the lateral interaction is not considered); beyond the cluster, all neurons are silent. The procedure let us to understand how lateral interaction generates clusters activated by specific stimuli (such as a light bar).

8.5.3 *Model of the tonotopic map*

The next phenomenon of interest to simulate is the tonotopic map formation found in all sensory systems and in many motor systems too. Kohonen (1987) simulated this topology, in which it is assumed that there is lateral inhibition. Because of that, for an input pattern

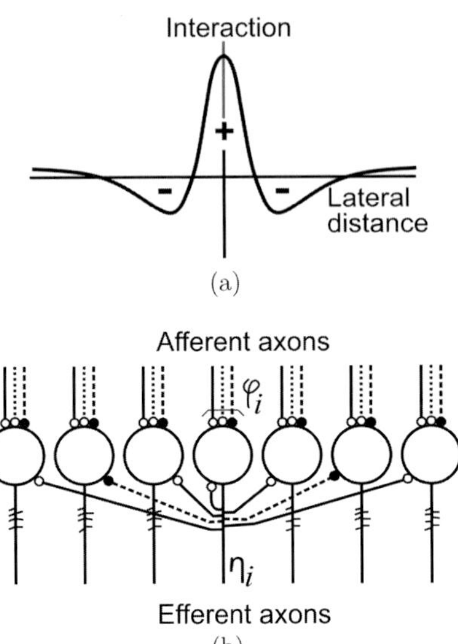

Figure 8.8: Mexican hat model. A, upper part, shows the synaptic weight value of the lateral interaction as a function of distance. Positive value: excitation. Negative value: inhibition. B, lower part, is a schematic representation of lateral connectivity, which may implement the function shown at A. Open (small) circles: excitatory synapses. Solid circles: inhibitory synapses. Dotted line: polysynaptic connection. The variable φi is the frequency of action potential of external axons and η_i stands for frequency of axon outputs action potential (Kohonen, 1987).

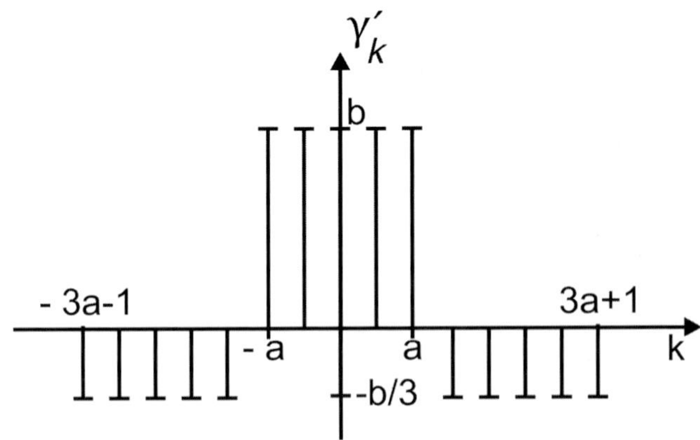

Figure 8.9: Graphic approximation of the Mexican hat function. The symbol γ_k represents values of neighbour neurons' synapses weight.

there is a cluster of a size of "p" neurons, centered in the neuron with maximum activity (maximum frequency discharge), called the *winning neuron*. In this way, the input vector $\boldsymbol{X}$, with m inputs, x_j, activates the Kohonen network and the neuron with maximum activity can be located using the minimum-distance Euclidean criterion. At iteration t, the *winning neuron* is called the *winner-takes-all* computed as the best matching neuron i_c, of all the "N" neurons of the network described by

$$i_c(t) = \min_i \|\boldsymbol{X} - \boldsymbol{W_i}\| = \min_i \left\{ \sum_{j=1}^{m} [x_j(t) - w_{ij}(t)]^2 \right\}^{1/2} \tag{8.7}$$

where x_j and w_{ij} are the jth elements of the vectors $\boldsymbol{X}$ and $\boldsymbol{W}_i$, respectively.

The synaptic weights $w_{ij}(t)$ are updated by Eqs. (8.3) and (8.5). Because here it is assumed that there is a cluster with the maximum output, all the p neurons inside the cluster have the output value $y_i = 1$ and 0 out of the cluster. Therefore, Eq. (8.5) takes the form of Eq. (8.8), that is,

$$\Delta w_{ij} = \begin{cases} \alpha(x_j - w_{ij}) \text{ if neuron } i \text{ is inside the cluster} \\ \qquad \text{centered at the neuron } i_c \\ 0 \qquad \text{if neuron } i \text{ is out of the cluster} \end{cases} \tag{8.8}$$

This simple model can simulate the tonotopic map formation found in all sensory and in many motor structures.

8.6 Motivational Processes and Attention

8.6.1 *The limbic system*

The limbic system is the brain place of subjective feelings associated with physiological states known as emotions. Although everyday feelings and emotions are as varied as happiness, surprise, anger, fear, and sadness, they share some common characteristics. The same forebrain structures that process emotional signals participate in a variety of complex brain functions, including rational decision-making, the interpretation and expression of social behavior, and even moral judgments. Historically, the higher order neural centers that coordinate emotional responses have been grouped under the rubric of the limbic system. Papez (1937) proposed a specific brain circuit to explain emotion. As time proceeded, the circuit for the control of emotional expression first elaborated suffered deep revision. Some of the structures of the original circuit (the hippocampus, for example) now appear to have little to do with emotional behaviour, whereas the amygdala, hardly mentioned by Papez, clearly plays a major role in the experience and expression of emotion. The actual limbic system includes parts of the orbital and medial prefrontal cortex, hypothalamus, ventral parts of the basal ganglia, the mediodorsal nucleus of the thalamus (a different thalamic nucleus than the one emphasized by Papez), and the amygdala. This set of structures, together with the parahippocampal gyrus and cingulate cortex, is generally referred to as the *limbic system* (Purves *et al.*, 2011).

8.6.2 *Motivation*

8.6.2.1 *Feeding: Prototypic motivational system*

What is being motivated towards something? What does it mean when somebody says feeling the inner drive to take a given action or project? Why do rats, people, and many other animals eat in bouts (meals) rather than continuously? How are they able to regulate not only total intake over a long period, but also within a meal? Is there a simple physiological explanation for these facts?

External factors such as taste (the evolutionary predictor of food quality), previous learning, habits, social situation, stress, emotions, and perhaps many others, all in a complex interplay, bring together a not fully understood set. Most omnivores, such as rats and guinea pigs — as well as human beings — eat not at fixed intervals of time, or at random times, but in more or less regular meals. To model this dynamic, the first step is to find the neuropsychological components that correspond to parts of the control circuit for feeding. It is known that endogenous factors that modify the size of an on-going meal are called *satiety signals* (SSs). Those factors are generated during and after a meal and provide information to the brain that inhibits feeding and leads to meal termination. The SSs are generated in the gastrointestinal tract and abdominal viscera, as well as in the oral cavity (Ritter *et al.*, 1994; Travers and Norgren, 1987). They provide information about mechanical (e.g., stomach stretch, volume) and chemical properties of food (e.g., via peptides such as *cholecystokinin, ghrelin,* and *peptide YY* [PYY]), which have been linked to short-term (within-a-day) feeding behaviors. Amylin and glucagon, which are secreted from the pancreatic islets during meals, also reduce meal size. SSs are relayed to the hindbrain, mainly to the *nucleus tractus solitarius* (NTS), either indirectly via nerves from the gastrointestinal tract, especially the *vagus* (e.g., *cholecystokinin* and *glucagon*), or else circulate via the blood and interact with local receptors in the hindbrain, as, e.g., *amylin* (Woods, 2004).

The hindbrain (mainly the NTS) also receives, via several *hypothalamic nuclei,* signals that reflect the fat mass of the body. Best known are the adiposity signals given by *leptin* and *insulin,* secreted into the blood in direct proportion to the amount of stored body fat. Leptin comes from fat cells (adipocytes) in direct proportion to the amount of stored fat and insulin is secreted by pancreatic β cells in response to increases of glucose. Moreover, basal insulin in the absence of elevated glucose, as well as every increment of insulin above baseline during meals is in direct proportion to total body fat or adiposity (Havel *et al.*, 1996).

Obese individuals show high basal insulin, whereas lean individuals have relatively low levels of circulating leptin. Both levels are good indicators of body fat, and both hormones are able to enter the brain from the blood and stimulate specific neural receptors. These two adiposity signals have been linked to long-term weight regulation, meaning over months and years (Marx, 2003). Some intestinal-related peptides are also long-term regulators. Rodents and humans with reduced PYY levels in response to food intake tend to obesity.

These data suggest that there is a large degree of redundancy in the orexigenic (appetite-stimulating) pathways, showing an evolutionary bias toward energy storage. Despite this redundancy, the neurophysiological pathways suggest that feeding is regulated by a feedback loop, where the hypothalamus provides the long-term regulatory input to the NTS that acts as the set point. It also receives SSs as feedback inputs, acting as short-term regulators.

Many areas of the brain are sensitive to long-term regulators. *Leptin* receptors have been found in *paraventricular nucleus* (PVN) and *lateral hypothalamic area* (LHA) neurons, implicating them as direct targets for regulation by circulating adiposity signals. PVN stimulation inhibits food intake, whereas the opposite is true of stimulation of the LHA (Fan *et al.*, 2004).

The NTS has been identified as a *satiety center*. It is feedback by inputs transmitted through there different pathways, the vagus, sympathetic fibers, and blood, and receives inputs from the hypothalamic pathways involved in energy homeostasis. In this way, the NTS could be functioning as a comparator in a feedback loop, and the hypothalamic inputs would work as the feedback loop set point. If that were the case, the NTS is a crucial component in the feeding regulation. It was found in rats that when the NTS is lesioned, they eat less and even may starve to death.

8.6.2.2 *Model of motivation*

This section shows a feedback model developed by Staddon and Zanutto (1997) for food intake regulation that readily may be interpreted in terms of existing neural data (Fig. 8.10).

As said earlier, SSs provide information about mechanical gastric characteristics as well as chemical properties of food. High-fiber (low-caloric) food will have a value of the satiation variable "X" similar to that of other food of comparable caloric value, but with greater volume; consequently, the stomach-stretch and volume components of SSs will increase faster than with other food. Thus, people will end a meal of high-fiber food sooner than a low-fiber meal. Fasting does cause changes in metabolism, which becomes more efficient in converting food into energy. An obvious possibility, therefore, is that a short diet that involves some fasting or much-reduced eating will only produce a temporary weight loss. Lost weight may be restored when the dieting regimen ends.

The role of external factors such as taste, learning and habits, social situation, stress, and emotion is to modulate the set point

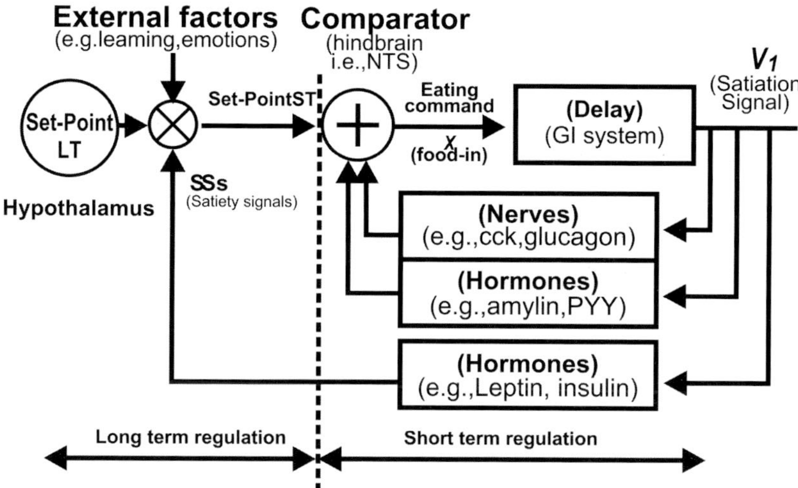

Figure 8.10: Block diagram of a model of long- and short-term regulation of feeding. Eating leads to delayed SSs. When the feedback SSs falls below the Set-Point$_{ST}$, the command is *eat*, otherwise, the command is *stop eating*. The hypothalamic Set-Point$_{LT}$, is fedback by adiposity signals and modulated by external factors (as learning) to generate the Set-Point$_{ST}$. Drawn by Gustavo Idemi.

for long term (Set-Point$_{LT}$), yielding an output that comprises Set-Point$_{ST}$ (Short Term Set Point), which, in turn, affects the initiation and termination of meals. In this way, a very tasty food would increase the Set-Point$_{ST}$ so that meal size would be greater, and vice versa, if food is not tasty.

In conclusion: Controlling feeding is equivalent to controlling the motivation for eating. In the model, ingested food inhibits eating, but the inhibitory effect is delayed; so, if one wants to lose (or not gain) weight, eat slowly or in small bites. In this way, you will give the process of satiation time to catch up, rather than outrunning it by eating fast.

The model explains both eating-rate regulation and the broad features of meal duration and timing. It explains why rats adapt to changes in reward size by adjusting meal size rather than inter-meal interval (IMI), and why the interruption of feeding affects primarily the first post-interrupted meal. It also explains the effects on eating rate when imposing a minimum inter-pellet interval as well as other operant-schedule constraints in a wide variety of closed-economy experiments. It also accounts quantitatively for the complexities of meal–inter-meal correlations. If Set-Point$_{ST}$ increases, the meal size will increase and IMI will decrease (and vice versa if Set-Point$_{ST}$ decreases). Hence, in short, simulations run by Zanutto and Staddon, 2007 made out three simple points, that is,

(1) Feeding regulation in rats is bang-bang rather than proportional control. Eating is regulated in an on–off fashion under most conditions rather than being proportional to the difference between Set-Point$_{ST}$ and SSs.

(2) The satiating effects of eating are delayed in a way that can be modeled by a simple first-order linear system.

(3) Long-term regulation (i.e., control of body weight) is separable from short-term regulation (i.e., control of meal pattern). The former is controlled by a Set-Point$_{LT}$ provided by the hypothalamic input, while the latter depends of the comparison of the Set-Point$_{ST}$ and the SSs that vary from minute to minute rather than over days.

8.6.3 *Attention*

We have experienced that our brain allows a few simultaneous thoughts. How are some thoughts favored over others? Attention is not well understood, but recent work has naturally evolved toward efforts at a more integrative understanding. The best-understood model is visual attention. It suggests that focusing attention arises from interactions between widespread cortical and subcortical networks that may be regulated via their rhythmic synchronization. Attention can be focused volitionally (intelligent behavior) by *top-down* signals derived from task demands, and automatically by *bottom-up* signals from salient stimuli, as brightly colored or fast moving objects (Miller and Buschman, 2013). The consensus across a large number of studies, including both animal electrophysiology and human imaging work, suggests that visual attention stems from a *frontal–parietal* control network acting on visual cortex. Two cortical areas are well studied. The *Frontal Eye Fields* are involved in locating a target presented among distractors or maintaining a representation of target location during a delay period. The second one is the *Lateral Intraparietal Cortex* involved in eye movement, wherein it was found that the electrical stimulation evokes saccades (quick movements) of the eyes (Arcizet *et al.*, 2011).

Neurophysiological studies suggest that top-down signals may originate from the frontal cortex. Microstimulation of the Frontal Eye Fields produces top-down attention-like modulation of visual area V4, crucial for visual object recognition (Moore and Armstrong, 2003). The hypothesis is that top-down signals add energy to these circuits, boosting the synchrony of an attended object.

It is well known that synchronized oscillations (the "brain waves") vary with attentional focus. They are the result of interactions between specific regions. A possible effect of this interaction is to improve information transmission (Miller and Buschman, 2013). When two brain areas oscillate in phase, they are more likely to influence each other than if they are out of phase, and so they are less likely to influence each other, as some kind of synchronization. If sensory neurons tuned to the same stimulus, they would synchronize their firing, and that stimulus would show up stronger in downstream areas. Local synchrony may help the brain to improve its signal-to-noise ratio while, at the same time, to reduce the number of spikes needed to represent a stimulus, thus improving the information transmission by removing these common 'noise' sources (Tiesinga *et al.*, 2002; Siegel and Konig, 2003).

8.7 Brain Organization and Neuronal Communication

8.7.1 *Cortical modulation*

The brain has a population of hundreds of billions of neurons, so the variance of possible neural circuitry is huge. In spite of such enormous and interconnected size, it has an organization that makes easier to study how the brain works. As said before, the spinal cord is well delimited; the posterior face handles perceptual signals, while the anterior side manages the motor actions. In the brain, instead, even though the posterior side mainly keeps a perceptual function, the frontal lobe is devoted to action and planning as shows (Fig. 8.11).

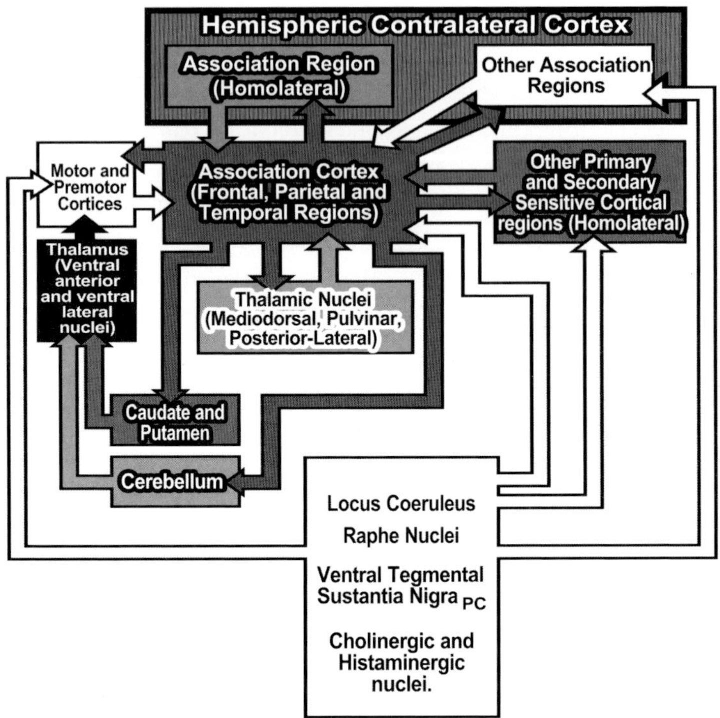

Figure 8.11: Main topological arrangements of neuroanatomy in the brain. The diagram shows bidirectional connection between different cortical regions and in the *thalamus* and *cortexes*. There are also unidirectional connections that leave from the *cortex*, enter in the *basal ganglia*, *cerebellum*, and *hippocampus*, and return to the *cortex*. Finally, the diagram shows diffuse projection over the brain originated in the *ventral tegmental area* and *sustantia nigra, pars compacta; locus coeruleus; rafe nucleus; cholinergic* and *histaminergic nuclei*. Drawn by Gustavo Idemi.

There are three major types of organizations (Edelman and Tononi, 2000).

8.7.1.1 *Organization with reciprocal connections between cortexes and cortico-thalamic structures*

Neurons within the same group interconnect in such a way that many of them respond simultaneously when triggered by an adequate stimulus; as said above, this phenomenon defines a *cluster*. Neuronal groups, placed in different sites but showing similar specificity, prefer mutual interconnection rather than with other groups. For example, as anticipated, neurons that visually respond to a vertical bar interconnect reciprocally and stronger than groups sensitive to other orientations. Moreover, groups nearby the visual field interconnect stronger than those that respond to distal positions. Separated and functionally different areas interconnect reciprocally. These bidirectional connections have been called *reentry* (Edelman and Tononi, 2000). In this way, and despite lacking central coordination regions, such areas can integrate themselves. The cortico-thalamic system gathers hundreds of functionally specialized areas, each with several thousands of neuronal groups, some processing motor responses, others related to planning and actual action, while still another group is devoted to abstract concepts.

8.7.1.2 *Organization of unidirectional connections between specialized cortical structures*

The cerebellum, the basal ganglia, and the hippocampus conform these structures. The cerebellum receives connections from the cortical levels, which after several synapses project to the thalamus and, back to the cortex again. Traditionally, the idea was that the cerebellum only controlled movement coordination and synchronization; however, it is involved in functions related to language. In turn, the cortex connects to the basal ganglia and, after several stages, projects to the thalamus and back to the cortex. These nuclei plan and execute complex movements and cognitive processes (as playing piano or developing a mathematical demonstration). The hippocampus receives inputs from different cortical areas, which project back to the cortex, in some cases to the same parts they came from. In this kind of particular and well-known arrangement, the function associated is with short- and long-term memory consolidation. These connections are unidirectional and show little lateral interaction. Apparently, they are involved in the execution of a variety of motor and complex cognitive routines. Most of them are isolated one from the other, helping in the speed and execution precision.

8.7.1.3 *Structures encoding the value system: Synaptic modulators*

The value system is a diffuse set of connections that originate in a relative small number of neurons in the brainstem and hypothalamus and then spread to most parts of the brain, modulating the activity of cortical neurons. The function of these widespread links is to distribute neurotransmitters to modulate the cortical neurons activity. Edelman and Tononi (2000) assign the diffuse projections of neurotransmitters as pertaining to the *value system*. It is defined as the information encoding the state and well-being of the whole organism, as well as the emotional responses to occurring events, such as novel and unpleasant stimuli to help the individual to environmental adaptation. This value system has been selected very early by natural selection to assure survival of the species. Significant neurotransmitters presents in that system are *dopamine* (DA) of the *ventral tegmental area* (VTA) and *sustancia nigra pars compacta* (SN_{pc}) neurons; *noradrenaline* (NA) of the *locus coeruleus* neurons; *serotonin* (5-HT) of the *Raphe Nuclei*, and *cholinergic* in two groups of projections, the *magnocellular basal forebrain*, and the *brainstem*. Besides, the *tuberomamillary nucleus* of the *posterior hypothalamus* is *histaminergic*.

The midbrain dopamine neurons encode a reward prediction error used to guide learning throughout the frontal and motor cortex and the basal ganglia. The noradrenergic system supplies NA throughout the CNS facilitating the processing of relevant or salient information. Substantial evidence indicates that the NA projection system regulates behavioral states and the dependent processing of sensory information. Tonic discharge correlates well with levels of arousal and demonstrates an optimal firing rate during good performance in a sustained attention task. Serotonin is an ancient neurotransmitter, and because of that, the probability of making contact creating new neuronal circuit throughout evolution was high. It is involved in the adaptive responses to environment changes. It mediates in emotional processing (the regulation of mood, appetite, and sleep), and in cognitive functions, including memory and learning. It is also involved in gastrointestinal regulation. *Acetylcholine* plays an important role in cognitive function, having impact in working memory, attention, episodic memory, and spatial memory function. *Histamine* is a transmitter involved in wakefulness and attention.

8.7.2 *Neuronal information*

The question of how we have higher brain functions as memory, cognition, or a simple perception, such as to see a color, has three domains. In first place, the subjective quality of individual experience (e.g., the redness of red, painful of pain), which is what the philosopher called *quale*; this is a subjective domain. The second, the physiological domain, is the result of brain functioning, and the third, the behavioral domain, is where data are reported and judged. The last two domains are verifiable between subjects. What occurs in the brain when a subject is reporting to see "red"? It is known that in that process specialized neural structures are activated. One is the *fusiform gyrus* (homologous to the inferotemporal cortex in monkeys); when this is lesioned, the individual loses the color awareness. Another structure is the lateral geniculate nucleus that contains the color-opponent neurons organized into three orthogonal axes. One become activated by wavelengths of red, and inhibited by green; another is activated by wavelengths of blue, and inhibited by wavelengths of yellow. The third axis responds to the amount of white light and is inhibited by darkness. To see graphically this effect, we can represent each axis corresponding to a set of neural groups whose firing can be 0 Hz when neurons are inhibited, 10 Hz when neurons are at its spontaneous firing, and 100 Hz to represent the maximum firing. When an individual sees pure red, for example, we would see 100 Hz in the red-green axis and 10 Hz on the blue-yellow and on the white-black axes. When this circuit is activated, the individual learned that this perception is called "red".

Given the complexity of the brain, it is necessary to find theories to explain experimental data in order to propose hypotheses. One way to study how the brain processes information is to analyze the electrical signals among neurons. This can be done in different ways, one very used is by recording action potentials with electrodes while animals perform a behavioral task.

Researchers found evidence that our brains use few cells to decode a given image (Quian Quiroga *et al.*, 2005). This does not mean that there is only one neuron encoding each image, instead, there is a group of neurons of specific areas storing many images, but not with the same neurons; an oversimplification of this process can be simulated by the Hopfield network (1982).

In some cases, instead of analyzing single neurons, an "average firing rate" is studied over many trials. This is a *spike-raster-plot*. Hence, neural signals in cortical areas that recognize faces, memories, or in bulbo-mesencephalic nuclei can be analyzed, where the prediction of conditioned stimulus (CS) is encoded. For example, the activation of face cells has been registered in the inferior temporal cortex of a rhesus monkey. The cell responds to different monkey faces, as long as they are complete and viewed from the front; the same cell also responds to a bearded human face but not as efficiently, and it does not elicit any response to irrelevant stimuli (Desimone *et al.*, 1984).

With the same technique, neurons involved in short-term memory (STM) near the *principal sulcus* of the frontal lobe during a delayed response task can be registered. For example, in an experiment with a monkey, it can be determined whether he is able to find where food is placed in one of two wells, after a screen is lowered for a standard time (when the wells are not on sight). After the screen is raised, the monkey is allowed to uncover only one well to retrieve the food. Normal monkeys learn this task quickly, usually performing at a level of 90% correct after less than 500 training trials. When the neurons in the *prefrontal cortex* are registered during the delayed response task, neurons begin firing when the screen is lowered and remain active throughout the delay period. When no food is presented, the same neuron is relatively silent (Goldman-Rakic, 1987).

There are small nuclei with a diffuse set of connections which originate in the brainstem and hypothalamus, spread to most parts of the brain, and they modulate the activity of the cortical neurons. In monkeys, dopamine neurons' output can be registered in a classical

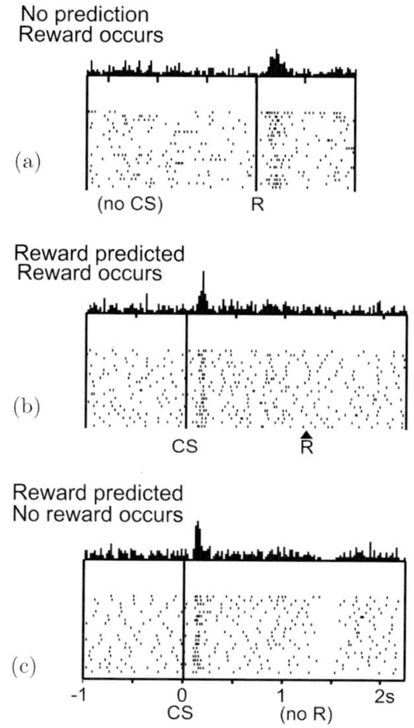

Figure 8.12: Changes in dopamine neurons' output when animals receive a small quantity of fruit juice into the mouth as liquid reward. The three figures are shown each one with the raster plot its histogram. The vertical axis shows 20 trials, where each dot represents an action potential. In the horizontal axis are shown at what time the stimuli are presented. (A) Before learning, a drop of fruit juice given in the absence of prediction. (B) The dopamine neuron is activated when the CS is presented, encoding the positive error in the prediction of reward (occurrence of juice). After learning, the CS predicts reward, and the reward occurs according to the prediction (no error). The dopamine neuron is activated by the reward-predicting process. (C) After learning, the CS predicts a reward, but if the reward fails to occur because of a mistake in the behavioral response of the monkey, the activity of the dopamine neuron is depressed exactly at the time when the reward would have occurred. CS and noCS are Conditioned and no CS, respectively. R and noR are Reward and no Reward, respectively (Schultz *et al.*, 1997). A similar figure is freely available in Internet and in several papers. Drawn by Gustavo Idemi.

conditioning task (Fig. 8.12). In that experiment, after pairing visual cues, the CS is followed (a few seconds later) by a reward, the *uncon-ditioned stimulus* (US). Here, the reward is a small quantity of fruit juice into the mouth. After pairing was repeated, dopamine neurons change the activation from the time of reward delivery, to the time of cue onset (Schultz *et al.*, 1997; Schultz, 1998).

The experiment shows that the firing rate of the dopaminergic structures: VTA ad SN_{pc} encode the prediction of the USs (Schultz *et al.*, 1997; Schultz, 1998,). Predictions give an animal time to learn behavioral reactions and can be used to improve the choices an animal will make in the future. The dopamine modulate the activity of the cortical neurons with the *value system* information.

A central question in neuroscience is how the nervous systems encode and process information. The most frequently studied process, perhaps because of the methodology involved, is when a neuron encodes information varying the firing rate (frequency of action potentials). In this case, the coding would have limitations from both, the information transmission and the metabolic perspective. In that case, the firing rate code would require that a neuron change (increase) considerably their firing rate to get capacity of representation. This requirement implies a very high metabolic cost, because the recovery of the ion gradient dissipated in the issuance of a spike is very expensive energetically (Laughlin *et al.*, 1998; Lennie and Place, 2003). From a bioengineering point of view, the information contained in the highly variable neuronal activity can be quantified by means of information theory, with commonly used measures of entropy and mutual information. The entropy measures the average amount of information needed to characterize a stochastic variable, while the mutual information measures the amount of information that is shared by two or more variables. Thus, the total capacity of information from a set of neurons can be estimated by calculating entropy, while information sharing group of neurons or between a group of neurons and a stimulus can be estimated by calculating mutual information (Effenberger, 2013).

8.8 Cognitive Functions: Learning and Behavior

8.8.1 *Learning: Classic and operant conditioning*

The behavior is an emergent property learned and controlled by the brain that occurs because of the individual–environment interaction.

John Locke (1632–1704), English physician and philosopher, in *An Essay Concerning Human Understanding* (1690) found one of the basic concepts of Experimental Psychology: *Human behavior is based on two desires, i.e., to avoid pain and to seek pleasure.*

The behaviour of all animals, from protists to humans, is guided by its consequences (Staddon, 2016). The bacterium obtains food by chemical gradient, a dog by asking for bones, a baby gets milk by sucking. This behavior is not learned, but when the baby grows, he/she will learn that getting good grades at school will be rewarding, and also he can be conditioned to buy things he might not need. Goal-oriented behaviors like these are examples of operant conditioning (OC, also known as instrumental conditioning). The latter is a process by which humans and animals learn to behave in such a way as to obtain rewards and avoid punishments (examples abound); even Charles Darwin, in the *Descent of Man*, published in 1871, described many of them.

Behavioral experiments suggest that learning finds a driving force in the expectation about the future and possible salient events, mainly reward and punishment. In *operant* and *classical conditioning*, the CS anticipates the US. As Staddon has pointed out, animals act as the CS allows them to elaborate an expectation or prediction of the US. As mentioned, Schultz (1998) found the neural substrates of prediction and reward. The OC is the prototypical learning responsible for much of the behavior of animals (including humans). Via this learning (or variants), for example, boys who "pee at night" are taught to control sphincters or to cure phobias, and less useful applications, such as in advertisements design or in the day of the week to pay the salary in order to increase productivity.

8.8.2 *Operant learning models*

Herein, we show how to simulate a real-time computational theory to find hypotheses in order to explain the most relevant behavioral paradigms of *Experimental Psychology*, such as simple operant learning, visual discrimination and perceptual and logical categorization. The model of Lew *et al.* (2008) is simulated with a biologically

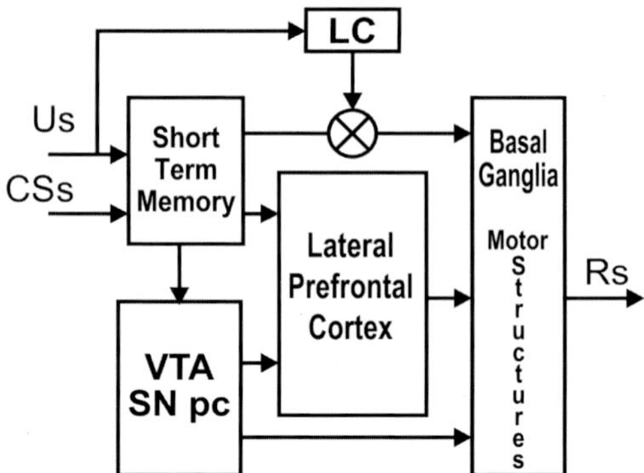

Figure 8.13: Oversimplified scheme modeling nervous structures and their connections between the CSs and US inputs and Rs outputs. VTA: ventral tegmental area. LC: locus coeruleus. SNpc: sustantia nigra pars compacta. US: unconditioned stimulus. CSs: conditioned stimuli. Rs: possible subject's responses. Drawn by Gustavo Idemi.

plausible neural network (Fig. 8.13), where the inputs are all the CSs and the US, and the outputs are all the possible responses "Rs" (i.e., press a lever, chose a path in a maze, walk or run).

STMs are computed from the inputs (CSs) and the US in later stages of the visual pathway, inferotemporal cortices, posterior parietal cortex, hippocampus and amygdala, as well as in ventromedial prefrontal cortex (Fuster and Jervey, 1982; Kesner *et al.*, 2001; Fiorillo *et al.*, 2003). Those STMs are inputs to VTA and SNpc in order to generate the US predictions, and both signals are inputs to the lateral prefrontal cortex, basal ganglia and motor cortices, and the pre-motor and motor cortices BG–PMC–MC (Cavada *et al.*, 2000). Since activity changes in BG, PMC, and MC occur at the same time course during operant learning, these structures are simulated as a single layer of neurons (Brasted and Wise, 2004). The outputs of the last structures are all possible motor responses. Finally, the *locus coeruleus* block represents the modulation exerted by this nucleus over direct input-output synapses.

The temporal difference algorithm (Sutton and Barto, 1990) simulated firing rates of dopamine neurons of the VTA and SNpc. Dopamine effects on PFC and BG–PMC–MC pyramidal neurons are related to modifications of synaptic efficacy via LTP and LTD

(Reynolds and Wickens, 2002; Pan *et al.*, 2005). The equations used in the simulation are based in those previously exposed. The computational simulation can be found in Lew *et al.* (2008) and in Rapanelli *et al.* (2015).

Signals that modulate cortices (VTA, SNpc, and LC) make up the *value system.* This model simulates the most relevant behavioral paradigms such as simple operant learning, *visual discrimination, delay matching to sample,* and *perceptual categorization,* but it cannot learn *equivalent classes.* To do that, it is necessary to include connections: first, the top-down modulation mechanisms found in monkeys during learning of association tasks and memory retrieval (Tomita *et al.*, 1999) and second, the effect of modulation of sensorial activity (Saalmann *et al.*, 2007; Taylor *et al.*, 2007; Ekstrom *et al.*, 2008). The model by Lew and Zanutto (2011) can simulate the previous experiment and equivalent relations. Because of that, the hypotheses of the model explain those experiments. Since this neural network is simple and works in real time, it is applicable to control autonomous systems like intelligent machines.

8.9 Neuroscientific Technologies in Bioengineering

Bioengineering is an interdisciplinary area for understanding processes and solve problems in biology, medicine, and veterinary using principles and methods of basic and engineering sciences. In this chapter, we have focused on the neural system and neural processes, which combined with technologies help solving neuroengineering problems. Just a quick browsing through the current literature offers the reader an overview of recent possibilities under study, such as in *neuromodeling*, neurorobotics and robotic surgical system, neuroelectromagnetic source imaging (NSI), neural prostheses, neuromechanical systems, prosthesis and computational biology, tissue engineering, brain–computer interface (BCI), exoskeletons, nanotechnology, and more. The list of new areas seems to be endless. Let us quickly browse what these new development try to accomplish.

Along this chapter, it was shown how to use *neuromodeling* to elaborate computational theories, to find hypotheses, and to explain higher functions of the brain. Another goal of *neuromodeling* is to find mathematical models that infer subject-specific mechanisms of brain disease from non-invasive measures of behavior and neuronal activity. These models aim at quantifying both physiological and computational principles that underlie adaptive cognition, such as aberrant learning and decision-making in individuals. A long-term goal is to use such models for a mechanistic re-definition of psychiatric and neurological diseases, leading to diagnostic classifications and individual treatment predictions. Areas that are more specific include the development of modeling techniques for inferring connectivity, synaptic plasticity, and neuromodulation from data obtained by techniques such as functional magnetic imaging (fMRI) or electroencephalography (EEG).

Neurorobotics, conceived as a combined study of neuroscience, robotics, and artificial intelligence, could be thought as the science and technology of embodied autonomous neural systems, understanding that neural systems include brain-inspired algorithms, computational models of biological neural networks, large-scale simulations of neural microcircuits, and actual biological systems. Such neural systems, embodied in machines, including robots, prosthetics, or wearable systems, are thinkable, indeed, at a smaller scale (micro-machines) and, at larger scales, as furniture and infrastructures. Neurorobotics is essentially based on the idea that the brain is embodied and the body is embedded in the environment. Therefore, most neurorobots are required to function in the real world, as opposed to a simulated environment.

NSI is devoted to modeling and estimating the spatiotemporal dynamics of the brain neuronal currents that generate the electric potentials and magnetic fields measured with non-invasive or invasive electromagnetic (EM) recording technologies. Unlike the images produced by fMRI, only indirectly related to neuroelectrical activity through neurovascular coupling, the current source density or activity images generated by NSI techniques are direct estimates of electrical activity in neuronal populations. There are algorithms that can localize simultaneously many active sources and even determine their variable spatial extents. The measured EM signals that are generated by the brain are thought to be primarily due to ionic current flow in the apical dendrites of cortical pyramidal neuron and the associated return currents throughout the volume conductor. Neurons that have dendritic arbors with closed field geometry (interneurons) produce no externally measurable signals. However, some non-pyramidal neurons do produce externally measurable signals. It is believed that some source localization methods can accurately image the activity of deep brain structures, such as the basal ganglia, amygdala, hippocampus, brain stem, and thalamus. On the other hand, there are also single neurons generating weak fields that may end up canceling each other. Thus, EM recordings are particularly suited for studying spatiotemporally coherent and locally synchronized collective neural dynamics (Ramirez *et al.*, 2011). Besides, EEG and magnetoencephalography provide an insight into neuronal processes in the brain in a real-time scale. Brain activity can be modeled in terms of a source distribution found by solving the bioelectromagnetic inverse problem. Such methods are particularly suitable to be used on modern highly parallel processing systems, such as widely available graphic processing units. These capabilities pave the way for online NSI. Pieloth *et al.* (2013) introduced a system that, according to its modular scheme, can be configured in a very flexible way using graphical building blocks.

Neural prostheses or neuroprosthetics appeared as a discipline related to neuroscience and biomedical engineering and concerned with developing prosthetic devices. They are sometimes contrasted with a BCI, which connects the brain to a computer rather than a device meant to replace missing biological functionality. Neural prostheses are a series of devices that can substitute a motor, sensory, or cognitive modality that might have been damaged because of an injury or a disease (cochlear implants provide an example). Through the replacement or augmentation of damaged senses, these devices intend to improve the quality of life for those with disabilities. Accurately probing and recording the electrical signals in the brain would better help understand the relationship among a local population of neurons that are responsible for a specific function. Neural implants are designed to be as small as possible, so as to minimize invasivity.

These implants typically communicate with their prosthetic counterparts wirelessly. Additionally, power is nowadays wireless received through the skin. The neuroprosthetic currently undergoing the most widespread use is the cochlear implant. Micera *et al.* (2010), in turn, focused on the least invasive interface, that is, transcutaneously, by using surface electrodes as interface between the stimulator and the sensory-motor system; say, a burst of short electrical pulses applied between pairs of electrodes over the skin, usually charge-balanced biphasic pulses to minimize tissue damage.

Nanorobots will be a great help in Neuroengineering, they would detect blockages in blood vessels or dissolve blood clot. They also detect foreign bodies (bacteria, viruses, and toxins) to help produce antibodies and detect cancer cells for diagnosis of diseases and even a treatment aimed at curing cancer. Nanorobots could be placed on damage tissue to assure proper repair. Moreover, in biotechnology, nanorobots could synthesize hormones, enzymes from amino acid. In a near future, bioengineers will have not only new developments in areas as biotechnology, nanotechnology, but also genetic information for each individual. It could be possible to know genetic predisposition to certain illnesses or health related behaviors. Hence, medicine will be much more personalized, and Bioengineering will have new tools and requirements for amazing developments.

8.10 Summary

The basics of the Nervous System were covered, first the neurophysiological principles were presented and, afterwards, the way to elaborate real-time models to simulate physiological experiments. We showed how high brain functions, such as learning and behavior, are emergent properties of plastic neural circuits. In addition, it was shown how these processes can be simulated with biological plausible models. Finally, a quick look through neuroscientific technologies was introduced.

References

Arbib MA. The metaphorical brain. Wiley, New York, 1972.

Arcizet F, Mirpour K, Bisley JW. A pure salience response in posterior parietal cortex. *Cereb Cortex* 21:2498–2506, 2011.

Bell SC. A system of dissections, explaining the anatomy of the human body, the manner of displaying the parts, and their varieties in disease: With plates. Mundell, Johnson, Longman and Rees, London, 1803.

Bienenstock EL, Cooper LN, Monro PW. A theory for the development of neuron selectivity. Orientation selectivity and binocular interactions in visual cortex. *J Neurosci* 2:32–48, 1982.

Brasted PJ, Wise SP. Comparison of learning-related neuronal activity in the dorsal premotor cortex and striatum. *Eur J Neurosci* 19:721–740, 2004.

Cainiello ER. Outline of a theory of thought processes and thinking machines. *J Theor Biol* 2:204–235, 1961.

Cavada C, Compañy T, Tejedor J, Cruz-Rizzolo RJ, Reinoso-Suarez F. The anatomical connections of the macaque monkey orbitofrontal cortex. A review. *Cereb Cortex* 10:220–242, 2000. doi:10.1093/cercor/10.3.220.

Cooper LNA. A possible organization of animal memory and learning. In: Lundquist B, Lundquist S, eds. Proceeding of the Nobel symposium on collective properties of physical systems. Academic Press, New York, 1973.

Cooper LN. Distributed memory in the central nervous system: Possible test of assumptions in visual cortex. In: Schmitt FO, Worden FG, Adelman G, Dennis SG, eds. The organization of the cerebral cortex. MIT Press, Cambridge, 1981.

Desimone R, Albright TD, Gross CG, Bruce C. Stimulus-selective properties of inferior temporal neurons in the macaque.*J Neurosci* 4(8):2051–2062, 1984.

Edelman GM, Tononi G. A universe of consciousness: How matter becomes imagination. Basic Books, New York, 2000.

Effenberger, F. A primer on information theory with applications to neuroscience. In: Computational medicine in data mining and modeling (April):60, 2013. http://arxiv.org/abs/1304.2333\nhttp://link.springer.com/chapter/10.1007/978-1-4614-8785-2_5.

Ekstrom LB, Roelfsema PR, Arsenault JT, Bonmassar G, Vanduffel W. Bottom-up dependent gating of frontal signals in early visual cortex. *Science* 321:414–417, 2008. doi:10.1126/science.1153276.

Fan W, Ellacott KLJ, Halatchev IG, Takahashi K, Yu P. Cholecystokinin-mediated suppression of feeding involves the brainstem melanocortin system. *Nature Neurosci* 7:335–336, 2004.

Farley BG, Clark WA. Simulation of self-organizing systems by digital computer. *IRE Trans Inform Theory* 4:76–84, 1954.

Fiorillo C, Tobler P, Schultz W. Discrete coding of reward probability and uncertainty by dopamine neurons. *Science* 299:1898–1902, 2003. doi:10.1126/science.1077349.

Funahashi K. On the approximate realization of continuous mappings by neural networks. *Neural Networks* 2:183–192, 1989.

Fuster JM, Jervey JR. Neuronal firing in the inferotemporal cortex of the monkey in a visual memory task. *J Neurosci* 2:361–375, 1982.

Galilei G. Dialogo sopra i due massimi sistemi del mondo Ptolemaico e Copernicano [In Italian, Dialog about the two most important systems of the world]. Galileo wrote this *magnus opus* between 1624 and 1632, printed in the latter year, 1632.

Goldman-Rakic PS (1987) Circuitry of primate prefrontal cortex and regulation of behavior by representational memory. In: Plum F, Mountcastle F, eds. Handbook of physiology. American Physiological Society, Washington DC, 373–517, 1987.

Grossberg S. Adaptive pattern classification and universal recording: I. Parallel development and coding of neural feature detectors. *Biol Cybern* 23:121–134, 1976.

Guyton AC. Textbook of medical physiology. 7th ed. WB Saunders, Philadelphia, 1986.

Havel PJ, Kasim Karakas S, Mueller W, Johnson PR, Gingerich RL. Relationship of plasma leptin to plasma insulin and adiposity in normal weight and overweight women: Effects of dietary fat content and sustained weight loss. *J Clin Endocrinol Metab* 81:4406–4413, 1996.

Hebb DO. The organization of behavior: A neuropsychological theory. John Wiley and Sons, New York, 1949.

Hodgkin AL, Huxley AF. A quantitative description of membrane current and its application to conduction and excitation in nerve. *J Physiol* 117(4):500–544, 1952.

Hopfield JJ. Neural network and physical systems with emergent collective computational abilities. *Proc Natl Acad Sci USA* 79:2554–2558, 1982.

Hubel DH, Wiesel TN. Receptive fields, binocular interaction and functional architecture in the cat's visual cortex. *J Physiol Lond* 160(1):106–154, 1962.

Kesner RP, Ravindranathan A, Jackson P, Giles R, Chiba AA. A neural circuit analysis of visual recognition memory: Role of perirhinal, medial, and lateral entorhinal cortex. *Learn Mem* 8:87–95, 2001. doi:10.1101/lm.29401.

Kohonen T. Self-Organization and associative memory. 2nd ed. Springer-Verlag, Berlin, 1987.

Laughlin SB, De Ruyter Van Steveninck RR, Anderson JC. The metabolic cost of neural information. *Nature Neurosci* 1:36–41, 1998. doi: 10.1038/236.

Lennie P, Place W. The cost of cortical computation. *Cur Biol* 13:493–497, 2003.

Lew SE, Rey HG, Gutnisky D, Zanutto BS. Differences in prefrontal and motor structures learning dynamics depend on task complexity: A neural network model. *Neurocomputing* 71:2782–2793, 2008.

Lew SE, Zanutto BS. A computational theory for the learning of equivalence relations. *Front Hum Neurosci* 5:113, 2011. doi:10.3389/fnhum.2011.00113.

Malsburg VD. Self-organization of orientation sensitive cells in the striate cortex. *Kybernetik* 14:85–100, 1973.

Marr D. A theory of cerebellar cortex. *J. Physiol* 202(2):437–470, 1969.

Marr D. A theory for cerebellar neocortex. *Proc Royal Soc Lond B* 176:161–234, 1970.

Marr D. Vision: A computational investigation into the human representation and processing of visual information. Freeman, San Francisco, 1982.

Marx J. Cellular warriors in the battle of the bulge. *Science* 299:846–849, 2003.

McCulloch WS, Pitts WA. A logical calculus of the ideas immanent in nervous activity. *Bull Math Biophys* 5:115–133, 1943.

Micera S, Keller T, Lawrence M, Morari M. Wearable neural prostheses. Restoration of sensory-motor function by transcutaneous electrical stimulation. *IEEE Eng Med Biol Mag* 29(3):64–69, 2010.

Miller EK, Buschman TJ. Cortical circuits for the control of attention *Curr Opin Neurobiol* 23:216–222, 2013.

Minsky M, Papert S. Perceptrons. MIT Press Cambridge, 1969.

Moore T, Armstrong KM. Selective gating of visual signals by microstimulation of frontal cortex. *Nature* 421:370–373, 2003.

Mountcastle VB. The problem of sensing and the neural coding of sensory events. In: Quarton GC, Melnechuk T, Schmitt FL, eds. The neurosciences. Rockefeller University Press, New York, 1967.

Nass MM, Cooper LN. A theory for the development of feature detecting cell in visual cortex. *Biol Cybern* 19:1–18, 1975.

Nicholls JG, Martin AR, Fuchs PA, Brown DA. From neuron to brain. 5th ed. Sinauer Associates Publishers, Sunderland, 2011.

Pan WX, Schmidt R, Wickens JR, Hyland BI. Dopamine cells respond to predicted events during classical conditioning: Evidence for eligibility traces in the reward-learning network. *J Neurosci* 25: 6235–6242, 2005. doi:10.1523/ *JNEUROSCI*.1478-05.

Papez JW. A proposed mechanism of emotion. *Arch Neurol Psychiat* 38:725–743, 1937.

Pellionisz A, Llinas R. Brain modelling by tensor network theory and computer simulation. The cerebellum: Distributed processor for predictive coordination. *Neuroscience* 4:323–348, 1979.

Perez R, Glass L, Schlaer RJ. Development of specificity in the visual cortex. *J Math Biol* 1:275–288, 1975.

Pieloth C, Knösche T, Maess B, Fuchs M. A System for online neuroelectromagnetic source imaging. *Biomed Tech* (Berlin) 7(Sep) pii, 2013. /j/bmte.2013.58.issue-s1-G/bmt-2013-4177/bmt-2013-4177.xml. doi: 10.1515/bmt-2013-4177.

Purves D, Augustine GJ, Fitzpatrick D, Hall WC, La Mantia AS, White LE. Neuroscience. 5th ed. Sinauer Associates Publishers, Sunderland, 2011.

Quian Quiroga R, Reddy L, Kreiman G, Koch C, Fried I. Invariant visual representation by single neurons in the human brain. *Nature* 435:1102–1107, 2005.

Ramirez AD, Ahmadian Y, Schumacher J, Schneider D, Woolley SMN, Paninski L. Incorporating naturalistic correlation structure improves spectrogram reconstruction from neuronal activity in the songbird auditory midbrain. *J Neurosci* 31(10):3828–3842, 2011.

Rapanelli M, Frick LR, Miguelez Fernández AM, Zanutto BS. Dopamine bioavailability in the mPFC modulates operant learning performance in rats: An experimental study with a computational interpretation. *Behav Brain Res* 280:92–100, 2015.

Reynolds JN, Wickens JR. Dopamine-dependent plasticity of corticostriatal synapses. *Neural Netw* 15:507–521, 2002. doi:10.1016/S0893-6080(02)00045-X.

Ritter S, Dinh T, Friedman M. Induction of Fos-like immunoreactivity (Fos-li) and stimulation of feeding by 2,5-anhydro-D-mannitol (2,5-AM) require the vagus nerve, *Brain Res* 646:53–64, 1994.

Rosenblatt F. The perceptron: A probabilistic model for information storage and organization in the brain. *Psych Rev* 65:386–408, 1958.

Rosenblatt F. Principles of neurodynamics: Perceptron and the theory of brain mechanisms. Spartan Books, Washington DC, 1961.

Saalmann YB, Pigarev IN, Vidyasagar TR. Neural mechanisms of visual attention: How top-down feedback highlights relevant locations. *Science* 316:1612–1615, 2007.

Schultz W. Predictive reward signal of dopamine neurons. *J Neurophysiol* 80:1–27, 1998.

Schultz W, Dayan P, Montague PR. A neural substrate of prediction and reward. *Science* 14(March):275(5306):1593–1599, 1997.

Sejnowski TJ, Koch C, Churchland PS. Computational neuroscience *Science* 241:1299–1306, 1988.

Siegel M, Konig P. A functional gamma-band defined by stimulus-dependent synchronization in area 18 of awake behaving cats. *J Neurosci* 23:4251–4260, 2003.

Staddon JER. Adaptive behavior and learning. 2nd ed., with a 1st ed in 1983, Cambridge University Press, Cambridge, 2016. ISBN: 9781107442900.

Staddon JER, Zanutto BS. Feeding dynamics: Why rats eat in meals and what this means for foraging and feeding regulation. In: Bouton ME, Fanselow MS, eds. Learning, motivation and cognition: The functional behaviorism of Robert C. Bolles. American Psychological Association, Washington DC, 131–162, 1997.

Steinbuch K. Die Lernmatrix. *Kybernetik* 1:36–45, 1961.

Sutton RS, Barto AG. Time derivative models of pavlovian reinforcement. In: Gabriel M, Moore J, eds. Learning and computational neuroscience: Foundations of adaptive networks. Bradfoud Books. MIT Press, Cambridge, 1990.

Taylor PC, Nobre AC, Rushworth MF. FEF TMS affects visual cortical activity. *Cereb Cortex* 17:391–399, 2007. doi:10.1093/cercor/bhj156.

Tiesinga PHE, Fellous JM, Jos JV, Sejnowski TJ. Information transfer in entrained cortical neurons. *Network* 13:41–66, 2002.

Tomita H, Ohbayashi M, Nakahara K, Hasegawa I, Miyashita Y. Top-down signal from prefrontal cortex in executive control of memory retrieval. *Nature* 401(6754):699–703, 1999.

Travers S, Norgren R. Gustatory neural processing in the hindbrain. *Annu Rev Neurosci* 10:595–632, 1987.

Valentinuzzi M. The organs of equilibrium and orientation as a control system. Harwood Academic Publishers, London and New York, 194, 1980.

Valentinuzzi ME, Saltor J. Lamarck, Darwin, Wallace, Ameghino: Revolutionary biology heritage showing quantitative and philosophical spots, even with social remarks, too. Does all this qualify as being called a Biological Engineering view? *IEEE Pulse Mag* in press.

Widrow B, Hoff ME. Adaptive switching circuits. IRE WESCON Convention Records, New York, Part 4, 96–104, 1960.

Woods SC. Gastrointestinal satiety signals I. An overview of gastrointestinal signals that influence food intake. *Am J Physiol Gastrointest Liver Physiol* 286: G7–G13, 2004.

Zanutto BS, Staddon JER. Bang-Bang control of feeding: Role of hypothalamic and satiety signals. *PLoS Comput Biol* 3(5):924–931; e97, 2007. doi:10.1371/journal.pcbi.0030097.

CHAPTER 9

MUSCULAR AND SKELETAL SYSTEMS

Mónica T. Miralles and Ignacio Ghersi

We look at it almost with worship, for we stand up, sit down, lie down, jump, run, move about, dance, search ... and yet ... how far behind the mind is!

Gustavo Idemi

Abstract

The first part of this chapter introduces the musculoskeletal system, from the perspective of bioengineering, tackling aspects of the osseous tissue, joints and muscles, tendons, and ligaments. The second part of the chapter focuses into human motion and the discussion and analysis of new means for acquiring and processing signals concerning human balance. As opposed to traditional means, accelerometry-based devices are especially described, as a promising alternative. Given the importance of risk of falls for the elderly population, basic components and signal analysis strategies on a study case are presented, in which an accelerometry-based device was used during clinical, qualitative assessments of risk of falls according to scales of performance. Finally, results from the definition of specific indexes through signal processing strategies are shown.

9.1 Introduction

When thinking about the musculoskeletal system, models of articulated segments, which have become highly popular from wood dummies used by artists, up to Sci-Fi battle androids in more advanced versions, jump immediately to our minds. Within this concept, the

305

immediate notion that we are a biological design for motion stands out. For the human body, the "dummy model" finds a first scientific counterpart in parametrized body-segment models, where the length of each segment is expressed as a percentage of body height (Winter, 2009). Together with information provided by anthropometric tables, these models allow estimating specific properties for each individual segment, for example, through characteristics such as the respective percentages of body weight. In this way, the estimated position of the center of mass (COM) for each segment is usually a basic initial calculation, resulting from considering the ratio between segment length and height, radius of gyration, density, and inertial moments. The sum of this knowledge has been incorporated into commercial motion-analysis systems.

The final goal of this chapter is to describe and analyze gesture performance, based on fundamental variables, such as body COM, through a certain time lapse. Thanks to modern motion capture systems, based on markers or other technologies, segments can be defined, and the progress of a gesture can be acquired for each segment through time. From these data, all the associated kinematics and some of the kinetics (velocity and acceleration, both linear and angular, joint amplitude, floor reaction forces, among other relevant information for the bioengineer) can be generated. These studies are further supplemented with multi-view records, electromyography and electrocardiography signals, among other relevant clinical data for every scenario. Finally, these data can be uploaded into dedicated software (such as OpenSim) to create and analyze derived models for dynamic simulations of motion (Delp *et al.*, 2007).

The musculoskeletal system aims at assuring equilibrium and balance, both during static and dynamic gestures and scenarios, whatever the complexity of the motion in question. From a basic analytical perspective, this requires the materialization of a support structure (skeleton), the presence of biological mechanisms (muscles, tendons, and ligaments) to allow displacements in the space of bone segments and, finally, an elaborate control system architecture, endowed with the standout feature of anticipation, or predictive capabilities (central nervous system [CNS] and its receptors).

Recent scientific advances allow us to conceive a continuity between the biological and the non-biological, progressively fusing natural and artificial technology. With new implants and prostheses, the human begins to overlap with biped humanoid robots, and the common ground between both grows day by day, reaching up to subjective aspects such as language or emotion (Johnson and Virgo, 2006; Lohmeier, 2010; Mavridis, 2014). Traditional engineering disciplines, as well as those constantly linked to innovation (such as bioengineering), move forward, converging from both extremes. The result of this synergy is an array of revolutionary products, including intelligent prostheses, 3D-printed artificial organs, among other developments that merge latest findings and applications of nanomaterials in this field (Bronzino and Peterson, 2006). Simultaneously, the emergence of increasingly specific journals, such as the *Journal of Biomechanical Engineering*,[1] reporting research results involving the application of mechanical engineering principles to the improvement of human health, joins an extensive scientific production in the area, as well as into new fields like Biomimetics.[2]

The fine, continuous automatic adjustment of body disposition and corrector postures for the muscles is the result of information perceived by three supplementary systems for information acquisition and of an array of effector systems for order transmission and interpretation. These systems are

(a) visual, for situational perception, location awareness and determination of changes between the body and its environment;
(b) auditory, through the vestibular system, providing information on body accelerations in three axes (semicircular conduits for

---

[1]See http://biomechanical.asmedigitalcollection.asme.org/journal.aspx

[2]The study of the formation, structure, or function of biologically produced substances and materials (such as enzymes or silk) and biological mechanisms and processes (such as protein synthesis or photosynthesis) especially for the purpose of synthesizing similar products by artificial mechanisms which mimic natural ones. See http://www.merriam-webster.com/dictionary/biomimetics. Term coined by Otto H. Schmitt (1913–1998).

angular accelerations, and otoliths for linear accelerations), and finally,

(c) the brainstem and the cerebellum, receiving proprioceptive information, mainly from the soles, and tendon–ligament and muscular sensors, acquiring information about body segments and their position and displacements, as well as about the actions of muscles involved in motion.

The array of effector systems is composed, in successive and supplementary stages, of different cerebellar zones, basal ganglia, and different regions of the cerebral cortex. The closed-loop interaction between these neural and musculoskeletal functions is beyond the reach of this chapter, but essential to the understanding of the neural-musculo-skeletal system.

The following section includes a brief introduction into fundamental concepts concerning the main components of the musculo-skeletal system, *from the perspective of a bioengineer*, in order to concentrate, in the second part of the chapter, on the central function of the musculo-skeletal system, which is that associated with equilibrium and human balance. A discussion and analysis of new means for acquisition and processing of motion-related signals, in particular for human balance in older adults with a higher risk of experiencing falls, closes the chapter.

9.2 Bones

Speaking of a biological structure for support (holding organs in fixed positions), protection and motion, requires to approach fields like material sciences, and the consequent characterization of the mechanical properties of biomaterials and living tissue (Fung, 1993; Ratner *et al.*, 1996). Unlike any mechanical test piece, the *form* and *structure* of the bones are the result of the mechanical strain to which they are subjected. That is, they are the direct result of a complex and dynamic set of loads due to action, always combined to a certain degree, with tensile and compressive efforts, but also with torsions, flexion, and shearing stresses. A paradigmatic example of this feature is represented by the different curvatures of the spine, in particular, lumbar lordosis, consequence of the bipedal human posture (Miralles *et al.*, 2005), not present in other primates.

Most bones come in mirrored pairs, and those which do not, such as vertebras or the pelvis, are symmetrical. This symmetry represents a well-known evolutionary optimization strategy, aimed at easing locomotion over any medium. For this purpose, it is also required that the support structure, as well as the whole body mass, be able to generate adequate reactions from contact surfaces, in order to perform vital motions such as walking, jumping or running, and their combinations (Novacheck, 1998; Norberg, 2015).

We have approximately 233 bones that weigh, together, an average of 9 kg. This support structure is relatively light, compared against the total body weight (which always imposes compressive forces), aimed at minimizing the energy expenditure required for its support and transportation, as well as increasing its resistance.

Bones can be short, long, flat, or irregular. They all share a common structure. They, broadly, have a compact tissue, covered by periosteum, and an internal region with spongy bone tissue, covered by endosteum. Given the interest and particular structure of long bones, only a brief presentation of features of a highest interest for a bioengineer for long bones, like the femur, will be detailed. From this basis, any other bone with its distinctive characteristics would require particular analysis for developing any application.

9.2.1 *Macrostructure*

From a macrostructural level, the shape of a mature long bone can be described as a shaft, called *diaphysis*, which shows expansions in its extremes, giving place to the *epiphyses*. Their extremes are covered by an articular cartilage, in such a way that a gliding surface, with a coefficient of dry friction that can be as low as 0.002, is generated. Their outer layer is a thin layer of cortical bone, which is continuous with the dense cortex of the diaphysis.

The whole outer surface of the bone, with the exception of the joint surface of the epiphyses, is covered by *periosteum*. The inner layer of the periosteum is called the *osteogenic* layer, since it includes the cells (*osteoblasts*) that allow bone growth, as well as cells that allow its destruction (*osteoclasts*, remodeling). The dense network of capillary blood and lymphatic vessels in the periosteum joins to the shaft through the nutrient foramen. On the other hand, the outer layer of the periosteum is fibrous, and formed by dense, irregular connective tissue. The periosteum provides attachment points for ligaments and tendons. At these points, the gripping structure between the periosteum and the bone, formed by collagen fibers, becomes more dense (Sharpey lines).[3] A membrane of connective tissue, called endosteum, covers the inner cavities of the bones, which overlays the trabeculae of *cancellous bone* of the *diaphyseal* and medullary cavities, as well as in the channels going through *compact bone*. This membrane also has osteoblasts and osteoclasts. The central part of the diaphysis contains the medullary cavity, where the bone marrow is located.

In order to appreciate the macrostructure of any bone piece, it is advisable to have a comprehensive anatomy atlas (McMinn and Hutchings, 1977), presenting different views of the bones in a 1:1 scale, including the position and insertion zones of different muscles and ligaments. This exercise contributes to comprehend the complex outer shape of the bones, which would seem apparently whimsical, but is justified by the physics covering the points in which forces and moments are applied to these structures. For example, great protuberances will anchor more active and powerful muscles.

It is interesting to highlight the great number of morphological variables that characterizes a long bone like the femur (Noble *et al.*, 1988). Sizes are not homogeneous, showing great standard deviations and range differences among subjects, which adds to a great variability in bone density. All these differences influence the fabrication of prostheses and implant design, especially to achieve proper tolerances. For example, the design of a stem for cemented hip prostheses requires no less than 17 stems of different sizes (Pruitt and Chakravartula, 2011). This knowledge is instrumental for the design

---

[3]Matrix of connective tissue consisting of bundles of strong collagen fibres connecting periosteum to bone.

of medical tools and surgical procedures, another fruitful field open to the work of bioengineers (Klausner *et al.*, 2014).

The anatomical model of a long bone can be simplified into a hollow cylinder, where the thickness of the wall is variable, growing from its ends to its center (diaphysis midpoint), reaching approximately one-half of the radius of the diaphysis at this point. This approximation represents a lightweight structure, with the same resistance of a similar dimensions compact tube. The *medullary cavity* is analogous to the neutral plane of a beam, since muscles and tendons pull from the outside of the bone and tend to bend it (flexion). The medullary cavity is the region in which compression and traction forces notably diminish, hence, there being cancellous bone in that region. The increased thickness in the midpoint of the cortical tissue is due to being precisely the place where compression and tension stresses display maximum values, on the sides of the medullary cavity, when the beam is loaded on one of its ends.

9.2.2 *Microstructure*

The microstructure of a long bone is different, according to whether it concerns to *cortical* or *cancellous tissue*. The *cortical* (or lamellar) *bone* is a composite material, whose microstructure can be described following Ham's Sketch (Fung, 1993). Within the bone matrix, a series of osteons or Haversian systems are inserted. They are the basic components that can be thought of as small cylinders, parallel to the longitudinal axis of the bone, at the center (Haversian canal) of which an artery, a vein, and a nerve are located, keeping the osteon alive.

These cylinders are formed of superimposed, concentric cylindrical shells called lamellae. Lamellae are formed of collagen fibrils (a mixture of aminoacids), parallel to each other, to which little mineral hydroxyapatite crystals are added ($3Ca_3(PO_4)_2 \cdot Ca(OH)_2$, about $-50 \times 50 \times 200$ Å^3). These fibers are angled with respect to the longitudinal direction of the osteon. On each successive layer, their angles are opposed, almost perpendicular, a design that endows the osteon, and statistically the bone, with a high resistance to torsion. The top view of an osteon is reminiscent of the transversal cut of the trunk of a tree, with its growth rings. Osteons are, in this way, comparable to little columns that carry a small part of the load to which the bone is put through. In the union of two successive lamellae, cavities called lacunae are located, which hold the osteocytes, cells with ramifications that go through micro-canals called canaliculi, connecting the lacunae of the osteon with each other and to the Haversian canal.

Osteons' centers connect to each other by transversal channels, the Volkmann's canals. This series of canals, in turn, and toward the periphery, link to the aforementioned periosteum network of vessels and nerves, and, close to the bone axis, to the medullary cavity. Such a circulatory system demonstrates the essential relevance of blood flow for this organ, sensitive to external stresses. Among the osteons, incomplete lamellae (or interstitial lamellae)[4] show up, and, in the end and surrounding the bone, external circumferential lamellae are found; they are in contact with the deep layer of the periosteum. Towards the center and ends of the bone, there is *cancellous tissue*, formed of cellular layers called trabeculae, oriented in such a way that they follow the bone force lines. They do not present osteons, only irregular lamellae, with osteocytes connected by canaliculi.

9.2.3 *Mechanical properties*

The main mechanical properties of the bones, which are composite materials (Young's modulus, shear modulus, viscoelastic properties, ultimate stress, strain, and failure), are a result of the aforementioned bone matrix structure, as well as of its composition. Bones are largely (70%) composed of calcium, phosphorus, oxygen and hydrogen (hydroxyapatite) as crystals, and collagen (organic material). This composition makes for considerably improved mechanical resistance, with respect to those of any isolated component (Table 9.1), offering high tension and flexion strengths.

Strain and deformation are measured through Poison's modulus, which measures the ratio between the differential length in the direction of the applied force, and the change in width over the

---

[4]See http://medical-dictionary.thefreedictionary.com/interstitial+lamella

Table 9.1: Young's modulus for isolated bone components and materials for industrial use (Fung, 1993).

Material	Collagen	Hydroxya-patite	Cortical bone (femur under tension)	Steel	Aluminum 6061-alloy[*]
Young's modulus	1.24 GPa	165 GPa	17–18 GPa (longitudinal) 12 GPa (transversal)	200 GPa	70 GPa

[*]It is a precipitation-hardened aluminum alloy, containing magnesium and silicon as its major elements. It shows good mechanical properties.

perpendicular direction to the force. For cortical bone, this ratio will be close to 0.26, which means that, during compression, the bone will shorten and widen. For shearing, the rigidity modulus applies, which, for metals, can have values of one-third of the aforementioned one.

For a uniaxial traction essay, and based on stress–strain curves, it can be found that dry, more brittle bone, experiences failure around 80 MPa, while fresh bone fails at a value lower than 60 MPa. Strain values are, respectively, 0.4% and 0.2%. In both cases, Hooke's law applies for a limited range of deformations, but Young's modulus is higher in tension than in compression. This is a consequence of the anisotropy and inhomogeneity of the bone matrix (Table 9.1). Tabulated values depend on a great number of variables, both concerning the bones and the experimental conditions.

Composite materials, such as bones, have been well studied. In ancient Greece, iron rods were hidden into long marble beams, similar to the case of reinforced concrete. Rubber is made of a resin (playing a similar role to that of collagen in the bone), to which carbon dust is added (analogous to the role of hydroxyapatite crystals) in order to improve the overall properties of the material. Bones are living tissue, with such a dynamic that, on a weekly basis, between 5% and 7% of our bone mass is recycled. They adapt their function through our time (growth and decay), and, for that purpose, they count on adapting mechanisms, such as growth, self-repair (i.e., the appearance of calluses after fractures), and resorption.

Long bones have specific growth mechanisms, allowing them to sustain their shape during this growth, and especially preserving the points of joint insertion. Usually, growth takes place through a balance between bone production in the outer wall (osteoblasts) and the disappearance of bone from the inner layers (by the action of osteoclasts), reducing excess bone or other substances carried away by blood. The process of bone remodeling happens between the endosteum and periosteum surfaces, where non-uniform, deposition and reabsorption processes occur for a single bone piece. This remodeling process is controlled by (a) hormonal regulation (calcium binding) and (b) stress-dependent processes (mechanical and gravitational force), within specific stress-ranges, knowledge of which is essential for the development of prostheses, implants and rehabilitation programs (see Chap. 7, section on calcium regulation).

Well-known examples of these two regulations are the loss of calcium (during prolonged rest or periods of weightlessness), the role played by daily efforts, among other relevant examples, confirming that subnormal stresses cause loss of bone strength, radiographic opacity, and size. It is assumed that bones fit within the principle of functional adaptation (Wolff's law),[5] in the sense that bones grow and remodel in response to stresses to which they are subjected, that is, function appears as a dynamic process according to conditions of use.

Meanwhile, the principle of maximum–minimum design focuses on an architectural optimal-design principle, concerning the addition of material where loads are concentrated, and the disappearance of material in the remaining areas. Much is known, but also much remains unknown about the cellular physiology of bones, in relation to the many functions of osteocytes. Bone deformations produce electric currents (piezoelectric activity), which is subject of research in relation to remodeling processes, since such current flow accelerates fracture healing. The extensive research of osseous tissue, particularly cortical and cancellous tissue of long bones (composition,

---

[5]Wolff's law, developed by the German anatomist Julius Wolff (1836–1902), states that bone in a healthy individual adapts to the loads under which is placed.

distribution, orientation and disappearance patterns through time, biomaterials for replacement, implants), has given place to multiple specialized fields of study and technological development, based on means and devices for research that are not within the scope of this chapter.

In general, bone fractures occur because of a sudden, intense impact. They can break into many pieces (comminuted fracture), and they often break due to twisting (torsion fracture), presenting a typical spiraled profile on the face of the fracture. In both cases, realigning (reduction) between both ends of the broken bone is needed. Fracture reduction can be manual (closed) or performed during surgery (open reduction), where elastoplastic nails can be used to help position and healing, among other implants (Paterson *et al.*, 2011).

Finally, immobilization is mandatory to allow the healing process, which, in turn, has a variable duration, depending on the case. Bones link thanks to joints, giving place to a complex repertoire of movements, as well as postures, resulting in comfort and stability.

9.3 Joints

Individual bones interact in different ways through joints. Their structure leads to their classification, which somehow explains how bones link to each other, and emphasizes their connective tissue. Joints classify as

(a) fibrous (dense, irregular connective tissue, rich in collagen fibers);
(b) cartilaginous (bones connected by cartilage), and
(c) synovial (bones are not directly linked, but through a synovial cavity with its capsule).

Joints can also be classified by the type and degree of motion they allow, resulting in three groups,

(a) synarthrosis (low to no mobility, fibrous tissue, i.e., cranial sutures);
(b) amphiarthrosis (slight mobility, cartilaginous tissue, i.e., intervertebral-disc joints); and
(c) diarthrosis (great mobility), which are synovial.

In mobile joints, contact surfaces with the bones are covered in cartilage, and their stronger linkage elements are the capsular ligaments. Within diarthroses, six particular types should be mentioned,

(a) Hinge joint: An articular surface is concave, sliding around a convex surface, allowing for flexion and extension (elbow, fingers).

(b) Modified Hinge joint: Allows for flexion, extension, rotation (knee).

(c) Rotary joint: These allow for rotation around a vertical axis (i.e., ulna and radius).

(d) Ball-and-socket joint: Sphere and sphere-like cavity, allowing for flexion, extension, abduction, adduction, circumduction, and rotation, over a major axis (shoulder and hip).

(e) Saddle joint: Both bones have a "saddle" surface, reciprocally adapted. It allows for flexion, extension, abduction, and slight rotation (i.e., carpal–metacarpal joint of the thumb).

(f) Condyloid joint: It allows for flexion, extension, abduction, adduction, and circumduction.

What is their composition? In synthesis, they count on four fundamental components: (a) dense fibrous tissue (collagen–elastin matrix), (b) synovial fluid (mainly composed by water and hyaluronic acid for lubrication), (c) synovial capsule, and (d) osseous tissue, embedded in a matrix of calcium and phosphorous minerals.

9.3.1 *Articular cartilage*

It presents three distinct zones, on which the particular architecture of collagen fibers stands out, that is,

(a) A superficial or sliding zone, formed by collagen fibrils, distributed in parallel with the surface, and presenting the highest percentage of water (around 80%) and intercellular space.

(b) A medial zone, representing a transition between the deep layer, in contact with the bone, and the sliding zone. Collagen fibers are more oblong, and of a larger diameter.

(c) A deep zone, presenting the lesser percentage of water (approximately 65%). Here, collagen fibers and chondrocytes are round, and perpendicular to the surface of friction.

As for its mechanical properties, cartilage is a viscoelastic material, meaning that it presents the combined behavior of a viscous fluid and that of an elastic solid. Its two fundamental properties are creep and stress relaxation. Creep refers to the fact that, upon the application of a constant compressive load in time, the result is a time-variable deformation (Mow *et al.*, 1980). Stress relaxation, on the other hand, occurs upon a constant deformation over time, giving place to a time-variable load. From a bioengineering perspective, when we think of contact surfaces, they are immediately associated to three phenomena: friction, lubrication, and wear. These concepts are dealt with by tribology, which synthesizes concepts of physics, chemistry, fluid, and solid mechanics, and it is named bio-tribology when applied to organic materials.

Surface friction can be classified as dry, limit, and hydrodynamic. Internal friction is another subject of interest. Wear is the loss of material due to chemical or mechanical factors. It can be (a) adhesive wear, when a material adds to another; (b) fatigue wear, when breaking limits are crossed and a part is released; (c) abrasive wear; and (d) corrosive wear, if material is released by action of chemical factors.

As for lubrication, it exists when a liquid film is interposed between two solid layers. Friction occurs, then, between solid and liquid surfaces. Lubrication can be (a) hydrostatic, when pressure is exerted by external means; (b) hydrodynamic, when pressure is exerted by the fluid itself, and this is the lubrication found in the cartilage. For this type of lubrication, Reynold's principles concerning contact surfaces have to be taken into consideration: They have to be rigid and non-porous, they must form a wedge at the interface, fluid viscosity must be constant (Newtonian fluid), sliding speed has to be high and load transmission between surfaces has to be small. However, these conditions are not met in the articular surface, since it is composed of porous surfaces in contact with each other, not formed as a wedge at the interface, over which the fluid is non-Newtonian, sliding speed is low for high loads, and inter-surface load transmission is high and variable. To sum up, articular-surface lubrication is elasto-hydrodynamic, by fluid flattening. Reading Mow *et al.* (1990) supplies a deeper exploration of the subject, even though it is over 25 years old.

The understanding of all these subjects is fundamental when facing the design of joint-replacement implants, osteosynthesis implants (particularly for hip, knee, humerus, and ankle), or the design of plates, nails, and screws for spinal fixation. Such design involves

(a) the detailed, biomechanical study of the joint; (b) a kinematic study, concerning axes and movement planes, acting forces and moments; and (c) a kinetic study, concerning individual acting elements and their contribution to motion.

Methods are also numerous, including finite-element simulation, stereo-photogrammetry, mechanical essays for soft parts or of joints on cadavers, biocompatibility tests in models, functionality tests in cadavers, among the most relevant. To finish the paragraph, let us underline that when a bone is in motion turning around a rotation axis, always a lever is a good model.

9.3.2 *Joint angles*

When a joint takes a position so that muscles are relaxed or under a state of minimal contraction, it is said that such joint is in a resting state. This is the starting point for any articular-motion analysis. Over the past decade, there has been an extensive production of different methods for the evaluation of joint angles, with relative advantages related to performance, cost, and usability (Rowe *et al.*, 2001; Nasseri *et al.*, 2007; Djuric-Jovicic *et al.*, 2011). The measurement of angles between body segments consists in measuring the spatial orientation of one body segment with respect to another adjacent to it, separated from the first by the studied joint, or possibly in relation to another reference, like the vertical line. Measuring lower limb joints is essential in the different stages of walking, as for the study of any other gesture (Whittle, 2007; Stergiou, 2004). This knowledge completes the evaluation of the active and passive mobility, the characterization of axial deviations, and finally, the search of abnormal movements. It is also one of the parameters used for the complete evaluation of a running cycle.

The current level of technological development has enabled the rise of new methods for measuring joint movements during any gesture, in particular, gait and running, with excellent results. For example, systems based on video-analysis algorithms, which isolate active markers that are fixed to the moving segments, and acquire information concerning the dynamics of such joint angle over a gesture are of interest as cost-effective alternatives (Mondani *et al.*, 2015).

9.4 Muscles, Tendons and Ligaments

As discussed in the previous section, ligaments join bones between each other, attached precisely to the joints. The functional unit of skeletal muscle is *bone–tendon–muscle–tendon–bone*. The functional unit of a ligament, instead, is *bone–ligament–bone*. From the perspective of the engineer, muscles are the engines (devices that provide mechanical work), adapted to different functions, with a high level of specificity, as in the case of the cardiac muscle. They represent 40% of the weight of a man and slightly less in women. In recent years, a great number of models were proposed (Yamaguchi, 2005). Tendons transmit force from muscles to bones in order to set them into motion. Muscle can be smooth or striated. The latter responds to voluntary control while the smooth type acts involuntarily (see chapter on the GIS).

This section will focus on muscles linked to the bones (skeletal muscles). In this way, the combination of muscle and bone allows us to resort to the classic lever model, applying the first, the second, or the third type, depending on where the support (rotation axis), load, and force (tendon insertion) are located. The determination of moments, link segment model, calls for the calculation of internal forces and external ones in a complex reverse-engineering problem (Caldwell *et al.*, 2004).

Muscles can only contract. They cannot push. Therefore, they need to act with another muscle, called, antagonistic, which allows reversing the movement, when relaxing the muscle that had previously contracted (agonistic). The most familiar case is the biceps and triceps, used during the motion of elevation of the forearm (biceps contraction), as well as for the return to their resting position, which takes place thanks to the contraction of the triceps. It is important not to lose sight of the fact that, at any moment, the whole body is working on maintaining the remaining muscles fixed in order to contribute to any intended movement. The secret behind the efficiency of the muscles lies with their relation to the bones, resulting therefore in a broad combination of movements and force ranges. This analysis can distinguish between internal forces (generated by the muscles), compressive forces (body weight), and external forces. In all cases, their measurements rely on their effects: displacement or deformation. External forces cause motion with respect to rotation centers (joints), compensated for by muscular forces to sustain a certain posture over time.

Lever arms are tabulated based on the knowledge of anatomical studies. They usually refer to small distances, which compensate great forces, possibly reaching many times that of the whole body weight, as in the case of the hip abductor muscles. The application point, characterizing any force, is fundamental here, since in this case we are talking about surfaces instead of points, and therefore, there will be an exertion of stress through that surface. Each, and depending on the type of force, will be characterized by a modulus measured in pascals under the international system of units (usually, in gigapascals, $1\,\text{GPa} = 10^9\,\text{Pa}$). Moreover, a bone is viscoelastic in itself, like the ligaments, in that the deformation depends not only on the type of effort but also on the speed at which this is applied; it is this property that determines if a ligament can break from a fast movement.

9.5 Motion Analysis

The musculoskeletal system provides main functions of support, protection, and mobility (see Section 9.2). Voluntary muscle control allows us to perform intensive work and carry loads, as well as to control slight facial expressions, ranging from the dexterity in our hands to the support and strength of our feet, devised for locomotion (Kapit and Elson, 2002). Limit ranges of motion are of use in ergonomics (refer to functional anthropometry, as described by Peebles and Norris (1998)).

The study of motion has greatly progressed and spread in the past years with many purposes, reaching up to consumer and entertaining systems, as well as research proposals for high-performance athlete optimization and the assessment of decreased motor control. The following technologies (some already mentioned in this chapter) are considered useful in providing information concerning fine and broad motion, say,

Machine-vision systems widely implemented from research to consumer-ready solutions. These systems use high-speed, single or multi-camera arrays, can achieve 2D or 3D reconstructions of the position of (active or passive) markers on the body, or estimate human posture from known body models, such as face, body, hand, or other parts (Murphy-Chutorian and Trivedi, 2009). Information from these systems can be integrated into the aforementioned simulation environments, like Open-Sim, and their resolution varies greatly among devices.

Force platforms are other well-known devices, used for research on human balance and equilibrium. They can be used to estimate the center of pressure (COP) during standing balance, or its trajectory while walking, under different configurations. The following sections will include a deeper analysis of force platforms for standing balance assessments.

Electromyography allows acquiring motor signals from muscular activity, and is a method with a variable, relatively higher level of invasiveness. Given the nature of this signal, conditioning and isolation of relevant information can be a topic of special concern for its applications.

MEMS (from Micro-ElectroMechanical System, see Sonje and Borde, 2014) have been widely integrated into navigation and consumer devices. MEMS sensors, such as accelerometers, gyroscopes, and compasses, can be integrated into portable, lightweight, and relatively low-cost devices, which can be attached to a body segment and measure motion through acceleration, angular velocity (gyroscopes), and full 3D orientation signals (by the combination of MEMS accelerometer and compasses). The notion of combining multiple sources of information, in such a way that one is able to compensate for the other, resulting in an overall superior solution, has been used in navigation research. This concept of *sensor fusion* also applies to motion analysis, where Inertial Measurement Units can be developed, through the combination of different types of MEMS sensors, and also supplemented by other means.

In relation to this overview, motion analysis strategies, implementing accelerometers, can be tackled through dedicated devices that measure linear acceleration among other variables (these may include gyroscopes and compasses under multiple configurations). Commercial and research-oriented systems have been developed for this purpose over the past years, aimed at analyzing signals from limbs or the trunk (Yang and Hsu, 2010; Aziz and Robinovitch, 2011). These systems can be oriented to long-term or short-term measurements, and acquire records from one or more sensors. Data can be transmitted through wired or low-power wireless means to a PC (sometimes through a dedicated base), or be stored and retrieved via a docking station, and specifically, graphical user interfaces (GUI), are mostly used to control and display the data. This system layout sums all the needed components to acquire and display relevant information concerning motion under portable conditions. Once an experimental scenario is defined, and signals are acquired, adequate signal processing strategies need to be deployed. The following section covers these stages in balance-oriented research.

9.6 Human Balance and Gait Assessment

Equilibrium and balance assessment is a topic of special interest in the area of human biomechanics. It is an essential condition required to perform any type of physical activity while standing and walking, and related to a diverse amount of movements, gestures, and postures involved in our daily life. The primordial need for a persistent state of equilibrium or balance arises in the intrinsic instability of the musculoskeletal system. As an inherently unstable structure, human body stable balance happens through the continuous actions of control, sensory, and regulatory systems in feedback loops. Its study and its changes lead to both a general and a more specific learning about the human species, say, concerning learning stages and evolution, or degenerative processes associated with age, for example (Winter, 1995).

Human balance conforms a subject of intensive, basic, and applied research. Numerous fields take part of such activity, where there is interest in human dynamics. Thus, the way opens for adapting and developing new technologies leading to novel devices for the assessment and also perfecting the development of tests for diagnosis and follow-up of control subjects, high-performance athletes, or subjects suffering from changes or degradations in the sensory systems associated with these tasks (Rudge and Bronstein, 1995).

Additionally, the field of research that focuses on the development and study of biped robots with human-like stepping motions may also benefit from a more precise knowledge about gait and balance. Moreover, it well provides new and relevant information to the field of human motion biomechanics, with several examples, including passive-dynamic walkers (Kuo, 1999; Collins *et al.*, 2005). In particular, the study and understanding of ataxias in this field is complex, and classic neurology tests are used to assess their degree or potential impact, with only slight modifications up until these days. There are many unanswered questions relative to human balance, and at different levels.

On one hand, biomechanical records for balance assessment usually come from classic force platforms or from elaborate tracking systems that rely on optic markers and high-speed, multi-camera video information for three-dimensional kinematics analysis (Robertson *et al.*, 2004). The subarea called *posturography*, defined as a strategy aimed at the assessment of static or dynamic posture through the analysis of a subject's COP, obtains information from force platforms (Peydro de Moya *et al.*, 2005). The truth is that, many decades later, the force platform appeared as not yet considered a proven instrument for diagnostics, so that there is no agreement regarding the usefulness of technique, either in the clinical or in the laboratory environments. Apparently, there is lack of consistency and repeatability in the measurements, as well as a lack of proper association of physiologic meanings to the traces obtained from these platforms, that is, from *stabilograms* (say, COP, trajectory length, mean position, and the like). This emphasizes the need to supplement these data acquisition sources with other systems, working simultaneously.

On the other hand and, many are the current mathematical tools that can be proposed for the treatment of acquired biomechanical data, as well as the models that support their use. Among these, those ranging from the application of classic laws of mechanics and usual treatments of periodic signals (Fourier transforms), to newer applications in the field of biology of tools developed for the analysis of complex, non-stationary signals (multi-scale time–frequency analysis), stand out as potential examples. In particular, for human balance assessment implementations, exploring applications of the *Dynamical Systems Theory* to the study of functional movement patterns is of interest. This latter theory considers that movement patterns come from the synergistic organization of the neuromuscular system, which depends on morphologic (biologic structures), biomechanical (laws of kinetics and kinematics), environmental (spatial–temporal events), and task-specific (gait, race, and jump) factors. Therefore, it assumes that movement patterns are the result of individual muscles and neural trajectories working collectively to achieve a functional output according to the restrictions of the system. This would suggest that variations in the form in which the neuromuscular system is organized, which translate into variability in the associated movements organization and execution, can be related to health instead of dysfunction, as considered by classic methods (Stergiou, 2004).

Facing these pending questions, information and data provided by the bibliography presents a wide variety of situations, in many cases of difficult reproduction due to expensive equipment involved. Another drawback recognizes a wide range of different working protocols, lack of specificity, and clearness as to the models that support them, uneven study groups, lack of a sufficiently large number of study cases to allow generalizations, and profusion of calculations without proper interpretation. In sum, a great need to reflect and articulate different variables appears as necessary in order to bring forward indexes of practical usefulness.

In spite of these difficulties, many opportunities appear in this field in the form of recent advances in technology and communications, which have brought forward new alternatives for the acquisition of relevant information for the assessment of human balance, gait, or other associated gestures. As previously stated, 3D-accelerometers can be used as activity monitors for human balance studies, providing access to a supplementary dataset to that related to the already mentioned force platforms and optic marker systems.

9.6.1 *Decrease of motor control and risk of falls: Clinical scales*

The loss or reduction of motor control, whatever its cause, significantly diminishes the ability to perform daily tasks, such as walking or climbing stairs. As age progresses, the loss of balance translates into frequent falls, with consequent associated bone fractures and other injuries. Populations are growing older, and falls are becoming a permanent problem of increasing importance for medical facilities and public health (United Nations, 2013). However, there is no clearness as to the prevalent main risk factors to enable a dedicated work leading to a reduction in fall risk. As populations age at unprecedented rates, falls stand out as relevant cause of temporary or permanent disability, and even death. Additionally, a main sign of high-level disability is associated with a higher risk of unintentional injuries, from road accidents, burns, and other causes, such as dementia, rheumatoid arthritis, and osteoarthritis (World Health Organization, 2011).

To evaluate falls risk, it is important to note that, while aging is a universal, evolutionary process, it is also individual, and the apparently specific group of "older adults" offers a greatly wide diversity of conditions and singularities. Many factors of different nature can be associated with a higher risk, and they can be extrinsic (environmental), intrinsic (degradations of the sensory or motor sub-systems), and behavioral (Hill and Wee, 2012). Degradations in balance capabilities can be the result of any neuro-musculoskeletal disorder, although the reliance of that control system as a whole on many input sources allows for compensations, when one of these is limited or reduced, so the possibility of these compensations must be taken into account when evaluating a subject's balance capabilities (Winter, 1995).

Fear of falling relates to the actual risk of falls, too. It may derive from self-imposed restrictions to daily-life activities as well as postural changes (Párraga Martinez *et al.*, 2010). While trying to measure such a thing is difficult, and may lead to many different tests possibly with large-ranging outcomes, the Falls Efficacy Scale International (FES-I) is hinted as a widely accepted evaluation tool (Delbaere *et al.*, 2010). It is a short self-evaluation that is used to predict rehabilitation results for geriatric populations, and is subject of many studies concerning its validity and limitations (Kempen *et al.*, 2007; Denkinger *et al.*, 2010; Delbaere *et al.*, 2010). Gait disorders and falls are well-known causes of morbidity and mortality in the elderly population, and they are a subject of growing interest for public health institutions (Baezner *et al.*, 2008; Srikanth *et al.*, 2009; World Health Organization, 2011). There are a number of clinical tests in this field, such as the Tinetti scale or the Dynamic Gait Index (DGI) assessment of risk of falls, where physicians score a subject based on assessments that factor step frequency, symmetry, continuity, deviations, and need of assistance (Shumway-Cook and Wollacott, 1995). Based on the aforementioned technologies, the interest in quantifying these tests through objective indexes will provide access to new and more uniform evaluations.

9.6.1.1 *Tinetti scale*

The Tinetti scale (Tinetti *et al.*, 1986) is a simple and reproducible means to evaluate the risk of falls for the elderly or for those suffering from balance dysfunctions. It is a scale associated with the observation of different stages, defined for the assessment of the subject's balance capabilities. Both standing balance and gait (straight walk over 4 m) are evaluated, with a total score of 28 points: 12 points form the gait phase and 16 for the balance phase. The complete test allows stratifying subjects in three groups, according to their general score, that is, low risk (28–25 points), medium risk (24–19 points), and high risk (0–19 points).

9.6.1.2 *DGI: Assessment of falls risk*

The DGI for gait assessment and risk of falls evaluation for the elderly consists on a sequence of exercises, most involving a walk on a straight path (two 5-meter walks, for example), with obstacles, or responding to commands from the evaluating physician (Shumway-Cook and Wollacott, 1995). A detailed description for a particular implementation of this test is divided in stages, as follows:

(1) Straight walk, at a comfortable speed for the subject
(2) Straight walk, with speed changes (normal, slower or faster, when commanded)
(3) Straight walk, with horizontal head turns (look left or right, when commanded)
(4) Straight walk, with vertical head turns (look up or down, when commanded)
(5) Straight walk, with pivot (180° turns, and keeping walking, when commanded)
(6) Straight walk, stepping over one small obstacle in the middle of the path
(7) Straight walk, passing around an obstacle and returning to a straight walk
(8) Stair climbing

The aim of these stages is to evaluate each subject's ability to respond to slight changes during a regular, straight walk. Physicians assign a score between 0 and 3 points for each stage: 0 point when the subject shows severe changes or the impossibility to perform the activity and 3 points when there is little to no difficulty in making such changes.

Each particular activity has some specific parameters that need to be accounted for in order to determine the score. In particular, activities 1–4 have certain properties that make them better suited for a joint evaluation. The highest possible score is 24 and scores of 19 or less have been related to a greater incidence of falls in the elderly (Miralles *et al.*, 2012).

9.6.2 *Accelerometry-based balance and gait analysis*

One of the immediate observable consequences when there is a level of problem in the motor control is the change in balance records (higher variability). As a consequence, the possibility of having some kind of assessment about risk of falls derived from new tools, including but also exceeding accelerometers, becomes an interesting challenge, and is currently widely explored, even attempting to use and validate current entertainment devices as rehabilitation tools (Reed-Jones *et al.*, 2012; Singh *et al.*, 2012).

Hence, current work suggests that gait features and risk of fall probability for elderly subjects can be enriched through the use of specific mathematical tools applied to balance patterns from accelerometers (Lee and Carlisle, 2011; Mizuke *et al*, 2009). Such results could lead to new protocols, supplementary to the traditional, qualitative ones used in the field of neurology. These clinical protocols for gait and balance assessment mostly involve measures of performance that are subjective (yet meaningful, since specialized personnel handle them). As previously stated, they consist partly in straight walks, as well as orthostatic balance assessments with open and closed eyes, getting up from and sitting on a chair, and in normal, straight walks while performing different tasks when commanded by the physician.

In short, the study of balance patterns from accelerometers and its subsequent mathematical treatment will bring forward protocols for the evaluation of falls risk through quantifiable, repeatable and reliable, performance indexes, opening the way for an objective measure of progress, or degradation in the motor system. Acquisition of such motion records and their general characteristics requires discussion.

Pictures from a portable system for accelerometry-based gait and balance research are shown in Fig. 9.1. A GUI (Graphical user Interface) running on the computer allows its user to control the sensing terminal and view the measured signals. All fits into the aforementioned description for these devices, that is, wearable, low-power data loggers and processors. In this particular case capable to measure acceleration in three axes at a rate of 100 Hz, through an analog 3D accelerometer with available ranges between $\pm 1.5\ A_{\mathrm{unit}}$, and $\pm 6\ A_{\mathrm{unit}}$, with $A_{\mathrm{unit}} = 9.8$ m/s^2. Other devices include different ranges, with digital-output alternatives, incorporating analog filtering and sampling stages within different sensing circuits. Programmable ranges and high sampling rates are necessary to open

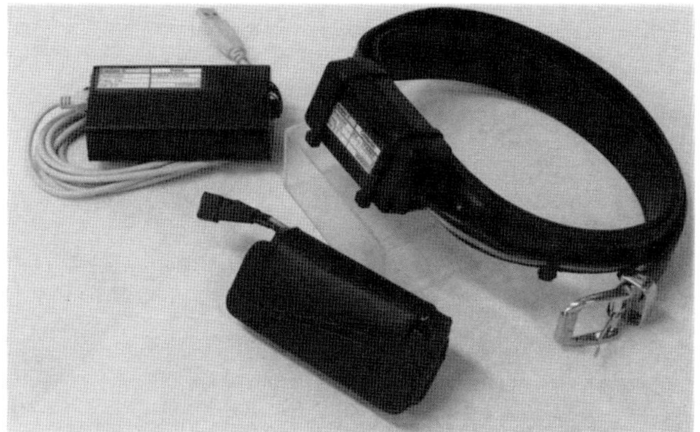

Figure 9.1: System components: The sensing terminal (top) is attached to a belt, which holds the battery pack (bottom right). The base appears on the left.

the way for implementations of these devices in the assessment of fine movements as well as other, more violent gestures (like vertical jumps), within different experiments involving static, semi-static, and dynamic balance tests. For the assessment of risk of falls in the elderly, however, a range of $\pm$ 1.5 A_{unit} is deemed sufficient, since their displacements tend to be less significant. Signals from all patients performing clinical tests were acquired under medical supervision, always under signed informed consent and approval from the ethics committee of the involved institutions.

The actual sensing device is placed in a small case, fixed to an adjustable belt, so that the device can be firmly positioned in place. It is designed to be placed at the widest part of the hips, at the middle of the subject's back, as an approximation of the body COM. This is a common placement for this type of devices and measurements, while other possible places are also in use (Mayagoitia *et al.*, 2001; Yang and Hsu, 2010). With this placement, the orientation of the accelerometer is such that its three axes coincide with the anterior–posterior (AP), vertical (VE), and medial–lateral (ML) axes of motion.

Information provided by the accelerometer forms a tridimensional vector, which can be represented in cartesian coordinates AP, ML, and VE (Fig. 9.2a) or through the magnitude of the composite vector A_{MOD} and two inclination angles, θ_{AP} and θ_{ML} (Fig. 9.2b), as

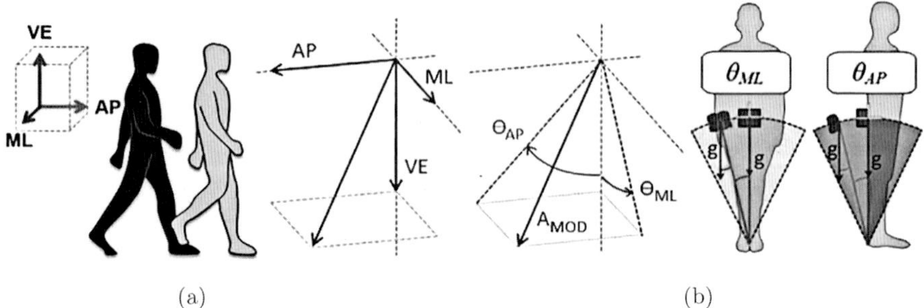

Figure 9.2: Two alternate representations for the 3D-acceleration vector (purple): (a) cartesian components AP, ML, and VE, more adequate for gait analysis, or (b) through A_{MOD} and θ_{AP} and θ_{ML}, suitable for standing balance assessments.

described in Miralles *et al.* (2012). Both representations can be used, according to the case, for data processing. The second alternative, which considers angular coordinates, usually improves the visualization of some static balance experiments.

9.6.2.1 *Calibration and device positioning*

Acceleration signals need calibration for proper interpretation. At least six parameters must be identified (considering a linear response for each axis), two for each of the sensing axes, say, zero-level and sensitivity. One possible approach for these six unknown constants is to use six equations. When a linear accelerometer as the one previously described is not moved, the compound acceleration that should measure would have a magnitude of approximately 9.8 m/s^2, that of the gravitational acceleration. This results in the following equation:

$$|A_{\text{gravity}}| = \sqrt{\left(\frac{x - Zx}{Sx}\right)^2 + \left(\frac{y - Zy}{Sy}\right)^2 + \left(\frac{z - Zz}{Sz}\right)^2} \qquad (9.1)$$

where x, y, and z, actual measured values, in volts;
Zx, Zy, and Zz, zero levels for each axis, in volts;
Sx, Sy, and Sz, sensitivity values for each axis, in V/(m/s^2);
A_{gravity}, compound acceleration for motionless sensing.

Six equations like (9.1) can be obtained, after placing the device in six different orientations, and an approximate solution to this system with six unknown variables can be derived. Since signals are inevitably somewhat noisy, an iterative approach estimates the set of values that brings the system to an optimal solution.

A similar approach for accelerometer calibration through six equations, corresponding to alignments with the full gravitational acceleration in the three axes and both directions for each axis, is proposed as an alternative, also applicable to 2D sensors (Mayagoitia *et al.*, 2002). Considering these extreme values, and assuming perfect alignment, zero-levels and sensitivities derive from the difference and mean between the maximum and minimum values detected by each axis, and considering that the magnitude of the difference between them is twice the gravitational acceleration.

When positioning the sensing device in the desired placement for a subject, careful effort must be made in order to make sure that the axes of the accelerometer are as aligned as possible with the actual expected AP, ML, and VE axes. In particular, this type of misalignments can affect the absolute offset values of measured acceleration signals, since the constant gravitational acceleration would result in a constant offset for these axes. It is not possible to fully eliminate this effect, but an appropriate standing balance record can help to provide an estimate of more accurate zero-levels for the device after positioning. Furthermore, when experiencing imbalance, or for other reasons, subjects can change their posture while walking, slightly shifting the device's orientation once again, and this could actually be a measure of performance or behavior (sudden or slow changes in "zero" levels over long periods of time). Another alternative is to actually consider the dynamics of the signal as their main source of information and not so much their offset, given the fact that the actual movements or changes in the signals is what actually matters for gait assessment, so the following sections will favor the use of relative instead of absolute levels for signal processing. Sensor fusion, based on added sensors (MEMS gyroscopes and compasses, as well as a higher number of accelerometers), is an alternative for the estimation of accurate orientations in space, as well as derived linear accelerations and improved accuracy in motion assessments.

9.6.2.2 Gait records

This section focuses on the signals provided by an acceleration-sensing device placed at an approximation of the COM, and some basic analytical strategies that are considered well suited for these signals. Figure 9.3 shows the averaged gait cycle signals for the AP, ML, and VE axes, corresponding to a steady-state segment of normal walk for a subject (28 years old). This was achieved by detecting feature points for the signals (AP or VE local maxima, e.g., when peaks are visible), and by cutting these records between two consecutive feature points to cover one full gait cycle (using prior knowledge about the characteristics of the signals). The derived segments of the

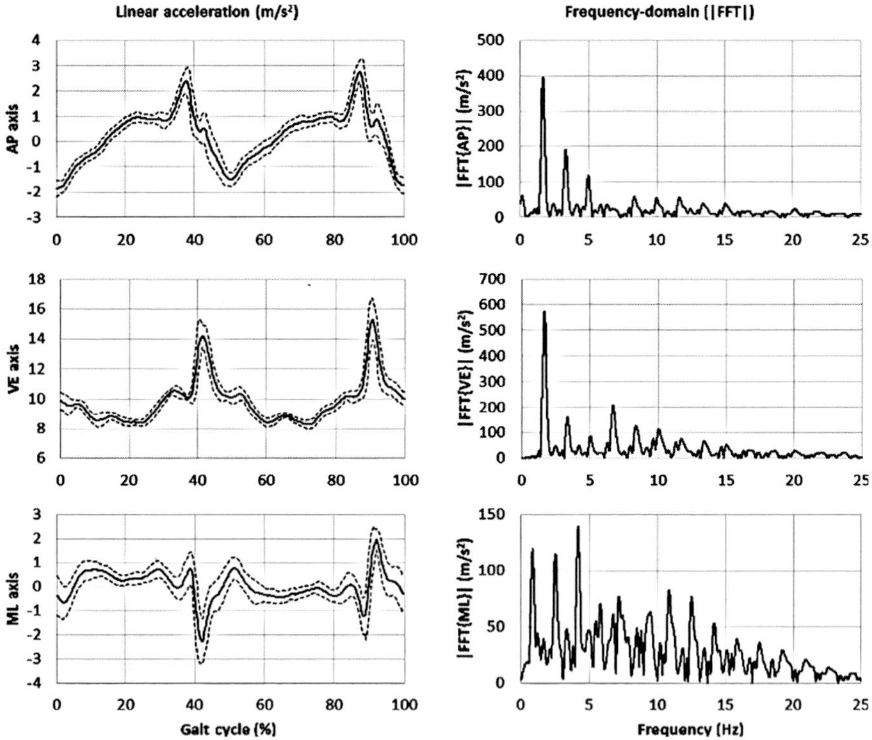

Figure 9.3: Left: Average gait cycle signals for one subject (waist sensor), over 20 cycles, with standard deviations plotted at each side of the mean points (dotted lines). AP, VE, and ML signals are shown. Right: Fast Fourier transform of the corresponding signal segments, indicating high periodicity at multiples of stepping frequency.

main record were later averaged for each point. A similar approach was used by Bugané *et al.* (2012). The original signal is, in fact, the succession of the segments that were cut as individual gait cycles, and later averaged.

Even as an individual case, it can be seen (and also expected) that the signals obtained from steady-state gait records be highly periodic, this hinting possible advantages of implementing signal analysis strategies in the frequency domain. The right side of Fig. 9.3 shows the absolute value of the windowed fast Fourier transform (FFT) of a segment of the three acceleration signals used for mean gait cycle evaluation, after offset subtraction. The result of an FFT of a real signal in time is a complex number, but the absolute value of the signal is initially considered as most important, so, when referring to the FFT, it should be implied that the $|\text{FFT}|$ is calculated within this section. A Hanning[6] window is applied to the time domain signal prior to the transform, as an example of a window function applied to the time-domain signal, useful for spectral analysis. Since signal dynamics can be of more interest than their static levels for gait evaluations, related to device orientation, this preprocessing applies to all signals converted to the frequency domain in this section.

The expected high periodicity can be detected in this new signal ($|\text{FFT}_{\text{AP}}|$), as a distinct peak at a frequency that can be related to a mean step frequency for that segment in the time domain. When the gait cycles become less repetitive or sudden changes occur (sudden imbalances, losses of concentration), the periodicity characteristics of the signal should also change. Therefore, a single $|\text{FFT}|$ from a straight walk record, as well as a continuous evaluation of the structure of the $|\text{FFT}|$ of a segment of the signal as the user walks, can provide information that could be of interest in the analysis of human balance.

Apart from AP signals, ML and VE signals are also of interest, as they show more variable structure, instead they are naturally

---

[6]The **Hann Function**, named after the Austrian meteorologist Julius von Hann (1839–1921), is a discrete window function given by $\omega(n) = (1/2)\{1 - \cos[2\pi n/(N-1)]\}$.

periodic when collected from gait records. The fundamental frequency in the ML axis of a straight walk should be one-half of the step frequency, associated with the main component of the FFT_{AP} for regular subjects. The hips shift to alternating sides once per step, taking two steps to form one full ML cycle. For many cases, the VE signal acquired through this device shows peaks, linked to the actual ankle contact and the following deceleration in the VE axis associated with the beginning of the next step, for each foot. There is an offset level in the VE signal of about 9.8 m/s^2, as shown in this figure, and that is because the plot is showing accelerations as measured by the device, which is sensitive to gravitational acceleration (which, in turn, mostly affects the VE axis when standing straight, even considering sensor misalignments). The dynamics of this signal should remain, however, mostly unaffected by this, while still considering possible changes in posture, as previously stated. VE and ML signals can be particularly variable in shape between different subjects (consider a VE signal for a subject that drags his feet when walking, or the ML signal of a subject with knee problems in one leg). Finally, it is the period of the ML signal that marks the size of a gait cycle, while two periods of the AP and VE signals can be observed within one gait cycle (one corresponding to each leg, only mirrored in the ML direction).

It should be underlined that some specifics of these signals are dependent on the physical properties of the sensing devices, so that some questions appear pertinent. For example: Are they prone to vibrations or oscillations when exposed to certain types of perturbations? Is it possible that they shift their position or spatial orientation in time, while measuring over long periods?

The limits and virtues of each specific device and implementation need to be taken into consideration when analyzing the signals they provide.

9.6.2.3 *Static balance records*

A different set of parameters of interest can be acquired from standing balance signals. Within these cases (i.e., orthostatic balance experiments, where subjects are asked to stand with their feet together and eyes open or closed), accelerations generated by subjects should normally be much lower in magnitude than the gravitational acceleration, which the proposed sensing device is sensitive to. This could enable the use of the accelerometer as an inclinometer for these tests. In this way, the accelerometer would gather information on the spatial orientation of the sensor, showing inclinations and changes in orientation of the subject during standing balance through θ_{AP} and θ_{ML}, over time, as well as sudden changes and peaks in A_{MOD}, due to more violent correction attempts (see Fig. 9.2b, corresponding to an inverted pendulum model).

The purpose of this interpretation of acceleration signals is to obtain an approximation to the signals and features presented by force platforms, which are considered useful for many purposes, such as balance assessments or adequate adaptation of prosthesis (Cuesta López and Lema Calidonio, 2009). However, a key difference between the information provided by these devices is of relevance.

Force platforms track the path followed by the COP of a subject over time, while the proposed acceleration-sensing device approximates accelerations of that subject's COM. While these two terms can be related to one another, they are not interchangeable. The COP is linked to the instant position of the vertical reaction force of the platform. Changes in the COP appear to compensate for changes in the COM, so as to maintain the standing balance (Winter, 2009). Therefore, the ground area covered by the COP should be larger than the one covered by the COM of one subject.

Through this model, and considering these differences, accelerometers can attempt to estimate traditional biomechanical parameters, analogous to those obtained from force platforms. Stabilograms from traditional force platforms track the path covered by the COP through characteristic indexes, such as COP path length, COP path mean radius and speed, internal density, anterior–posterior, and medial–lateral ranges of displacement of the COP. In the case of accelerometer implementations, the proposed approach can be used to approximate similar parameters from the angle–angle plots that can be formed through θ_{AP} and θ_{ML}, for example, in order to analyze the behavior of subjects during standing balance experiments.

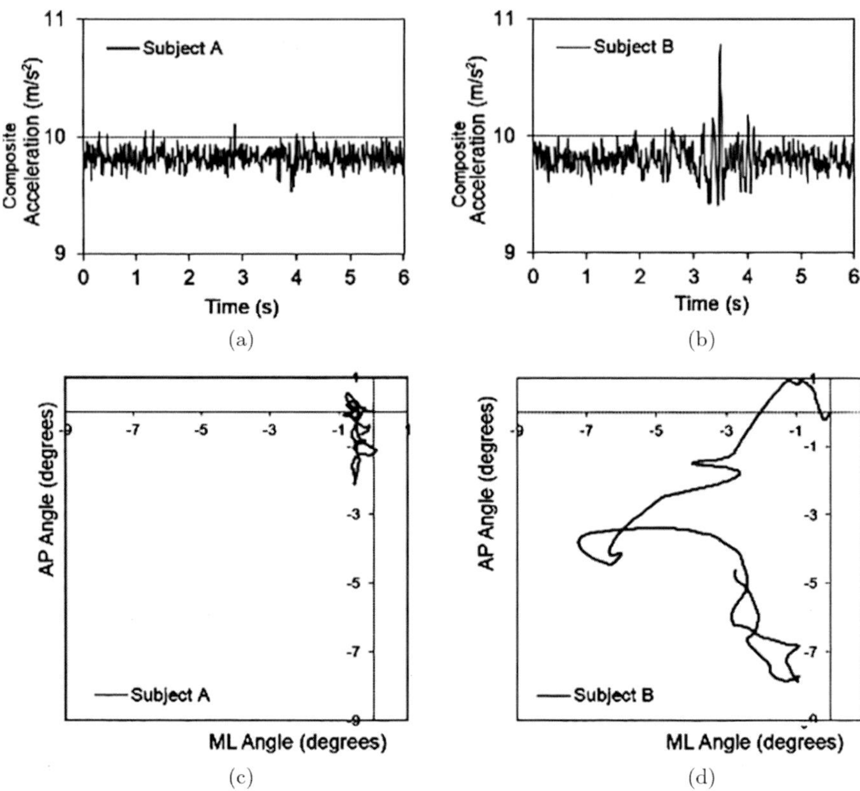

Figure 9.4: Standing balance analysis for subjects A (left), and B (right). (a), (b) Unfiltered composite acceleration magnitude, in m/s². (c), (d) Angle–angle plots of filtered anterior–posterior (AP) versus medial–lateral (ML) displacement from the beginning of the record. Positive AP and ML angles correspond to leaning forward and to the right, respectively.

Figure 9.4 shows results of a standing balance test, gathered with the proposed acceleration-sensing device from two subjects over 60 years old, standing with their feet together, referred to as A and B. Records of 6 seconds are compared.

Figure 9.4a and 9.4b shows the composite acceleration magnitude over time for A and B (Miralles *et al.*, 2012). While A showed an almost flat response, a peak in Fig. 9.4b indicates that subject B may have experienced imbalance and attempted to regain control. Figure 9.4c and 9.4d shows analyses for the same records displayed

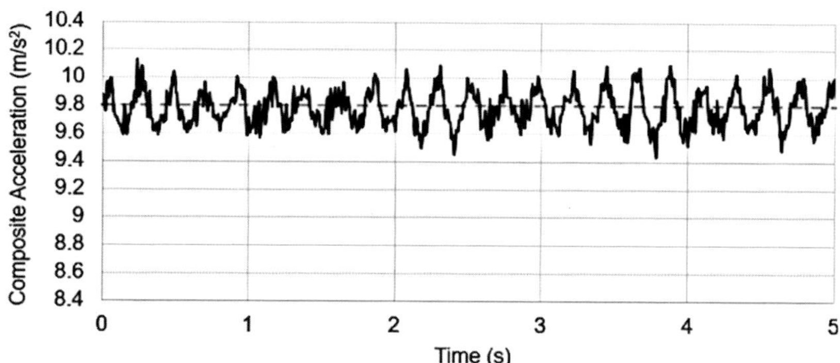

Figure 9.5: Standing balance signal on a subject experiencing lumbar tremors: composite acceleration magnitude. Tremor frequency is visible in the time-domain, and identified at 4.4 Hz through FFT analysis.

in Figs. 9.4a and 9.4b, and shown as angle–angle trajectories from the beginning of each record, in both directions. Angle values at the beginning of the records were considered as reference levels. In order to better estimate these trajectories, both angular signals were filtered with a finite impulse response filter, with a cutoff frequency of 3 Hz. Once again, A showed better control, with an AP sway (range) of 2.7°, and an ML sway of 0.9°, unlike B, who showed significant displacements, with an AP and ML sways of 8.8°and 7.2°, respectively.

Figure 9.5 shows a particular case from a standing balance test. This record was acquired from a subject who was diagnosed with lumbar tremors, and the sensor clearly identified a particular behavior from the standing balance record. Based on a 6-s signal (A_{MOD}, shown in Fig. 9.5), its |FFT| resulted in an isolated peak at a frequency of 4.4 Hz. This emphasizes the potential role of accelerometers as capable of providing additional information in fields beyond that of balance assessment (Miralles *et al.*, 2012).

The attempt to associate acceleration records to those of force platforms is here explored as an indication of potential implementations of these devices in the field of human balance assessment beyond that of direct movement measurements. The actual sensitivity of the accelerometer devices in these implementations is completely device dependent, and the sensitivity of force platforms is, however, a tough match for these new devices. Examples shown here are extreme cases, selected to illustrate different results and types of information that can be acquired from these untraditional devices (specifically, the alternate use of the device as a two-axis inclinometer), and the possibility to compare its information to that from force platforms, considering their differences (Mayagoitia *et al.*, 2001). Sensor fusion strategies can be of value in acquiring relevant, highly precise versions of these signals.

9.6.3 *Signal-processing strategies/case studies*

As described, the aforementioned clinical tests result in comprehensive estimations of risk and response. Given the complexity of the human balance process, a comprehensive analysis becomes almost mandatory. However, since the scoring procedure for these tests is mostly qualitative, a unified measure would be needed for comparisons between tests over long periods of time, or among different patients and physicians. Under basic guidelines from these protocols, the final decision as to what makes "slight" or "severe" imbalances or changes in pattern depends on each specialist. A unified standard would also be necessary for reliable statistical assessments of large groups, necessarily relying on different specialists and locations for data acquisition.

These reliable sources are meant to supplement, and not replace clinical tests, by being taken into account by physicians while evaluating patients, therefore combining their knowledge with new tools. Furthermore, precise signal-processing strategies may lead to finer assessments of conditions and evolution of patients.

For all further cases, referred signals were acquired from patients over 60 years of age, experiencing variable risk of falls, while they performed the aforementioned clinical tests. Refer to cited bibliography for further details on subjects, patient subsets and protocols. The purpose of these analyses concerned two initial objectives (1) and (2) toward a final research goal (3), that is,

(1) To learn more about particular properties of the acceleration signals involved in gait and standing balance tests.

(2) To propose indexes as a supplementary measure of features that physicians find of interest while evaluating a subject's performance, as new sources of objective information.

(3) To implement means for feature selection and patient classification.

Through a joint analysis, general characteristics and features for gait acceleration records can be interpreted in relation to the scores given by physicians over the gait phase of the Tinetti test.

Additionally, records from the initial four stages of the DGI assessment of risk of falls can be evaluated through a moving-window strategy, calculating indexes and parameters with biomechanical meaning within these moving windows and evaluating changes in these parameters. These two signal processing strategies are explored as follows.

9.6.3.1 *Tinetti scale for risk of falls assessment: Analysis and general response*

The gait phase of the Tinetti test consists of two 4-meter-long straight walks, and results in general scores ranging from 0 to 12 points, and covering deviations, imbalances, step symmetry and continuity, posture and step length and height, for example. In order to evaluate general signal characteristics and their relationship with a good or poor performance in this test, a frequency-domain analysis over records of fixed length can be considered as a good initial source of information, given the periodicity characteristics of the gait signals that were covered previously. For these purposes, initial and final transients should be excluded from these signals, leaving a regular signal (depending on the subject) corresponding to the gait regime. The following signal processing strategy is a way for comparing these signals between subjects, that is,

Average step frequency estimation (F_{avg}). Experimental results show that information given by the sensor in the AP axis is the most reliable signal out of the three provided by the sensor for normal gait evaluations: AP signals are the most consistent, and their FFT have a peak at a fundamental frequency ($F_{\mathrm{max}}\{|\mathrm{FFT_{AP}}|\}$), in most cases directly linked to the actual mean stepping frequency (Fig. 9.3). Only when a very significant asymmetry between left and right steps occurs, that peak takes place at approximately one half of the actual mean step frequency (representing a full gait cycle frequency), and those particular cases can be detected and compensated for by comparing the three axes (Miralles *et al.*, 2012). Other significant deviations in posture during the gait record can also generate a low-frequency peak component. For general purposes, the proposed feature for average step frequency is defined in the following equation:

$$F_{\mathrm{avg}} = F_{\max}\{|\mathrm{FFT}_{\mathrm{AP}}|\} \tag{9.2}$$

The average step frequency estimate can be evaluated for each subject, using their AP signals, after applying an offset subtraction and Hanning window to the signal segments.

Frequency-normalized FFT. Since it is difficult to compare signals from different subjects, a frequency-domain compensation can be used to facilitate this task. An ideal subject, walking in the exact same manner, but at two different stepping frequencies, should show similar signals in the frequency domain, except for the fact that one of them would be a stretched version of the other one (the slower walk record will be a "stretched" version of the faster record, without considering changes in magnitude). Therefore, if the actual average step frequency could be isolated from these signals in the frequency domain, a step frequency-independent analysis could be performed. After calculating the $|\mathrm{FFT}|$ for the three signals provided by the accelerometer, these three results are normalized by re-scaling their x-axes (frequency), dividing them by F_{avg} ($f_{\mathrm{N}} = f/F_{\mathrm{avg}}$). As a result, a frequency-normalized FFT is obtained for each of the three $|\mathrm{FFT}|$ obtained from steady-state gait records. The frequency-normalized $|\mathrm{FFT}|$ for each axis is described by Eq. (9.3), and an example is shown in Fig. 9.6.

$$\mathrm{FFT}_{\mathrm{Nf}}(f_{\mathrm{N}}) = |\mathrm{FFT}(f/F_{\mathrm{avg}})| \tag{9.3}$$

The result of the two previous signal processing stages is a value of F_{avg} and a set of three frequency-normalized $|\mathrm{FFT}_{\mathrm{Nf}}|$, one for

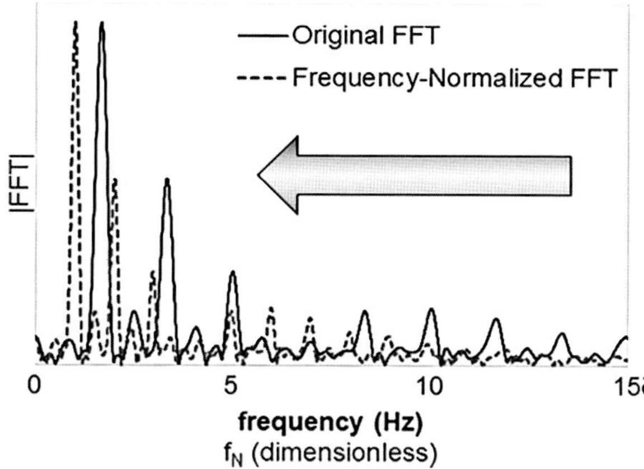

Figure 9.6: Frequency-normalized |FFT| of an AP signal.

each signal from each subject (be them in cartesian or angular coordinates). Since these new results are now normalized, they can be compared between subjects. This strategy was implemented by Miralles *et al.* (2012) in order to detect general response characteristics from 51 subjects over 60 years of age performing the gait phase of the Tinetti test, according to their score in that phase. Scores from that phase (0–12) were divided in three sub-groups, representing poor, medium, and good performance, and each normalized signal from their gait records (5 s measurements of steady-state walk) was averaged out within each sub-group. Finally, three averages were obtained for each signal as shown in Fig. 9.7, indicating key features corresponding to a good performance, as listed below:

- Most of the spectral information considered of interest for these subjects was concentrated within the first 6–10 multiples of the average step frequency estimate.
- Values of Fig. 9.7a for $f/f_{\mathrm{med}} = 1$ can be proposed as an indirect measure of general energy expenditure for the exercise (this value decayed for the sub-groups with lower scores in the test).
- The ratio between the values of Fig. 9.7b for $f/f_{\mathrm{med}} = 1$ and Fig. 9.7c for $f/f_{\mathrm{med}} = 0.5$ can be proposed as an indication of how much effort is actually directed to moving forward (corresponding

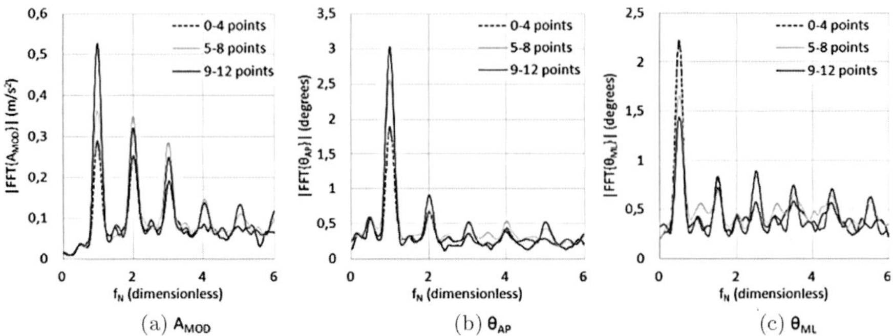

Figure 9.7: Sub-group averages, according to Tinetti-scale gait phase scores, for normalized signals: (a) $|\text{FFT}\{A_{\text{MOD}}\}|$, (b) $|\text{FFT}\{\theta_{\text{AP}}\}|$, and (c) $|\text{FFT}\{\theta_{\text{ML}}\}|$.

to higher values of this ratio) and how much is dedicated to lateral sways.

These results show how information provided by the accelerometer can actually translate into features that link, at least initially, to performance in the gait phase of the Tinetti test, therefore pointing toward potential uses of these devices for evaluations of risk of falls.

Gait symmetry is another characteristic available from accelerometry. Step symmetry can be roughly defined as the similarity between parts of the gait signal that correspond to left and right steps. In order to quantify this characteristic from gait signals, *autocorrelation* is a useful tool. Autocorrelation of a discrete-time signal $(r_{xx}[j])$, compares a signal with a time-shifted copy of itself. Its algebraic expression, for a discrete sequence $x[n]$, starting in $n = 0$, zero mean value and size N (in samples), is shown in the following equation:

$$r_{xx}[j] = \sum_{n=-N}^{N} x[n]x[n-j], \quad N < j < N \tag{9.4}$$

An advanced version of this regular autocorrelation is described by Eq. (9.5), that is, a dimensionless version of Eq. (9.4), normalized to a value of 1 for $j = 0$, and divided by $N - |j|$, to compensate for null coefficients that appear as the signals are shifted. While this form of autocorrelation eliminates the downward slopes from $j = 0$, resulting in Eq. (9.4) due to the growing amount of null coefficients, it also sets a limit to the reliability of its value, for $|j|$ approaching the limits ($|j| \approx N$).

$$rn_{xx}[j] = \frac{N}{r_{xx}[0]} \frac{r_{xx}[j]}{(N - |j|)}, \quad N < j < N \tag{9.5}$$

This implementation of autocorrelation for step symmetry assessment relies on the fact that autocorrelation is higher when the original signal and its time-shifted copy are more similar. For a periodic signal, this happens at shift values corresponding to multiples of the step period, close to $1/F_{\mathrm{avg}}$ (Yang *et al.*, 2011). As a result, the autocorrelation of the AP signal, for example, will likely resemble a wave. In the case of VE signals, this autocorrelation would emphasize peaks, and the ML signal would have relevant peak values that can be negative and positive, given the difference between its fundamental frequency and that of the other axes. An example is shown in Fig. 9.8.

In general, it can be assumed that the value of the first local maxima from $j = 0$, for AP and VE signals, corresponds to a relative displacement of one step period between signals, therefore corresponding to a measure of how similar these signals are when there is a shift of one step. This would represent a measure of similarity between a succession of steps made with the left and the right leg, as long as subjects always perform alternate steps and do not skip stepping on one foot (which is not likely to be the case). The remaining ML signal has a local minimum at the time where the first local maximum locates, for AP and VE signals. It can be proposed, then, that these first local maxima could be in a way related to step symmetry and regularity, be it on their own or in relation to other features, like the second local maximum (Mizuike *et al.*, 2009; Yang *et al.*, 2011; Miralles *et al.*, 2012).

A signal-processing strategy for gait symmetry measurements through this first local maximum was applied to 50 subjects over 60 years of age, performing the Tinetti scale (Miralles *et al.*, 2012). The same records that were previously used for spectral analysis (after cutting initial and final transients) were used for autocorrelation analysis.

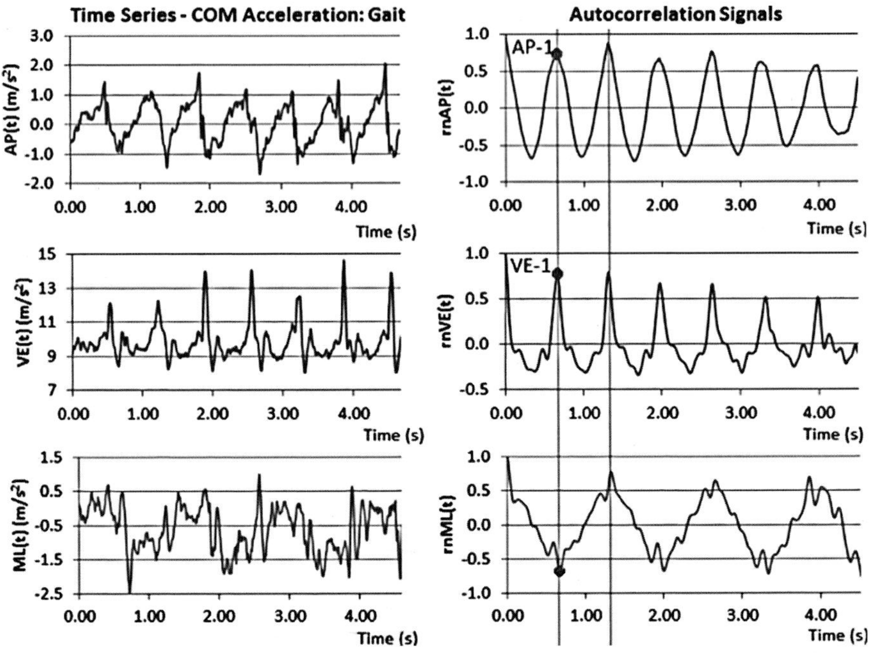

Figure 9.8: Autocorrelation maximum values between alternate steps. Left: acceleration signals (AP, VE, and ML) for gait evaluations. Right: respective autocorrelations, with emphasized peaks. The first maximum values for AP and VE axes (AP-1 and VE-1, respectively) are relevant to the clinical scores of the Tinetti test.

VE and AP first local-maxima from autocorrelation signals were analyzed according to the scores that each subject received from the physicians concerning step symmetry (0 or 1), as shown in Miralles *et al.* (2012). Higher values of these two features should correspond to more symmetric steps. While not all of the cases could be isolated through these parameters, these estimations of step symmetry could discriminate thresholds for determining highly asymmetrical cases. It should be noted that this measure of symmetry is a continuous one, unlike the test score, which is a Boolean value. However, this value, provided by the physicians, can encompass more than one variable, which may or may not be detectable by the accelerometer. It should be re-emphasized that the proposal for the addition of this type of devices is meant to be supplementary, and not just a replacement to previously available measures of performance.

9.6.3.2 *Dynamic gait index: Feature extraction*

In this section, a quantitative approach to the evaluation of different DGI stages is presented. Unlike the previous approach, this test requires to not only evaluate general performance characteristics but also to evaluate the progression of certain features related to the gait cycles over time, while dynamics are set into play, specifically changes in speed and in balance during the straight walk that conforms most stages. As a result, a *moving-window feature extraction* process is proposed, so that its results allow us to see how much these features change over time, during each stage of the test (Miralles *et al.*, 2013).

In order to obtain actual indexes of performance for these tests, which are of variable length, it is convenient to calculate performance features in a smaller window of the whole length of the records, as proposed in Fig. 9.9. The highlighted window signal (shaded rectangle in the figure) is processed in order to obtain a value for the proposed features, then the window is moved one sample to the right, and the process is repeated. As the window is displaced, a new feature signal is obtained, Feature(t), representing the evolution of that signal over time, as shown in the figure.

Changes in these signals over time should be representative of the performance for each subject during a certain stage, given the features that are to be proposed for these tests. The size of the window should be smaller than the full record, long enough to include an even number of steps, and short enough to allow for a feature signal of relevance. A fixed window four times the average size step period

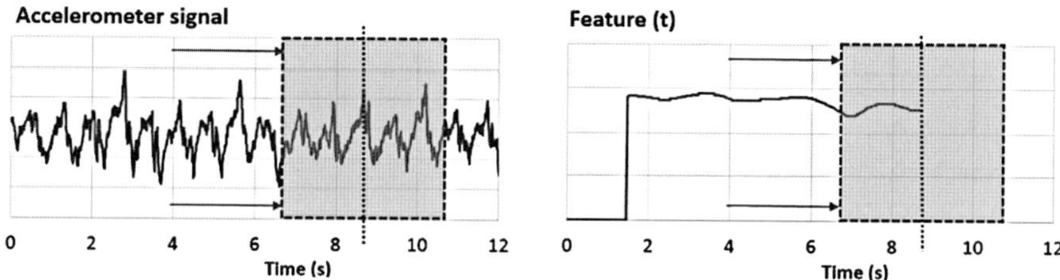

Figure 9.9: Moving-window feature extraction strategy for a signal. As the window (shaded) is moved over the signal (left), a time chart for the feature, Feature (t) is created (right).

from the normal walk, as gathered for the Tinetti test, is chosen in this case, while other alternatives are available.

When physicians evaluate a subject while performing the DGI assessment, they follow a series of general guidelines that point out what could be related to a specific score, from 0 to 3 (Shumway-Cook and Wollacott, 1995). Most stages (particularly 1–4) are scored according to evaluations of walking speed or fluctuations in balance. In relation to these characteristics, three features can be selected as sources of performance information for stages 1–4 of this test, estimating walking speed (linked to step frequency and stride length) and changes in balance (which can be roughly associated to changes in the shape of the signal, independently of its amplitude and fundamental frequency). These features are described in detail as follows (Miralles *et al.*, 2013):

Average step frequency estimate (F_{avg}), as described for the Tinetti test, but with additional considerations (incorporating ML feature frequency and smoothing, in order to work with more dynamic experiments, resulting in an improved step frequency estimate.

Average energy expenditure per step factor (E_{step}). In order to estimate the subject's energy expenditure for each acceleration signal ($a[n]$), E_{avg} is defined in Eq. (9.6). The proposed, derived factor for per step measurements, (E_{step}) is defined in Eq. (9.7).

$$E_{avg} = \frac{1}{W} \sum_{n=n_0}^{n=n_0+W-1} (a[n] - \text{mean}(a[n]))^2 \qquad (9.6)$$

$$E_{step} = \frac{E_{avg}}{F_{avg}} \qquad (9.7)$$

An alternate version of Eq. (9.7), E'_{step} is proposed, calculated after applying the window function. Therefore, this estimate is smoother and more suitable for further analysis.

Signal shape evolution factor (FFT_{evol}). This feature represents changes in the actual shape of the records over time, when compared to the average performance of a normal walk. The first step in order to obtain this feature is to calculate the frequency-normalized $|\text{FFT}|$

for the three signals provided by the accelerometer, as indicated for the Tinetti test processing strategy. Results are then normalized by re-scaling the y-axis with the following square root:

$$\sqrt{E_{\mathrm{avg}}}$$

This normalized spectrum is defined as

$$\mathrm{FFT_N}(f_\mathrm{N}) = \frac{|\mathrm{FFT}(f/F_{\mathrm{avg}})|}{\sqrt{E_{\mathrm{avg}}}} \tag{9.8}$$

This new feature should be independent of both the average step frequency and its average magnitude, therefore being only indicative of the shape of the signal, its structure and composition in the frequency domain. In order to compare two normalized spectra (e.g., a spectrum from any stage $|\mathrm{FFT1_N}|\,(f_\mathrm{N})$ to the average from the normal walk $|\mathrm{FFT2_N}|\,(f_\mathrm{N}))$, a mean square error is used in the following equation:

$$\mathrm{FFT}_{\mathrm{evol}} = \frac{1}{N_{\mathrm{bins}}} \sum_{f_\mathrm{N}=0}^{f_\mathrm{N}=f_\mathrm{N}\,\mathrm{max}} (|FFT1_\mathrm{N}|[f_\mathrm{N}] - |FFT2_\mathrm{N}|[f_\mathrm{N}])^2 \tag{9.9}$$

where N_{bins} is the number of bins for the normalized functions. This final feature is to be calculated for all three signals from a record, comparing each $\mathrm{FFT_N}$ to its mean counterpart for a normal walk. Higher values reflect a larger difference between signal shapes. This strategy can also be used to compare the average response of a normal walk to smaller segments of that walk record, so this does not imply that the feature cannot be used for normal straight walk assessments. f_{max} will necessarily be less than one half of the sampling frequency.

Seven signals can be obtained through the application of these indexes to the three acceleration signals provided by the sensing device: $F_{\mathrm{avg}}(t)$, $E'_{\mathrm{step}}\mathrm{AP}(t)$, $E'_{\mathrm{step}}\mathrm{ML}(t)$, $E'_{\mathrm{step}}\mathrm{VE}(t)$, $\mathrm{FFT}_{\mathrm{evol}}\mathrm{AP}(t)$, $\mathrm{FFT}_{\mathrm{evol}}\mathrm{ML}(t)$, and $\mathrm{FFT}_{\mathrm{evol}}\mathrm{VE}(t)$, for example, when working with an orthogonal component-representation of the vector. This set of features separates measures of the fundamental frequency, intensity and shape of each signal, as indicated in Fig. 9.10. While these parameters are theoretically independent, their relationship to each other depends on each subject.

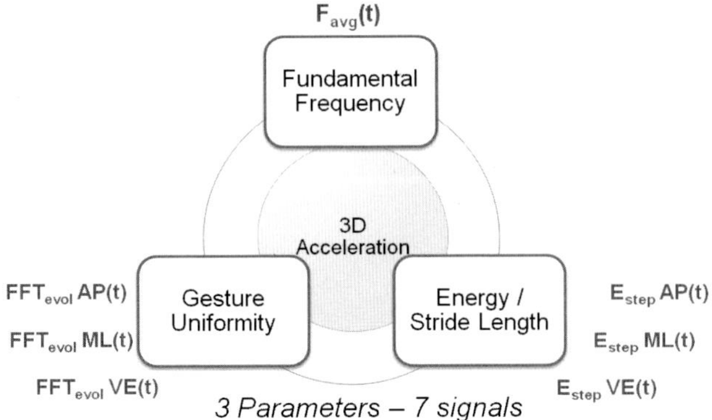

Figure 9.10: Signal acquisition strategy for dynamic gait assessments. Instead of generating one index or a single transform from a fixed record, other experiences require to assess the evolution over time of certain parameters, resulting in the derivation of signals from the original 3D acceleration components. Changes in these seven signals are proposed as indicative of performance.

Now that these features have been defined, a relationship as to what should be expected of them during each stage can be established. Based on the "clinical" evaluations for a good response by the subjects Table 9.2 translates these conclusions to the features that have been defined for gait progress evaluations. During an ideal straight normal walk, there should be no great changes in speed (no great changes in step frequency or in the average energy expenditure per step in any direction) or in the balance structure of the subject, which is intended to be measured through $\mathrm{FFT_{evol}}(t)$ for the three coordinates. For walks with speed changes, however, there should be significant changes in either or both $\mathrm{F_{avg}}(t)$ and $E'_{\mathrm{step}}(t)$, with only moderate changes in $\mathrm{FFT_{evol}}(t)$. This implies that basic statistical evaluations over these new feature signals (mean value, maximum, and standard deviation) could be of interest as objective indexes of performance during the tests.

An initial evaluation of these features, indexes, and their biomechanical meaning was presented, and compared to the clinical results of their evaluations (Miralles *et al.*, 2013). As a result, averaged results per score provided coherent tendencies to the ones presented in Table 9.2, and the most reliable of these seven features was

Table 9.2: Expected relationship between accelerometry-based features and performance in DGI index assessment.

Feature	Normal walk	Walk with changes in speed	Walk with head turns
$F_{\mathrm{avg}}(t)$	Constant	Significant changes	Constant
$E_{\mathrm{step}}(t)$			
$FFT_{\mathrm{evol}}(t)$	Constant	Slight to moderate changes	Constant

identified. Additionally, both supplementary features that aim at measuring walking speeds (step frequency and average energy expenditure per step) provided compensating information for cases when one of the features could not clearly identify these sub-groups in average.

9.6.3.3 *Subject classification*

After defining relevant, potential indexes of performance, the final task in gait or balance assessment is the classification of a subject, according to the expected results from records acquired during clinical tests, into groups with a low, sometimes moderate, or high risk of falling. This stage in signal processing relies on *feature subset selection* strategies, as well as in the consolidation of *classifiers*, based on the features with the highest performance. These classifiers can be binary (good/poor response), or rely on continuous measures of performance for patient assessment. AdaBoost, neural networks, as well as linear discriminant analysis and support vector machines have been implemented for these tasks (Aziz and Robinovitch, 2011; Barth *et al.*, 2012). A specific challenge associated with motion signals is the process of achieving generalization from a sometimes-limited subject group.

Concerning the aforementioned signals and indexes obtained from the first two stages of the DGI assessment, a modified AdaBoost and Leave-One-Out Cross-Validation strategy was implemented (Ghersi *et al.*, 2014). Based on this study over a relatively small population, results provided preliminary high classification rates, in particular, for the case of walks with speed changes. Besides, they show promising results for the case of a normal walk (stage 1), when taking into consideration the biomechanical meaning of the derived indexes, hinting at the relevance of these indexes and their potential to classify subjects into good or poor performance.

9.7 Summary

This chapter presented the basic topics concerning the musculoskeletal system. It aims to stimulate further reading and research on those subjects the reader finds intriguing, since each is an extensive field in itself, with an abundance of high-complexity models and simulations. This has been the origin of a preliminary introduction to the second part, centered on equilibrium and human balance, a highly sensible subject of social concern, in particular when directed at the assessment of risk of falls for older adults. Currently, the subject area envisions new modeling, simulations, and experimentation, based on novel means and advanced motion analysis devices. The road covered by this second portion of the chapter has been aimed at presenting different stages required to propose a signal-processing strategy toward performance classifiers for gestures and movements, considering human gait as an application example. Countless alternative parameters could be chosen and calculated, resulting in derived indexes and signals from records coming from one or more sensors. Data interpretation would depend on the experiment and gesture that is to be understood, as well as on the sensing device and the test subject (age, mobility), for any signal processing stage. As a final note, a thorough dedicated study should be devoted to the control system involved in voluntary motion, reaching into the CNS, allowing us to point, in a comprehensive manner to a neural-musculoskeletal system.

References

Aziz O, Robbinovitch SN. An analysis of the accuracy of wearable sensors for classifying the causes of falls in humans. *IEEE Trans Neural Syst Rehabil Eng* 19(6):670–676, 2011.

Baezner H, Blahak C, Poggesi A, Pantoni L, Inzitari D, Chabriat H, Erkinjuntti T, Fazekas F, Ferro JM, Langhorne P, O'Brien J, Scheltens P, Visser MC,

Wahlund LO, Waldemar G, Wallin A, Hennerici MG, LADIS Study Group. Association of gait and balance disorders with age-related white matter changes. *Neurology* 70(12):935–942, 2008.

Barth J, Sünkel M, Bergner K, Schickhuber G, Winkler J, Klucken J, Eskofier B. Combined analysis of sensor data from hand and gait motor function improves automatic recognition of Parkinson's disease. *EMBC Ann Intl Conf IEEE* 2012:5122–5125, 2012. doi 10.1109/EMBC.2012.6347146.

Bronzino JD, Peterson DR. Tissue engineering and artificial organs. 3rd ed. CRC Press, Boca Raton, 2006.

Bugané F, Benedetti MG, Casadio G, Attala S, Biaggi F, Manca M, Leardini A. Estimation of spatial-temporal gait parameters in level walking based on a single accelerometer: Validation on normal subjects by standard gait analysis. *Comput Methods Programs Biomed* 108(1):129–137, 2012. doi: 10.1016/j.cmpb.2012.02.003.

Caldwell GE, Robertson GE, Whittlesey SN. Forces and their measurement. In: Research methods in biomechanics. 1st ed., Chapter 4. Human Kinetics, Champaign, 73–102, 2004.

Collins S, Ruina A, Tedrake R, Wisse M. Efficient bipedal robots based on passive-dynamic walkers. *Science* 307(5712):1082–1085, 2005.

Cuesta López LF, Lema Calidonio JD. "CgMed": Diseño y construcción de plataforma para determinar posición del centro de gravedad en bipedestación (in Spanish, Design and construction of platform for center of gravity determination in standing position). *Rev Ingen Bioméd* (Colombia) 3(6):26–36, 2009.

Delbaere K, Close JCT, Mikolaizak AS, Sachdev PS, Brodaty H, Lord SR. The Falls Efficacy Scale International (FES-I). A comprehensive longitudinal validation study. *Age Ageing* 39(2):210–216, 2010.

Delp SL, Anderson FC, Arnold AS, Loan P, Habib A, John CT, Guendelman E, Thelen DG. OpenSim: Open-Source software to create and analyze dynamic simulations of movement. *IEEE Trans Biomed Eng* 54(11):1940–1950, 2007.

Denkinger MD, Igl W, Lukas A, Bader A, Bailer S, Franke S, Denkinger CM, Nikolaus T, Jamour M. Relationship between fear of falling and outcomes of an inpatient geriatric rehabilitation population-fear of the fear of falling. *J Am Geriatric Soc* 58(4):664–673, 2010.

Djuric-Jovicic MD, Jovicic NS, Popovic DB. Kinematics of Gait: New method for angle estimation based on accelerometers. *Sensors* 11(11):10571–10585, 2011. doi:10.3390/s111110571.

Fung YC. The meaning of the constitutive equation. In: Biomechanics, mechanical properties of living tissues. 2nd ed., Chapter 2. Springer, New York, 23–65, 1993a. ISBN 3-540-97947-6.

Fung YC. Bone and cartilage. In: Biomechanics, mechanical properties of living tissues. 2nd ed., Chapter 12. Springer, New York, 500–538, 1993b. ISBN: 3-540-97947-6.

Ghersi I, Álvarez F, Miralles MT. Classification of performance in risk-of-falls assessment based on accelerometer data and feature boosting. IFMBE

Proc, vol. 49, VI Latin American Congress Biomedical Engineering CLAIB, Paraná, Argentina 29, 30 & 31 October, 607–610, 2014. http://link. springer.com/chapter/10.1007%2F978-3-319-13117-7_155.

Hill KD, Wee R. Psychotropic drug-induced falls in older people: A review of interventions aimed at reducing the problem. *Drugs Aging* 29(1):15–30, 2012.

Johnson FE, Virgo K. The bionic human: Health promotion for people with implanted prosthetic devices. Human Press, New Jersey, 720, 2006.

Kapit W, Elson LM. The anatomy coloring book. 3rd ed. Pearson Education, San Francisco, CA, 400, 2002. ISBN 9780805350869.

Kempen GI, Todd CJ, Van Haastregt JC, Zijlstra GA, Beyer N, Freiberger E, Hauer KA, Piot-Ziegler C, Yardley L. Cross-cultural validation of the Falls Efficacy Scale International (FES-I) in older people: Results from Germany, the Netherlands and the UK were satisfactory. *Disabil Rehabil* 29(2):155–162, 2007.

Klausner S, Szriber M, Miralles MT. Bioengineering and arthroscopy: Knotless anchor. *IFMBE Proc.*, vol. 49; and VI Latin American Congress Biomed Eng (CLAIB), Paraná, Argentina, 29–31 October, 711–714, 2014. http://link.springer.com/chapter/10.1007/978-3-319-13117-7_181.

Kuo AD. Stabilization of lateral motion in passive dynamic walking. *Int J Robot Res* 18(9):917–930, 1999.

Lee RYW, Carlisle A. Detection of falls using accelerometers and mobile phone technology. *Age Ageing* 40(6):690–696, 2011.

Lohmeier S. Design and realization of a humanoid robot for fast and autonomous bipedal locomotion. Technische Universität München, Lehrstuhl für Angewandte Mechanik, Munich, Germany, 214 pp., 2010. http://mediatum.ub.tum.de/doc/980754/980754.pdf.

Mavridis N. A review of verbal and non-verbal human–robot interactive communication. *Cornell University Library*, 2014. http://arxiv.org/pdf/1401.4994.pdf.

Mayagoitia RE, Lötters JC, Veltink PH, Hermens H. Standing balance evaluation using a triaxial accelerometer. *Gait Posture* 16(1):55–59, 2001.

Mayagoitia RE, Nene AV, Veltnik PH. Accelerometer and rate gyroscope measurement of kinematics: An inexpensive alternative to optical motion analysis systems. *J Biomech* 35(4):537–542, 2002.

McMinn RMH, Hutchings RT. Color atlas of human anatomy. Year Book Medical Pub., London, 352 pp., 1977. There are several editions and with other authors, too.

Miralles MT, Ghersi I, Vecchio R, Paterson R, Pérez Akly M, Ferrando M, Paterson A, Álvarez F. Comprehensive feature extraction for objective Dynamic Gait Index assessment of risk of falls in the elderly. *J Phys Conf Series* 477, Conference 1; 012029; 10, 2013. doi:10.1088/1742-6596/477/1/012029.

Miralles Marrero RC, Miralles Rull I, Puig Cunillera M. Columna vertebral. In: Masson S, ed. Biomecanica Clínica de los Tejidos y las Articulaciones del Aparato Locomotor [In Spanish, Vertebral Column; Clinical Biomechanics of Tissues and the Locomotor Apparatus Joints]. 2nd ed., Chapter 11. Barcelona, Spain, 177–201 pp., 2005.

Miralles M, Vecchio R, Ghersi I, Paterson R, Pérez Akly M, Ferrando M, Paterson A, Álvarez F. Estudios de equilibrio en pacientes con riesgo de caída a partir de datos de acelerometría en tres ejes [in Spanish, Balance assessments in patients facing risk of falls, based on 3D-accelerometry data]. *Proc XIV Jornadas Internacionales Ingeniería Clínica y Tecnología Médica*, City of Paraná, Province of Entre Ríos, Argentina, 2012. http://bioingenieria.edu. ar/grupos/geic/biblioteca/j2012/Documentos/Trabajos/T12TCAr14.pdf.

Mizuike C, Ohgi S, Morita S. Analysis of stroke patient walking dynamics using a tri-axial accelerometer. *Gait Posture* 30(1):60–64, 2009.

Mondani M, Ghersi I, Miralles MT. Video-analysis interface for angular joint measurement. *Proc XX Congreso Argentino de Bioingeniería* [in Spanish, XX Argentine Congress of Bioengineering]. SABI, San Nicolás, Argentina, 2015.

Mow VC, Kuei SC, Lai WM, Armstrong CG. Biphasic creep and stress relaxation of articular cartilage in compression: Theory and experiments. *J Biomech Eng* 102(1):73–84, 1980.

Mow VC, Ratcliffe A, Woo SL-Y. Biomechanics of diarthrodial joints. Vols. 1 and 2. Springer-Verlag, New York, 1990. ISBN: 978-1-4612-8015-6.

Murphy-Chutorian E, Trivedi MM. Head pose estimation in computer vision: A survey. *IEEE Trans Pattern Anal Mach Intell* 31(4):607–626, 2009.

Nasseri N, Hadian MR, Bagheri H, Talebian S, Olyaei G. Reliability and accuracy of joint position sense measurement in the laboratory and clinic; utilizing a new system. *Acta Medica Iranica* 45(5):395–404, 2007.

Noble PC, Alexander JW, Lindahl LJ, Yew DT, Granberry WM, Tullos HS. The anatomic basis of femoral component design. *Clin Orthop Relat Res* (235):148–165, 1998.

Norberg JD. Biomechanical analysis of race walking compared to normal walking and running gait. *Theses and Dissertations–Kinesiology and Health Promotion*; Paper 20, University of Kentucky, Kentucky, 184, 2015. http:// uknowledge.uky.edu/khp_etds/20.

Novacheck TF. The biomechanics of running. *Gait Posture* 7(1):77–95, 1998.

Párraga Martinez I, Navarro Bravo B, Pretel FA, Denia Muñoz JN, Elicegui Molina RP, López-Torres Hidalgo J. Miedo a las caídas en personas mayores no institucionalizadas [In Spanish, Fear to fall in older non institutionalized people]. *Gaceta Sanitaria* 24(6):453–459, 2010.

Paterson R, Del Sel G, Paterson A, Miralles MT, Del Sel N. Elastoplasticity analysis of the nails used in long bone fractures. J Phys: Conf Ser, vol. 332, conference 1, 8th Argentine Bioengineering Conference (SABI 2011) and 7th Clinical Engineering Meeting 28–30 September, Mar del Plata, Argentina, 2011. http://iopscience.iop.org/article/10.1088/1742-6596/332/1/012002.

Peebles L, Norris B. Adult data: The handbook of adult anthropometric and strength measurements. Data for design safety. Department of Trade and Industry, London, 332, 1998.

Peydro de Moya MF, Baydal Bertomeu JM, Vivas Broseta MJ. Evaluación y rehabilitación del equilibrio mediante posturografía [In Spanish, Evaluation and rehabilitation of equilibrium by means of posturography]. *Rehabilitación* 39(6):315–323, 2005.

Pruitt LA, Chakravartula AM. Mechanics of biomaterials: Fundamental principles for implant design. 1st ed. Cambridge University Press, Cambridge, 696 pp., 2011. ISBN 978-0521762212.

Ratner BD, Hoffman AS, Schoen FJ, Lemons JE. Biomaterials science: An introduction to materials in medicine. Academic Press, San Diego, 484 pp., 1996. ISBN 9780125824613.

Reed-Jones RJ, Dorgo S, Hitchings MK, Bader JO. WiiFit^{TM} Plus balance test scores for the assessment of balance and mobility in older adults. *Gait Posture* 36(3):430–433, 2012.

Robertson DGE, Caldwell GE, Hamill J, Kamen G, Whittlesey SN. Research methods in biomechanics. 1st ed. Human Kinetics, Champaign, 309, 2004. ISBN 0-7360-3966-X and ISBN 978-0736039666.

Rowe PJ, Myles CM, Hillmann SJ, Hazlewood ME. Validation of flexible electrogoniometry as a measure of joint kinematics. *Physiotherapy* 87(9):479–488, 2001.

Rudge P, Bronstein AM. Investigations of disorders of balance *J Neurol Neurosurg Psychiatry* 59(6):568–578, 1995. doi:10.1136/jnnp.59.6.568.

Shumway-Cook A, Wollacott M. Motor control: Theory and practical applications. 1st ed. Williams and Wilkins, Michigan, 475 pp., 1995. ISBN 0683077570 and ISBN 9780683077575.

Singh DK, Rajaratnam BS, Palaniswamy V, Pearson H, Raman VP, Bong PS. Participating in a virtual reality balance exercise program can reduce risk and fear of falls. *Maturitas* 73(3):239–243, 2012.

Sonje SU, Borde SV. Micro-Electromechanical systems (MEMS). *Int J Mod Eng Res* 4(3):102–105, 2014.

Srikanth V, Beare R, Blizzard L, Phan T, Stapleton J, Chen J, Callisaya M, Martin K, Reutens D. Cerebral white matter lesions, gait, and the risk of incident falls: A prospective population-based study. *Stroke* 40(1):175–180, 2009. doi:10.1161/STROKEAHA.108.524355.

Stergiou N. (2004) Innovative analyses of human movement: Analytical tools for human movement research. Human Kinetics, Champaign, 344 pp., 2004. ISBN 0-7360-4467-1.

Tinetti ME, Williams TF, Mayewski R. Fall Risk Index for elderly patients based abilities *Am J Med* 80(3):429–434, 1986.

United Nations, Department of Economic, Social Affairs, Population Division. Characteristics of the older population. In World population ageing. Chapter IV. United Nations, New York, 2013. http://www.un.org/en/development/ desa/population/publications/pdf/ageing/WorldPopulationAgeing2013.pdf.

Whittle M. Gait analysis: An introduction. 4th ed. Butterworth-Heinemann, Michigan, 255 pp., 2007. ISBN 0750688831 and 9780750688833.

Winter DA. Human balance and posture control during standing and walking. *Gait Posture* 3(4):193–214, 1995.

Winter DA. Biomechanics and motor control of human movement. 4th ed. Wiley, Ontario, Canada; 384 pp., 2009. ISBN 978-0-470-39818-0.

World Health Organization. World report on disability. World Health Organization Press, Geneva, Switzerland, 363 pp., 2011. ISBN 978 92 4 068825 4 (ePub), ISBN 978 92 4 356418 0 (print). http://www.who.int/disabilities/world_report/2011/report.pdf.

Yamaguchi, GT. Dynamic modeling of musculoskeletal motion: A vectorized approach for biomechanical analysis in three dimensions. Springer Science & Business Media, New Delhi, India; 257 pp; ISBN 0387287043 and ISBN 9780387287041.

Yang C, Hsu Y. A review of accelerometry-based wearable motion detectors for physical activity monitoring. *Sensors* (Open-Access Journal) 10(8):7772–7788, 2010.

Yang C, Hsu Y, Shih K, Lu J. Real-time gait cycle parameter recognition using a wearable accelerometry system. *Sensors* 11(8):7314–7326, 2011.

CHAPTER 10

THE TIMEKEEPING SYSTEM: A KEY ORGANIZATIONAL ELEMENT

Veronica Sandra Valentinuzzi and Gisele Akemi Oda

There is a time for everything ... and a season for every activity under the heavens.

Ecclesiastes 3:1

Abstract

Life has evolved in a cyclic environment, and consequently, during evolution the time dimension was incorporated in the inner organization of every single biological structure. This incorporation is evident in the biological rhythms detected in the whole phylogenetic range, from unicellular organisms to humans, and at all organizational levels within an organism, from molecular to behavioral variables. "Doing the right thing at the right time" enhances reproduction, health, survival, and longevity. Biological clocks generate such rhythms, afferent pathways synchronize them with the cyclic environment, and efferent pathways convey this time information to the organism as a whole. In mammals, the main biological clocks are localized in the hypothalamus, the suprachiasmatic nuclei, which receive direct specialized inputs from the retinas allowing synchronization to the external night–day maintaining the organism in tune with its environment. A particular and refined set of photoreceptors allows this light perception. A network of peripheral clocks in other neural, as well as non-neural tissues throughout the organism, receives this time information assuring precise temporal relationship between every system within the organism, the so-called internal temporal order. When considering any biological system, time of sampling should always be tracked, and the use of basic and simple chronobiological methods considered so avoiding misinterpretations.

10.1 Organisms Live in a Cyclic Environment

Life on Earth evolved in an environment characterized by rhythmic fluctuations. The spinning of the Earth around its own axis (rotation) generates the 24-h day–night cycle, with all the physical (light–dark [LD], temperature, etc.) and biotic (food availability, predation, social interaction, and others) daily changes that take place. The elliptic movement of the Earth around the Sun (translation) and the inclination of the Earth generate seasonal cycles every 365.25 days, with their more or less dramatic changes, according to latitudes.[1] The moon movement around the Earth and its rotation around its axis induce different gravitational pulls over the oceans every 12.4 h. During full and new moon, its alignment with the sun increases the gravitational pull, so that the high tide is the highest and the low tide is the lowest. This results in the lunar month of 28.53 days, which in addition to this effect on the tides, determines nocturnal light intensity cycles. These environmental cycles have their counterparts in the periods of rhythms of all living organisms.

The first living forms were probably exposed to damaging solar radiation and/or daily drastic increases in free oxygen levels. This daily variation has probably driven certain cellular processes to occur only during the more benign hours leading to the first adaptive regular timing event. This 24-h timing structure resulted so important for survival that was probably selected with basic elements conserved along the full phylogenetic scale. In effect, rhythmicity appears in the simplest prokaryotic unicellular organisms up to fungi, plants, invertebrates, and vertebrates. Today, it is known that clock mechanisms

---

[1]In the 2nd century BC, Hipparchus of Nicaea (*ca.*?190–*ca.*?120 BC), astronomer, geographer, and mathematician, measured the time required for the Sun to travel from an equinox to the same equinox again. He found the length of the year to be 365.24667 days (say, 365.25 days, rounding it out).

along the whole phylogenetic scale, although species specific, possess conserved basic mechanistic components (Dunlap, 1999).

10.2　Time, Biological Rhythms and Chronobiology

Time is a nonspatial continuum in which events occur in apparently irreversible succession from past to present and beyond to the future. A biological rhythm is the periodic occurrence or variation of a biological function or event. According to Isaac Newton, we are only capable of perceiving relative time, which is a measurement of perceivable objects in motion, like the Moon or Sun (Rynasiewicz, 2014). From these movements, we infer the passage of time. In the same way, we can infer the mechanism of the biological clocks through the regular expression of biological variables at defined intervals. Chronobiology (*chronos*: time; *bios*: life, and *logos*: study) is the study of the temporal organization of living organisms (Halberg, 1969). In other words, it refers to the study of biological rhythms, the mechanisms that generate them, and their relationship with the environmental cycles. For long, it was assumed reasonably that biological rhythms were direct responses to the periodic stimuli of our cyclic environment. This concept remained until Jean Jacques d'Ortous de Mairan (1678–1771), a French astronomer, conducted a simple but innovative experiment in 1729 with the leaf movements in a "sensitive" heliotropic plant, *Mimosa pudica*. This plant opens its leaves during the day and folds them at night. De Mairan showed that night-day environmental stimulus was not necessary to generate the periodic leaf movements by placing the plant in total darkness. Even under these constant conditions, the leaves opened and folded daily. The persistence of a daily rhythm in the absence of environmental time cues was thereafter fully demonstrated (Bunning, 1960). However, the fact that circadian timekeeping is an endogenous property of animals and plants was not widely accepted until 250 years later. Chronobiology, as a formal discipline, was consolidated only in 1960, as a result of the Cold Spring Harbor Symposium on Quantitative Biology on Biological Clocks organized by one of the fathers of chronobiology, Colin Pittendrigh. This was a milestone meeting

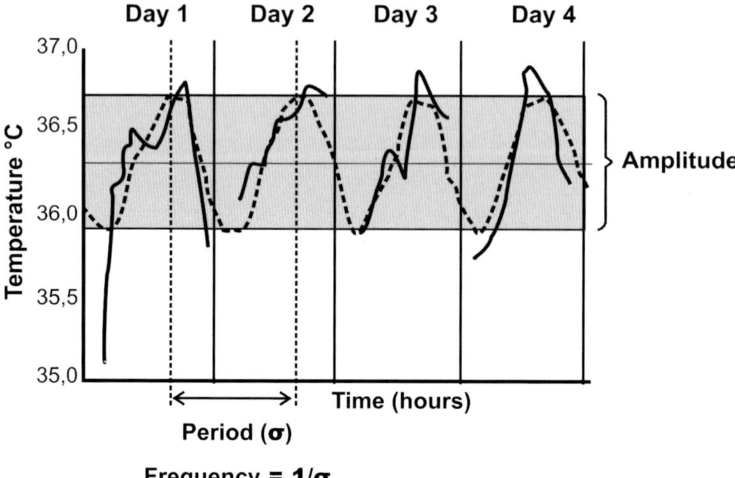

Figure 10.1: A sinusoidal curve (dashed curve) can be fitted to the body temperature data obtained at different time points in humans. The interval between the occurrences of the reference points, in this case the *acrophase*, defines the period. The reciprocal is the frequency, the number of cycles per interval. The horizontal line is the mean temperature value and the range from maximum to minimal temperatures in the wave defines the amplitude of the rhythm (adapted from Marques *et al.*, 2003).

where chronobiological concepts were defined, its own mathematical and statistical methods were incorporated, and chronobiology began to be accepted as a scientific area of knowledge (Chovnick, 1960; DeCoursey, 2004).

10.2.1 *Properties of rhythms*

The concept of rhythm encompasses parameters of waves (Fig. 10.1). Every wave has a minimal (*batyphase*) and maximal (*acrophase*) point (see http://www.circadian.org/dictionary.html). The range from maximum to minimal values defines the amplitude. The period is the interval between the occurrences of two reference points, which commonly is the *acrophase*. Its reciprocal, frequency, is the number of cycles per unit of time. Phase is any defined point in the wave or also a major area within the wave with a common characteristic (e.g., the active or inactive phase, the dark or light phase). The relationship of two mutually synchronized rhythms is expressed in terms of the phase angle difference.

10.2.2 *Chronobiology and homeostasis*

From the beginnings of chronobiology, ample discussions were generated between the newly idea of a generalized rhythmicity in contraposition to the idea of the constancy of the internal environment, strongly rooted in the physiologist's academia being the central idea of physiology courses. The concept of internal environment and its regulation has its roots in the French physiologist Claude Bernard (1813–1878). Subsequently, Cannon (1929) refined this idea and coined the term homeostasis (*homo*, equal, and *stasis*, static; see also Chapter 5, on the Renal System). The "steady state" of an organism, with no emphasis in the temporal dimension, was considered the essence of physiology. Fluctuations of variables were understood as perturbations that the system quickly had to correct through homeostatic mechanisms. On the other hand, chronobiology highlighted the regular rhythmic variation, rescuing its value in the context of adaptation to a cyclical world. Currently, conciliation of both concepts is easier than believed at the time. As Moore-Ede *et al.* (1984) pointed out, Cannon (1929) specifically mentioned that even the most tightly regulated variables may oscillate. He actually defined homeostasis as the process that regulates a physiological variable within certain limits, but that the variable may oscillate between those limits, and the limits themselves may change in response to some special demand. A conciliatory point of view is that of reactive and predictive homeostasis proposed by Moore-Ede (1986). The first refers to homeostatic mechanisms initiated in response to perturbations of the system and the second to those initiated in anticipation to rhythmic predictable changes. Another integrative view is the concept of *rheostasis*, the regulation of physiological variables in a changing environment (Mrosovsky, 1992).

10.2.3 *Different types of biological rhythms*

All physiological variables are rhythmic unless proved otherwise. Biological rhythms of various periodicities are ubiquitous. They may be externally imposed, internally generated, or a combination of these two. Up to now, we have been considering rhythms with 24 h intervals called circadian (*circa* = about, and *dies* = day), which are the most obvious and well known. However, other types of rhythms with different periods abound. In a general classification, other than the circadian rhythms, there are those that occur at intervals shorter than 24 h called *ultradian,* as is the case of the pulsatile liberation of hormones, the sleep stages, electroencephalogram (ECG), respiration, or heartbeat. The third category of rhythms is those with intervals longer than 24 h called *infradian,* as is the case of annual migration and reproduction in seasonal animals. Within this *infradian* category, other examples with different periods are *circalunar* (about 28 days) or *circaseptal* (about 7 days). However, not all seems to be directly related to an environmental cycle. For example, the week-long duration rhythms have never been shown to be related to the human-determined week nor has the menstrual cycle been related to the 28 day-long lunar cycle, despite the coincidence in the intervals (if this were not the case, all women would be synchronized in their menstrual cycle with the lunar cycle marking the pace!). Circadian rhythms are the most obvious and best-studied rhythms, and it is on these that we will continue to concentrate. Figure 10.2 shows diagrammatic examples of circadian rhythms in humans.

10.2.4 *Activity rhythm as a widely used marker of the timing system*

We mentioned above the temperature rhythm as a precise and representative circadian rhythm. Another good and obvious example is the activity/rest rhythm, often associated with the sleep–wake cycle. This is the main rhythm used to define a species as being nocturnal, diurnal, or crepuscular, depending on the phase of the day during which activity is expressed (Aschoff, 1960). Each activity allocation is associated with morphological and physiological adaptations related to the temporal niche occupied (e.g., big sensitive eyes in nocturnal animals). A fourth type of activity pattern is *cathemeral* (from the Greek "through the 24 h day"), which is a term applied when an animal is neither prescriptively nocturnal nor diurnal nor crepuscular

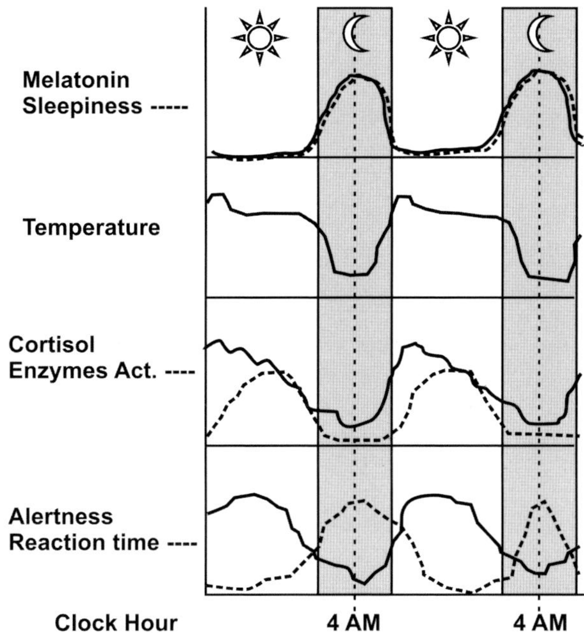

Figure 10.2: Diagrammatic examples of daily rhythms in humans. The level of the variable is on the left axis and the abscissa represents two complete day–night cycles. 4 am is a critical time point in which melatonin level peaks (in plasma, or saliva or as metabolite in urine; used as an index of body clock time), and sleepiness is maximum (not in vain is melatonin called the "sleep hormone"). Additionally, at 4 am, temperature, cortisol levels (considered one of the "activity hormones"), and subjective alertness are at their minimum level while, coincidently, reaction time is the longest. The activity of diverse digestive and detoxification enzymes are at their lowest level during the night, consequently not a good time for snacking! (adapted from Moore-Ede *et al.*, 1984; Arendt 2010).

but, instead, irregularly active at any time of night or day, according to the prevailing conditions at that moment (Tattersal, 2006). Humans are typically diurnal being active during the light phase of the day–night cycle and inactive or sleeping during the dark phase (Czeisler *et al.*, 1980). However, in the current modern world with electricity, internet, and 24 h-open companies and stores, in many cases, speaking of nocturnal or even cathemeral individuals is not far-fetched (Foster and Wulff, 2005). Keep in mind that accompanying and supporting this behavioral rhythm, the majority of all other behavioral, physiological, and biochemical rhythms express this rhythmicity, too, each with its own pattern and phase relationship with the other variables and with the external environmental cycles.

Activity is measured through diverse devices depending on the species being studied. Motion-detectors (burglar detectors), intra-abdominal transmitters, stabilimeter-cages, photo-electric devices, perch-hopping in birds, accelerometers, wrist actographs in humans, rocking actimeters in certain insects, and running-wheels (see Fig. 10.3) are some examples. Wheel-running response has been demonstrated for a wide variety of mammals, from rodents to large carnivores and even primates, as well as other vertebrates (Meijer and Robbers, 2014). This is a non-invasive very precise way of measuring activity rhythms and as a great side effect, it additionally

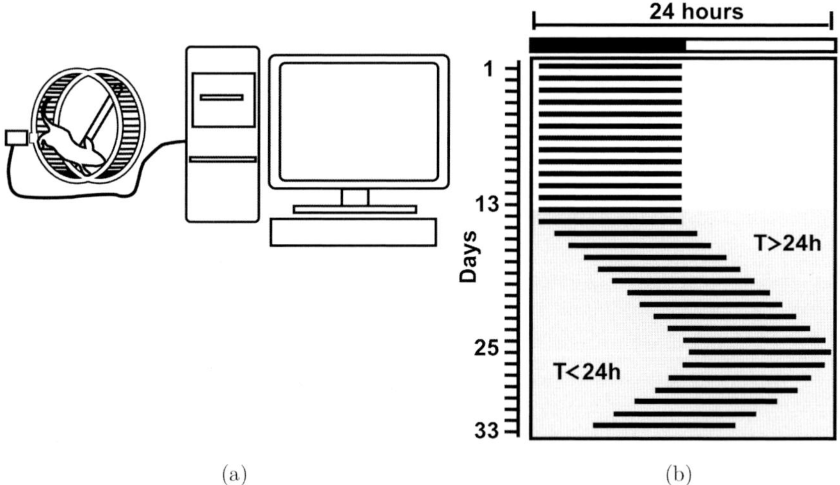

(a)   (b)

Figure 10.3: (a) Setup for monitoring wheel-running activity continuously. During the active phase of the species in study, the animal jumps on the wheel, in this case in the dark phase. A computer records each turn as a pulse that is stored and collected over time. (b) Schematic actogram. Each horizontal line represents a 24-h day. Successive days appear vertically (a total of 33 days). The bar on the top shows the LD cycle to which the animal is exposed, 12 h dark and 12 h light. The LD cycle appears also on the actogram with gray and white background. Horizontal black lines represent activity; lack of lines represent inactivity (adapted from Tomotani, 2011).

provides encaged captive animal with an enriched environment with the opportunity to exercise (which they actually do intensely).[2] We will use this type of behavioral measure to evaluate the main characteristics of circadian rhythms. A setup for measuring activity rhythm is shown in Fig. 10.3 as well as its representation in the form of an actogram, the most common mode to visualize rhythms.

In the case of Fig. 10.3b, for the first 13 days, this animal was submitted to a LD cycle and, as a nocturnal animal, it only ran in the wheel during the dark phase. On day 14, the lights were turned off, and left so for the rest of the experiment. The animal kept expressing the activity rhythm, however, with a different period. In this case and just for didactic purposes, a free-run longer than 24 h is represented until day 25, that is, activity starts every day a bit later; after day 25 the period is represented as being shorter than 24 h, meaning that activity starts every day a bit earlier.

10.3 Fundamental Properties of Circadian Rhythms

A measured rhythm is considered circadian if it meets three major characteristics (Johnson *et al.*, 2004). In the schematic representation of an actogram shown in Fig. 10.3b, it is possible to visualize two of them, as explained below.

10.3.1 *Entrainment to external cycles*

As mentioned before, there are diverse 24 h cycles in the environment, however, the most reliable is the Earth's LD cycle. It has been present from the beginnings of life and indefectibly occurs day after day and, consequently, through evolution, it has become the main temporal anchorage of organisms to the world's time niche. One of the fathers of chronobiology, Jürgen Aschoff, coined the term *Zeitgeber* (from German, *zeit*, time; *geber*, one that gives) meaning "time

---

[2]There is an extensive literature on the benefits of wheels for captive animals (Schroeder *et al*, 2012; Leise *et al*, 2013). This daily voluntary exercise increases life expectancy, improves life quality, promotes healthier and calmer animals, and increases cognitive performance.

giver", to describe such periodic environmental cycles. Other terms are *synchronizer* and *entraining agent*.

In order to be adaptive, a rhythm should be capable of entrainment to the Earth's cycles. This means that, by daily adjustments, it should acquire the same period of the *Zeitgeber* and maintain a stable phase relationship with it. In Fig. 10.3b, the first 14 days represent the activity rhythm of an animal in a standard LD cycle, with its activity concentrated during the dark phase and no activity during the light phase (nocturnal animal). The period of the activity rhythm is equal to that of the synchronizer (24 h) and a stable phase relationship is evident. In this example, activity onset (a good phase reference for the activity rhythm) occurs every day a few minutes after the light is off (a good reference point for the LD cycle). In this case, the animal expresses a negative phase relationship with respect to the LD cycle. This synchronization process becomes clearer when the time giver is no longer present.

10.3.2 *Endogenous nature of circadian rhythms*

Since De Mairan's first observations in the sensitive plant, it is known that biological rhythms persist in constant conditions. In Fig. 10.3b, when the animal is free of environmental time cues on day 13, the activity rhythm persists with a different period, which is species specific. In this schematic representation, a period longer than 24 h is displayed first, where every daily activity begins a bit later than in the previous cycle. This is the case of rats and certain primates, including humans. On day 25, a rhythm with a period shorter than 24 h is represented, that is, activity begins every day earlier than in the previous cycle as occurs in mice.

When no environmental cycle is present, it is said that the biological clock is *free running*. That points out to an own endogenous period, never exactly 24 h, but around 24 h. Circadian (*circa*, about; *dies*, a day) is a word coined by the chronobiologist Franz Halberg from two Latin words, "circa", meaning "about", and "dies", meaning "a day". Hence, it reflects the slight deviation of almost all free-running rhythms from the theoretical 24 h. For environmentally related pacemakers, the equivalent terms are *circalunar*

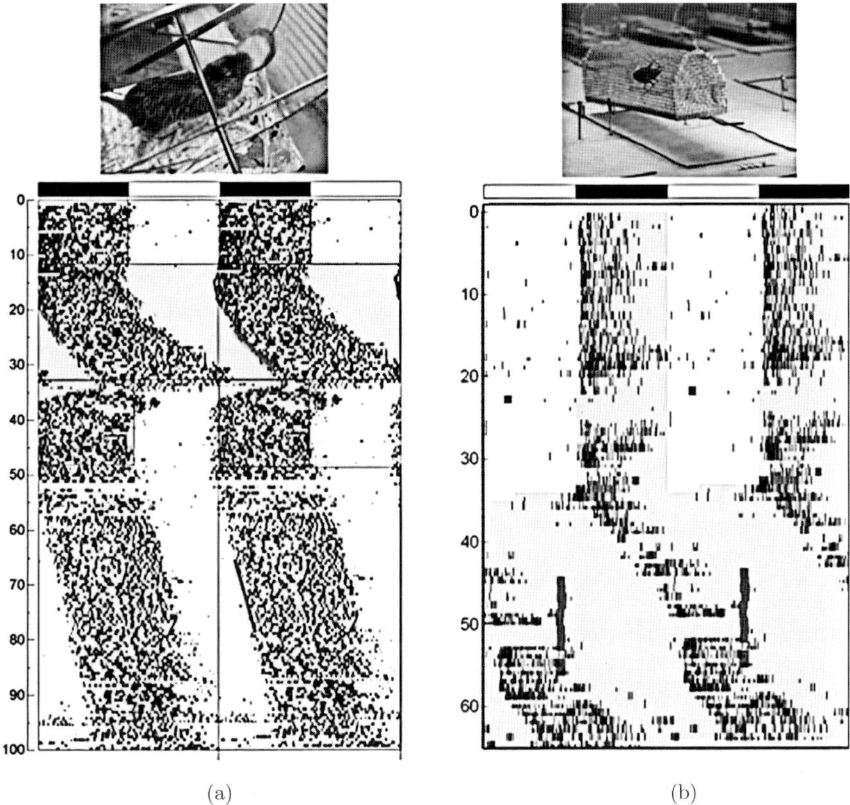

(a)   (b)

Figure 10.4: Double plotted actograms of two different species exposed to different light conditions as shown in the white and gray backgrounds. (a) Subterranean rodent (*Ctenomys* sp.) in a running wheel exposed to a LD cycle for 12 days followed by 20 days of constant darkness, after which LD was reinstated for 18 days and, finally, constant light for the rest of the experiment. Lines through onsets of activity were used to calculate free-running periods (from Valentinuzzi *et al.*, 2009). (b) A kissing bug (*Triatoma infestans*) in a rocking actimeter exposed to a LD cycle and then to constant darkness. Small squares indicate when the insect was fed a blood meal, after which a temporary decrease in activity occurs. Vertical rectangles indicate daily exposure to a host (a chicken); a cycle that was able to entrain the kissing bugs activity rhythm (from Valentinuzzi *et al.*, 2013).

for endogenous lunar cycles, and *circannual* for endogenous annual rhythms.

Figure 10.4 shows examples of real actograms of two different species. In this case, the actograms are double plotted, which means that 48 h is represented in the horizontal axis instead of only 24 h. To the left, a subterranean rodent (*Ctenomys* sp.) in a cage with a running wheel was exposed to different light conditions, as indicated in the background color of the actogram. A LD cycle for 12 days followed by 20 days of constant darkness, after which LD alternation was reinstated for 18 days and, finally, constant light for the rest of the 50 days of the experiment. To the right, a kissing bug (*Triatoma infestans*), vector of the parasite *Trypanosoma cruzi*, causative agent of Chagas disease, in a rocking actimeter. The insect was exposed first to a LD cycle during 25 days and then to constant darkness. As the insect walks across the cage's midline, the actimeter slightly re-balances (as a seesaw) closing an electric circuit recorded as a pulse of activity. Note how the rhythms free-run with a period longer than 24 h when the synchronizing environmental cycle is eliminated (when in constant dark or constant light conditions).

10.3.3 *Temperature compensation*

Biochemical and physiological processes are sensitive to changes in temperature. They accelerate with increases in temperature, while decelerate with decreases in it. For instance, the rate of oxygen consumption in animals can duplicate or even triplicate when ambient temperature increases $10°C$ (Schmidt-Nielsen, 1997). To calculate the effect of changes in temperature on the rate of any biochemical process the temperature coefficient Q_{10} is used, that is,

$$Q_{10} = \left(\frac{R_2}{R_1} \right)^{10/(T_2 - T_1)} \tag{10.1}$$

where R is the rate of the variable being measured and T is the temperature. Q_{10} is a unitless quantity, as it is the factor by which a rate changes. In essence, it is the rate between two values measured with a $10°C$ difference (Schmidt-Nielsen, 1997). In this way, if a value duplicates with an increase in $10°C$, it would have a Q_{10} equal to 2. If it triplicates, Q_{10} would be 3, and so on. The rate of most physiological and biochemical processes changes two- or threefold with each $10°C$ change in temperature ($Q_{10} = 2$–3).

Considering that circadian oscillations arise in cycles of gene expression and protein synthesis (as explained later), it would be expected that these biochemical reactions would respond in the same way to temperature change. If the mechanisms of rhythm generation had a $Q_{10} = 2$, for instance, an increase in $10°C$ would make the oscillator faster, which would result in a very short period, maybe close to 12 h. Following the same line of thought, a decrease in $10°C$ would result in a considerable lengthening of this period. This would jeopardize synchronization to the environmental cycles and the temporal allocation of physiological and behavioral processes to the adequate phase. When the circadian periods of different rhythms, in different species are measured at different temperatures, Q_{10} values vary from 0.85 to 1.3. In other words, circadian period is maintained essentially the same despite temperature changes (Pittendrigh 1954; Hastings and Sweeney, 1957). The mechanism of temperature compensation, however, is yet poorly understood. A hypothesis states that time-keeping could be the net result of two reactions, one producing a substance that inhibits the other, and both reactions speeding up with increasing temperature (Rensing and Ruoff, 2002). The bottom line is that, as it occurs with human-made clocks, temperature compensation is essential for biological clocks to be reliable timekeepers (Pittendrigh 1954).

10.4 Circadian Timing System

The circadian pacemaker system is constituted by three major components (Fig. 10.5),

(1) the biological clock, generator of circadian rhythms;
(2) afferent pathways, or neural inputs, into the clock from other parts of the brain;
(3) efferent pathways from the clock to target tissues, which couple their internal rhythmicity to that of the clock.

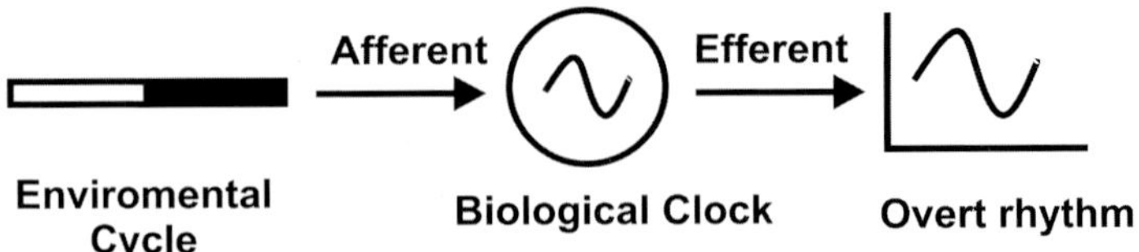

Figure 10.5: Diagram of the main components of the circadian timing system. From left to right, an environmental cycle (e.g., a LD cycle) is detected by the organism and transmitted through afferent pathways to the biological clock(s). From here the time information is conveyed to the rest of the organism through efferent pathways resulting in the expression of overt rhythms.

10.4.1 *Biological clocks*

Much of the knowledge about circadian clocks stands on observations of overt rhythms. These rhythms are not the clock itself, but the observable "hands of the clock". In mammals, there are central master clocks located in the hypothalamus, at the suprachiasmatic nuclei (SCN, Klein *et al.*, 1991; Weaver, 1998). As mentioned in Chapter 7, the hypothalamus — a functionally ubiquitous portion of the diencephalon — is a main center controlling the majority of body functions. Therefore, the location of the biological clocks in such a strategic localization seems very meaningful. The SCN are a pair of structures, about the size of a grain of rice, located at each side of the third ventricle, a fluid-filled space in the midline of the brain (Fig. 10.6). They derive their name from their position just above the optic chiasm, which is a large bundle of axons formed by the convergence of the two optic nerves as they pass from the retinae to the visual brain cortices. Formed by about 10,000 neurons, each nucleus consists of a shell and a core. The shell lies mostly dorsal and medial (DM), whereas the core lays ventral and more lateral (VL). Each area has different cytoarchitectures, neurotransmitters, and different input/output pathways, revealing differential functionalities. For instance, it is the VL portion that receives projections from the eye's retinas. Glutamate is the neurotransmitter involved in synchronization to light (Moore 1993; Golombek and Rosenstein, 2010; Canteras *et al.*, 2011).

The neurons that compose these nuclei have the capacity to generate circadian rhythmicity due to a mechanism of negative feedback

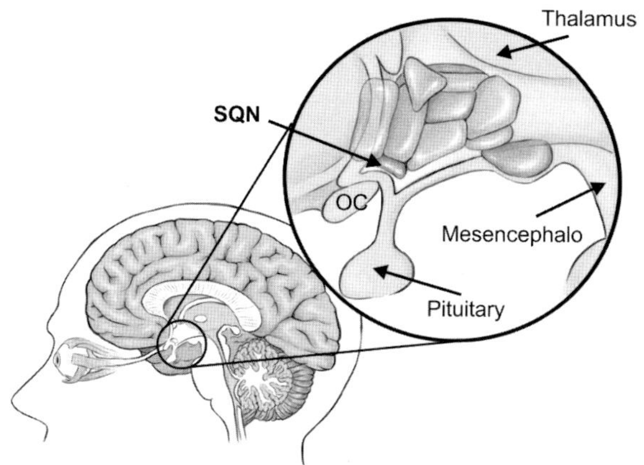

Figure 10.6: Localization of the biological clock in the hypothalamus of the human brain. OC: optic chiasm; SQN: suprachiasmatic nuclei (adapted from Moore-Ede *et al.*, 1984 and Lent, 2002; redrawn by Gustavo Idemi).

loops of *gene expression* and protein synthesis. The common elements in the design of the circadian oscillatory loop are four core components, two transcriptional activator proteins (the positive elements named CLOCK, and BMAL1, in the case of mammals), and two transcriptional repressors (the negative elements, named Periods, PERs, and Cryptochromes, CRYs, in the case of mammals) (Fig. 10.7). This activation and repression cycle is the clockworks, the core of the clock, the quartz crystal taking circa 24 h to be completed (Dunlap 1999; Lowrey and Takahashi, 2011).

How does this bridge to physiology? This core clock directs the cycling of thousands of mRNAs in the genome (Zhang *et al.*, 2014). The heterodimeric CLOCK and BMAL1 complex not only activate and promote the expression of the clock genes that constitute the main oscillatory mechanism but also activate other genes. We have 25,000 genes and up to half of these undergo circadian oscillations. These rhythmic genes generate proteins rhythmically and these proteins regulate a myriad of cellular functions, in a rhythmic fashion as well.

Clock is a common figure of speech, and even though a master clock seems to be in the hierarchical top of the circadian organization, more than one clock exists in an organism. Circadian clocks can be

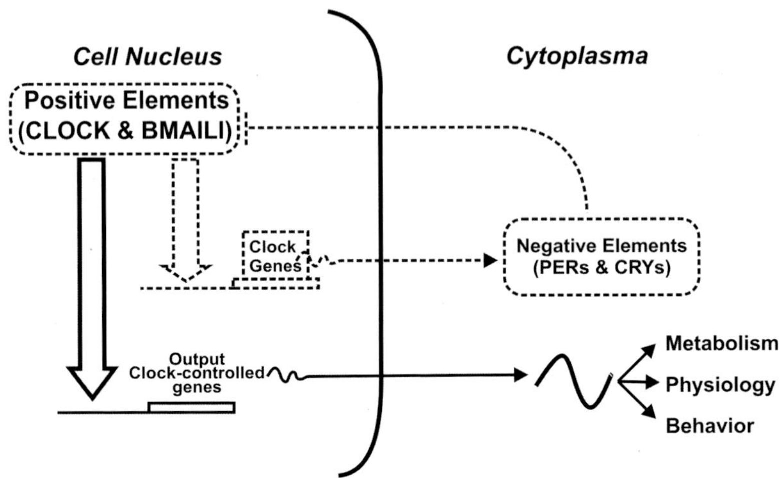

Figure 10.7: Basic elements in the design of a circadian oscillatory loop (in dashed lines) and the rhythmic cellular output to the rest of the organism (modified from Dunlap, 1999) in mammals.

systems, instead of a single anatomical unit. In fact, each of our cells contains a molecular clock, with clock genes and proteins controlling its own functioning as well as its interactions with other cells and tissues (Buijs and Kalsbeek 2001; Schibler *et al.*, 2003; Yoo *et al.*, 2003; Lamia *et al.*, 2008; Dibner *et al.*, 2010; Mohawk *et al.*, 2012). Overall, this organization allocates specific biological functions to optimal times of the day.

All the cells that are oscillating individually synchronize through other body rhythms generated by the SCN. Several important internal synchronizers have been recognized, such as the daily rhythms of body temperature (Brown *et al.*, 2002; Buhr *et al.*, 2010), glucocorticoids (Le Minh *et al.*, 2001), and the plasma level of melatonin (Arendt, 2010) (see Fig. 10.2). These rhythms, generated and controlled by the master SCN, set the timing of cells, and ultimately tissues and organs, to be active or inactive as specific phases (Reppert and Weaver, 2002). Like an orchestra director, the SCN marks the pace of every single cell in the organism through direct and indirect pathways. However, in many occasions, this internal synchronization may be disrupted, as mentioned later.

10.4.2 *Afferent pathways and synchronization to light*

Circadian clocks are entrained by environmental cycles resulting in synchronization of physiology and behavior with the environment. The LD cycle is the most important factor in synchronizing the endogenous circadian rhythms in animals and humans. The importance of this *zeitgeber* is reflected in the neural connections between the retina (photosensitive structure of the eye) and the SCN, the *retinohypotalamic tract* (RHT) (Moore and Lenn, 1972), as the most obvious afferent connections of the circadian system (Figs. 10.6 and 10.8). Light reaching the eyes every day is transduced into a neural signal and this information travels through the RHT reaching the biological clock. Let us underline that this transduction occurs in photoreceptors that are independent from the traditional visual cones and rods. These are a subset of melanopsin-containing retinal ganglion cells (RGCs), in the mammalian retina, that respond to light and connect to the brain's clock (Berson *et al.*, 2002; Panda *et al.*, 2003). Interestingly, the whole wavelength spectrum of light (white light) is not necessary for synchronization; evidence indicates that short wavelengths between 460 and 480 nm (blue) have the most powerful resetting effects (Foster *et al.*, 1991; Skene and Arendt, 2006). Blind human subjects due to retinal destruction, including the melanopsin-containing RGCs, display free running of behavioral and physiological rhythms with periods longer than 24 h, underlining the importance of light as a synchronizer. Under such conditions, peak levels of different rhythmic variables may be achieved at any time during the day or night.

Despite the important role of light as a synchronizer, keep in mind that many other non-photic environmental cues can entrain the clock as well, if they occur in a time predictable pattern. See, for example, Fig. 10.4b, in which a kissing bug was synchronized by a daily 1-h exposure to a chicken, a main food source for this hematophagous insect (Valentinuzzi *et al.*, 2013). Temperature cycles, feeding schedules, or social cues are other examples (Dunlap *et al.*, 2004). In humans, social cues become especially relevant. Non-photic neural input pathways to the clock are many, deriving

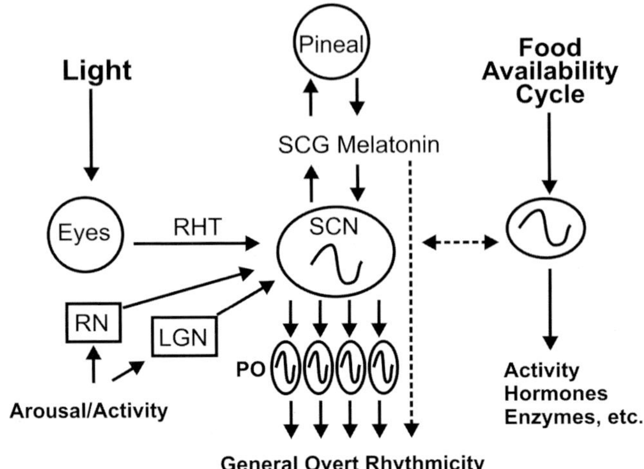

Figure 10.8: Schematic representation of some of the components of the circadian timing system and their interactions. SCG: superior cervical ganglion; RN: raphe nuclei; LGN: lateral geniculate nuclei; SCN: suprachiasmatic nuclei; RHP: retino-hypotalamic tract. PO: peripheral oscillators.

from diverse brain areas (thalamus, brain stem, hypothalamus, limbic system), but the better understood are the ones from the Raphe Nuclei and Lateral Geniculate Nuclei. These structures relay information about the general arousal state, which is easily modulated by diverse non-photic stimuli (Mrosovsky, 1996). Synchronization to food availability cycles is an important area of research that has clearly revealed other input pathways and oscillators specifically sensible to this particular type of environmental cycle (Mistlberger, 1994; Stephan, 2002) (Fig. 10.8).

10.4.3 *Efferent pathways*

The SCN convey the endogenous temporal information (in free-running conditions) or the synchronized temporal information (when in a cycling environment), to the rest of the organism through many nervous and neurohormonal paths that guarantee adequate arrival of rhythmic information to brain structures and body organs. The strategic location of the master clock in the hypothalamus allows a short distance timing control of the main functions. The SCN sends many neural connections to different brain areas. Most of them terminate directly within the hypothalamus and few others reach other more distant brain structures. A relevant example is the neural connections to the sleep centers in the reticular formation of the midbrain, pons, and medulla controlling all the variables related to the sharp and clear sleep–wake rhythms (Foster and Kreitzman, 2005). The neurotransmitters secreted rhythmically in these nerve endings are important time markers for the different brain structures they innervate, including the hypothalamic nucleus that control the endocrine system (Kalsbeek *et al.*, 1996). Through its connection with several key homeostatic centers in the hypothalamus, the SCN conveys, in principle, circadian regulation of organs through the nervous pathways that comprise the autonomous nervous system (Bartness *et al.*, 2001; Buijs and Kalsbeek, 2001; Buijs *et al.*, 2003).

The hypothalamus is extensively innervated by the clock and this, in turn, is intimately linked to the anterior pituitary gland ("master gland" of the body). Considering that this is the site of release of a battery of neurohormones and hormones that collectively influence all cells and affect virtually all physiologic processes, the pervasive effect of "time" on the organism becomes evident. All the functions controlled by this master gland become rhythmic. Rhythmicity is thus amplified, as both the nervous and the neuroendocrine axis project into the whole organism conveying their particular functions in a time-regulated manner. One important link that involves the SCN regulation of both the peripheral and central nervous system is the control of the *pineal gland* and its circadian release of *melatonin.*

The *human pineal gland*[3] is a very small structure attached by a stalk to the postero-dorsal aspect of the *diencephalon* (see Chapter 7). Under the SCN's command, the pineal's main product, *melatonin* (N-acetyl-5-methoxytryptamine), is secreted rhythmically into the

---

[3]**René Descartes** (1596–1650), the French mathematician and philosopher, identified the pineal gland as the seat of the human soul. He believed that the rational soul made contact with the body at the pineal gland (mind–body interface). It appeared to him to be the only organ in the brain that was not bilaterally duplicated. With the advent of empirical science, the pineal gland was widely believed to be a vestigial, non-functional organ as recently as 1950. Since then, intense research has clearly established it as an integral and important component of the neuroendocrine system.

systemic circulation. The retinohypothalamic tract (RHT) and its connection to the clock play a critical role in entrainment of this melatonin circadian rhythm. This is accomplished via a long and circuitous multi-synaptic pathway, passing through the superior cervical ganglion, by which light information that reaches the SCN through the RHT ends up reaching the pineal (Moore and Klein, 1974; Klein and Moore, 1979; Lucas *et al.*, 1999).

The production of melatonin within the pinealocytes (the secretory unit of the pineal) requires the uptake of the amino acid tryptophan from the circulation, which, in turn, is obtained mainly from ingested proteins. This amino acid is converted through a series of steps, first to serotonin and then to melatonin. Two of the enzymes essential for this conversion, hydroxyindol-*O*-methyltransferase and *N*-acetylserotonin, show circadian rhythmicity in their concentration and activity level (Moore and Klein, 1974; Klein and Moore, 1979).[4]

Melatonin's function for the circadian timing system provides a humoral time cue for the organization of seasonal and circadian rhythms. The pineal gland secretes melatonin with a marked daily rhythm, peaking at night — it has been called the 'darkness hormone'. In animals that depend on day length to time their seasonal physiology, the length of melatonin secretion signals the length of the night (Carter and Goldman, 1983). In humans, its circulating concentration is high from 21:00 to 07:00 h (with individual variations) and is used as the primary output marker of the internal clock (Arendt, 2010). The peak secretion occurs at 04:00 h, closely associated with the nadir of core body temperature, alertness, and performance (Fig. 10.2). Exposure to light of sufficient intensity and spectral composition, when melatonin levels are highest, quickly causes suppression of further melatonin production (Redlin, 2001). Suppression is detectable at 30–50 lux and maximum suppression occurs between 1000 and 2000 lux (sunlight can attain 100,000 lux).

---

[4]Assays that quantify NAT and HIOMT activity are often used in legal medicine to determine the approximate time of death. The activity of these two enzymes at the time of autopsy is higher in the pineal gland of humans who died during the night than in those who succumbed during the day.

Melatonin consumption has the ability to induce sleepiness, change circadian phase, and entrain free-running rhythms (Arendt, 1999; Lockley *et al.*, 2000). However, its effects depend on the circadian phase that it is administered (correct timing). In many countries, melatonin capsules can be purchased off the counter, however, care should be taken since its long-term effects in an indiscriminate use are not completely studied. The abundance of melatonin-binding sites in several brain areas, including the brains clock (SCN), is a strong indicator that this hormone is an important coupler of circadian rhythms and endocrine functions (Dubocovitch, 2007).

10.5 Biological Value of Rhythms

Whereas structure in terms of cells, tissues, organs, and systems gives form and spatial organization, the biological clock system provides temporal organization. Harmonization with the environment refers to an adequate coincidence of rhythmic organic functions with the cyclic environmental events according to the activity-allocation characteristics of the organism in study. A good phrase for this phenomenon is *Do the right thing at the right time* (a statement certainly valid for each human life, too). Appropriate timing refers to the moment in which doing something (foraging, looking for mate, reproducing, taking care of its progeny, and defending territory) happens with the highest efficiency and minimal risk (Daan, 1981).

A representative example for this in humans is related to the biological activity of liver detoxification enzymes (Tahara and Shibata, 2016). Food, particularly vegetables and fruits, contain toxic elements. These do not refer to human-made pesticides, fertilizers, and others (even though these also enter the equation), but toxins that the plants naturally produce to avoid being eaten. During evolution, organisms, including humans, have developed enzymes, mainly located in the liver to neutralize these toxins (Rosbash, 2014). Simultaneously, there are enzymes dedicated to DNA repair, which may be damaged by these toxins. These DNA repairing enzymes also show a sharp diurnal expression (Sancar *et al.*, 2010). Maximum activity occurs during the day (Fig. 10.2) and, if correctly synchronized before regular meals, they are ready to enter in action before food full of toxins enters the organism. The consequences of eating at a phase that these enzymes are inactive, during the night, become obvious, and this is associated with the reasons jetlag and shift work affects the gastrointestinal system.

Endogenous rhythms provide the organism the ability to *anticipate* cyclical challenges of the environment (Enright, 1970). By the time sunrise occurs, for instance, melatonin levels are down, cortisol and temperature are up (see Fig. 10.2), and so the organisms are already in an adequate physiological and behavioral state to start day activities.

Maintenance of an internal synchronization among the diverse internal functions of an organism (*internal temporal order*) is essential for the efficient and healthy workings of the body (Moore-Ede and Sulzman, 1981). Virtually, every bodily function has been shown to exhibit circadian rhythms with remarkable precision and stability in healthy individuals. The period and phases of these physiological parameters must be synchronized with respect to each other as well as with the environmental cycles in order for them to have an optimal effect on body organs and systems. Regular alteration of peaks and troughs is now known to be essential for normal function. Evidence showing that pathological states are associated with abnormal rhythms has accumulated, and there is nowadays substantial support for the hypothesis that disorders of temporal organization may be involved in the etiology of various diseases (Scheer *et al.*, 2009).

The concept of internal temporal order may be easily understood when making an analogy with the functioning of a machine, usually called *engine fine tuning*. For example, to work correctly, a car must have a precise spatial format and arrangement (cylinder, pistons, crankshaft, etc.). In addition, it is fundamentally important that every event occur in the right sequence and with the correct timing. Only in this way, will the motor function and do so correctly, smoothly, and efficiently, without unnecessary wear. Why is it that we, humans, find it obvious that the temporal order of a machine is essential for its functioning (many rapidly run to a mechanic upon the slight noise in the engine car!); however, it is difficult to accept that our bodies respond to the same need of temporal organization? When our rhythms are in synchrony, life flows easily, we have more energy, tend to view things more positively; then, we are more socially connected and find life more satisfying. These considerations lead us to the next item related to human health.

10.6 Importance of Clocks in Human Health

Alterations in the synchronization to external time cues and in the internal synchronization between variables, as well as other changes such as in the endogenous period or sensitivity to *zeitgebers*, lead indefectibly to pathological states. Disorders of temporal organization have been shown to be involved in the etiology of various diseases.

10.6.1 *Alterations in environmental time cues: Jet-lag and shift-work*

At the beginning of this chapter, it was said that all organisms, including *Homo sapiens*, have evolved in an environment orchestrated by a 24-h LD cycle. Our system is then adapted to this temporal environment; however, in the last 150 years (an instant in the evolutionary time scale), technological advances drastically changed our temporal niche (Foster and Wulff, 2005; Rajaratnam and Arendt, 2001). Transmeridean flights and shift working schedules are the main cause of chronobiological malaise. Millions of people are subjected to abrupt shifts in environmental time cues when they fly across time zones (transmeridean flights). This produces the known **Jet-Lag** syndrome defined as a temporary loss of synchrony between an abruptly shifted sleep–wake cycle and the local time (Waterhouse *et al.*, 1997). Symptoms are fatigue during the day time and inability to sleep satisfactorily at nighttime, unpleasant digestive feelings, such as loss of appetite and indigestion, cognitive distress such as inability of concentrating, also loss of motivation, increased irritability, headaches, and others. According to the number of time zones crossed, this produces phase shifts in many rhythmic variables, including melatonin rhythm, that take a few days to re-entrain to the post flight LD cycle (Dunlap *et al.*, 2004).

Research in circadian systems and pineal gland has led to practical applications. For example, elite athletes, by following simple chronobiological prescriptions (as light exposure, melatonin ingestion, eating, and exercise habits), all at a predetermined and carefully selected time of day, can adapt themselves in just two days to as much as a 12-h jet lag. Hence, they have better chances for success in high-performance competitions, as demonstrated by several authors (Reilly *et al.*, 2007; Samuels, 2012).

The set of jet lag symptoms is similar to night worker's malaise, the second most important cause of internal and external desynchrony of the circadian timing system in our modern world (*shift work schedules*). While most people can tolerate an occasional adjustment to a new time zone without too much discomfort, repeated shifts to rest-activity schedules presents a much greater stress. Airline pilots, hospital staffs, police, army members, and many industries personnel are examples. The symptoms are the same as in jet lag, however, the long-term effects on health are worse considering the constant effort of the system to adjust to new schedules. Increased incidence of cancer and heart deceases as well as decreased longevity has been reported in long-term shift workers (Waterhouse *et al.*, 1992). The consequences of disrupted schedules are much more widespread than the health of the individual concern. Human errors, for example, by pilots or air traffic controllers contribute to many aircraft accidents, and part of the cause undoubtedly is incomplete adjustment to their shift work schedule (Waterhouse and DeCoursey, 2004; Foster and Kreitzman, 2005).

10.6.2 *Insufficient environmental time cues*

This is particularly an underestimated issue in hospitals. In intensive-care units, patients are monitored very closely and the room lighting is often maintained at a constant level at all times. If they are conscious, the patients' social contacts, with nurses, for example, occur equally during night and day. Even meals may have no circadian schedule, if all metabolic requirements are given continuously by vein. These patients are thereby deprived of external temporal cues (Mirmiran and Ariagno, 2000). Health consequences have become obvious in many studies. The organism fails to function optimally and consequently recovery may be seriously delayed.

Insufficient *zeitgeber* strength in older people is also an issue. With age, less activity and less social interactions are common. This behavioral change added to neurodegenerative deterioration of the circadian system such as a decrease in the sensitivity to light and other entraining agents may determine misalignment in internal rhythms (Turek *et al.*, 2000). In some Nursing Homes and Assisted Living Facilities, an effort in enhancing synchronizers, such as light treatments, scheduled social, and physical activities, are already being implemented.

10.6.3 *Disorders of internal timekeeping*

The circadian timing system may occasionally malfunction, even in individuals living in a regular 24-h day–night schedules. On one side, known diseases clearly report a time of day propensity as nighttime asthma, or higher death rates due to early morning cardiovascular disease and stroke. On the other hand, there are also states that are characterized by altered circadian expression such as the delayed or advanced sleep phase syndrome, as well as non-24-h sleep–wake cycles (especially in the blind). Finally, some psychiatric disorders have been related with circadian disorders (Wehr *et al.*, 1979).

Pathologies can occur in the time cue reception part of the system, or the pacemaker function or the coupling mechanism between oscillators. Most of these are related to alterations in the sleep–wake process. With simple sleep diaries, or more sophisticated EEG recordings in Sleep Disorder Clinics, diseases are classified according to phase, period and amplitude disorders. Delayed Sleep Phase Insomnia (DSPI) is characterized by difficulty of the patient to fall asleep at night followed by a corresponding difficulty in awakening in the morning (Fig. 10.9). Several hypothesis about their origins have been discussed, such as changes in the period of the pacemaker driving the sleep–wake cycle or, more probably, a low-amplitude response to light

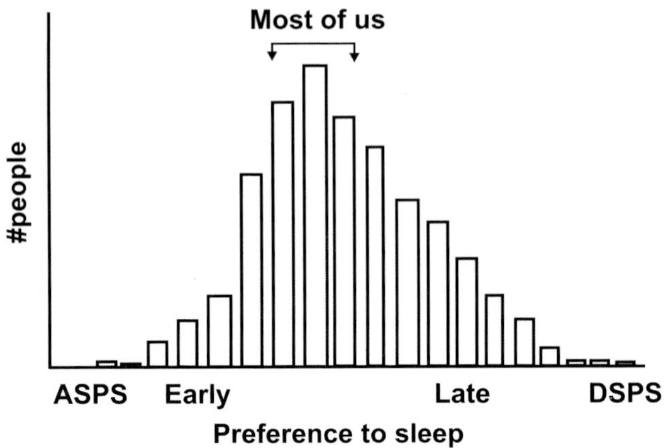

Figure 10.9: Within a given population, when individuals are grouped according to preferred time to sleep, a Gaussian distribution is obtained with a bias towards the right, late time sleep. The majority of the population can adapt to wakening up within an ample range (intermediate chronotype). Morning and evening type people are on both sides of this range. Extreme morning and evening type people are out of these ranges, characterizing chronobiological diseases such as Advance Sleep Phase Syndrome and Delayed Sleep Phase Syndrome, respectively (adapted from Roenneberg *et al.*, 2004).

(related to a change in the sensitivity of the clock to light) (Weitzman *et al.*, 1981).

Psychiatric disorders are in many occasions related to circadian disorders. For example, schizophrenic patients as well as depressed people show a phase advance, going to sleep early and waking up very early (Wehr *et al.*, 1979). The relationship between psychiatric diseases and circadian disruption has led to etiological questions: Are circadian disruptions a consequence of the psychiatric disease or are they the cause? In many cases, the second may be true since treating psychiatric patients with chronobiological manipulations of daily routines improves symptoms significantly. It has been shown that lithium salts, a common anti-depressive medicine, lengthens the endogenous circadian period (Benedetti, 2012). Treatments of certain psychiatric diseases through chronobiological approach have been used. Benign methods as simple as daily activity schedule manipulations, light therapy, or melatonin are promising, avoiding all the collateral effects of the strong psychotic drugs commonly used (Wehr *et al.*, 1979). However, this is not an approach that the multinational multimillionaire pharmaceutical companies approve!

10.6.4 *Chronotypes*

With all this discussion, it is of interest to mention an important chronobiological phenotype of humans that certainly affects health, performance, social interactions, as well as many other aspects of human's life. It is part of common knowledge that some people like to stay up late at night, whereas others are at their best early in the morning (Fig. 10.9). The extreme cases are easiest to identify. The extreme "lark" arises at 4 or 5 a.m., and does his best work in the morning becoming tired and less able to concentrate by 8 or 9 at night. In contrast, the extreme "owl" has an enormously hard time getting up in the morning and prefers to sleep until noon, then slowly gains momentum during the day, reaching its maximum efficiency at night, probably staying awake until 2 or 3 a.m. The majority of the human population is in the middle, called *intermediate*. In any selected population, the distribution of these behavioral patterns can be quantified using a questionnaire developed by Horne and Ostlberg (1976); this questionnaire can be accessed in Internet and you can determine your own chronotype (see www.cuclock.org). It presents questions designed to evaluate the person's relative performance at different times of day, and it mainly concentrates on questions of timing preferences for different activities (Roenneberg *et al.*, 2003). The type of chronotype you will have depends on three mains factors. There is a genetic basis in which polymorphisms seem to be the case, in which different genes contribute (Toh *et al.*, 2001). Age is a strong determinant, with teenagers and young adults being usually of the evening type, while young children and elderly adults are morning type (Roenneberg *et al.*, 2007). Light exposure can also be determinant in this characterization. For instance, morning-like people prevail in equatorial areas where sunrise occurs early. In contraposition, in northern cold areas the evening chronotype dominates (Fig. 10.9) (Roenneberg *et al.*, 2004).

10.7 Summary

The pervasive influence of time in biology and specifically in humans has been discussed. The cyclic nature of the Earth's environment has incorporated time organization in the living matter since the primordial of life. A clear anatomically and functionally defined circadian system is part of our organism. The adaptive value of this system is evident as well as the negative effects that result from its malfunctioning due to diverse diseases; also because of the around-the-clock structure of the society we live in. The time domain should always be considered for a complete understanding, control and intervention in any biological system, and Bioengineering should not be an exception. Recall, for example, the relatively new techniques available to follow up different species' behaviors, either in the controlled restricted laboratory or in the open field, along with the application of mathematical tools to process and interpret the records.

References

Arendt J. Jet-lag and shift work: Therapeutic use of melatonin. *J Royal Soc Med* 92:402–405, 1999.

Arendt J. Shift work: Coping with biological clocks. *Occup Med* 60:10–20, 2010.

Aschoff J. Exogenous and endogenous components in circadian rhythms. *Cold Spring Harb Symp Quant Biol* 25:11–28, 1960.

Bartness TJ, Song CK, Demas GE. SCN efferent to peripheral tissues: Implications for biological rhythms. *J Biol Rhythms* 16:196–204, 2001.

Benedetti F. Antidepressant chronotherapeutics for bipolar depression. *Dialog Clin Neurosci* 14(4):401–411, 2012.

Berson DM, Dunn FA, Takao M. Phototransduction by retinal ganglion cells that set the circadian clock. *Science* 295:1070–1073, 2002.

Brown SA, Zumbrunn G, Fleury-Olela F, Preitner N, Schibler U. Rhythms of mammalian body temperature can sustain peripheral circadian clocks. *Curr Biol* 12(18):1574–1583, 2002.

Buhr ED, Yoo SH, Takahashi JS. Temperature as a universal resetting cue for mammalian circadian oscillators. *Science* 330(6002):379, 2010. doi: 10.1126/science.1195262.

Buijs RM, Kalsbeek A. Hypothalamic integration of central and peripheral clocks. *Nat Rev Neurosci* 2:521e526, 2001.

Buijs RM, van Eden CG, Goncharuk VD, Kalsbeek A. The biological clock tunes the organs of the body: Timing by hormones and the autonomic nervous system. *J Endocrinol* 177:17e26, 2003.

Bunning E. Opening address: Biological Clocks. *Cold Spring Harb Symp Quant Biol* 25:1–9, 1960.

Cannon WD. Organization for physiological homeostasis. *Physiol Rev* 9:399–431, 1929.

Canteras NS, Ribeiro-Barbosa ER, Goto M, Cipolla-Neto J, Swanson LW. The Retinohypothalamic tract: Comparison of axonal projection patterns from four major targets. *Brain Res Rev* 65(2):150–183, 2011.

Carter DS, Goldman BD. Antigonadal effects of timed melatonin infusion in pinealectomized male Djungarian hamsters (*Phodopus sungorus sungorus*): Duration is the critical parameter. *Endocrinology* 113(4):1261–1267, 1983.

Chovnick A. Biological clocks. *Cold Spring Harb Symp Quant Biol* 25:1–524, 1960.

Czeisler CA, Weitzman ED, Moore-Ede MC, Zimmerman JC, Knauer RS. Human sleep: Its duration and organization depend on its circadian phase. *Science* 210:1264–1267, 1980.

Daan S. Adaptive daily strategies in behavior. In: Aschoff J, ed. Handbook of behavioral neurobiology. Biological rhythms. Springer, New York, pp. 275–298, 1981.

DeCoursey PJ. Overview of biological timing from unicells to humans. In: Dunlap JC, Loros JJ and DeCoursey PJ, eds. Chronobiology: Biological timekeeping. Sinauer Associates Publishers, Sunderland, pp. 3–24 2004.

Dibner C, Schibler U, Albrecht U. The mammalian circadian timing system: Organization and coordination of central and peripheral clocks. *Annual Rev Physiol* 72:517e549, 2010.

Dubocovitch ML. Melatonin receptors: Role on sleep and circadian rhythm regulation. *Sleep Med* 8(Suppl 3):34–42, 2007.

Dunlap JC. Molecular basis for circadian clocks. *Cell* 96(2):271–290, 1999.

Dunlap JC, Loros JJ, DeCoursey PJ. Chronobiology: Biological timekeeping. Sinauer Associates Publishers, Sunderland, 406, 2004. ISBN 0-87893-149-X.

Enright J. Ecological aspects of endogenous rhythmicity. *Annu Rev Ecol Syst* 1:221–238, 1970.

Foster RG, Kreitzman L. Rhythms of life. Yale University Press, London, 2005.

Foster RG, Provencio I, Hudson D, Fiske S, De Grip W, Menaker M. Circadian photoreception in the retinally degenerate mouse (rd/rd). *J Comp Physiol* A 169(1):39–50, 1991.

Foster RG, Wulff K. The rhythm of rest and excess. *Nat Rev Neurosci* 6:407–414, 2005.

Golombek DA, Rosenstein RE. Physiology of circadian entrainment. *Physiol Rev* 90:1063–1102, 2010.

Halberg F. Chronobiology. *Annu Rev Physiol* 31:675–725, 1969.

Hastings J, Sweeney B. On the mechanism of temperature independence in a biological clock. *Proc Natl Acad Sci USA* 43:6119–6121, 1957.

Horne JA, Ostberg O. A self-assessment questionnaire to determine morningness-eveningness in human circadian rhythms. *Int J Chronobiol* 4(2):97–110, 1976.

Johnson CH, Elliott J, Foster R, Honma K, Kronauer R. Fundamental properties of circadian rhythms. In: Dunlap JC, Loros JJ and DeCoursey PJ, eds. Chronobiology: Biological timekeeping. Sinauer Associates Publishers, Sunderland, pp. 67–105, 2004.

Kalsbeek A, van Heerikhuize JJ, Wortel J, Buijs RM. A diurnal rhythm of stimulatory input to the hypothalamo–pituitary–adrenal system as revealed by timed intrahypothalamic administration of the vasopressin V1 antagonist. *J Neurosci* 16:5555–5565, 1996.

Klein DC, Moore RY. Pineal N-acetyltransferase and hydroxyindole-Omethyltransferase: Control by the retinohypothalamic tract and the suprachiasmatic nucleus. *Brain Res* 174:245–262, 1979.

Klein DC, Moore RY, Reppert SM (Eds). Suprachiasmatic nucleus: The mind's clock. Oxford University Press, New York, 467 pp., 1991.

Lamia KA, Storch KF, Weitz CJ. Physiological significance of a peripheral tissue circadian clock. *Proc Natl Acad Sci NY* 105:15172e15177, 2008.

Leise TL, Harrington ME, Molyneaux PC, Song I, Queenan H, Zimmerman E, Lall GS, Biello SM. Voluntary exercise can strengthen the circadian system in aged mice. *Age* 35:2137–2152, 2013.

Le Minh N, Damiola F, Tronche F, Schütz G, Schibler U. Glucocorticoid hormones inhibit food-induced phase-shifting of peripheral circadian oscillators. *EMBO J* 20(24):7128–7136, 2001.

Lent R. Cem Bilhões de Neurônios: Conceitos Fundamentais de Neurociência. Editora Atheneu, São Paulo, Brasil, 698 pp., 2002.

Lockley SW, Skene DJ, James K, Thapan K, Wright J, Arendt J. Melatonin administration can entrain the free-running circadian system of blind subjects. *J Endocrinol* 164:R1–R6, 2000.

Lowrey PL, Takahashi JS. Genetics of circadian rhythms in mammalian model organisms. *Adv Genet* 74:175–230, 2011. doi:10.1016/B978-0-12-387690-4.00006-4.

Lucas RJ, Freedman MS, Munoz M, Garcia-Fernandez JM, Foster RG. Regulation of the mammalian pineal by non-rod, non-cone, ocular photoreceptors. *Science* 284:505–507, 1999.

Marques MD, Golombek D, Moreno C. Adaptação temporal. In: N Marques e L Menna-Barreto, eds. Cronobiologia: Princípios e Aplicações. 3rd ed. Editora da Universidade de São Paulo, São Paulo, Brazil, 435 pp., 2003.

Meijer JH, Robbers Y. Wheel running in the wild. *Proc R Soc B* 281:20140210, 2014.

Mirmiran M, Ariagno RL. Influence of light in the NICU on the development of circadian rhythms in preterm infants. *Semin Perinatol* 24:247–257, 2000.

Mistlberger RE. Circadian food anticipatory activity: Formal models and physiological mechanisms. *Neurosci Biobehav Rev* 18:171–195, 1994.

Mohawk JA, Green CB, Takahashi JS. Central and peripheral circadian clocks in mammals. *Annu Rev Neurosci* 35:445–462, 2012.

Moore RY. Organization of the primate circadian system. *J Biol Rhythms* 8(Suppl.):S3–S9, 1993.

Moore RY, Klein DC. Visual pathways and the central neural control of a circadian rhythm in pineal serotonin N-acetyltransferase activity. *Brain Res* 71:17–33, 1974.

Moore RY, Lenn NJ. A retinohypothalamic projection in the rat. *J Comp Neurol* 146:1–9, 1972.

Moore-Ede M. Physiology of the circadian timing system: Predictive *versus* reactive homeostasis. *Am J Physiol* 250:R737–R752, 1986.

Moore-Ede M, Sulzman FM. Internal temporal order. In: Aschoff J, ed., Biological rhythms: Handbook of behavioral neurobiology. Vol. 4. Plenum Press, New York, pp. 215–241, 1981.

Moore-Ede MC, Sulzman FM, Fuller CA. The clocks that time us. Physiology of the circadian timing system. 1st ed. Harvard University Press, Cambridge, MA and London, 448 pp., 1984. ISBN/ISSN 0-674-13581-4.

Mrosovsky N. Rheostasis: The physiology of change. Oxford University Press, New York/Oxford, 183 pp., 1992.

Mrosovsky N. Locomotor activity and non-photic influences on circadian clocks. *Biol Rev* 71:343–372, 1996.

Panda S, Provencio I, Tu DC, Pires SS, Rollag MD, Castrucci AM, Pletcher MT, Sato TK, Wilshire T, Andahazy M, Kay SA, Van Gelder RN, Hogenesch JB. Melanopsin is required for non-image-forming photic responses in blind mice. *Science* 301:525–527, 2003.

Pittendrigh CS. On temperature independence in the clock system controlling emergence time in *Drosophila. Proc Natl Acad Sci USA*, 40:1018–1029, 1954.

Rajaratnam SM, Arendt J. Health in a 24-hr society. *Lancet* 358:999–1005, 2001.

Redlin U. Neural basis and biological function of masking by light in mammals: Suppression of melatonin and locomotor activity. *Chronobiol Int* 18:737–758, 2001.

Reilly T, Atkinson G, Edwards B, Waterhouse J, Akersted T, Davenne D, Lemmer B, Wirz-Justice A. Coping with jet-lag: A position statement for the European College of Sport Science. *Eur J Sport Sci* 7(1):1–7, 2007.

Rensing L, Ruoff P. Temperature effect on entrainment, phase shifting, and amplitude of circadian clocks and its molecular bases. *Chronobiol Int* 19:807–864, 2002.

Reppert SM, Weaver DR. Coordination of circadian timing in mammals. *Nature* 418:935–941 (29 August); 935e941, 2002. doi:10.1038/nature00965.

Roenneberg T, Kuehnle T, Juda M, Kantermann T, Karla Allebrandt, Gordijn M, Merrow M. Epidemiology of the human circadian clock. *Sleep Med Rev* 11:429–438, 2007.

Roenneberg T, Kuelnle T, Pramstaller PP, Ricken J, Havel M, Guth A, Merrow M. A marker for the end of adolescence. *Curr Biol* 14(24):R1038–R1039, 2004. doi:10.1016/j.cub.2004.11.039.

Roenneberg T, Wirz-Justice A, Merrow M. Life between clocks: Daily temporal patterns of human chronotypes. *J Biol Rhythms* 18(1):80–90, 2003.

Rosbash M. Understanding Circadian Rhythms: Understanding Sleep Disorder. Conference at CLS-Coalition for the Life Sciences, 2014. It is a Youtube. https://www.youtube.com/watch?v=mJ8ZOCHVrmI&ebc=ANyPxKqIl1E6 pnVHY_Jltk9tGy-uvWoe3yUM1ofTz2DjUIN3nRPLyDdONc11vTls8CFl3lu N0uOReLD4fc4uLQErVIQuv6khxg; Rosbash, director of the Brandeis

National Center for Behavioral Genomics, has spent more than three decades studying circadian rhythms, the built-in biological clock that governs functions such as sleep and wakefulness, metabolism and hormone levels, in organisms as simple as fruit flies and as complex as humans.

Rynasiewicz R. Newton's views on space, time, and motion. In: Edward N. Zalta, ed. The stanford encyclopedia of philosophy (Summer edition), ed 2014. http://plato.stanford.edu/archives/sum2014/entries/newton-stm/.

Samuels CH. Jet lag and travel fatigue: A comprehensive management plan for sport medicine physicians and high-performance support teams. *Clin J Sport Med* 22:268–273, 2012.

Sancar A, Lindsey-Boltz LA, Kang TH, Reardon JT, Lee JH, Ozturk N. Circadian clock control of the cellular response to DNA damage. *FEBS Lett* 584(12):2618–2625, 2010. doi 10.1016/j.febslet.2010.03.017.

Scheer F, Hilton MF, Mantzoros CS, Shea SA. Adverse metabolic and cardiovascular consequences of circadian misalignment. *Proc Natl Acad Sci USA* 106:4453–4458, 2009.

Schibler U, Ripperger J, Brown SA. Peripheral circadian oscillators in mammals: Time and food. *J Biol Rhythms* 18(3):250–260, 2003.

Schmidt-Nielsen K. Animal physiology: Adaptation and environment. Cambridge University Press, New York, 607 pp., 1997.

Schroeder AM, Truong D, Loh DH, Jordan MC, Roos KP, Colwell CS. Voluntary scheduled exercise alters diurnal rhythms of behaviour, physiology and gene expression in wild-type and vasoactive intestinal peptide-deficient mice. *J Physiol* 590(23):6213–6226, 2012.

Skene DJ, Arendt J. Human circadian rhythms: Physiological and therapeutic relevance of light and melatonin. *Ann Clin Biochem* 43:344–353, 2006.

Stephan FK. The "other" circadian system: Food as a Zeitgeber. *J Biol Rhythms* 17(4):284–292, 2002.

Tahara Y, Shibata S. Circadian rhythms of liver physiology and disease: Experimental and clinical evidence. *Nat Rev Gastroenterol Hepatol* 13:217–226, 2016. Online 24 February; doi:10.1038/nrgastro.2016.8.

Tattersall I. The concept of cathemerality: History and definitions. *Folia Primatol* 77:7–14, 2006.

Toh KL, Jones CR, He Y, Eide EJ, Hinz WA, Virshup DM, Ptacek LJ, Fu YH. An hPer2 phosphorylation site mutation in familial advanced sleep phase syndrome. *Science* 291:1040–1043, 2001.

Tomotani BM. Aftereffects of field entrainment and the activity phase of the subterranean rodent tuco-tuco (*Rodentia: Ctenomyidae*). Master Sciences Thesis, Digital Library of Theses and Dissertations, University of São Paulo, 2011, http://www. teses.usp.br/teses/disponiveis/41/41135/tde-23042012- 105030/publico/Barbara_Tomotani_CORRIG.pdf.

Turek FW, Scarbrough K, Penev P, Labyak S, Valentinuzzi VS, Van Reeth O. Aging of the mammalian circadian system. In: Takahashi JS, Turek FW, Moore RY, eds. Circadian clocks, handbook of behavioral neurobiology. Vol. 12, pp. 292–317, 2000. Kluwer Academic/Plenum Publishers, New York, NY, 770 pp., 2000. ISBN 0-306-46504-3.

Valentinuzzi VS, Amelotti I, Gorla DE, Catalá S, Ralph MR. Circadian entrainment by light and host in the Chagas desease vector, *Triatoma infestans*. *Chronobiol Int* (Online) 1–11, 2013. doi:10.3109/07420528.2013.846352.

Valentinuzzi VS, Oda GA, Araújo JF, MR Ralph. Circadian pattern of wheel-running activity of a South American subterranean rodent (*Ctenomys cf knightii*). *Chronobiol Int* 26(1):14–27, 2009.

Waterhouse JM, DeCoursey PJ. The relevance of circadian rhythms for human welfare. In Dunlap JC, Loros JJ, DeCoursey PJ, eds. Chronobiology: Biological timekeeping. Sinauer Associates Publishers, Sunderland, pp. 325–356, 2004.

Waterhouse JM, Folkard S, Minors DS. Shiftwork, health and safety: An overview of the scientific literature 1978–90. The Stationery Office, London, 1–31, 2, 1992.

Waterhouse JM, Reilly T, Atkinson G. Jet Lag. *Lancet* 350:1611–1616, 1997.

Weaver DR. The suprachiasmatic nucleus: A 25-year retrospective. *J Biol Rhythms* 13:100–112, 1998.

Wehr TA, Wirz-Justice A, Goodwin FK, Duncan W, Gillin JC. Phase advance of the circadian sleep–wake cycle as an antidepressant. *Science* 206:710–713, 1979.

Weitzman ED, Czeisler CA, Coleman RM, Spielman AJ, Zimmerman JC, Dement W, Pollak CP. Delayed Sleep Phase Syndrome: A chronobiological disorder with sleep-onset insomnia. *Arc Gen Psychiatry* 38(7):737–746, 1981. doi:10.1001/archpsyc.1981.01780320017001.

Yoo SH, Yamazaki S, Lowrey PL, Shimomura K, Ko CH, Buhr ED, Siepka SM, Hong HK, Oh WJ, Yoo OJ, Menaker M, Takahashi JS. PERIOD2: LUCIFERASE real-time reporting of circadian dynamics reveals persistent circadian oscillations in mouse peripheral tissues. *Proc Natl Acad Sci NY* 101(15):5339–5346, 2003.

Zhang R, Lahensa NF, Ballancea HI, Hughes ME, Hogenesch JB. A circadian gene expression atlas in mammals: Implications for biology and medicine. *Proc Natl Acad Sci NY* 11(45):16219–16224, 2014.

CHAPTER 11

BIOSENSORS AND NANOBIOSENSORS

Rossana E. Madrid, Rosana Chehín, Ting-Hsuan Chen
and Anthony Guiseppi-Elie

*So, naturalists observe, a flea
has smaller fleas that on him prey;
and these have smaller still to bite 'em,
and so proceed ad infinitum.*
from **On Poetry: A Rhapsody** (1733),

by Jonathan Swift (1667–1745)

Abstract

Biosensors, sensing systems that use biological recognition, physicochemical transduction, and companion instrumentation, have been widely used to analyze and quantify different analytes. The application of biological entities to develop selective sensors has grown exponentially since the late 1990s, and the worldwide biosensor industry is now worth billions of dollars and tens of thousands of papers have been published in the area. Biosensors can be applied in medicine, pharmacology, food and process control, environmental monitoring, defense, and security. The biggest market is in the medical area, particularly the glucose biosensors for people with diabetes. The latest trend is to apply biosensors for personalized medicine. The modification of electrode surfaces with biological molecules (enzymes, antibodies, nucleic acids, organelles, whole cells, tissues, etc.) has now evolved with the use of nanostructures. The emergence of nanotechnology promises to enhance the performance of biotransducers through interface engineering. Among the nano-enabled enhancements, we find preferential orientation of biomolecules onto

391

device surfaces, improved bioelectrocatalytic performance, direct electron transfer, and the consequential increased sensitivity and miniaturization. Recently, it has been suggested that amyloids, a class of protein aggregates, originally associated with human diseases, could be used for the bio-nanotechnological applications, particularly in the development of biosensors.

This chapter introduces the concept of biomedical biosensors, discusses the various transduction principles, the various biological agents used in biorecognition, the engineering challenges in fabrication and systems applications, and the new frontier opportunities for both fabrication and applying biosensors.

11.1 Introduction

In all living beings, there are several systems to detect, analyze, and quantify different analytes, which are very important for life. These detecting systems could be chemical or physical ones. Chemical systems, for example, are those that use specialized receptors, which normally detect chemical analytes. Examples of this type of systems are taste and smell. Taste detects the food flavor and other materials that pass across the tongue and through the mouth. Gustatory cells, located primarily on the surface of the tongue and adjacent portions of the pharynx and larynx, interact with dissolved chemicals. Different tastes bind to specific receptors that release neurotransmitters to afferent fibers causing action potential firing. Smell has a more complex mechanism, but also has olfactory receptor cells, so allowing the body to recognize chemical molecules in the air through inhalation. Odorant molecules stimulate the receptors and activate chemical mechanisms that result also in an action potential.

Physical systems, on the other hand, have a different mechanism in the sense that the stimulus is a physical variable, such as light or sound. Light, for example, enters the retina and excites a special type of neuron called a photoreceptor cell. A local potential begins there and travels to larger neurons, where action potentials must be created for the signal to have enough strength to reach the central nervous system. Visual information is processed in the occipital lobe, specifically in the primary visual cortex. In the case of hearing, a mechanical vibration of the tympanic membrane, caused by sound reaching the external ear, proceeds to the auditory ossicles in the middle ear; from there, the disturbance continues to the cochlea, where specific organs are deflected as the waves of fluid travel due to the first vibration. Bipolar sensory neurons located in the center of the cochlea monitor the information from these receptor cells and pass it onto the brain. These systems use chemical or physical receptors to detect the analyte. These receptors, or its biomimetic expression, can be used in an artificial device, to perform the same operation. It is possible to use, for example, different enzymes to detect specific analytes, or an antibody, to detect its antigen. On the other side, there are the physical sensors. They may detect and quantify different physical or chemical potential magnitudes, such as pressure, current variation, flow, temperature, dissolved oxygen, and pH. They are called *transducers*, because they transform some kind of energy into another. Say, a temperature transducer converts thermal energy into an electrical resistance change, which, in turn, can be measured electrically. The integration of these two types of elements, the physicochemical and the biorecognition ones, is present in all life forms and give rise to new type of sensors called *biosensors*.

The present chapter addresses the key concepts in the area of biosensors. It describes the different types of physicochemical transduction modes used in biosensors, the methods used to integrate the biological recognition component with the physicochemical component to produce a biotransducer, and the challenges of engineering the interface for stability and reproducibility. Recent developments in the field are also dealt with. This includes the application of nanotechnology, hydrogels, and biopolymers to increase the biosensor's sensitivity and its analytical performance.

11.2 Detectors, Sensors, Transducers and Actuators

The first contact between the electronic and biological worlds is the *biotransducer*. This device "transforms" the energy commanding the biological phenomenon to be measured into another type of energy, which can be registered, processed, and displayed. In some cases, simple electrodes can do it, but in other, more specific or complex

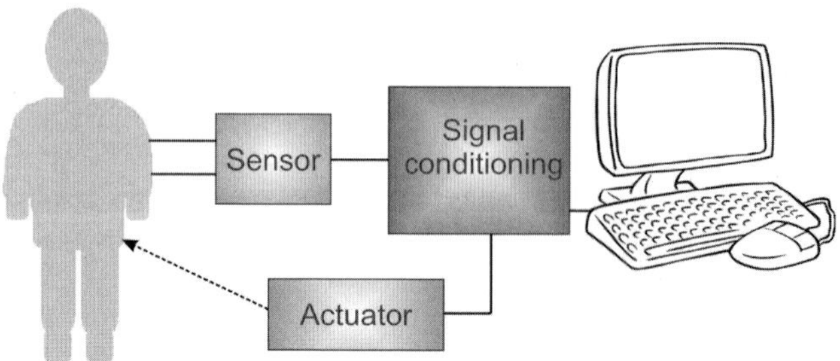

Figure 11.1: The measurement system.

transducers or sensors are necessary (Madrid *et al.*, 2010). The development of biosensors had a great impact, since the level and variety of signals that can be detected have exponentially grown.

When a transducer converts a measurable quantity (pressure, liquid flow, temperature change, or other) to an electrical variable (voltage or current), it is called a *detector*. *Sensors*, instead, are distinguished from detectors in that the electrical output is updatable in response to changing conditions and are quantifiable. When the transducer converts an electrical signal into another form of energy, such as force, light, and mechanical movement, its name is *actuator*.

Figure 11.1 depicts schematically a measurement system, where the transducer is the first element in contact with the sample. The detected signal is thereafter amplified or modified to some extent. On the other side, an actuator is needed.

Electrodes are the simplest transducers. They transform ionic currents, usually present in biological media, into electronic ones, which are recorded, amplified, modified, and displayed in the electronic device. There are different types of sensors and transducers available in the marketplace, and the choice of which one to use depends upon the quantity being measured or controlled. They can be classified according to different characteristics, such as transduction principle, measured magnitude, location of the sensor, output signal generation, or other characteristics. Table11.1 summarizes different classifications and gives some examples.

In the area of sensors and transducers, there are two important concepts to take into account: *transducible property* and *transduction*

Table 11.1: Transducers classification. More extensive in Pallàs-Areny and Webster (2001).

Output signal generation	Transduction principle	Measured magnitude	Location of the sensor
Active or generative (*Thermocouples*)	**Resistive** (*Thermistor*) **Inductive** (*LVDT*) **Capacitive** (*Microphone*) **Electromagnetic** (*Electromagnetic flowmeter*) **Ultrasonic** (*Ultrasound transducer*) **Electrochemical** (*pH electrode*) **Optical** (*Photodiode*) **Others. . .**	**Temperature** (*Thermistor, thermocouple*) **Pressure** (*Strain gauge, LVDT*) **Force** (*Strain gauge, dynamometer*) **Flow** (*Ultrasound flow transducer, electromagnetic flowmeter*) **Velocity** **Acceleration** (*Accelerometer*) **Oxygen** (*Clark electrode*) **Ionizing radiation** (*Ionization chamber, solid-state detectors*) **Others. . .**	**Noninvasive/ noncontact** (*Blood gases*)
Passive (*Thermistor*)			**Noninvasive/ contact** (*Temperature, oxymetry*) **Invasive/ short-term** (*Catheter*) **Invasive/ long-term** (*Pacemakers, implantable electrodes*)

principle. The former refers to those properties that characterize the biological phenomenon and that allow it to be measured. For example, the movement of the arterial wall, say, to measure pressure. The latter, instead, is the basic idea used by a transducer to register the former. Examples are the variation of a thermistor resistance with temperature, or the Doppler effect to measure blood flow. Almost all these devices can be used as a physical transducer in a biosensor. The biological receptor is normally immobilized over this physical transducer.

11.3 Biosensors

Biosensors are systems comprising a biotransducer and associated instrumentation. A biotransducer, as mentioned above, is the combination of a physical transducer and a biological recognition. Figure 11.2 shows a generalized schematic representation.

The *"physicochemical transducer"* of the biotransducer is one of the different kinds of sensors described previously. It can be a physical transducer (metallic electrode, thermistor, and piezoelectric) or a chemical one (pH or oxygen electrode). The bioreceptor can be an enzyme, an antibody, an antibody fragment, nucleic acid, a sub-cellular fragment, a cell, microorganisms, some kind of tissue or other biological molecule, which normally is called the *biorecognition element*. This bioreceptor selectively recognizes the specific chemical information or *analyte* in the sample.

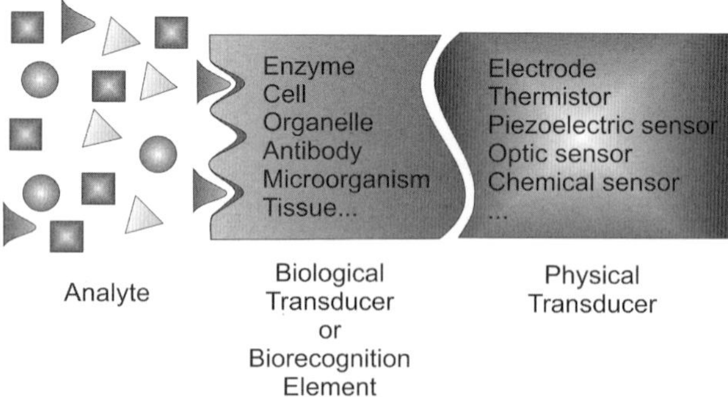

Figure 11.2: The biotransducer. Its parts are the *biological recognition* or *biorecognition element* and the *physical transducer*.

Biosensors not only help in the understanding of biological processes, but also have many biomedical and industrial applications. Glucose, pregnancy, and urea sensors are commonly found in daily life and hospital practice, but an increasing number of new applications emerge every day. In the early 1960s, Clark and Lyons reported the possibility of using an enzyme membrane to detect urea or glucose by means of a pH or oxygen electrode, respectively (1962). At the end of that decade, Updike and Hicks (1967) introduced the first enzyme electrode, and they trapped glucose oxidase (GOD) in a polyacrylamide gel. This was the first biosensor that made use of an oxygen Clark's electrode to detect glucose. It was based on the property of GOD to catalyze the reaction between glucose and oxygen, as described by

$$\text{Glucose} + O_2 Gl \xrightarrow{\text{Glucose oxidase}} 1 + H_2O_2 \qquad (11.1)$$

The electrode oxidizes the glucose of the sample into gluconic acid by using the oxygen of the solution. The sensor detects the reduction in oxygen partial pressure, indirectly detecting the glucose concentration of the sample. This new type of sensor gave rise to different types of biosensors by combining existing sensors, such as amperometric, potentiometric, thermal, piezoelectrical, optical, and biological molecules, such as enzymes, antibodies, cells, and chemical receptors, among others.

A few years later, Guilbault and Lubrano (1973) developed a new type of glucose sensor, which uses a metallic electrode for the oxidation of peroxide. In Eq. (11.1), we can see that when glucose is oxidized, one of the products of the reaction catalyzed by the enzyme is peroxide. By polarizing the working electrode at 0.9V vs a Ag/AgCl reference electrode, the H_2O_2 is oxidized to give a current proportional to the H_2O_2 present in the sample, that is,

$$H_2O_2 \xrightarrow{\sim 0.9 \text{ V}} O_2 + 2H^+ + 2e^- \qquad (11.2)$$

New electrochemical biosensors were then developed. In general, the information decoded by the bioreceptor is transformed in some kind of signal. This transduction could take place by different techniques, such as potentiometry, amperometry, or impedance measurements,

Table 11.2: Biosensor classification.

According to the bioreceptor	According to the immobilization method	According to the transduction principle
Enzymatic	Membrane	Optical
Immunological	In volume	Mass
Microbiological	Electromagnetic	Thermal
Whole cell	Adsorption	Electrochemical

but also by optical, piezoelectrical, and thermometrical techniques, among others. Electrochemical biosensors currently dominate the field, but are focused mainly on metabolite monitoring. In the case of immunosensors, the transduction is carried out principally using optical means. However, both transducers find utility across the whole field, along with piezoelectric, thermometric, magnetic, and micromechanical transducers (Turner, 2013).

Biosensors can be classified based on the biological receptor, the transduction principle used, or the immobilization method used (Table 11.2).

A brief description of each type of biosensor according to the transduction principle is now offered. The main objective aims at an overview of some of the technologies used. After that, the different type of bioreceptors will follow.

11.3.1 *Optical biosensors*

Optical chemical sensors use optical transduction techniques to yield analyte information. The most widely used techniques used in optical chemical sensors are optical absorption and luminescence, reflectance, or fluorescence emissions that can occur in a wide range of the electromagnetic spectrum, including the ultraviolet, visible, or near-infrared ranges (Borisov and Wolfbeis, 2008). Other types of optical sensors are based on other spectroscopic or optical parameters, such as changes in refractive index and reflectivity (McDonagh *et al.*, 2008). The optical principles have not changed over the years; they are basically the same now, but the transduction systems have

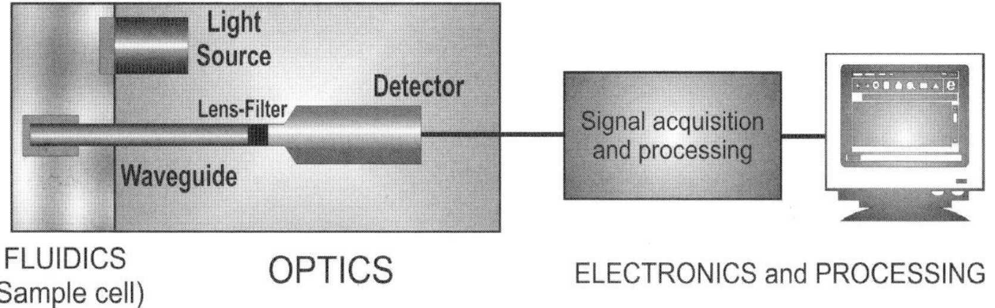

Figure 11.3: Schematic diagram of an optical biosensor. It is composed of three main parts: the optics box, the fluidic system, and the electronics, to acquire and process the signal.

changed a lot with the advent of new technologies such as microfluidics and nanotechnology.

Optical biosensors can be classified as direct sensors and reagent-mediated ones. Biosensors where the analyte has optical properties, such as luminescence or absorbance, belong to the first type. When a second-mediated reagent with optical properties is used to quantify the analyte, it refers to a reagent-mediated sensor (McDonagh *et al.*, 2008). Generally, these reagents are dye molecules that are sensitive to the analyte. These types of biosensors are used when the analyte has no optical properties, which is generally the most common case. The instrumentation used in optical biosensors by and large includes three blocks: an *optical* part, the *fluidics*, and the *electronic* section (Fig. 11.3).

The OPTICS includes the *Waveguide*, such as an optical fiber or planar waveguides; a *Detector*, and in some cases where fluorescence must be excited, there is a *light source* that depends on the wavelength required to excite the fluorophore. In the coupling between the waveguide and the detector there are generally filters and lenses. The detector can be a photodiode or a photomultiplier when using an optical fiber, and CCD or CMOS cameras, or photodiodes arrays in case of planar waveguides. If using a light source, this can be solid state laser diodes, which have a collimated light beam and a uniform wavelength.

The FLUIDICS is necessary to introduce or remove the sample, and finally, the ELECTRONICS, captures the signal and processes it, to show it to the user.

Optical fibers are plastic or glass waveguides that transmit light and are immune to electromagnetic and radio frequency interferences. They are excellent devices to sense light from bio- and chemoluminescence or fluorescence. They act as "guides" of light that enters at one end through the fiber to the other end. The light remains confined in the core because the coating has a lower refractive index and also the incidence angle is greater than the limiting angle, which is called total internal reflection. The latter is the complete reflection of a ray of light within a medium, such as water or glass, from the surrounding surfaces back into the medium

$$\theta_c = \text{arcsen}\left(\frac{n_{\mathrm{m}}}{n_{\mathrm{wg}}}\right)\frac{-b \pm \sqrt{b^2 - 4ac}}{2a} \quad \text{for } n_{\mathrm{m}} < n_{\mathrm{wg}} \qquad (11.3)$$

Such a phenomenon occurs if the angle of incidence surpasses a certain limiting angle, called the critical angle, and is calculated with Eq. (11.1), where n_{m} is the refractive index of the external medium, and n_{wg} is the refractive index of the waveguide. Unfortunately, the total reflection is not perfect. Even though the entire incident wave is reflected back into the originating medium, there is some penetration into the second medium at the boundary, giving rise to an important side effect, the appearance of an evanescent wave beyond the boundary surface. This wave travels along the boundary between the two materials. It propagates parallel to the optical fiber and has an exponential attenuation in the perpendicular direction (Fig. 11.4).

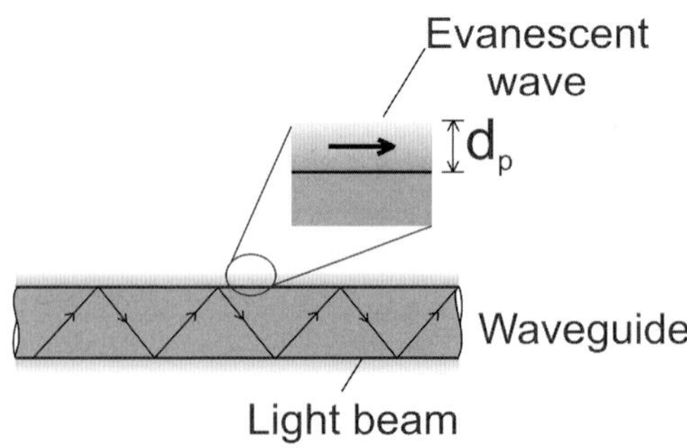

Figure 11.4: Evanescent wave in a waveguide.

The evanescent waves quickly decay beyond the boundary surface in an exponential way. The electric-field strength of an unperturbed evanescent wave at a point (x, z) above the boundary can be derived from Fresnel's equations and is given for s polarization by (Meixner *et al.*, 1994),

$$E(x, z) = E_o e^{(-z/d)} e^{i(2\pi/\lambda)x} \tag{11.4}$$

where

$$|E_o|^2 = \frac{4n_{wg}^2 \cos^2 \theta}{n_{wg}^2 - n_m^2} |E_i|^2, \quad \lambda = \frac{\lambda_o}{n_{wg} \sin \theta},$$

$$d = \frac{\lambda_o/n_m}{2\pi[(n_{wg}/n_m \sin \theta)^2 - 1]^{1/2}}$$

Besides, E_i, λ_o, and θ are the amplitude of incident wave, the vacuum wavelength, and the angle of incidence, respectively.

The distance from the boundary is very short, and the evanescent wave only has effect up to nanometers, so the exponential decay of the electromagnetic field confines the transducible optical signals to a very short distance from the surface. They can be used to detect variations of the optical properties of chemical or biological films immobilized on the fibers. Kronick and Little were the first to use in 1975 the evanescent waves to excite fluorescent immunoassays (1975). They attached haptens (small molecules that can elicit an immune response only when attached to a large carrier, such as a protein) to a planar wave guide and detected binding to evanescently excited fluorescent antibodies. The first evanescent wave fiber optic immunosensor was developed by Hirschfeld in 1984 (Ligler and Taitt, 2008). The selectivity of these evanescent waves was exploited for the design of a great variety of biosensors, such as immunosensors or deoxyribonucleic acid (DNA) biosensors.

Optical biosensors have several advantages over those that use other signal transduction methods. No direct electrical connection to the transduction system is required. Therefore, they are immune to many interfering electrochemical and electromagnetic effects that plague sensors that use these transduction methods. In addition, the availability of high-quality and miniature light sources continue to decrease the size and price of optical biosensors. In particular, biosensors based on evanescent wave excitation have the advantage of surface-specific detection, but this could also be a disadvantage, depending on the application, because interactions that occur away of the evanescent field are not significantly detected. In optical biosensors, the response time depends on the rate of molecular interactions, but not on the rate of the signal transduction. In this sense, there is a critical flux where the mass transport can limit the target union to the immobilized recognition molecule. Fluorescent labels are widely used, because they are simple to use, stable, and have more quantum efficiency; nevertheless, there is a poor coupling efficiency of the generated fluorescence back into the fiber modes. Future perspectives of optical biosensors are promising. The optical biosensor is only as good as the recognition element it uses, and new types of recognition molecules, for example, DNA and peptide aptamers,[1] are being developed. However, some technological hurdles must be overcome before this new receptors begin replacing traditional antibodies in applications where most antibodies are incompatible (Ligler and Taitt, 2008). Besides, the advent of smaller, low-cost lasers, and detectors permits the capability to perform multiple assays in parallel, and this is only limited by the biochemical tests that can be simultaneously performed. The emergence of microfluidics provided new designs and advantages, such as the use of very low-sample volumes, miniaturization, and new applications. The lab-on-a-chip is a new technology that includes optical biosensors and can overcome many of the problems of mass transport, increasing response time and simultaneous detection of many analytes in a single-miniaturized system.

11.3.2 *Piezoelectric biosensors*

These kinds of biosensors use a piezoelectric substrate, which reacts electrically in response to a deformation used as a transduction method to quantify a specific analyte. Piezoelectricity describes the generation of electrical charges on the surface of a solid caused by a

---

[1]Aptamers (Latin *aptus*, fit; Greek *meros*, part) are oligonucleotide or peptide molecules that bind to a specific target molecule.

stress or a strain and vice versa. The direct piezoelectric effect arises when a mechanical stress, made on the surface of the piezoelectric crystal, induces a positive or negative displacement in the lattice elements that manifests itself as a dipole moment. The resulted electric field develops an electric potential on the insulated electrodes. In inverse piezoelectric effect, the crystal is deformed as a function of both the polarity of the voltage applied and the direction of the polarization vector (Pramanik *et al.*, 2013). Quartz crystals and other special materials provide the combination of mechanical, electrical, chemical, and thermal properties that make them of commercial importance. There is wide research in this area, and many piezoelectric materials that have been used as transducing or sensing elements, such as ceramics (Lec, 2001), polymers (Li *et al.*, 2014), and composites (Pramanik *et al.*, 2013). However, this successful research has not yet led to wide commercial success because it is still difficult, in practical terms, the handling of ceramic quartz crystals for various sensing applications owing to its fragile nature and other physical characteristics.

The quartz crystal microbalance (QCM) is a well-known mass-sensitive device widely used in the development of biosensors due to its high-temperature frequency stability and mass sensitivity (Liang *et al.*, 2013). Sauerbrey demonstrated the linear relationship (Eq. (11.5)) between the deposited mass and the resonance frequency displacement in a piezoelectric crystal (1959):

$$\Delta f = -\frac{2f_0^2 \Delta m}{A\sqrt{\rho\mu}} = -S_f \Delta m \qquad (11.5)$$

where Δf and Δm are the frequency shift and mass change, respectively, f_0 is the fundamental frequency, A is the excitation electrode area, ρ and μ are the density and shear modulus of the quartz crystal, respectively, and S_f the Sauerbrey constant. The expression $\sqrt{\rho\mu}$ is also called mechanical impedance. Piezoelectric biosensors are very sensitive, being capable of detecting weight variations in the picogram range. The resonant frequency changes as a result of mass change associated with binding between an analyte and its complementary molecule immobilized onto receptor electrode surface. When

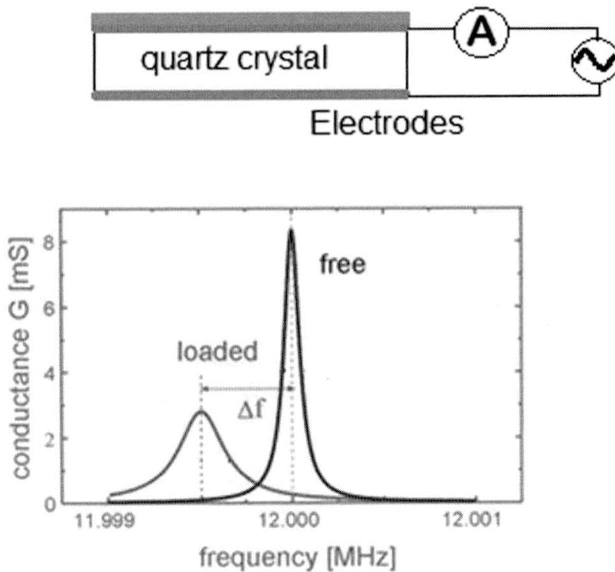

Figure 11.5: Quartz crystal microbalance. Conductance response of a crystal when it is loaded.

the quartz crystal is loaded, the resonant frequency decreases. Figure 11.5 shows the conductance variation of the quartz crystal when loaded.

The frequency variation is quite small. This was why bioanalytical applications arose when new oscillator circuits capable of operating resonators in fluids along with circuits able to solve those small frequency changes were developed. Some examples of applications with this kind of biosensors are the detection of interaction between DNA or RNA complementary chains, specific recognition of protein ligands, virus, bacteria, cells and immunological detections. They are more associated with the detection of binding molecules, rather than to the measurement of a reaction product, as in the case of an enzymatic biosensor. In this sense, the biosensor response is influenced by interfacial phenomena, viscoelastic properties of the adhered biomaterial, surface charge of the adsorbed molecules, and surface roughness. The sensitivity of the biosensor depends on the thickness of the crystal; the thinner the crystal, the greater the sensitivity. For example, the sensitivity of a 5-MHz crystal is 0.057 Hz cm^2/ng. The mechanical vibration is very sensitive to any change on the surface, and due to oscillation, the adsorbed mass experiences an acceleration of 10^6 g, with an amplitude variation in air of 10–20 nm, and in water of 1–2 nm.

The adsorption of biomolecules and cells can be evaluated by measuring the resonant frequency with an oscillator connected to a frequency counter, or by a passive method. The latter consists of applying a frequency near the resonant one with a frequency generator and to monitor the acoustic impedance. The parameters can be fitted to an electrical equivalent circuit, determining the mass of the charge and the dissipated energy. In the development of QCM-based biosensors, the quality of the bioreceptor immobilization method is very important because it determines the specificity and sensitivity of the biosensor. In this sense, a homogeneous surface coverage, which minimizes nonspecific adsorption, determines a good sensitivity. To minimize the detection limit, high-resonant frequency crystals must be used.

The most widely used applications of this type of biosensors are the piezoimmunosensors, which detect the union of antigens to antibodies confined in the surface of a crystal. The resonant frequency decreases when the antigen–antibody binding is produced.

11.3.3 *Thermal biosensors*

This type of biosensor uses the heat absorption or liberation during biochemical reactions as a transducible property for the development of the biosensor. This heat absorption and liberation is proportional to the molar enthalpy and to the total number of molecules created in the biochemical reaction. The following equation relates these two parameters

$$Q = -n_p(\Delta H)Q = C_p(\Delta T) \tag{11.6}$$

where Q is the total heat, n_p is the products moles, ΔH the molar enthalpy change, and C_p is the caloric capacity of the system.

The temperature change registered by the biosensor is

$$\Delta T = \frac{-\Delta H \cdot n_{\mathrm{p}}}{C_{\mathrm{p}}} \tag{11.7}$$

The measurement of the heat absorbed or liberated equals the sum of all enthalpy changes of the reaction mixture (Ramanathan and Danielsson, 2001). Enzymatic reaction enthalpies are in the range of -10 to -100 kJ/mol and cause local temperature changes of a few mKs (Lammers and Scheper, 1999). This temperature change can be measured by different transduction principles. The concept of heat absorption or liberation of a homogeneous conductor with different temperatures is explained by the Thomson effect. When a current exists in a conductor with a nonhomogeneous temperature, there is an absorption or liberation of heat in the conductor (Pallàs-Areny and Webster, 2001). The liberated heat is proportional to the current and it is a reversible effect. When the electrons go from cold points to hot points, heat is absorbed. If the electrons reverse direction, heat is released. There is a report of a bi-nanowire-based thermal biosensor for the detection of salivary cortisol using the Thomson effect (Lee *et al.*, 2013). This biosensor uses cortisol antibodies as bioreceptors and a single bismuth nanowire as the conductor where the heat variation is detected by the Thomson effect. This biosensor is based on the fact that all biological reactions are exothermic, so a temperature difference occurs on the nanowire as a result of the binding of biomolecules. This temperature variation leads to a potential difference due to the Thomson effect. The cortisol concentrations are then determined through the measurement of the voltage difference via the antibody–antigen reaction of cortisol.

Other types of thermal principles could be used. Thermocouples, for example, have a transduction principle based on the combination of the Thomson effect and the Peltier effect. The last one is also a reversible effect, and establishes that there is a heating or cooling of a junction of two different metals when an electric current exists through it (Pallàs-Areny and Webster, 2001). Wang *et al.* reported a microelectromechanical systems (MEMS) differential thermal biosensor integrated with microfluidics for metabolite measurements in two modes (2008). The MEMS device consists of two identical freestanding polymer diaphragms, resistive heaters, and a thermopile between the diaphragms. The polymer-based microfluidic chambers allow sensitive measurement of small volumes of liquid samples. Enzymes for a specific analyte were immobilized on microbeads packed in the chambers. When a sample solution containing the analyte was introduced to the device, the heat released from the enzymatic reactions with the analyte was detected by a thermocouple. It uses a reference chamber, and the temperature difference between the sample and reference chambers was detected by the thermocouple integrated on the diaphragms. The thermocouple consists of thin-film chromium and nickel junctions. The temperature difference induces a voltage in the thermocouple, and this is the output signal of the biosensor. With this kind of biosensor, a sensitivity with respect to a glucose concentration of approximately 2.2 $\mu V/mM$ was obtained, with a resolution of about 0.025 mM.

Another type of the transduction principle can be used for the development of thermal biosensors. One of them is the variation in resistivity of a semiconductor material due to temperature variations. These devices are the thermistors, which exhibit high sensitivities for monitoring of enzymatic conversions. Miniaturized sensors are simply developed with the great advance in semiconductor technology. In recent years, several simple, inexpensive thermistor devices for the determination of different analytes have been developed (Mishra *et al.*, 2014). Enzymes have been shown to be ideally suitable as biorecognition elements for thermometric measurements, unlike immunological thermal biosensors, of which there are very few reports in the literature. In general, thermal biosensors have the advantage of using very simple instrumentation and procedures. The assays are normally reagent-less and are independent of the color or turbidity of the sample (Yakovleva *et al.*, 2013).

11.3.4 *Electrochemical biosensors*

An electrochemical biosensor has as a physical transducer an electrode and its transduction principle is electrochemical.

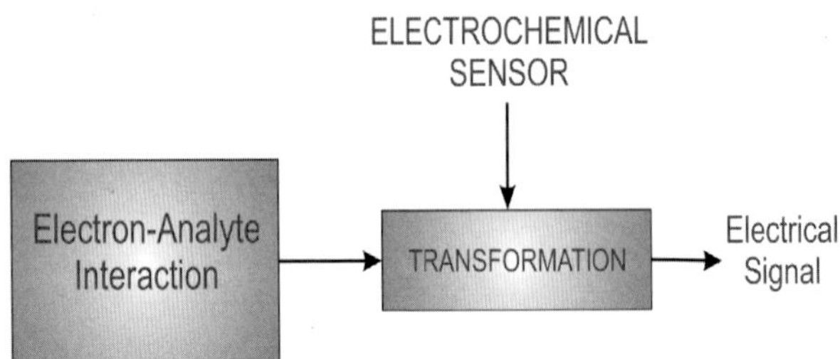

Figure 11.6: Electrochemical interaction and transformation of the signal.

Electrochemistry implies an interaction between an electrode and the analyte, with a consequent movement of charge that may transfer charge (faradic) from the electrode to the other phase, or there may be accumulation of charge at the interface (capacitative). There is a redox reaction at the electrode. The electrochemical transducer (Fig. 11.6) registers such transformation of ions into electrons to give a useful electrical signal.

During this process, chemical changes take place at the electrodes. These electrode reactions and/or the charge transport serve as the basis of the sensing process. An electrochemical transducer is classified according to the specific electrochemical principle used. They can be potentiometric, amperometric, impedimetric (conductometric), and ion charge or field effect (Thévenot *et al.*, 2001). Their operating and measurement principles are summarized in the next sections.

11.3.4.1 *Potentiometric biosensors*

In this type of biosensor, the primary signal arising due to the interaction between the analyte and the biomolecule is a change in an electrode potential. This is measured at zero current and with respect to a suitable reference electrode. The reference electrode is required to provide a constant half-cell potential against which the change is measured. The working electrode develops a variable potential depending on the activity or concentration of the analyte. The electrodes are, thus, the simplest transducers, which transform ionic currents into electronic ones. The contact zone between the electrode and the biological sample is known as the electrode–electrolyte interface (EEI) (Madrid *et al.*, 2010). Distribution perturbations of the ionic chemical potential, at this interface, alter the interfacial potential.

All electrochemical measurements involve charge transfer at the EEI. This EEI generates an impedance and a DC potential (the half-cell potential). The change in potential is related to the activity of the species in solution, also with its concentration if it is sufficiently low, in a logarithmic manner. The potential difference at the interface is related to the activities of species i in sample phases (s) and in the electrode phase (β) through the Nernst equation (Wang *et al.*, 2008)

$$E = E_0 + \frac{RT}{Z_{\mathrm{i}}F} \ln \frac{a_{\mathrm{i}}^{s}}{a_{\mathrm{i}}^{\beta}} \qquad (11.8)$$

where E_0 is the standard electrode potential of the sensor electrode, a_i is the activity of the ion, R is the universal gas constant, T stands for the absolute temperature, F is the Faraday constant, and Z_i is the ion valence. For low-ionic concentration in the solution ($<10^{-3}$ M), the activities could be replaced by the concentration of the redox species. The Nernst equation cannot accurately predict half-cell potentials for solutions in which the total ionic concentration exceeds about 10^{-3} M (Lower, UC Davis-ChemWiki, no date).

Potentiometric sensors can be classified into three types: metallic electrodes, membrane sensors (*solid*, such as the pH glass electrode, and *polymeric*), and ion-sensitive field-effect transistors (ISFET). The last one can be of different types: ChemFET (with a crystalline or polymeric membrane), EnFET (enzymatic membrane), and ImmunoFET (with an antibody–antigen bioreceptor). Potentiometric sensors based on FETs are electronic semiconductor devices, which are modified to be used as potentiometric sensors of analytes. The basic component is a metal-oxide semiconductor field-effect transistor (MOSFET), which is based on the original field-effect transistor introduced in the 1970s (Fig. 11.7).

When a voltage is applied between the drain and the source, this voltage finds p–n unions inversely polarized, so there is no current flow. An inversion layer at the semiconductor–oxide interface acts

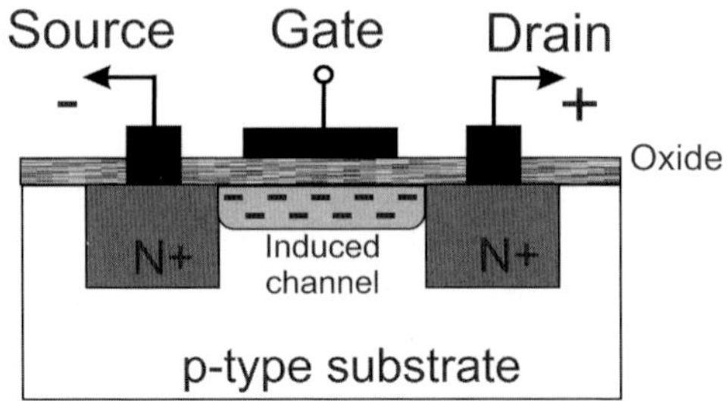

Figure 11.7: The MOSFET. It consists of a source and a drain, two highly conducting n-type semiconductor regions, which are isolated from the p-type substrate by reversed-biased p–n diodes. A metal gate covers the region between the source and the drain, but is separated from the semiconductor by the gate oxide.

as a conducting channel. An inversion charge can be induced in the channel by applying a suitable gate voltage relative to other terminals. If a positive potential is applied on the Gate, the "field effect" appears and the positive carriers are repelled by the field, and a conduction channel is formed between the drain and the source (Ytterdal *et al.*, 2003). The current is then

$$i_D = K_1 \cdot (V_G - V_\mathrm{T})^2 \qquad (11.9)$$

The onset of strong inversion is defined in terms of a threshold voltage V_T being applied to the gate electrode relative to the other terminals. This threshold voltage is one of the most important parameters characterizing metal–insulator–semiconductor devices and can be adjusted through ion implantation. If the metallic gate is eliminated and replaced by an ion-selective membrane, an ISFET is created. The oxide of the MOSFET gate is in direct contact with the solution. Figure 11.8 schematizes an ISFET.

It is possible to detect the potential variation at the insulator–solution interface, thermodynamically defined by considering the surface of a non-hydrated layer and an ideal insulating layer but ion selective, which is the ion-selective electrode. This is similar to the pH glass electrode, with a hydrated glass layer in contact with an insulating dry glass layer beneath, so forming the ion-sensitive electrode (Silva Cárdenas, 2008). The Nernst equation (Eq. (11.6)) is the fundamental working equation. In the present case, the change

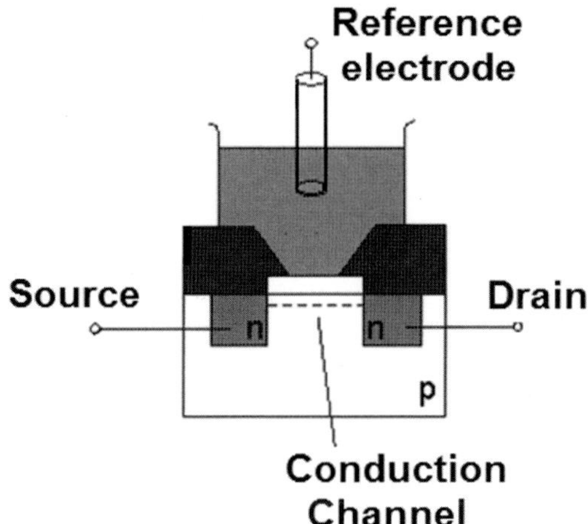

Figure 11.8: ISFET — Ion-selective field-effect transistor. The gate is replaced by an ion-selective membrane.

of voltage is given by

$$\Delta V = K + \frac{0.05916}{z} \log a = K + S \log a \qquad (11.10)$$

where S is the sensor sensitivity. The ISFET operating principle is based on pH sensitivity of the insulating inorganic layer in a direct contact with the solution, which can be SiO_2, Si_3N_4, Al_2O_3, or other. The last one offers a chemical sensitivity in the range of 53–57 mV/pH, very similar to that of the Si_3N_4, which is in the range of 52–56 mV/pH, but markedly different to the range of the SiO_2, which is between 29 and 36 mV/pH. The phenomena that explain the behavior of the ISFET are complex due to physical and chemical processes involved (Silva Cárdenas, 2008). Potentiometric biosensors generally integrate enzymatic material with glass ion selective electrodes, metallic electrodes, or solid state transducers (ISFETs). The enzymes used usually generate reaction products feasible to be detected with this type of sensors. Some examples are H^+, NH_4^+, NH_3, CO_2, or CN^-.

11.3.4.2 *Amperometric and voltametric biosensors*

Another electrochemical method is amperometry, where the current, $I(A)$, in response to an applied voltage $E(V)$, is the signal of interest.

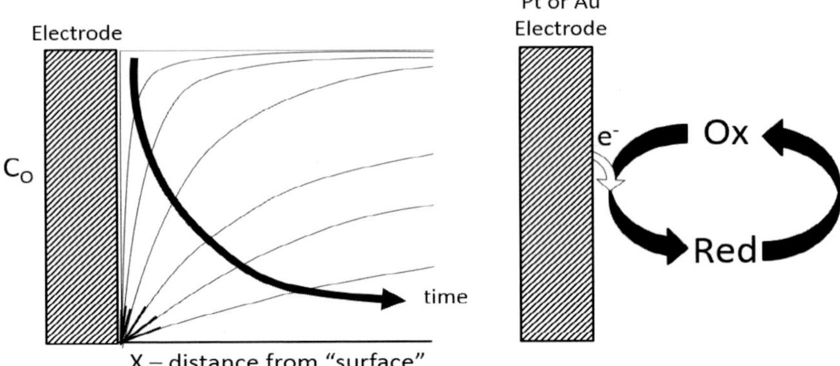

Figure 11.9: Amperometric method. Molecules at the interfaces are involved in redox reactions at the EEI of a metallic electrode, with the corresponding charge transfer.

In this case, the current is proportional to the concentration of the analyte and is given by the Cottrell equation (Eq. (11.9)). There are redox reactions at the interface between the electrolyte and the metallic or semiconductor electrode, and as a result, charge transfer occurs. The direction of the electron flow depends on the redox properties of the analyte and can be controlled with the voltage applied to the working electrode. The magnitude of the current and its dependence upon concentration depends on the size and shape of the electrode (Fig. 11.9)

$$i = nFAD\frac{C}{\sqrt{\pi Dt}} \tag{11.11}$$

where,

i = current (A)
n = number of electrons (to reduce/oxidize one molecule of analyte, for example)
F = Faraday constant (96,485 C/mol)
A = area of the (planar) electrode in cm^2
C = initial concentration of the reducible analyte in mol/cm^3
D = diffusion coefficient for species in cm^2/s
t = time in s.

An amperometric biosensor could use bipolar or tripolar measurements. In the first case, only two electrodes are used and both

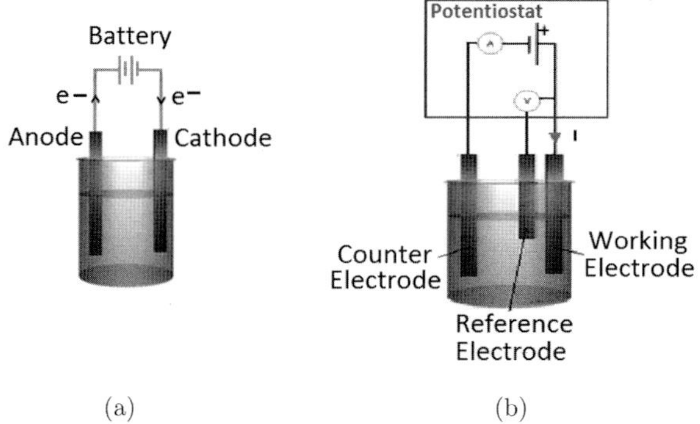

Figure 11.10: Amperometric measurements showing: (a) Bipolar configuration. (b) Tripolar configuration. The working electrode is the biotransducer.

are "active". In the tripolar case, there is a reference electrode (RE), generally a Ag/AgCl one, a counter electrode (CE) of large surface area (1000 $\times$ WE of Pt or stainless steel), and the working electrode (WE), which is the biosensor. The area of the CE is much larger than the working electrode to diminish its contribution to the IEE impedance (Madrid *et al.*, 2010). Figure 11.10 shows both methods.

The potential to be applied to an amperometric measurement is determined by the species to be measured. Each species undergoes oxidation or reduction at a characteristic potential. Voltammetry can also be performed by varying the working potential in a controlled way and by sweeping through this potential. For the first scan, the resulting peak current magnitude is directly proportional to the analyte concentration, $C(M)$, but is also proportional to the square root of the scan rate (ν), as shown in the following equation:

$$i_{\text{peak}} = 0.446nFAC\sqrt{\frac{nFD\nu}{RT}} \tag{11.12}$$

Amperometric biosensors are those that have shown the greatest progress due to its extensive application in the field of clinical analysis. Significant examples of this progress are biosensors for the measurement of analytes in whole blood and serum (Mayorga Martínez, 2009). In the case of amperometric enzymatic biosensors, different methods have been developed for electron transfer between the actions of the enzyme and the amperometric transducer, which allows

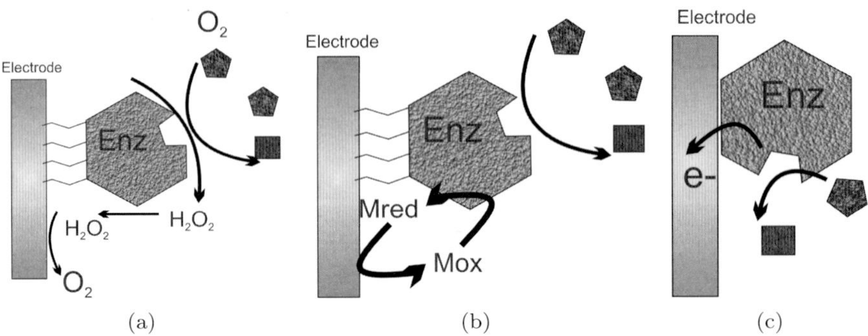

Figure 11.11: Operating principle of (a) first generation, (b) second generation, and (c) third generation of amperometric biosensors.

us to differentiate at least three generations of amperometric biosensors (Fig. 11.11).

First generation of biosensors is those based on the measurement of one of the products or cofactors of the enzymatic reaction. The normal product of the reaction diffuses to the transducer and causes the electrical response. These products must show electrochemical properties. The first biosensor of this type was Clark's biosensor to detect glucose, wherein the consumption of oxygen, a reactant in the biochemical conversion of glucose to gluconolactone, is cathodically measured at an oxygen electrode. Another example is the Guilbault and Lubrano biosensor, which anodically measures glucose by applying a proper potential to electro-oxidation of H_2O_2, a product of the same reaction. The anodic detection of peroxide, although more sensitive than the cathodic determination of oxygen of Clark's electrode, presents many analytical difficulties. For example, the analytic reaction depends on the pH and the potential to be applied is very high, between 0.6 and 0.9 V vs Ag/AgCl at pH 7.0. This potential makes the biosensor also responsive to other molecules, interferents, which oxidize at the same potential (e.g., ascorbic acid, citric acid, uric acid, or paracetamol). This drawback has been partially solved in multiple designs using perm-selective membranes. A better solution to solve this problem led to the use of *redox mediators*; they gave rise to the second generation of amperometric biosensors (Mayorga Martínez, 2009).

The *Second Generation of Biosensors* involves specific "mediators" between the reaction of interest and the transducer in order to generate improved electrochemical response. It is also known as *electronic mediation.* The mediator is responsible for the electron transfer between the active site of the enzyme and the electrode surface, thus avoiding the dependence on the natural enzyme co-substrate. The operating voltage corresponds now to the oxidation potential of the mediator, which is chosen to be much smaller than that used in first generation biosensors, reducing in this way the influence of interferents. The mediators must have the following properties:

(i) react in a rapid way with the active site of the enzyme to diminish the competition with the natural cofactor of the enzyme;
(ii) have a heterogeneous transfer of reversible electrons;
(iii) have a low regeneration potential at the electrode to reduce the effect of interferents;
(iv) be independent of pH;
(v) be stable in both the oxidized and reduced form;
(vi) not be reactive with O_2 in the reduced form, and
(vii) must be innocuous.

Mediators could be used free in solution or co-immobilized with the enzymes. Some examples of widely explored mediators include ferrocyanide/ferricyanide, 1,4-benzoquinone, ferrocene, Prussian blue, methylene blue, Au nanoparticles, and conducting salts, among others (Wang *et al.*, 2008).

Third-generation biosensors are those wherein the reaction itself causes a direct electrochemical response and no product or mediator diffusion is involved. There is a direct electron transfer between the cofactor associated active site of the enzyme and the electrode. Such biosensors show greater selectivity, since they work very close to the intrinsic potential of the enzyme's cofactor, being less exposed to possible interferents. Some problems that could appear in this kind of biosensor are that the direct transfer of electrons between proteins and the electrode is very slow, irreversible, if not impossible. It decreases exponentially with the distance, and also, proteins normally denature in contact with the surface. Nonetheless, a direct transfer could be achieved if the enzyme is properly oriented over the electrode and also, the distance between the redox site and the electrode decreased (comparable with the tunnel effect). It is quite difficult for some enzymes that have the redox site in its interior, but there are some enzymes, for example, the Horseradish Peroxidase, which has its redox site on the surface. Advances in nanotechnology and other very small-scale fabrication technologies have played an important role in the improvement of this third generation biosensing systems (Senveli and Tigli, 2013). These nanomaterials, not only can improve the charge transfer between the active site and the electrode, but also could act as effective immobilization matrices. Moreover, due to their intrinsic and unique features (large surface areas, electrocatalytic properties, controlled morphology, and structure), in combination with the biomolecules contribute to improving bioelectrode performance in terms of sensitivity and selectivity (Walcarius *et al.*, 2013).

11.3.4.3 *Conductimetric and impedimetric biosensors*

These types of biosensors are based on the measurement of temporal changes in electrolyte conductivity or changes in the biotransducer resistance and use DC interrogation. The total impedance of the system can also be measured, and in this case, the excitation signal is an alternating current. In this context, the mechanisms of charge transfer between molecules could provide fundamental knowledge about the behavior of some electrochemical systems. Electrochemical measurements provide the tools to measure this charge transfer, since these chemical systems could be modeled as a set of parallel and serial circuit elements connected in such a way that they permit to study their behavior.

The conductivity change is linear with the concentration of ions, but the ionic concentration is often not linear with the concentration of the electrolyte. Moreover, conductance is not specific for one ion type. For electrolytes, the conductance has a complex dependence upon electrolyte concentration and is given by the Fuoss–Onsager conductance theory (1978) in high relative permittivity solvents, for example, water. For a given set of conductivity values (c_j, Λ_j, $j = 1, \ldots, n$), for each electrolyte in the solvent, the conductance is given by the first of the following equations:

$$\Lambda = P\Lambda_o[(1 + R_X) + E_L] \tag{11.13}$$
$$P = 1 - \alpha(1 - \gamma)$$
$$\gamma = 1 - K_A c\gamma^2 f^2$$

with P and γ defined by the two lower relationships of the set of Eq. (11.11).

In the latter three equations, Λ_o is the limiting molar conductance, R_X stands for the relaxation field effect, E_L represents the electrophoretic counter current, α is the fraction of contact pairs, γ stands for the fraction of solute present as unpaired ion, K_A is the association constant for the ions that form the electrolyte, and f means the activity coefficient.

For simple Ohmic systems,

$$E = I \bullet R \tag{11.14}$$

where E is the potential calculated when the current I passes through a system whose resistance to the current pass is R. The conductivity is the reciprocal of R

$$G = \frac{1}{R} \quad \text{so } E = \frac{1}{G} \tag{11.15}$$

The measurement method to choose depends on where the change is visualized. If the resistance or conductance variation is at the electrode–electrolyte interface, the tripolar method is preferred, because it allows one to measure the EEI at the working electrode. If the variation is in the bulk, the bipolar method is used. In both cases, a frequency sweep is generally carried out, called the electrochemical (tripolar) of electrical (bipolar) impedance spectroscopy (EIS), to know at which frequency the maximum variation of impedance, resistance, or conductance occurs. The impedance of an electrode–electrolyte interface can be modeled with the Warburg-enhanced Randles equivalent circuit (Fig. 11.12).

The Randles model, frequently used to model the EEI, is a parallel of the charge-transfer resistance in series with the Warburg

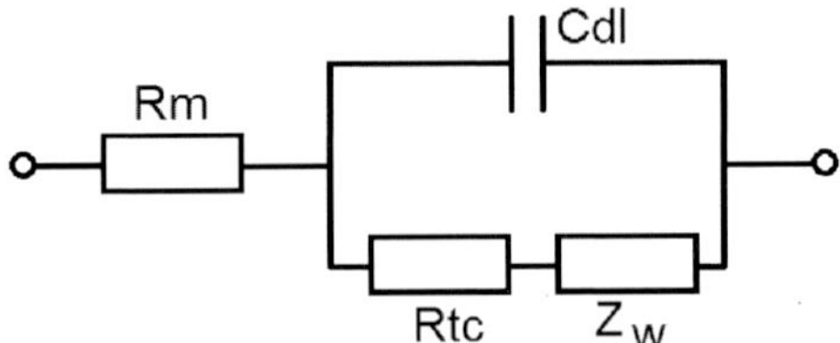

Figure 11.12: Randles electrical model for the electrode–electrolyte interface impedance. Rm: medium resistance; Cdl: double-layer capacitance; R_{tc}: charge-transfer resistance; Z_w: Warburg impedance.

impedance (mass transfer), and the double-layer capacitor. This parallel is in series with the solution or medium resistance. The charge-transfer resistance (R_{tc}) considers the physical hindrance faced by the electrons when they move from and to the electrodes. The Warburg impedance models the diffusion of the reactive species through the boundary layer, but it can be considered negligible in a high ionic buffer solutions, which is normally the case with biosensors. The solution also presents a resistance in the electrochemical system. The resistance of an ionic solution depends on the ionic concentration, type of ions, temperature, and the geometry of the area of electrodes (Madrid *et al.*, 2010).

In the case of biosensors, the charge-transfer resistance (R_{tc}) of the biosensor–medium interface is used to calculate the concentration of the analyte. The disadvantage of this method is that it requires multiple measurements to adjust the model, and that consumes many minutes to have low errors. The method is not amenable for on-line use. Mayorga Martínez *et al.* (2010) developed a new transduction method, called the chrono-impedance method, which allows determining the concentration of an analyte in the sample in real time. The technique consists of determining first, the optimum parameters of potential and frequency to be used to measure impedance, by means of electrochemical measurements. Normally, the optimum DC potential determined is the oxidation potential of the analyte or mediator, depending on the type of biosensors used. After that, an alternate low potential of the same frequency is superimposed to the DC potential in order to measure impedance. The method allows measuring online the analyte concentration. This technique was evaluated in first and second-generation biosensors with good results (Mayorga Martínez *et al.*, 2010 and 2011).

11.3.4.4 *Future trends in electrochemical biosensors*

Until the late 1990s, electrochemical biosensors have generated a large number of scientific publications, demonstrating a greater scientific interest to this kind of biosensor than for the other types, and although quite old, they still hold a very promising future. These devices are clearly preferred because they are robust, simple, and inexpensive to manufacture, they have high specificity, high sensitivity, and a wide range of linearity, without forgetting its fast response and flexibility (Iqbal *et al.*, 2012). On one hand, the equipment needed for signal acquisition, such as potentiostats, are inexpensive, easy to manufacture, miniaturize, and maintain. On the other hand, over the past decade, the main advances and improvements in the sensitivity of electrochemical biosensors have resulted from the incorporation of nanomaterials into functionalized electrodes to develop the nanobiosensors (Walcarius *et al.*, 2013). Another trend for the future research on electrochemical biosensors is the development of biosensors for *in vivo* analysis and continuous testing with high sensitivity and very low detection limit, which would allow the development of noninvasive sensors for clinical use; the latter aspect, so far, is now only in the research stage.

11.4 Bioreceptors

Bioreceptors are those molecules that provide for molecular recognition of the analyte. This system could be a biomolecule such as an enzyme, antibodies, antibody fragments, organelles, a cell, microorganisms, tissue, or other, and it is the part of the biotransducer which provides the specificity to the biomolecular recognition of events. Bioreceptors are classified as affinity bioreceptors and catalytic ones.

11.4.1 *Affinity bioreceptors*

Affinity bioreceptors have the characteristic of recognizing other biological molecules with high binding affinity and selectivity. This

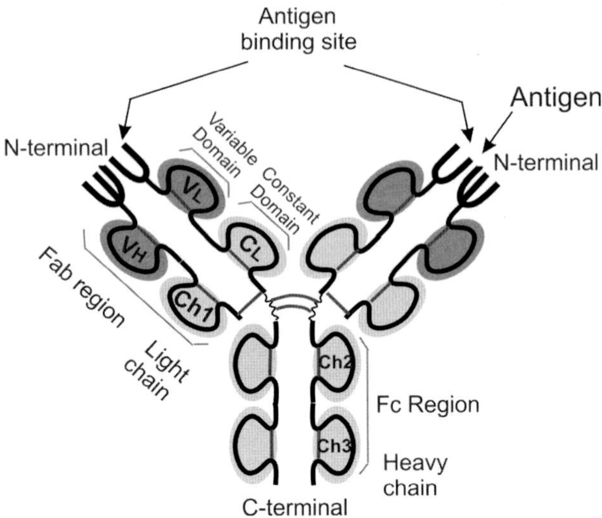

Figure 11.13: Antibody structure.

bioreceptors do not involve reaction product release, substrate consumption or detectable charge-transfer between the coupled molecules (there may be slight variations in potential, but hardly detectable by the methods previously described). Instead, they do involve the coupling of molecules, like antigen–antibody (immunological bioreceptors), and hybridization of complementary DNA (genetic bioreceptors), or nucleic acid or peptide aptamers. Antibodies are glycoproteins which are in the globulin fraction of blood serum. There are five classes of immunoglobulin IgG, IgA, IgM, IgD, and IgE. Figure 11.13 shows the structure of an antibody.

The basic structure of an antibody consists of four polypeptide chains: two heavy chains and two light chains, equal to each other and linked by disulfide bonds. As in any protein, an amino-terminal extreme (N-terminal) is distinguished in each chain and also a carboxylic extreme (C-terminal) in the third chain. The N-terminal fraction is characterized by varying sequences (V), both in the light chains (VL) and in the heavy ones (VH); the rest of the molecule has a relatively conserved and constant structure (C). These fragments are the antigen-binding fragment (F_{ab}), and is a region on the antibody that binds to antigens.

The high affinity in these molecules is due to non-covalent forces, such as hydrogen bonding, electrostatic or hydrophobic interactions, and attractive or repulsive Van der Waals forces. The interaction of all of these forces defines the affinity, and this affinity lies in the different conformations of the F_{ab} region. Little changes in the structure of the antigen can lead to big changes in the affinity. The antibody–antigen complex is a dynamic equilibrium.

$$Ab + Ag \leftrightarrow AbAg \text{ and the complex affinity constant is } K = \frac{[AbAg]}{[Ab][Ag]}$$

The immunosensors devices are able to give an analytical signal from the immune complex formation event. They combine immune recognition agents (antigen or antibodies) that confer selectivity, with a transducer, which confers sensitivity and converts the immunological recognition event into an electronic signal. The main advantage of the immunosensors derives from the proximity between the biological material and the transducer, which allows detecting changes in a more sensitive manner. The specificity and affinity of the antigen–antibody interaction determine the selectivity and sensitivity of the immunosensor and the possibility of regeneration.

In practice, a complex with a very high affinity is required to achieve an appropriate sensitivity. In this case, the subsequent dissociation for regeneration is difficult; therefore, they are usually disposable systems. However, in practice, the antibodies can be regenerated by dissociation of the complex through the application of different agents that change the conditions of the environment promoting the dissociation. This change might be a change of pH, although this sometimes can damage the structure of the antibody.

11.4.2 *Genetic bioreceptors*

The detection of specific sequence of nitrogenous bases allows analytical measurements for different purposes. DNA is a very long molecule with a large capacity to store information. It is rolled and compacted into the nucleus of living cells (say, human cells, virus, or bacteria). The DNA contains information coded for the template synthesis of proteins. DNA is a very long unbranched polymer. The monomeric

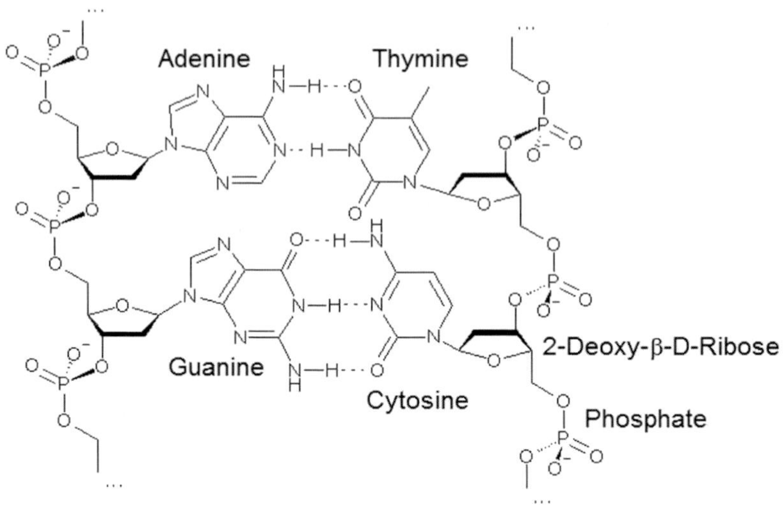

Figure 11.14: DNA molecule formed by repetitive units of 2-deoxyribose in two strands presented as a spiral, and bonded each other with nitrogenous bases pairs.

units are called the nucleotides and the polymer is known as the polynucleotide. The macromolecule is composed of two complementary nucleotides chains. The molecular components of the polymer are 2-deoxyribose (a five atoms sugar), nitrogenous bases: purine (adenine and guanine) and pyrimidine (thymine and cytosine) and monohydrogen phosphate groups (Fig. 11.14).

The monocatenary linear polymer is formed by repetitive units of molecules of 2-deoxyribose, bonded to each other by phosphodiester bonding. This link is formed via a bond between the –OH in position 5′ of the pentose sugar molecule with an –OH in position 3′ of another molecule. The phosphate group and the deoxyribose are the structural molecules of the chain. The DNA molecule is comprised of two polymeric chains or complementary strands, which are presented as a double helix or spiral, and coiled in parallel form, as a spiral staircase, where the steps are the nitrogenous bases pairs –adenine and thymine, guanine and cytosine bonded by hydrogen bridges. This structure is called plectonemic. The nitrogenous bases are integrated in pairs, according to the principle of complementarity: Adenine (A) binds to Thymine (T) and Guanine (G) binds to Cytosine (C) (Fig. 11.15).

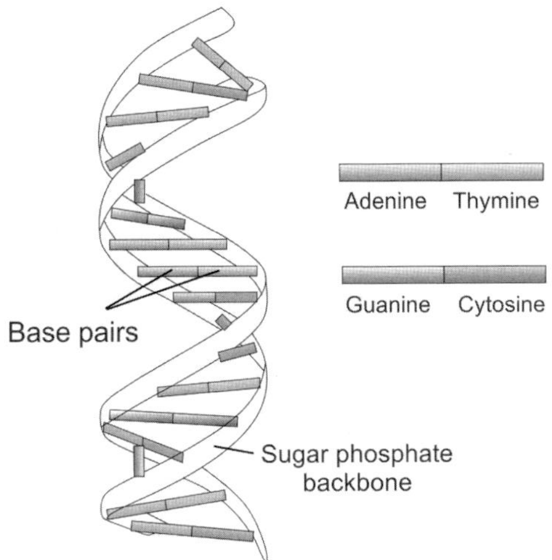

Figure 11.15: DNA molecule. Two complementary chains adopting a double helix form, typical of the DNA molecule.

The structure of the nitrogenous bases is such that A binds with T (by means of three H-bridges) and C binds with G (with two H-bridges). The order of these pairs along the helix determines the genetic information and controls the development of living beings. Each DNA molecule consists of coded segments or genes, which carry out the hereditary instructions to produce proteins that control living processes. Some other characteristics of this molecule are that both chains extend in opposite directions, and that it has the same diameter in its entire length.

Genosensors (or DNA biosensors) are devices, which respond to a hybridization event that is the binding of a DNA chain with its complementary strand. This type of biosensor possesses a high selectivity and specificity and the DNA analysis with them represents the most recent determinations in biosensors. A single-stranded DNA (ssDNA) is immobilized on the sensor surface. The methods used to immobilize DNA strands are similar to that of immobilization of enzymes. The sensor is subjected to a solution of the ssDNA target. If the sequence is complementary, the DNA target is captured in the recognition layer and the resulting hybridization signal is transduced into a useful electrical signal. The hybridization could be measured by direct or indirect DNA detection.

Direct detection of DNA is based on reduction or oxidation of DNA molecules. The use of gold, TiO_2- and polymer-coated electrodes has been reported for electrochemical impedance spectroscopy measurements (Bala *et al.*, 2015; Galán *et al.*, 2015; Kokkinos *et al.*, 2015; Zhu *et al.*, 2015), to mention a few, but showing only a very small number of relevant publications. With this kind of biosensor, it is possible to detect simple polymorphisms through the charge-transfer resistance (R_{tc}).

In the case of indirect DNA detection, electroactive indicators are used together with DNA double-stranded (dsDNA). "Sandwich" hybridization with probes marked with electroactive markers or enzymes is used for this type of transduction. The electroactive indicators are mainly polymers, organic dyes, metal complexes, ferrocene, and derivatives. Nanoparticles have shown a great potential to be used as labels in DNA sensors and immunosensors. Gold nanoparticles followed by quantum dots have been the most reported labels in various biosensing applications (Walcarius *et al.*, 2013). Developments of DNA biosensors and DNA microarrays have progressed tremendously as reported by several investigators (Singh, 2011).

11.4.3 *Catalytic bioreceptors*

Catalytic bioreceptors are those receptors where a catalytic reaction generates the signal. Examples are the enzymes, microorganisms, tissues, and animal and plant cells. During the last few years, enzymes were the most used bioreceptors for the design of biosensors. They are biological catalysts, which increase the speed of a chemical reaction by diminishing the activation energy necessary to transform a substrate in a product. As all catalysts, they have the property to recover its native form after the reaction.

The use of enzymes as bioreceptors has the following advantages (Alegret *et al.*, 2004):

- High sensitivity to a substrate or family of substrates
- Action mechanism well characterized

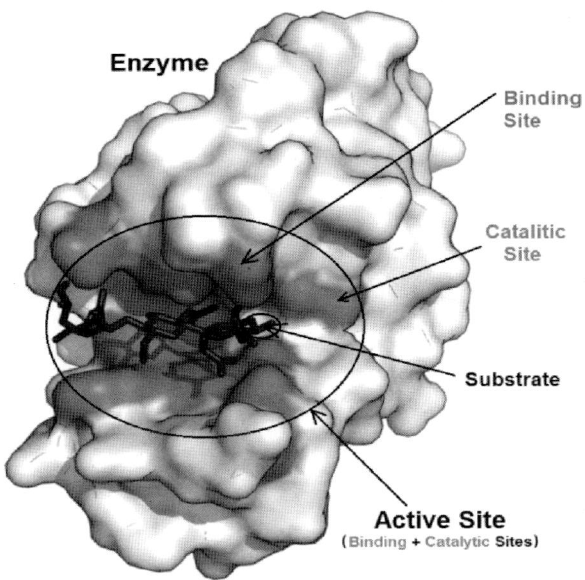

Figure 11.16: Enzyme molecule structure. (Adapted from https://en.wiki2.org/wiki/Active_site)

- Sensitive and fast catalytic mechanisms
- Feasible to use with different transduction methods
- Availability of different types, origin, activity and stability of enzymatic material
- Possibility of using the enzymes as selective for one type of substrate or by enzymatic inhibition by other analytes
- Possibility of reusing the enzyme
- Availability of pure enzymes at moderate price.

The enzymes present a hydrophobic cavity called the *active site*, where the transformation of the substrate in product takes place (Fig. 11.16). The substrate binds to the active site via weak interactions, such as hydrophobic forces, van der Waals unions, or by hydrogen bridges. The surface of the active site is composed of amino acid residues, whose substituent R-groups bind to the substrate and participate to catalyze the chemical transformation. Often, the active site covers the substrate and kidnaps it completely from the solution

(Nelson and Cox, 2005). Some residues from the active site binds temporarily to the substrate and some catalyze the reaction.

Some enzymes supplement their functional groups with metals or organic molecules (cofactors) that contribute to expand the number of reactions to catalyze. The cofactor is a non-protein chemical that is necessary to support the reaction. Some examples are Flavin adenine dinucleotide (FAD), Flavin mononucleotide (FMN), NAD^+, or oxygen. Metals can act as catalysts, stabilizers for transient structures, or electron carriers. The organic molecules which act as cofactors may be permanently attached (prosthetic group) or not (coenzyme) to the rest of the enzyme molecule (Alegret *et al.*, 2004). The classical curve for a reaction catalyzed by an enzyme is presented in Fig. 11.17. This curve represents the relationship between the reaction velocity and the concentration of the substrate, and it can be algebraically expressed by the Michaelis–Menten equation(11.13).

$$v = \frac{V_{\max}[S]}{K_{\mathrm{m}} + [S]} \tag{11.16}$$

In this equation, two important parameters characterizing the enzymatic reactions appear. The first one is $V_{\max}$, which is observed when

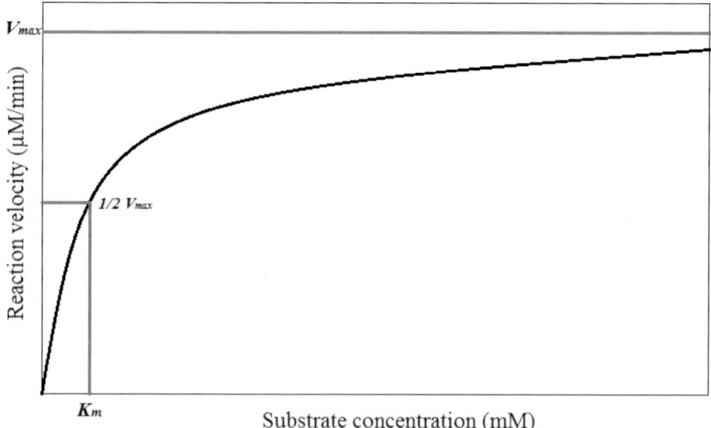

Figure 11.17: Reaction rate vs substrate concentration for a reaction catalyzed by an enzyme. $V_{\max}$: maximum speed, K_{m}: Michaelis–Menten constant.

all enzyme molecules of the reaction are in the form of the ES complex (intermediate complex in the enzymatic reaction which results from the binding of the enzyme with the substrate) and the concentration of enzyme is small. The complex is necessary for the transformation of the substrate to product. Under these conditions, the enzyme is saturated with its substrate so that a further increase in substrate concentration will have no effect on the catalytic rate. This condition occurs when the substrate concentration is high enough so that the entire free enzyme population has become the ES complex (Nelson and Cox, 2005).

The other important parameter is K_M, which is called the Michaelis–Menten constant. Practically, K_M can be defined as the substrate concentration at which the reaction rate is half of V_{max}: $V = V_{max}/2$. In other words, K_M is the substrate concentration at which half the enzyme active sites are occupied by substrate molecules. Therefore, K_M can be used as a relative measure of the affinity of the substrate for the enzyme and evaluates the catalytic performance of an enzyme.

The last important parameter of a catalytic reaction is k_{cat}; it represents the maximum rate at which an enzymatic reaction can proceed at a fixed concentration of enzyme and infinite availability of the substrate (Copeland, 2000). Also called the *turnover number*, k_{cat} is the maximum number of substrate molecules converted to product per enzyme molecule per second. In the area of biosensors, these 3 parameters are very important because they characterize the power amplification of a signal. In a biosensor, K_M is called the apparent $K_M(K_M^{app})$. It is an apparent constant, because the enzyme is not free and pure in solution, but is immobilized on the sensor, and in general with other molecules, like mediators, so K_M^{app} measures the enzyme–substrate kinetics of the whole biosensor (Mayorga Martinez *et al.*, 2010).

Enzymes are classified into six main classes (Table 11.3).

Enzymes known as oxidoreductases catalyze oxidation or reduction reactions by transferring hydrogen or electrons. They are ideal for coupling with amperometric transducers. The most popular redox

Table 11.3: Enzyme classification.

Enzyme	Reaction
Transferases	Catalyze the transfer of a chemical group from one substrate to another
Hydrolases	Catalyze hydrolysis reactions
Lyases	Catalyze groups additions to double bonds or double bonds formations by removing groups
Isomerases	Catalyze the inter-conversion of isomers
Ligases	Catalyze the formation of C–C, C–S, C–O and C–N bonds by condensation reactions coupled to ATP hydrolysis
Oxidoreductases	Catalyze redox reactions, that is, transfer of hydrogen or electrons from one substrate to another

enzymes used in the development of enzyme biosensors are oxidases, dehydrogenases, and peroxidases.

11.4.3.1 *Oxidases*

These enzymes catalyze the transfer of hydrogen from the substrate to molecular oxygen. They have a prosthetic group strongly bound to the apoenzyme. This prosthetic group is a Flavin derivative and consists of Flavin mononucleotide (FMN) or of Flavin adenine dinucleotide (FAD), which are groups conferring redox properties to the active center of enzymes of this class. They catalyze reactions of the following type, where S is the substrate and P the product,

$$S + \frac{1}{2}O_2 \qquad P + H_2O$$
$$S + O_2 \qquad P + H_2O$$

11.4.3.2 *Deshydrogenases*

These enzymes catalyze the transfer of hydrogen from the substrate to a redox acceptor A (not molecular O_2) or vice versa. It requires coenzyme NAD^+ or $NADP^+$ to catalyze the reaction. They catalyze reactions of the following type:

$$S + AP + AH$$

11.4.3.3 *Peroxidases*

Peroxidases are a large family of enzymes that typically catalyze a reaction of the form,

$$ROOR' + \text{electron donor}(2e^-) + 2H+ \rightarrow ROH + R'OH$$

For many of these enzymes the optimal substrate is hydrogen peroxide, but others are more active with organic hydroperoxides such as lipid peroxides. The nature of the electron donor is very dependent on the structure of the enzyme. In nature, peroxidases are associated with oxidoreductases as the hydrogen peroxide formed by the latter may be deleterious to tissue. This strategy is also useed in multi-enzyme biotransducers.

11.5 Surface Functionalization for Bioreceptors Immobilization

The physical or chemical attachment of the bioreceptor to a solid, insoluble support is called bioimmobilization. Biommobilization of bioreceptors plays an essential role in any biosensor. The principal reasons are

(i) to establish a locus (specific location) for the biochemical reaction;

(ii) to enhance reusability of the expensive bio-reagent and hence reduce bioreceptor cost;

(iii) to reduce cost of separations of the bioactive reagent from the other reactants and products;

(iv) to confer the bioactive property to the surface onto which the bioactive reagent is immobilized;

(v) to enable ready automation of the bioanalytical system;

(vi) to allow for continuous operation and to integrate into feedback control systems;

(vii) to minimize waste and effluents;

(viii) to control bioreceptor properties (activity and stability); and

(ix) to serve as a model system for natural, *in vivo*, membrane-bound systems.

                    R. E. Madrid et al.

The immobilized bioreceptor ensures the selective attachment, concentration, and quantitative differentiation of target molecules (analytes) from other interfering materials. For example, in enzyme-linked immunosorbent assays (ELISAs), which are used to quantify the amount of antigen or hapten in a sample, requires the immobilization of protein target or capture antibody on a plastic substrate to selectively solicit the corresponding signals in absorbance or fluorescence (optical format). In addition, for biosensors on miniaturized devices, the immobilization of bioreceptors is also essential to concentrate the target analyte to the substrate such that the detection, for example, electrochemical reaction, can allow the efficient collection of electrons reflecting the target molecules. Thus, reproducible control of bioreceptor immobilization is considered as one of the determinants for the success of biosensors.

There are four major mechanisms for bioreceptor immobilization: physicochemical adsorption, the use of self-assembled monolayers (SAMs) that exploit gold–sulfur bonds (thiols), or organo-slanes, covalent crosslinking, and the use of protein-G or biotin–streptavidin binding. Adsorption is the simplest process, where molecules spontaneously adhere to the solid surface, for example, sensing area of a biosensor, based on physical forces such as van der Waals forces or electrostatic attractions. Because this force is surface charge dependent, the polarity of the molecule (or its hydrophilic/hydrophobic properties), and how it matches with the solid surface, will strongly influence the binding strength. ELISA is a good example using physical adsorption for quantifying the amount of proteins (Lillehoj *et al.*, 2010). For direct ELISA, it requires the adsorption of protein-based analytes onto the solid surface, and usually the hydrophobic surface of plastic substrate is required. Later, an enzyme-conjugated antibody is used to label the protein analyte, yielding a colorimetric signal via enzymatic reactions following incubation with substrate and dye. Similarly, for sandwiched ELISAs, the capture antibody is first coated onto the surface using physical adsorption prior to the capture of target protein. This type of ELISA provides more specific binding of protein analytes to the surface, and reduces background noise. The amount of target protein of interest will be revealed following incubation with the enzyme-conjugated secondary antibody by the subsequent absorbance or fluorescence assay.

Notably, although physical adsorption is widely applied in many types of biosensors without a complicated surface chemistry, it may result in denaturation of protein-based bioreceptors. It is because the use of van der Waals forces or electrostatic forces relies on the polarity of molecules. Since protein is composed of both polar and non-polar residues, adsorption may force the residue that has higher affinity with the substrate surface to contact with the substrate surface, resulting in conformational change of the protein structure and hence its function. In addition, adsorption is not specific, which may lead to an increased background noise in biosensing.

In contrast, gold–sulfur bonds, mediated by the sulfhydryl functional group in thiols, are another immobilization approach for biosensing devices. Typically, gold–sulfur bonds are used for bioreceptor modified with a sulfhydryl group and immobilized on the gold surface. This gold–sulfur bond is a covalent interaction between the sulfhydryl group and the gold surface through a deprotonation process (Häkkinen, 2012). Because this covalent bonding is specific and has a strong bond strength similar to that of the gold–gold bond, gold–sulfur bond provides a more reliable and stable immobilization method for bioreceptors as compared with physical adsorption.

For biosensing applications, gold–sulfur bond was applied in gold nanoparticles (AnNPs)-based biosensors. Depending on the dispersion and agglomeration of nanoparticles, AnNP solution can exhibit a colorimetric signals representing the amount of target molecules (Rosi and Mirkin, 2005). For example, a detection of target DNA/RNAs was reported based on this phenomenon. Oligonucleotide probes with sequence complementary to a target oligonucleotides were immobilized onto AnNPs such that the hybridization between target molecules and oligonucleotide probes can lead to the agglomeration of AnNPs and result in a change of solution color from pink to purple (Rosi and Mirkin, 2005). Importantly, the immobilization of oligonucleotide probes is based on gold–sulfur bonds, where the oligonucleotide probes were thiolated on one terminus and kept another terminus free in the solution. Thus, when encountering the target oligonucleotide, AnNPs were connected via DNA hybridization and formed sandwiched structures, AuNPs–targets–AuNPs, which enlarge the characteristic size of nanoparticles without steric hindrance and change the solution color.

Self-assembled monolayers (SAMs) are other representative applications of gold–sulfur bonds. Typically, SAMs are organic molecules having a head group, tail, and functional end group. The head group can be the sulfhydryl group, which has a strong affinity to the gold substrate and anchors to it, while the tail groups that carry functional end group are far from the substrate. As such, the surface property of the gold substrate can be tailor-made based on the covered SAMs because its property will be overridden by the functional groups. In an example of biosensing application, SAMs were used to covalently immobilize proteins. Using an optical fiber coated with gold film, the optical fiber was modified with the carboxylic group using SAMs with the carboxylic group. When protein was added, the SAMs allowed the formation of covalent peptide bounds between the carboxylic group and the amino group of the protein (Tseng *et al.*, 2009). This immobilized protein can serve as the detection probes for subsequent recognition of antibody biomarker, thus enabling a continuously in-situ immune sensing. Organo-silanes such as (3-aminopropyl) triethoxysilane, 3-glycidoxypropyltrimethoxysilane, and 3-cyanopropyl dimethylchlorosilane also find favor in presenting terminal reactive functional groups as the silanol condensation reaction increases molecular weight results in adsorption onto and subsequent condensation with surface –OH groups on biochip substrates such as glass, silicon dioxide, and silicon nitride.

Biotin–streptavidin binding is one of the strongest non-covalent binding reactions known in nature. Streptavidin, a protein purified from the bacterium *Streptomyces avidinii*, possesses a tetrameric quaternary structure, providing four binding sites that accept biotins. For each streptavidin monomer, there is a β-barrel structure containing a biotin binding-site, and the strong binding strength is due to the high complementary shape between the binding pocket and the biotin. Because it is a non-covalent bond, depending on only hydrogen bonds, van der Waals force-mediated contacts, and hydrophobic interactions, the biotin–streptavidin bond has very high association rate and tolerance for a variety of reaction conditions, which provide great flexibility for different biosensing environment.

The most typical example is to use biotin–streptavidin binding for immobilization of biosensing probes on transducer surfaces. Using self-assembled monolayer of mercaptoundecanoic acids, terminated by a carboxyl group, this SAM was able to form an amine–PEO_2–biotin labels at the gold substrate (Wei *et al.*, 2010). Thus, this modified gold surface with biotin was allowed us to coat with streptavidin via biotin–streptavidin bonds, and the unused binding sites of the streptavidin were used to accept biotinylated DNA oligonucleotides, that were later used as the bioreceptor for recognizing target DNA/RNA oligonucleotides in the solution. Based on this scheme, the biotinylated DNA oligonucleotides can be immobilized on the gold substrate, and an electrochemical reaction can be conducted after hybridizing with target oligonucleotide. Alternatively, biotin–streptavidin binding can also be used in the modification of suspended particles (Ho *et al.*, 2005; Li *et al.*, 2015). Specifically, streptavidin was coated on micro- or nanoparticles, so allowing the attachment of single-strand oligonucleotide probes for nucleic acid detection. Based on this modification, the target oligonucleotide creates particle–particle connections via hybridization in juxtaposition. Besides, the particle–particle connections can be visualized through the change of fluorescence color due to the mixing of nanoparticles with different fluorescent color, or visualized by the subsequent suppression of coffee ring, where non-spherical particle agglomerates are able to suppress the coffee ring effect, after evaporation of solution making the microscale connections visible in macroscale aspect (Li *et al.*, 2015).

In conclusion, these variable types of bioreceptor immobilization may be suitable in different situations due to their different binding strength and association rate. Physical adsorption is the simplest procedure in surface preparation and has strong binding strength, providing great flexibility in immobilizing bioreceptors. However, the non-specificity and potential for denaturation of protein bioreceptors may be a concern. Gold–sulfur interactions establish covalent bonds where higher specificity and strong bonding strength can be achieved. However, gold–sulfur bonds are compatible with only gold surface; the association is time consuming ($\sim$16 hrs), which may limit the overall performance. In contrast, biotin–streptavidin binding provides extremely strong binding force, simple and rapid association, and high specificity, offering great tolerance in a variety of applications. However, the immobilization of either biotin or streptavidin is required before conducting the biotin–streptavidin bonds, which is non-trivial and may need SAMS based on gold–sulfur bonds, organosilane or physical adsorption. Thus, the developer needs a strategic plan to select the immobilization method based on one or combination of approaches which best suits their application.

11.6 Nanotechnology Applied in Bioengineering

Nanotechnology, as it is named, refers to knowhow that precisely manipulates and benefits from unique properties of objects at the nanoscale. Nanomaterials has become attractive because of,

(1) their size which yields a correspondingly large surface-to-volume ratio;
(2) unusual chemical binding activities with implications for enhanced sensing mechanism; and
(3) overall structural robustness (Wei *et al.*, 2010; Elghanian *et al.*, 1997; Zhan *et al.*, 2010; Cao *et al.*, 2009; Taton *et al.*, 2000; Xu *et al.*, 2007).

Bioengineering indeed has a long history of using bio-conjugation and labeling technique, for example, antibody labeling, for the past decades (Baudhuin *et al.*, 1989), which allows visualization and analysis of biomarkers into the sub-cellular and molecular levels. In recent decades, as alternatives to these molecular labeling techniques, the development of nanotechnology focuses on new synthesis, fabrication, or measurement of nanomaterials with tailored size, shape, and composition, thereby allowing exquisite control of the material properties (Rosi and Mirkin, 2005). For example, nanoparticles and quantum dots with specific size and geometry enables customized fluorescent, absorbance, or light-scattering properties. This has significantly broadened the spectrum of signals for detection, which is essential for differentiation of multiple analytes in a single bioassay, or multiplexed bioassay.

On the other hand, materials on the nanoscale may also exhibit extraordinary properties. For example, graphene or carbon nanotubes, a layer of carbon molecules with hexagonal arrangement in two-dimensional or tubular sheet, was found to offer superior electrical conductivity and mechanical strength. Such nanostructure has also led to intriguing properties such as intrinsic peroxidase catalysis or highly efficient fluorescence quenchers (Song *et al.*, 2010; Wang *et al.*, 2010; Zhao *et al.*, 2011). Additionally, characterization platforms and techniques for surface modification and patterning have enabled spatial layout of biomacromolecules and molecules on these surfaces. Altogether, the combinatory advances derived from nanotechnology has opened up new possibilities for implementation of new assays with improved signal transduction compared to traditional approaches such as PCR or ELISA.

11.7 Nanomaterials for Biosensors

Reported by Mirkin and coworkers in 1996 (Mirkin *et al.*, 1996) and 1997 (Rosi and Mirkin, 2005), the most well-known nanomaterials for biosensors may be AuNPs. Based on the sequence of the oligonucleotide probes modified on AuNPs, the resulted particle dispersion or assembly of AuNPs in response to target analytes may exhibit unusual optical and melting properties, which later inspired the hybridization-based colorimetric assay (Fig. 11.18). Specifically, using AuNPs with diameter of 13–30 nm, AuNPs were surface functionalized with two types of oligonucleotide designed to hybridize in juxtaposition with one target DNA. Thus, the presence of target DNA would induce the aggregation of AuNPs, resulting in a color change of bulk solution as a consequence of particle surface plasma and scattering properties. This phenomenon is particularly interesting because it allows macroscale visualization of the particle assembly

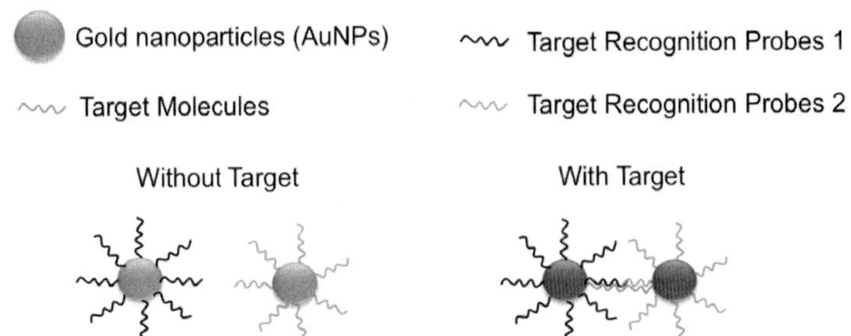

Figure 11.18: The use of probes modified by AuNPs to detect target molecules would induce the aggregation of AuNPs, resulting in a change of bulk solution color from red to purple.

events, which enables the detection of nucleic acid into a type of "litmus test". With the compatibility of visual inspection or quantification by spectrophotometry, this colorimetric assay via nanoparticle assembly exemplifies the use of nanomaterials for biosensors in a simple and rapid strategy.

Interestingly, this hybridization-based AuNPs aggregation also shows a sharp melting transition, which provide improved selectivity for detections (Elghanian *et al.*, 1997). Specifically, measured by the absorbance difference between single-strand and double-strand DNA, DNA solution usually exhibits a smooth DNA melting profile with the increase of solution temperature. In contrast, the AuNPs modified with oligonucleotide probes was observed with a sharp transition of absorbance. This unanticipated phenomenon may be due to the enhanced contrast in absorbance resulted from the dissociation of AuNPs, and the structurally aligned DNA duplexes as the tethers between AuNPs which may lead to coherent melting events. More importantly, this sharp melting transition enabled the differentiation of DNA with single mismatches which was hardly achievable by microparticles (Elghanian *et al.*, 1997). It suggests that the hybridization-induced particle assembly offers simple and economical approaches for biosensor, and has great selectivity owing to the sharp melting transition derived from dense loading of oligonucleotides and the aligned DNA duplexes at the nanoscale. Based on the unique advantages, numerous methods based on AuNPs were developed to detect DNA, protein, and metal ions (Elghanian *et al.*, 1997; Wei *et al.*, 2010; Xia *et al.*, 2010). Also, this unique binding event via oligonucleotides on AuNPs was recently evolved into a format of lateral flow strip. Using capture oligonucleotides immobilized on a paper-based flow strip, a self-driven capillary flow would carry the AuNPs and target molecule, and AuNPs would be retained by the capture oligonucleotides when target is present, resulting in a visible bar that can provide fast, colorimetric readout for DNA detection (Rastogi *et al.*, 2012).

However, although promises were shown, these colorimetric methods based on AuNP assembly are subject to the limit of detection (LOD), which usually falls into the nM level because a large amount of target oligonucleotides were required to induce AuNPs aggregation. Effort has been made using 50 nm particles for increased density of oligonucleotide probes, and the LOD was achieved between 5 nM and 50 pM (Hurst *et al.*, 2006). However, it is still incomparable with the conventional fluorophore-based assays and limits the potential applications.

To improve its sensitivity, the AuNP-based assays have been modified with enzymatic amplification. For example, a concept was developed using cyclic enzymatic signal amplification and hairpin aptamer probe to detect target molecules, adenosine triphosphate (ATP) (Li *et al.*, 2012). Specifically, when ATP was absent, a short linker oligonucleotide would link AuNPs, resulting in the bulk solution as purple color. However, when ATP was present, the hairpin aptamer probe was opened after binding with the target, forming an aptamer–target complex. Notably, this complex was allowed to hybridize with the short linker oligonucleotides, resulting in a duplex that contains a specific recognition site which can be enzymatic cleaved by nicking endonuclease. Thus, as the short linker oligonucleotides was cleaved and released from the aptamer–target complex, the freed aptamer probe–target complex can then restart the cleavage of other short linker oligonucleotides. Therefore, the cyclic amplification greatly reduced the short linker oligonucleotides while target molecule was present, resulting in an enhanced difference in colorimetric readouts.

Similarly, horseradish peroxidase (HRP), an enzyme widely used in producing colored products for conventional assays such as ELISA and immunohistochemistry, was also used in AuNPs-based assays. For example, in the format of lateral flow test strip, with AuNPs conjugated to HRP, the presence of target molecule would retain the AuNPs with HRP conjugates (Qin *et al.*, 2015), leading to an enhanced colorimetric readout that significantly amplified the signals and potential of substantial improvement by enzymatic amplifications.

Although sensitivity was improved, enzymatic amplification requires specific facility for additional chemical reaction or delicate protocols, which may not be suitable for biosensors in miniaturized format or simplified settings. Alternatively, as the colorimetric signals are produced by the light scattering of AuNPs, the amplification can also be achieved by attachment of extra AuNPs (Qin *et al.*, 2015) or gold deposition combined with locked nucleic acids (Rastogi *et al.*, 2012), where LOD was improved to 0.4 nM for short oligonucleotides (Rastogi *et al.*, 2012) or 2.9 fg ml^{-1} for CEA (Qin *et al.*, 2015). On the other hand, special optical setups such as using dynamic light scattering to measure the hydrodynamic size of AuNPs before and after particle aggregation (Dai *et al.*, 2008), or directly counting single AuNPs suspended in solution using flash-lamp darkfield microscopy (Yuan *et al.*, 2012). In the method of flash-lamp darkfield microscopy, considering that AuNPs are moving rapidly and randomly in the solution, the flash-lamp was used to apply ultrashort pulse of exposure which snapshot hundreds of AuNPs, allowing statistically a large enough number of counts for convincing analysis. Based on the special optical setup, the LOD reached 5 pM by dynamic light scattering (Dai *et al.*, 2008) and 0.1 pM by flash-lamp darkfield microscopy (Yuan *et al.*, 2012).

In addition to AuNPs, researchers also found that the oligonucleotide can itself be a powerful nanomaterial for biosensing. As mentioned above, based on the pairing nature, single-strand oligonucleotides with complementary sequences were widely used as the recognition probes for nucleic acid-based biomarkers. In addition, oligonucleotides can also serve as an aptamer which can recognize and bind with protein biomarkers. Molecular beacon is a representative example of oligonucleotide-based biosensors (Tyagi and Kramer, 1996). Typically, the nucleotide sequence of the molecular beacon is designed with a loop region in the center and two stem regions at both termini. The loop region is designed with sequence complementary to the target DNA or RNA, while the two stem regions are complementary with each other. Thus, when the target molecule is absent, the two stem regions at both termini would hybridize, making the entire molecular beacon as a shape of hairpin. Note that a fluorescent dye and a quencher may be covalently bound to both ends of the molecular beacon. Thus, when the molecular beacon is closed as the hairpin, the fluorescent dye and the quencher are close, which quenches the fluorescent emission. In contrast, if the target molecules are encountered, the molecular beacon would hybridize with the target, which opens the hairpin and the fluorescence light is observed. Based on this concept, molecular beacon has been used for the detection of gene expression in live cells (Li and Ho, 2008). Such hybridization-based fluorescence quenching, combined with aptamer that can bind with specific protein, it was also developed into protein biosensors for thrombin (Wen *et al.*, 2013). Moreover, as the molecular beacon usually suffers, due to the efficiency of fluorescence quenching (Dubertret *et al.*, 2001), the quencher may be replaced by AuNPs, or gold-coated substrate, as they offer much higher efficiency for the fluorescence quenching, and so resulting in a higher single base mismatch selectivity compared to conventional molecular beacons (Cheng *et al.*, 2011; Zhang *et al.*, 2010).

To further improving the sensitivity, oligonucleotides were also used for a new enzyme-free amplification based on the hybridization DNA circuit. Two typical approaches of DNA circuits were reported: hybridization chain reaction and catalytic hairpin assembly (Dirks and Pierce, 2004; Yin *et al.*, 2008). For both types of DNA circuits, single-strand DNA targets were used as the triggers to initiate a series of opening and assembly of DNA hairpins (Fig. 11.19). Subsequently, the opening and assembly of DNA hairpins result in either elongated double helix composed by alternating hairpin 1 and

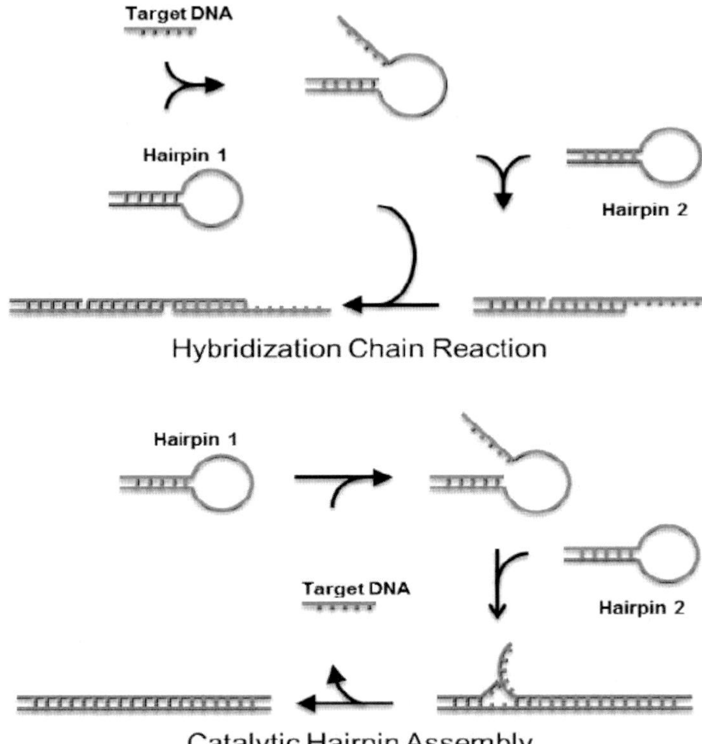

Figure 11.19: Two types of DNA circuits, hybridization chain reaction and catalytic hairpin assembly, for signal amplification.

hairpin 2 (hybridization chain reaction) or a vast amount of double helix formed by hybridized hairpin 1 and hairpin 2 (catalytic hairpin assembly). Therefore, after the hybridization DNA circuit, a small amount of target was converted into double helix structures with bulky size or a larger amount of copies, which significantly enhances the sensitivity of detections, therefore an isothermal and enzyme-free amplification were achieved. Notably, compared with catalytic hairpin assembly where the target is continuously needed for each cycle of the catalytic hairpin assembly (Yin *et al.*, 2008), hybridization chain reaction only need target in the initial condition to trigger the reaction (Dirks and Pierce, 2004), creating more reliability of the circuit operation and signal amplification.

11.8 Self-assembling Peptides and Proteins as Nanomaterials: Promising Use in Biosensors

The requirement to discover novel materials capable of exploiting the properties of nanostructured world is nowadays growing exponentially. In fact, several scientific disciplines made huge effort to study and reproduce under laboratory conditions the natural peptide/protein self-assemblies process, since they proved to be versatile, cheap, and shape-size controlled nanomaterials for future technological applications. In contrast to established "top-down" fabrication techniques, protein self-assembly is emerging as a "bottom-up" approach for fabricating novel nanostructured materials.

11.8.1 *Mimicking the strategy of nature in self-assembling processing*

Even though in nature, the most popular form in proteins/peptide is characterized by a well-defined folded structure which remains in soluble state called folded or native ($\mathbf{N}$) (Dunker *et al.*, 2002), some cellular function are developed by proteins in unfolded ($\mathbf{U}$) or aggregated ($\mathbf{A}$) state.

Once the polypeptide chain emerges newly synthesized from ribosomes, without any secondary or three-dimensional structural element (in U state), aqueous environment push most proteins to a folding process that results in order to hide hydrophobic residues, generating new non-covalent bonds reducing the ΔG of the system. The $\mathbf{N}$ state has an almost a unique and defined tridimensional conformation but only little structural changes related with the physiology activity of the protein are allowed. However, it is worth noting that some proteins called IUPs (intrinsically unordered proteins) do not have any folding process and thus lacks any fixed or ordered three-dimensional structure, however, they play important physiological roles in their $\mathbf{U}$ state (Dyson and Wright, 2005).

Surprisingly, the folded state is by and large only 5–10 kcal/mol more stable than the unfolded one (Pace, 1990) and thus the $\mathbf{N}$ state is frequently considered as metastable. Even for proteins that successfully fold, small fluctuations in the medium can push the $\mathbf{N} \rightleftharpoons$

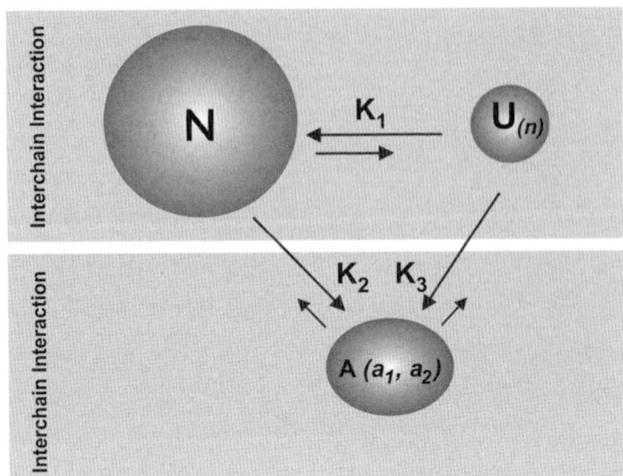

Figure 11.20: Schematic representation of protein folding equilibrium. N is the protein population in native, U in unfolded and A in aggregated states. The different aggregation states a_1 (amorphous) and a_2 (amyloid) are also indicated. Circle surfaces are only indicative of the different population sizes but they do not represent any real system. Adapted from Torres-Bugeau *et al.* (2011). Amyloid fibril is the name given to the insoluble nanostructures formed by a particular self-assembly of normally soluble protein or peptide monomers into fibrous aggregates with a highly organized repeated structure characterizable by biophysics techniques. This term is nowadays strongly coined to refer to this protein aggregates although etymologically make references to sugars.

U equilibrium to protein misfolding and aggregation. In fact, proteins can leave the **N** state with relative facility reaching the **A** state, which is energetically favourable.

From both the states **N** and **U**, all proteins are capable of undergoing self-assembly processes depending on the physicochemical of the environment. This process represents the generation of novel interchain interactions between different polypeptide chains and complete folding reaction of a protein (Fig. 11.20). Protein aggregation is a frequent occurrence event in nature and has the ability to self-assemble into a variety of structures that are capable of performing desired functions in the biological world.

It is important to note that currently we can distinguish mainly two types of aggregate states (Fig. 11.20): the amorphous one without any order (a_1) and a highly organized aggregate called amyloid (a_2). Amyloid fibrils are one such self-assembling biological structure, formed when native or IUPs misfold into insoluble fibrous quaternary structures. Amyloid self-assemble structure has been extensively studied due to its role in neurodegenerative disorders. In fact, scientific community is nowadays taking in mind the huge potential of this kind of protein aggregates as novel nanomaterials for nanotechnological supports (Mankar *et al.*, 2011).

11.8.2 *Bio-inspired nanoassemblies: The bright side of amyloid aggregation*

Nature has many examples of amyloid self-assembly that could be valuable in order to understand how is the better way to exploit this important properties of peptides for biotechnological proposes. In addition to self-assembly, the topography of amyloid fibrils also recommends their use as nanomaterials. They naturally exist in the nanometre scale, with widths of generally 7–10 nm and lengths of up to several microns (Mankar *et al.*, 2011). These lengths are usually highly heterogeneous, but could be controlled by experimental conditions.

Among the properties that make protein aggregates an excellent choice for the development of new materials and devices, we can mention

(1) Strong mechanical properties: By using atomic force microscopy and spectroscopy (AFM) Smith *et al.* (2006) have studied mechanical properties of insulin amyloids with real important conclusions. Amyloid fibrils have been shown to have a strength comparable to steel and are flexible, with a stiffness similar to that of silk. The amyloids fibrils were also found to be stable up to 100°C temperature (Meersman and Dobson, 2006).

(2) Low production cost: It has been proposed that amyloid fibril formation is a generic property of all proteins, and if the correct environment conditions are found, every protein can reach this aggregation state (Chiti and Dobson, 2006). In this way, the study of conditions in which proteins related to human disease aggregates in amyloids grew rapidly (Brange *et al.*, 1997; Murphy, 2002; Uversky, 2008), but also a number of proteins

that are not known to form fibrils in disease or nature such as insulin (Nielsen *et al.*, 2001), wheat proteins, hen egg lysozyme (Krebs *et al.*, 2000), α-lactoglobulin (Pearce *et al.*, 2007; Sardar *et al.*, 2014) milk proteins, and protein from lactoserum (WHEY) (Thorn *et al.*, 2005) have been induced to form fibrils *in vitro*.

(3) Functionalization: the chemistry of the different amino acid side chains can be exploited to add functionality to fibrils (Hartgerink, 2004) via the formation of crosslinks between mature fibrils and enzymes of interest. The nanoscale properties of amyloids provide them with a large surface-to-volume ratio allowing for the possibility of a high functionalization packing density. Figure 11.21 shows a schematic representation of different modes of amyloid fibril functionalization.

In fact, amyloids could be considered a quite sophisticated support for enzyme immobilization or drug carrier since they are not only a high stable chemical and mechanical nanosupport, but are easily functionalizable due to the presence of amino acid lateral chains. On the other hand, amyloids have poor electrical characteristics, which have prevented their direct use in electrical circuits. However, this significant challenge was overcome by using metal coating techniques, and the construction of nanowires to enable the electrical connection

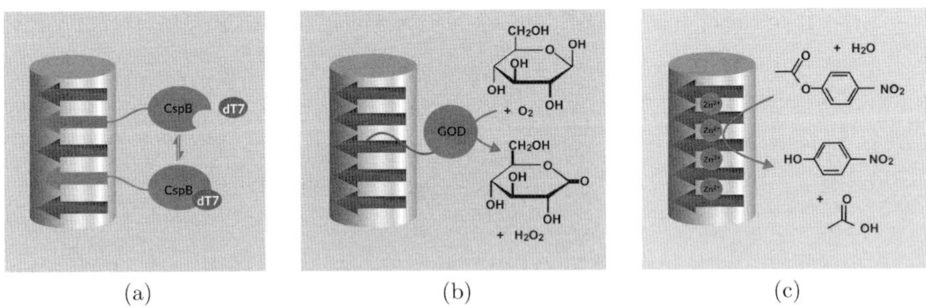

(a)   (b)   (c)

Figure 11.21: (a) Display of cold-shock protein B (CspB) on fibrils achieved by biotechnological fusion to a fibril-forming polypeptide (N-terminal domain of PABPN1 protein). (b) Immobilization of GOD on bovine insulin fibrils by glutaraldehyde crosslinking. (c) Korendovych, DeGrado and co-workers showed that Zn^{2+}-binding fibrils can hydrolyze pNPA into nitrophenole and acetic acid. Figure adapted from Aumüller and Fändrich (2014).

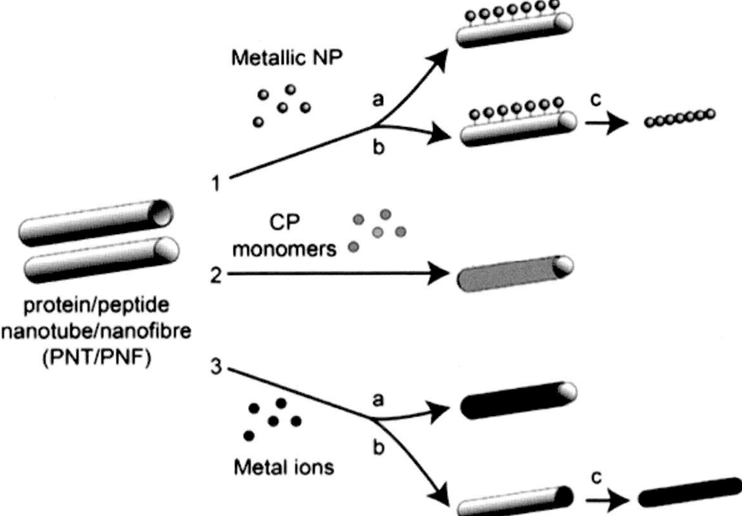

Figure 11.22: The most common ways in which protein/peptide nanotubes/nanofibres (PNTs/PNFs) are used for the templating of conductive nanowires.

was then possible (Scheibel *et al.*, 2003). Once the fibers are formed, it has a higher than average chemical stability, the metal coating procedure allows us to produce nanowires with good electrical conductivity and linear I–V curves.

In summary, protein/peptide nanotubes (PNTs), and nanofibres (PNFs), could be used as template molecules in a number of different ways. Figure 11.22 shows a schematic representation of the three most common ways; by modification with metallic nanoparticles (NPs), conducting polymers (CPs), and metal ions: (1) The PNT/PNF is modified with metallic nanoparticles (NPs). This may occur such that the PNT/PNF itself is decorated with metallic NPs (a), or with the PNT/PNF being used as a "sacrificial" template for the formation of a chain of metallic NPs (b). (2) The PNT/PNF is modified with CP monomers, with the polymerisation reaction being carried out in the presence of the PNT/PNF, forming a coating of CP on the surface of the PNT/PNF. (3) The PNT/PNF is modified with metal ions. (a) A coating of metal may be formed on the exterior of the PNT/PNF or (b) on the interior of a PNT. In some cases, (c), the PNT can then be removed, leaving a thin metallic nanowire.

11.9 Hydrogels in Biosensors

11.9.1 *Hydrogels*

Engineering control of the abio–bio interface of the biotransducer continues to be a major challenge in the development and commercialization of biosensor systems. The principal constituent of the abio–bio interface is the biorecognition element and much of the engineering control is to preserve its biorecognition activity upon exposure to the analyte, its matrix and the many possible interferences and insults associated with its practical use. Early materials of biosensor construction were generally abundant, commercially available, and of industrial origin; for example, plasticized polyvinyl chloride (PVC) and Nafion®. These have been supplemented or supplanted by polymers of biological origin, for example, chitosan and cellulose acetate. Recently, hydrogels have grown in favor. Hydrogels are 3-D crosslinked polymers of highly hydratable monomers which when exposed to water, physiological buffers of physiological conditions, swell to bring into balance the tendency of the monomer to dissolve, and the retractive force of the polymer network. Such polymers have become a mainstay in biological and biomedical applications (Ottenbrite *et al.*, 2010) due to their desirable physicochemical properties made possible by the controllable high degree of hydration (Brahim *et al.*, 2003; Porter *et al.*, 2007), cell, and tissue biocompatibility (Phelps *et al.*, 2012), high potential for pre- and post-synthetic molecular engineering (Abraham *et al.*, 2005; Aucoin *et al.*, 2013), architecture control, and potential for multifunctional action.

The hydrogel, when used as a coating or membrane layer on the biotransducer, serves as a site for the immobilization of the biorecognition element. Three advantages are immediately apparent. The first is the presentation of a 3D environment for the immobilization of a larger number of immobilized biorecognition elements per unit of plan area of the biotransducer compared to adsorption, covalent tethering to a self-assembled monolayer (SAM) or crosslinking.

The extent of this difference will depend upon the nano- through meso- to micro-porosity of the hydrogel membrane. The second is the highly hydrated milieu for the retention of the native conformation of biorecognition elements. The extent of this value contribution depends upon whether the biorecognition element is covalently tethered to the hydrogel network of physically entrapped and on the physicochemical interactions between the immobilized biorecognition element and the hydrogel's primary repeat unit structure. The third is that the molecular recognition by the biomolecule may be linked to any if the stimuli stimuli-responsive properties of the hydrogel itself (Kopeček and Yang, 2007) thereby effectively making the hydrogel part of the transducer. Two disadvantages are also immediately apparent. The first is the mass transfer barrier presented by the hydrogel. The may include a diffusion barrier to the movement of analytes and secondary reagents but may also contribute a membrane potential that opposes charged species. The second is the occupancy of active sites of the transducer by the hydrogel. For piezoelectric transducers, this may be viscoelastic damping while for electrochemical transducers this may be a reduction in effective surface area for charge-transfer reactions. Figure 11.23 shows a schematic for the biotransducer using a hydrogel membrane layer.

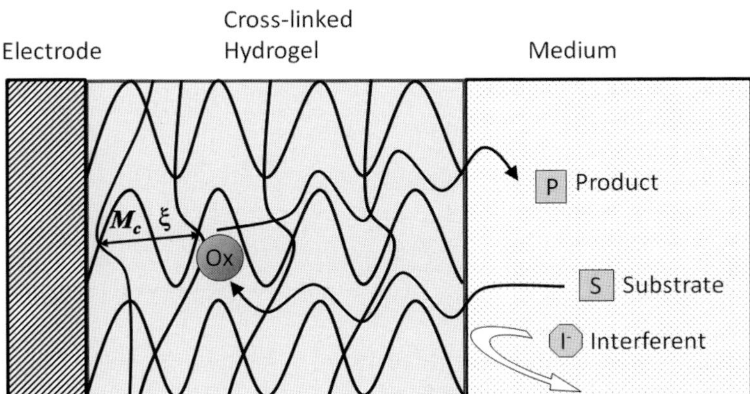

Figure 11.23: Schematic illustration of a crosslinked hydrogel membrane supported on a transducer (electrode) showing hosting of an oxidoreductase enzyme (Ox), substrate (S) diffusion, product (P) effusion, interferent (I) suppression, Mc = molecular weight between crosslinks and ξ = polymer mesh size of the polymer network.

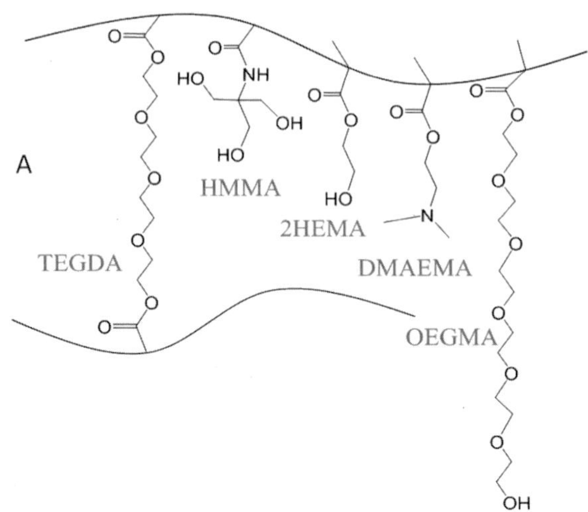

Figure 11.24: A. Schematic illustration of a poly(MEMA)-based hydrogel used in an enzyme hosting electrochemical biosensor.

A prototypical hydrogel is crosslinked poly(2-hydroxyethyl methacrylate) (pHEMA) (Fig. 11.24), which was first used in the development of contact lenses because of its optical clarity, dimensional stability, matched mechanical properties, and its biologically benign nature (Wichterie, 1961). Since the 1960s, hydrogels have expanded in molecular repeat unit composition, architecture, and applications to have a broad technological impact in areas such as molecular theranostics (Yoon *et al.*, 2012), biomolecular detection (Choi *et al.*, 2012), biosensor-based molecular diagnostics (Bromberg *et al.*, 2012), drug-free macromolecular therapeutics (Wu *et al.*, 2010), and tissue engineering and regenerative medicine (Rustad *et al.*, 2012). Moreover, hydrogels are continually being molecularly engineered to serve as stimuli-responsive polymers (Guiseppi-Elie *et al.*, 2002) within their particular end-use environments by exploiting various sense and respond mechanisms (Hoffman and Crocker, 2009; Kirschner and Anseth, 2013; Wilson and Guiseppi-Elie, 2013). More recently, feedback control has been integrated into bioresponsive hydrogels creating synthetic analogs to tightly regulated metabolic pathways.

The relationship between immobilized biorecognition elements and polymer structure is still an active area of research (Hoffman, 2013; Kost and Langer, 2001). The effects of hydrogel swelling dynamics; solute transport properties; *in situ* kinetics of enzymatic and/or nucleic acid (NA), aptamer, antibody binding recognition reactions; ionic interactions; and environmental (pH, temperature, ionic strength) changes can alter the performance of a biologically responsive system (Brandl *et al.*, 2010; Qiu and Park, 2001; Wang *et al.*, 2010; Wilson and Guiseppi-Elie, 2013). These factors greatly influence polymer design and could have unforeseen consequences on the reactivity and stability of the immobilized biorecognition element.

11.9.1.1 *Biologically responsive hydrogels*

Biologically responsive hydrogels are a class of hydrogels conferred with the recognition of biological molecules that are linked to the stimuli-responsive properties of the hydrogel (Kopeček and Yang, 2007; Wilson and Guiseppi-Elie, 2013) including their swelling and de-swelling dynamics (Ottenbrite *et al.*, 2010). In this way, the hydrogel is conferred with the specificity of the biomolecule and may be part of the biotransducer transduction. Hydrogels may be used in diagnostic biomedical biosensors (Kotanen *et al.*, 2012; Tanase *et al.*, 2014; Viter *et al.*, 2011). Among these smart diagnostic biosensors are ones wherein the hydrogel, via its conferred biomolecular recognition entity, reacts with a biological entity through catalysis or binding and produces an integrated physicochemical response (Bhat *et al.*, 2013; Chung *et al.*, 2008). The response seeks to restore equilibrium or a pseudo-steady state within the hydrogel (Wilson and Guiseppi-Elie, 2013). Thus, a pH-dependent response pursuant to the appearance of gluconic acid from glucose, the result of immobilized GOD activity, may lead to a change in hydration (Guiseppi-Elie *et al.*, 2002) and hence impedance of a hydrogel membrane that sits astride a pair of interdigitated microelectrodes (Yang *et al.*, 2011). In all cases, the hydrogel response is governed by its polymeric network morphology.

Polymer network morphology successfully describes the orientation polymer chains and the void spaces that co-exist among polymer chains. Together the network and the voids describe transport and mechanical properties and influence the movement of biological species within the milieu (Guiseppi-Elie *et al.*, 2012; Munoz-Pinto *et al.*, 2009; Zustiak *et al.*, 2010). The quantitative parameters that describe the network morphology include the void fraction (ε), the tortuosity (τ), the molecular weight among crosslinks ($\bar{M}_c$), the elastic modulus (E), and the mesh size (ξ). Polymer swelling theory, originally developed by Flory and Rehner (1943), enables the determination of ξ based on polymer solvation (Peppas and Merrill, 1977) (see Eq. 11.15),

$$\xi = v_{2,s}^{-1/3} \left(\frac{2C_n \bar{M}_c}{M_r} \right)^{1/2} l \qquad (11.17)$$

In the latter relationship, C_n is the Flory characteristic ratio, M_r is the molecular weight of the repeat unit, $v_{2,s}$ is the polymer volume fraction in the swollen state, and l is the bond length projection along the polymer backbone. The determination of $\bar{M}_c$ can also be achieved through mechanical methods by relating applied stress or modulus to $\bar{M}_c$ by extensions of rubber elasticity theory (Peppas and Merrill, 1977; Safranski *et al.*, 2011). Void fraction can be determined by gravimetric or mercury porosomitry (Karageorgiou and Kaplan, 2005) and tortuosity can be determined from transport analysis (Faraji *et al.*, 2011). Combing polymer swelling theory with transport equations based on hydrogel morphology enables the definition of zones of performance in biologically responsive hydrogels. Peppas *et al.* provide a more robust treatment of polymer swelling theory and its effects on transport (Peppas, 2000; Slaughter *et al.*, 2009).

11.9.2 *Electroconductive hydrogels for biosensor fabrication*

Electroconductive hydrogels (ECH) combine the one-dimensional electrical conductivity of inherently conductive polymers (ICPs) with the dynamic swelling characteristics of hydrogels (Brahim and Guiseppi-Elie, 2005; Guiseppi-Elie, 2010). These polymeric hybrid materials go well beyond the advantages given above for hydrogels.

They are particularly useful as they offer a method means to confer biospecificity during the biosensor fabrication process. Electroactive monomer, for example, pyrrole, thiophene, aniline, or any of their analogues contained within the hydrogel cocktail that is cast onto an electrode, or imbibed/partitioned into the pre-formed electrode-supported hydrogel, or held within the bathing solution, may be electropolymerized at the hydrogel/electrode interface and in so doing entrain and cause to be entrapped and immobilized biorecognition molecules within the hydrogel. Biorecognition molecules such as enzymes, nucleic acids, antibodies, DNA, and peptide aptamers are generally net negatively charged and so may be electrostatically entrained during the polymerization reaction. Alternatively, as an enhancement and to improve reproducibility, the biomolecule may be monomerized (i.e., conjugated to an electroactive monomer) or made a dopant (e.g., conjugated to a sulfonate) to render them more readily entrained (Kotanen *et al.*, 2014). This may be used to great advantage in immobilizing different biorecognition molecules on closely spaced electrode features of multi-analyte biosensors. ECHs of polypyrrole and poly(HEMA) have been shown to have excellent interference suppression properties, particularly for anionic interferents such as ascorbate and citrate (Brahim *et al.*, 2002).

11.9.3 *Hydrogels for implantable biosensors*

Synthetic and synthetic-hybrid polymeric hydrogels are being molecularly engineered to meet the demanding requirements of *in vivo* performance. The biosensor's biorecognition membrane may be made biomimetic to its surrounding host environment. Biomimetics seeks inspirations from nature to design practical materials and systems that can imitate structure and function of native biological systems (Sarikaya *et al.*, 2003). Hydrogels may be readily rendered biomimetic (Ottenbrite *et al.*, 2010) with the view to guiding cellular responses via biochemical or mechanical cues (Hoffman and Crocker, 2009) while effectively hosting enzymes in a three-dimensional tissue-like milieu (Guiseppi-Elie, 2006). Hydrogels modified with polyethylene glycol (PEG) and methacryloyloxyethyl phosphorylcholine (MPC) have been shown to be useful for resisting protein adsorption

(Nederberg *et al.*, 2006), limiting cell proliferation, and maintaining high cell viability *in vitro* (Abraham *et al.*, 2005). PEG is a commonly used for inhibition of extracellular matrix protein adsorption to materials (Lee and Lin, 2002). Phosphorylcholine is a cell–cell recognition moiety found as the bioactive polar head group of phospholipids that comprise the exoplasmic leaflet of cell plasma membranes (Ishihara, 2000). Adsorption and denaturation of host proteins onto materials will trigger cell recruitment, production of cytokines, and fibrous encapsulation of the foreign material (Ratner *et al.*, 2012). *In vivo* testing of poly(hydroxyethyl methacrylate) [p(HEMA)] hydrogels containing pendant groups of PEG and MPC has shown a significant reduction of the initial inflammatory response, decreases in fibrous encapsulation and a reduced foreign body response following intramuscular implantation (Abraham *et al.*, 2005; Guiseppi-Elie, 2011). Including low mole fractions of 0.5 and 10 mol.% for PEG and MPC, respectively, into a poly(HEMA) hydrogel yields a 64% reduction in adsorption of fibronectin and collagen compared to poly(HEMA) (Abraham *et al.*, 2005). Polysaccharides such as the dextrans, the alginates and hyaluronic acid, a degradation product of the ECM, may likewise be incorporated into hydrogels to confer *in vivo* biocompatibility (Censi *et al.*, 2009). Hybrid gel–ECP systems have demonstrated enhanced cell proliferation (Justin and Guiseppi-Elie, 2009) compared to the ECP or hydrogel only materials (George *et al.*, 2005).

11.10 Conclusions

The imitation of biological entities to develop selective sensors has grown exponentially since the late 1990s, and the biosensors industry is now worth billions of dollars in the world and tens of thousands of papers have been published in the area. The biggest market is in the medical area, particularly the glucose sensors for the self-management of people with diabetes. From the transduction point of view, electrochemistry has come to dominate the development of biosensors for diagnostics, while optical techniques have found their niche principally in R&D. To complete the picture concerning transduction strategies, advances in acoustic resonance devices are certainly worthy of note, but both thermometric and magnetic transduction have failed to have any serious practical impact to date. The modification of electrode surfaces with biological molecules (enzymes, antibodies, whole cells, tissues, etc.) has now evolved with the use of nanostructures. The application of nanotechnology allows obtaining, for example, preferential orientation of biomolecules onto electrode surfaces, improved bioelectrocatalytic performance, miniaturization, increased sensitivity, between many other interesting properties. Nanomaterials with at least one of their dimensions ranging in scale from 1 to 100 nm display unique and remarkably different property as compared to its bulk because their nanometer size gives rise to high reactivity and other enhanced beneficial physical properties (electrical, electrochemical, optical, and magnetic) (Singh, 2011). These features directly impact on the development of next-generation biosensors, with the ability to detect concentrations as low as nano- or picograms, which will surely lead to the development of noninvasive biosensors. Analyzing the current state-of-the-art, it is thought that the future of biosensors lies in the integration of portable and label-free systems, making use of nanotechnology and electronics. Also the advent of microfluidic systems may benefit the portability, low sample volumes and functionality of these sensors. It is expected then that these point-of-care devices to be commercialized in the future 5 or 10 years, which will revolutionize the field of self-care and biosensors research.

The authors are grateful to Dr Martín Lucas Zamora, for his contribution on the item 11.4.3 about Catalytic Bioreceptors.

References

Abraham S, Brahim S, Ishihara K, Guiseppi-Elie A. Molecularly engineered p(HEMA)-based hydrogels for implant biochip biocompatibility. *Biomaterials* 26(23):767–778, 2005.

Alegret S, Valle M, Merkoçi A. Sensores Electroquímicos: Introducción a los Quimiosensores y Biosensores?: Curso Teórico-Práctico. University Autónoma Barcelona, Spain, 176 pp., 2004.

Aucoin H, Wilson AN, Wilson A, Ishihara K, Guiseppi-Elie A. Release of potassium ion and calcium ion from phosphorylcholine group bearing hydrogels. *Polymers* 5(4):1241–1257, 2013.

Aumüller T, Fändrich M. Protein chemistry: Catalytic amyloid fibrils. *Nat Chem* 6(4):273–274, 2014.

Bala A, Pietrzak M, Górski L, Malinowska E. Electrochemical determination of lead ion with DNA oligonucleotide-based biosensor using anionic redox marker. *Electrochim Acta* 180:763–769, 2015.

Baudhuin P, vander Smissen P, Beavois S, Courtoy J. Molecular interactions between colloidal gold, proteins, and living cells. In: Hayat MA, ed. Colloidal gold: Principles, methods, and applications. Vols 1–2, Academic Press, Cambridge, 1989.

Bhat A, Hoch AI, Decaris ML, Leach JK. Alginate hydrogels containing cell-interactive beads for bone formation. *FASEB* 27(12):4844–4852, 2013.

Borisov SM, Wolfbeis OS. Optical biosensors. *Chem Rev* 108:423–461, 2008.

Brahim S, Guiseppi-Elie A. Electroconductive hydrogels: Electrical and electrochemical properties of polypyrrole-poly(HEMA) composites. *Electroanalysis* 17(7):556–570, 2005.

Brahim S, Narinesingh D, Guiseppi-Elie A. Interferent suppression using a novel polypyrrole-containing hydrogel in amperometric enzyme biosensors. *Electroanalysis* 14(9):627–633, 2002.

Brahim S, Narinesingh D, Guiseppi-Elie A. Synthesis and hydration properties of pH-sensitive p(HEMA)-based hydrogels containing 3-(trimethoxysilyl)propyl methacrylate. *Biomacromolecules* 4(3):497–503, 2003.

Brandl F, Kastner F, Gschwind RM, Blunk T, Tessmar J, Göpferich A. Hydrogel-based drug delivery systems: Comparison of drug diffusivity and release kinetics. *J Control Release* 142(2):221–228, 2010.

Brange J, Andersen L, Laursen ED, Meyn G, Rasmussen E. Toward understanding insulin fibrillation. *J Pharm Sci* 86(5):517–525, 1997.

Bromberg A, Jensen EC, Kim J, Jung YK, Mathies RA. Microfabricated linear hydrogel microarray for single-nucleotide polymorphism detection. *Anal Chem* 84(2):963–970, 2012.

Cao C, Li X, Lee J, Sim SJ. Homogenous growth of gold nanocrystals for quantification of PSA protein biomarker. *Biosens Bioelectron* 24(5):1292–1297, 2009.

Censi R, Vermonden T, van Steenbergen MJ, Deschout H, Braeckmans K, De Smedt SC, van Nostrum CF, di Martino P, Hennink WE. Photopolymerized thermosensitive hydrogels for tailorable diffusion-controlled protein delivery. *J Control Release* 140(3):230–236, 2009.

Cheng Y, Stakenborg T, Van Dorpe P, Lagae L, Wang M, Chen H, Borghs G. Fluorescence near gold nanoparticles for DNA sensing. *Anal Chem* 83(4):1307–1314, 2011.

Chiti F, Dobson CM. Protein misfolding, functional amyloid, and human disease. *Annu Revi Biochem* 75:333–366, 2006.

Choi NW, Kim J, Chapin SC, Duong T, Donohue E, Pandey P, Broom W, Hill WA, Doyle PS. Multiplexed detection of mRNA using porosity-tuned hydrogel microparticles. *Anal Chem* 84(21):9370–9378, 2012.

Chung I-M, Enemchukwu NO, Khaja SD, Murthy N, Mantalaris A, García AJ. Bioadhesive hydrogel microenvironments to modulate epithelial morphogenesis. *Biomaterials* 29(17):2637–2645, 2008.

Clark LC, Lyons C. Electrode systems for continuous monitoring in cardiovascular surgery. *Ann N Y Acad Sci* 102(1):29–45,1962.

Copeland R. Enzymes: A practical introduction to structure, mechanism, and data analysis. Wiley-VCH, 416 pp., 2000. ISBN: 0-471-35929-7.

Dai Q, Liu X, Coutts J, Austin L, Huo Q. A one-step highly sensitive method for DNA detection using dynamic light scattering. *J Am Chem Soc* 130(26):8138–8139, 2008.

Dirks RM, Pierce NA. Triggered amplification by hybridization chain reaction. *Proc Natl Acad Sci USA* 101(43):15275–15278, 2004.

Dubertret B, Calame M, Libchaber AJ. Single-mismatch detection using gold-quenched fluorescent oligonucleotides. *Nat Biotechnol* 19(4):365–370, 2001.

Dunker AK, Brown CJ, Lawson JD, Iakoucheva LM, Obradović Z. Intrinsic disorder and protein function. *Biochemistry* 41(21):6573–6582, 2002.

Dyson HJ, Wright PE. Intrinsically unstructured proteins and their functions. *Nat Rev Mol Cell Biol* 6(3):197–208, 2005.

Elghanian R, Storhoff JJ, Mucic RC, Letsinger RL, Mirkin CA. Selective colorimetric detection of polynucleotides based on the distance-dependent optical properties of gold nanoparticles. *Science* 277(5329):1078–1081, 1997.

Faraji AH, Cui JJ, Guy Y, Li L, Gavigan CA, Strein TG, Weber SG. Synthesis and characterization of a hydrogel with controllable electroosmosis: A potential brain tissue surrogate for electrokinetic transport. *Langmuir* 27(22):13635–13642, 2011.

Flory PJ, Rehner J. Statistical mechanics of cross-linked polymer networks I. Rubberlike elasticity. *J Chem Phys* 11:512–520, 1943.

Galán T, Prieto-Simón B, Alvira M, Eritja R, Götz G, Bäuerle P, Samitier J. Label-free electrochemical DNA sensor using "click"-functionalized PEDOT electrodes. *Biosens Bioelectron* 74: 751–756, 2015.

George PM, Lyckman AW, LaVan DA, Hegde A, Leung Y, Avasare R, Testa C, Alexander PM, Langer R, Sur M. Fabrication and biocompatibility of polypyrrole implants suitable for neural prosthetics. *Biomaterials* 26(17):3511–3519, 2005.

Guilbault GG, Lubrano GJ. An enzyme electrode for the amperometric determination of glucose. *Analytica Chimica Acta* 64(3):439–455, 1973.

Guiseppi-Elie A. Biomimetic hydrogels for in vivo biosensor biocompatibility. *Polymer Prepr* 47(2), 2006.

Guiseppi-Elie A. Electroconductive hydrogels: Synthesis, characterization and biomedical applications. *Biomaterials* 31(10):2701–2716, 2010.

Guiseppi-Elie A. An implantable biochip to influence patient outcomes following trauma-induced hemorrhage. *Anal Bioanal Chem* 399(1):403–419, 2011.

Guiseppi-Elie A, Brahim SI, Narinesingh D. A chemically synthesized artificial pancreas: Release of insulin from glucose-responsive hydrogels. *Adv Mater* 14(10):743–746, 2002.

Guiseppi-Elie A, Dong C, Dinu CZ. Crosslink density of a biomimetic poly(HEMA)-based hydrogel influences growth and proliferation of attachment dependent RMS 13 cells. *J Mater Chem* 22(37):19529–19539, 2012.

Häkkinen H. The gold–sulfur interface at the nanoscale. *Nat Chem* 4(6):443–455, 2012.

Hartgerink JD. Covalent capture: A natural complement to self-assembly. *Curr Opin Chem Biol* 8(6):604–609, 2004.

Ho Y-P, Kung MC, Yang S, Wang T-H. Multiplexed hybridization detection with multicolor colocalization of quantum dot nanoprobes. *Nano Lett* 5(9):1693–1697, 2005.

Hoffman AS. Stimuli-responsive polymers: Biomedical applications and challenges for clinical translation. *Adv Drug Deliv Rev* 65(1):10–16, 2012.

Hoffman BD, Crocker JC. Cell mechanics: Dissecting the physical responses of cells to force. *Annu Rev Biom Eng* 11:259–288, 2009.

Hurst SJ, Lytton-Jean AKR, Mirkin CA. Maximizing DNA loading on a range of gold nanoparticle sizes. *Anal Chem* 78(24):8313–8318, 2006.

Iqbal MA, Gupta SG, Hussaini SS. A review on electrochemical biosensors: Principles and applications. *Adv Biores* 3(4):158–163, 2012.

Ishihara K. Bioinspired phospholipid polymer biomaterials for making high performance artificial organs. *Sci Technol Adv Mater* 1(3):131–138, 2000.

Justin G, Guiseppi-Elie A. Electroconductive blends of Poly(HEMA-co-PEGMA-co-HMMAco-SPMA) and Poly(Py- co-PyBA): In Vitro Biocompatibility. *J Bioact Compat Polym* 25(2):121–140, 2009.

Karageorgiou V, Kaplan D. Porosity of 3D biomaterial scaffolds and osteogenesis. *Biomaterials* 26(27):5474–5491, 2005.

Kirschner CM, Anseth KS. Hydrogels in healthcare: From static to dynamic material microenvironments. *Acta Mater* 61(3):931–944, 2013.

Kokkinos C, Prodromidis M, Economou A, Petrou P, Kakabakos S. Quantum dot-based electrochemical DNA biosensor using a screen-printed graphite surface with embedded bismuth precursor. *Electrochem Commun* 60:47–51, 2015.

Kopeček J, Yang J. Hydrogels as smart biomaterials. *Polym Int* 56(9):1078–1098, 2007.

Kost J, Langer R. Responsive polymeric delivery systems. *Adv Drug Deliv Rev* 46(1–3):125–148, 2001.

Kotanen C, Karunwi O, Guiseppi-Elie A. Biofabrication using pyrrole electropolymerization for the immobilization of glucose oxidase and lactate oxidase on implanted microfabricated biotransducers. *Bioengineering* 1:85–110, 2014.

Kotanen CN, Moussy FG, Carrara S, Guiseppi-Elie A. Implantable enzyme amperometric biosensors. *Biosens Bioelectron* 35(1):14–26, 2012.

Krebs MR, Wilkins DK, Chung EW, Pitkeathly MC, Chamberlain AK, Zurdo J, Robinson CV, Dobson CM. Formation and seeding of amyloid fibrils from wild-type hen lysozyme and a peptide fragment from the beta-domain. *J Mol Biol* 300(3):541–549, 2000.

Kronick MN, Little WA. A new immunoassay based on fluorescence excitation by internal reflection spectroscopy. *J Immunol Methods* 8:235–240, 1975.

Lammers F, Scheper T. Thermal biosensors in biotechnology. *Adv Biochem Eng Biotechnol* 64:35–67, 1999.

Lec RM. Piezoelectric biosensors: Recent advances and applications. *Proceedings of the 2001 IEEE International Frequency Control Symposium and PDA Exhibition (Cat. No.01CH37218)* 215:419–429, 2001.

Lee S, Hyun Lee J, Kim M, Kim J, Song M-J, Jung H-I, Lee W. Bi nanowire-based thermal biosensor for the detection of salivary cortisol using the Thomson effect. *Appl Phys Lett* 103:143114, 2013.

Lee W-F, Lin W-J. Preparation and gel properties of Poly[hydroxyethyl-methacrylate-co-poly(ethylene glycol) methacrylate] Copolymeric hydrogels by photopolymerization. *J Polym Res* 9(1):23–29, 2002.

Li J, Fu H-E, Wu L-J, Zheng A-X, Chen G-N, Yang H-H. General colorimetric detection of proteins and small molecules based on cyclic enzymatic signal amplification and hairpin aptamer probe. *Anal Chem* 84(12):5309–5315, 2012.

Li N, Ho C-M. Aptamer-based optical probes with separated molecular recognition and signal transduction modules. *J Am Chem Soc* 130(8):2380–2381, 2008.

Li Y, Zhao Z, Lam ML, Liu W, Yeung PP, Chieng C-C, Chen T-H. Hybridization-induced suppression of coffee ring effect for nucleic acid detection. *Sensors Actuators B Chem* 206:56–64, 2015.

Li X, Zhao B, Li S, Wu X, Ren W, Shi P, Ye Z. Resonance behavior of piezoelectric polymer diaphragms for biosensors. *Ferroelectrics* 459(1):38–45, 2014.

Liang J, Huang J, Ding S, Ueda T. Flow-injection-based miniaturized quartz crystal microbalance. *Sensors Mater* 25(7):519–526, 2013.

Ligler F, Taitt C. Optical biosensors. Today and tomorrow. 2nd ed. Elsevier Science, Amsterdam, 712, 2008. ISBN: 9780444531254.

Lillehoj PB, Wei F, Ho C-M. A self-pumping lab-on-a-chip for rapid detection of botulinum toxin. *Lab Chip* 10(17):2265–2270, 2010.

Lower S. Electrochemistry 4: The Nernst Equation – Chemwiki (no date). http://chemwiki.ucdavis.edu/Analytical_Chemistry/Electrochemistry/Electrochemistry_4%3A_The_ Nernst_ Equation.

Madrid RE, Treo EF, Herrera MC, Mayorga Martínez CC. Electrodes. In: Splinter R, ed. Handbook of physics in medicine and biology. CRC Press, USA, 548 pp., 2010.

Mankar S, Anoop A, Sen S, Maji SK. Nanomaterials: Amyloids reflect their brighter side. *Nano Rev* 2:6032, 2011.

Mayorga Martínez CC. Aplicaciones Biomédicas de la Espectroscopía de Impedancia No Lineal (In Spanish, Biomedical Applications of Non-Linear Impedance Spectroscopy). PhD Dissertation, Universidad Nacional de Tucumán, Argentina, 2009.

Mayorga Martinez CC, Treo EF, Madrid RE, Felice CJ. Evaluation of chrono-impedance technique as transduction method for a carbon paste/glucose oxidase (CP/GOx) based glucose biosensor. *Biosens Bioelectron* 26(4):1239–1244, 2010.

Mayorga Martinez CC, Treo EF, Madrid RE, Felice CC. Real-time measurement of glucose using chrono-impedance technique on a second generation biosensor. *Biosens Bioelectron* 29(1):200–203, 2011.

McDonagh C, Burke CS, MacCraith BD. Optical chemical sensors. *Chem Rev* 108(2):400–422, 2008.

Meersman F, Dobson CM. Probing the pressure-temperature stability of amyloid fibrils provides new insights into their molecular properties. *Biochim Biophys Acta* 1764(3):452–460, 2006.

Meixner AJ, Bopp MA, Tarrach G. Direct measurement of standing evanescent waves with a photon-scanning tunneling microscope. *Appl Optics* 33(34):7995–8000, 1994.

Mirkin CA, Letsinger RL, Mucic RC, Storhoff JJ. A DNA-based method for rationally assembling nanoparticles into macroscopic materials *Nature* 382(6592):607–609, 1996.

Mishra GK, Sharma A, Deshpande K, Bhand S. Flow injection analysis biosensor for urea analysis in urine using enzyme thermistor. *Appl Biochem Biotechnol* 174(3):998–1009, 2014.

Munoz-Pinto DJ, Bulick AS, Hahn MS. Uncoupled investigation of scaffold modulus and mesh size on smooth muscle cell behavior. *J Biomed Mater Res Part A* 90(1):303–316, 2009.

Murphy RM. Peptide aggregation in neurodegenerative disease. *Annu Rev Biomed Eng* 4:155–174, 2002.

Nederberg F, Watanabe J, Ishihara K, Hilborn J, Bowden T. Biocompatible and biodegradable phosphorylcholine ionomers with reduced protein adsorption and cell adhesion. *J Biomater Sci Polym Ed* 17(6):605–614, 2006.

Nelson DL, Cox MM. Lehninger principles of biochemistry. 4th ed. W.H. Freeman and Company, New York, 1216 pp., 2005. ISBN 0-7167-4339-6.

Nielsen L, Khurana R, Coats A, Frokjaer S, Brange J, Vyas S, Uversky VN, Fink AL. Effect of environmental factors on the kinetics of insulin fibril formation: Elucidation of the molecular mechanism. *Biochemistry* 40(20):6036–6046, 2001.

Ottenbrite RM, Park K, Otano T. Biomedical applications of hydrogels handbook. 1st ed. Springer Science & Business Media, New York, 432 pp., 2010.

Pace CN. Conformational stability of globular proteins. *Trends Biochem Sci* 15(1):14–17, 1990.

Pallàs-Areny R, Webster JG. Sensors and signal conditioning. 2nd ed. John Wiley & Sons, New York, 608 pp., 2001. ISBN: 978-0-471-33232-9.

Pearce FG, Mackintosh SH, Gerrard JA. Formation of amyloid-like fibrils by ovalbumin and related proteins under conditions relevant to food processing. *J Agric Food Chem* 55(2):318–322, 2007.

Peppas N. Hydrogels in pharmaceutical formulations. *Eur J Pharm Biopharm* 50(1):27–46, 2000.

Peppas NA, Merrill EW. Crosslinked poly(vinyl alcohol) hydrogels as swollen elastic networks. *J Appl Polym Sci* 21(7):1763–1770, 1977.

Phelps EA, Enemchukwu NO, Fiore VF, Sy JC, Murthy N, Sulchek TA, Barker TH, García AJ. Maleimide cross-linked bioactive PEG hydrogel exhibits improved reaction kinetics and cross-linking for cell encapsulation and in situ delivery. *Adv Mater* 24(1):64–70, 2012.

Porter TL, Stewart R, Reed J, Morton K. Models of hydrogel swelling with applications to hydration sensing. *Sensors* 7(9):1980–1991, 2007.

Pramanik S, Pingguan-Murphy B, Osman NAA. Developments of immobilized surface modified piezoelectric crystal biosensors for advanced applications. *Int J Electrochem Sci* 8(6):8863–8892, 2013.

Qin C, Wen W, Zhang X, Gu H, Wang S. A double-enhanced strip biosensor for the rapid and ultrasensitive detection of protein biomarkers. *Chem Commun* 51(39):8273–8275, 2015.

Qiu Y, Park K. Environment-sensitive hydrogels for drug delivery. *Adv Drug Deliv Rev* 53(3):321–339, 2001.

Ramanathan K, Danielsson B. Principles and applications of thermal biosensors. *Biosens Bioelectron* 16(6):417–423, 2001.

Rastogi SK, Gibson CM, Branen JR, Aston DE, Branen AL, Hrdlicka PJ. DNA detection on lateral flow test strips: Enhanced signal sensitivity using LNA-conjugated gold nanoparticles. *Chem Commun* 48(62):7714–7716, 2012.

Ratner BD, Hoffman AS, Schoen FJ, Lemons JE. Biomaterials science: An introduction to materials in medicine. 3th ed. Academic Press, Elsevier, Oxford, UK, 484 pp., 2013. ISBN-13: 978-0-12-374626-9.

Rosi NL, Mirkin CA. Nanostructures in biodiagnostics. *Chem Rev* 105(4):1547–1562, 2005.

Rustad KC, Wong VW, Sorkin M, Glotzbach JP, Major MR, Rajadas J, Longaker MT, Gurtner GC. Enhancement of mesenchymal stem cell angiogenic capacity and stemness by a biomimetic hydrogel scaffold. *Biomaterials* 33(1):80–90, 2012.

Safranski DL, Weiss D, Clark JB, Caspersen BS, Taylor WR, Gall K. Effect of poly(ethylene glycol) diacrylate concentration on network properties and in vivo response of poly(β-amino ester) networks. *J Biomed Mater Res Part A* 96(2):320–329, 2011.

Sardar S, Pal S, Maity S, Chakraborty J, Halder UC. Amyloid fibril formation by β-lactoglobulin is inhibited by gold nanoparticles. *Int J Biol Macromol* 69:137–145, 2014.

Sarikaya M, Tamerler C, Jen AK-Y, Schulten K, Baneyx F. Molecular biomimetics: Nanotechnology through biology. *Nat Mater* 2(9):577–585, 2003.

Sauerbrey G. Verwendung von Schwingquarzen zur Wägung dünner Schichten und zur Mikrowägung [In German, Use of quartz crystal units for weighing thin films and microweighing].*Z Physik* 155(2):206–222, 1959.

Scheibel T, Parthasarathy R, Sawicki G, Lin X-M, Jaeger H, Lindquist SL. Conducting nanowires built by controlled self-assembly of amyloid fibers and selective metal deposition. *Proc Natl Acad Sci USA* 100(8):4527–4532, 2003.

Senveli S, Tigli O. Biosensors in the small scale: Methods and technology trends. *IET Nanobiotechnol* 7(1):7–21, 2013.

Silva Cárdenas BC. Desarrollo y Validación de un Sistema Basado en ENFET para Aplicación en Diálisis (In Spanish, Development and Validation of a System Based on ENFET for Dialisis Application). PhD Dissertation, Universidad Autónoma de Barcelona, Bellaterra, Barcelona, Spain, 2008.

Singh RP. Prospects of nanobiomaterials for biosensing. *Int J Electrochem* 2011:1–30, 2011.

Slaughter BV, Khurshid SS, Fisher OZ, Khademhosseini A, Peppas NA. Hydrogels in regenerative medicine. *Adv Mater* 21(0):3307–3329, 2009.

Smith JF, Knowles TPJ, Dobson CM, Macphee CE, Welland ME. Characterization of the nanoscale properties of individual amyloid fibrils. *Proc Natl Acad Sci USA* 103(43):15806–15811, 2006.

Song Y, Qu K, Zhao C, Ren J, Qu X. Graphene oxide: Intrinsic peroxidase catalytic activity and its application to glucose detection. *Adv Mater* 22(19):2206–2210, 2010.

Tanase CP, Albulescu R, Neagu M. Application of 3D hydrogel microarrays in molecular diagnostics: Advantages and limitations. *Expert Rev Mol Diagn* 11(5):461–464, 2014.

Taton TA, Mirkin CA, Letsinger RL. Scanometric DNA array detection with nanoparticle probes. *Science* 289(5485):1757–1760, 2000.

Thévenot DR, Toth K, Durst RA, Wilson GS. Electrochemical biosensors: Recommended definitions and classification. *Biosens Bioelectron* 16(1–2):121–131, 2001.

Thorn DC, Meehan S, Sunde M, Rekas A, Gras SL, MacPhee CE, Dobson CM, Wilson MR, Carver JA. Amyloid fibril formation by bovine milk kappa-casein and its inhibition by the molecular chaperones alphaS- and beta-casein. *Biochemistry* 44(51):17027–17036, 2005.

Torres-Bugeau CM, Borsarelli CD, Minahk CJ, Chehín RN. The key role of membranes in amyloid formation from a biophysical perspective. *Curr Protein Pept Sci* 12(3):166–180, 2011.

Tseng Y-T, Yang C-S, Tseng F-G. A perfusion-based micro opto-fluidic system (PMOFS) for continuously *in-situ* immune sensing. *Lab Chip* 9(18):2673–2682, 2009.

Turner APF. Biosensors: Sense and sensibility. *Chem Soc Rev* 42(8):3184, 2013.

Tyagi S, Kramer FR. Molecular beacons: Probes that fluoresce upon hybridization. *Nat Biotechnol* 14(3):303–308, 1996.

Updike SJ, Hicks GP. The enzyme electrode. *Nature* 214(5092):986–988, 1967.

Uversky VN. Alpha-synuclein misfolding and neurodegenerative diseases. *Curr Protein Pept Sci* 9(5):507–540, 2008.

Viter R, Starodub N, Smyntyna V, Kusevitch A, Sitnik J, Buk J, Macak J. Immune biosensor based on Silica Nanotube Hydrogels for rapid biochemical diagnostics of bovine retroviral leukemia. *Procedia Eng* 25:948–951, 2011.

Walcarius A, Minteer SD, Wang J, Lin Y, Merkoçi A. Nanomaterials for biofunctionalized electrodes: Recent trends. *J Mater Chem B* 1(38):4878–4909, 2013.

Wang L, Liu M, Gao C, Ma L, Cui D. A pH-, thermo-, and glucose-, triple-responsive hydrogels: Synthesis and controlled drug delivery. *React Funct Polym* 70(3):159–167, 2010.

Wang L, Sipe DM, Xu Y, Lin Q. A MEMS thermal biosensor for metabolic monitoring applications. *J Microelectromech Syst* 17(2):318–327, 2008.

Wang Y, Xu H, Zhang J, Li G. Electrochemical sensors for clinic analysis. *Sensors* 8(4):2043–2081, 2008.

Wei F, Lam R, Cheng S, Lu S, Ho D, Li N. Rapid detection of melamine in whole milk mediated by unmodified gold nanoparticles. *Appl Phys Lett* 96:133702, 2010.

Wen JT, Ho C-M, Lillehoj PB. Coffee ring aptasensor for rapid protein detection. *Langmuir* 29(26):8440–8446, 2013.

Wichterie O. Process for producing shaped articles from three-dimensional hydrophilic high polymers. US patent US2976576, 1961. http://www.google.com/patents/US2976576.

Wilson AN, Guiseppi-Elie A. Bioresponsive hydrogels. *Adv Healthcare Mater* 2(4):520–532, 2013.

Wu K, Liu J, Johnson RN, Yang J, Kopecek J. Drug-free macromolecular therapeutics: Induction of apoptosis by coiled-coil-mediated cross-linking of antigens on the cell surface. *Angew Chem* 49(8):1451–1455, 2010.

Xia F, Zuo X, Yang R, Xiao Y, Kang D, Vallée-Bélisle A, Gong X, Yuen JD, Hsu BBY, Heeger AJ, Plaxco KW. Colorimetric detection of DNA, small molecules, proteins, and ions using unmodified gold nanoparticles and conjugated polyelectrolytes. *Proc Natl Acad Sci USA* 107(24):10837–10841, 2010.

Xu X, Georganopoulou DG, Hill HD, Mirkin CA. Homogeneous detection of nucleic acids based upon the light scattering properties of silver-coated nanoparticle probes. *Anal Chem* 79(17):6650–6554, 2007.

Yakovleva M, Bhand S, Danielsson B. The enzyme thermistor-A realistic biosensor concept. A critical review. *Ana Chim Acta* 766:1–12, 2013.

Yang L, Guiseppi-Wilson A, Guiseppi-Elie A. Design considerations in the use of interdigitated microsensor electrode arrays (IMEs) for impedimetric characterization of biomimetic hydrogels. *Biomed Microdev* 13(2):279–289, 2011.

Yin P, Choi HMT, Calvert CR, Pierce NA. Programming biomolecular self-assembly pathways. *Nature* 451(7176):318–322, 2008.

Yoon HY, Koo H, Choi KY, Lee SJ, Kim K, Kwon IC, Leary JF, Park K, Yuk SH, Park JH, Choi K. Tumor-targeting hyaluronic acid nanoparticles for photodynamic imaging and therapy. *Biomaterials* 33(15):3980–3989, 2012.

Ytterdal T, Cheng Y, Fjeldly T. Device Modeling for Analog and RF CMOS Circuit Design. In Device modeling for analog and RF CMOS circuit design. John Wiley & Sons, 306 pp., 2003.

Yuan Z, Cheng J, Cheng X, He Y, Yeung ES. Highly sensitive DNA hybridization detection with single nanoparticle flash-lamp darkfield microscopy. *Analyst* 137(13):2930–2932, 2012.

Zhan Z, Cao C, Sim SJ. Quantitative detection of DNA by autocatalytic enlargement of hybridized gold nanoprobes. *Biosens Bioelectron* 26(2):511–516, 2010.

Zhang Y, Tang Z, Wang J, Wu H, Maham A, Lin Y. Hairpin DNA switch for ultrasensitive spectrophotometric detection of DNA hybridization based on gold nanoparticles and enzyme signal amplification. *Anal Chem* 82(15):6440–6446, 2010.

Zhao C, Song Y, Ren J, Qu X. A DNA nanomachine induced by single-walled carbon nanotubes on gold surface. *Biomaterials* 30(9):1739–1745, 2011.

Zhu X, Li J, He H, Huang M, Zhang X, Wang S. Application of nanomaterials in the bioanalytical detection of disease-related genes. *Biosens Bioelectron* 74:113–133, 2015.

Zustiak SP, Durbal R, Leach JB. Influence of cell-adhesive peptide ligands on poly(ethylene glycol) hydrogel physical, mechanical and transport properties. *Acta Biomater* 6(9):3404–3414, 2010.

CHAPTER 12

THE BIOLOGICAL AMPLIFIER

Enrique Spinelli and Federico N. Guerrero

Like a spying glass, it magnifies tiny electrical bioevents

Abstract

Biopotentials are essential signals that are immersed in the body like underwater currents in an ionic ocean. Electrodes allow their detection by transducing these ionic phenomena into electrical signals, feasible to be processed by electronic circuits. Thus, we do not have direct access to biopotentials but through variable (and sometimes unpredictable) electrode–skin impedances. Because of this, when available as electrical signals, biopotentials become faint and vulnerable to noise and interference sources. One approach to bypass this problem is to use invasive techniques such as skin abrading or percutaneous needles insertion, which bring us closer to the electrical source, thus reducing the impedance. These solutions try to find out a solution compromising the patients' comfort, instead of working on electronic design. It is a task for bioengineers to design bioamplifiers able to acquire high-quality signals using noninvasive techniques while dealing with high-electrode impedances. This chapter describes the characteristics of biomedical signals available at the electrodes, the influence of external interference sources, and the desired bioamplifiers' features to deal with. Classic and current amplifier design and new tendencies are herein presented, showing circuits for both traditional wet electrodes and capacitive electrodes (CEs). The latter is an emerging technology that does not require electrolytes (gel, liquid, or paste), which allow picking up biopotentials without any skin preparation, even through dielectric films or cotton clothes. A note should be made about the term "bioamplifier" itself. When analog electronics

463

were the predominantly available tool, biopotential measurements were improved by enhancing the amplifier design. This remains true, but in a time when digital electronics and integrated systems are ever cheaper and more efficient, we think about the bioamplifier in connection with a broader class of devices which encompass the digital domain. In this chapter we discuss *biopotential acquisition systems*, because the characteristics of data converters and mixed-signal devices have become an integral part of a "bioamplifier" design.

12.1 Introduction

The "human machine" needs many signals to work properly. They are essential to control and to synchronize the multiple processes that are right now occurring inside our body. Many of these signals are generated autonomously in some organs to maintain vital functions, such as the coordination of the cardiac or respiratory muscles, whereas others are voluntarily generated in the brain. These signals travel through nerves to multiple destinations, for example, to move the fingers that write the final point of this paragraph.

As described in Chap. 2, biopotentials originate in excitable cells, such as nervous and muscular cells. They populate throughout our body, and are an excellent means to diagnose the "human machine". Their appearances, spectral contents, propagation velocities, among other features, provide valuable information about the functioning of our body. Because of that, physiologists and physicians have been trained for decades to infer the behavior of the human organism from these signals.

12.1.1 *The biopotential signal model*

Bioelectric sources can be modeled as current sources (Fig. 12.1a). When we acquire biopotentials on the skin or scalp, we observe the potentials they produce flowing on the surrounding tissue that works as a *volume conductor* (Plonsey and Barr, 2007). The conductivities in the volume conductor are quite high, defining impedances (Z_{VC}) as low as a few hundred ohms. So, by applying Thevenin theorem to transform the current source, the signal under the skin can be seen as a voltage generator with low-output impedance (Fig. 12.1b).

However, these signals are still far from being manageable using our electronic instruments.

Inside our body, signals are carried by ionic currents that must be transduced to *electronic* currents in order to be measurable with electronic equipment. This process occurs in the interface between the skin and the electrodes placed to pick up the biopotentials. When standard wet electrodes are used, a metallic electrode is placed on the skin and an electrolyte is added between the electrode and the skin. This simple and usual procedure creates an interface in which complex electrochemical processes take place. A detailed analysis and model of the skin–electrolyte–electrode interface can be found in Valentinuzzi (2004), but for our electronic design purposes it suffices to consider a simplified electrical model. This model can be represented by an *electrode impedance* Z_E and an *electrode offset voltage* v_E, as shown in Fig. 12.1b. This DC source, product of the half-cell potential generated by the contact of metal and electrolyte, can be as high as hundreds of millivolt (several orders of magnitude above the biopotential signal amplitude) and imposes one of the main constrains for biopotential amplifier (BA) design. Electrode impedance

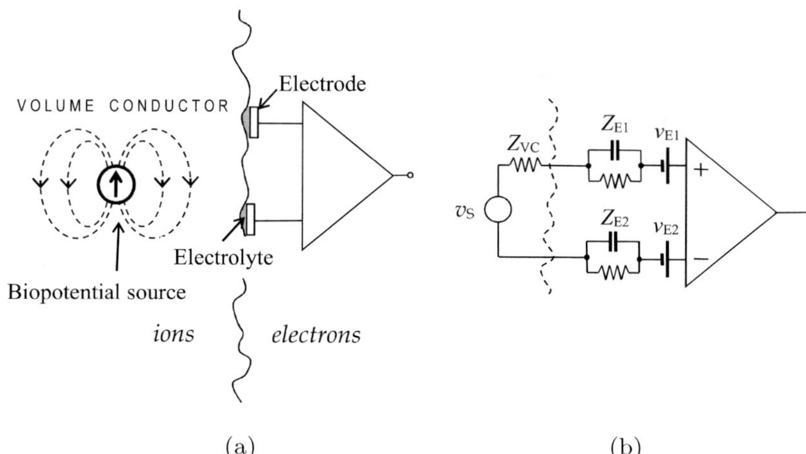

(a)   (b)

Figure 12.1: (a) Biopotential sources can be modeled as current sources. (b) They produce potential differences across the volume conductor (surrounding tissues) shown at the surface of the body as voltage sources v_S, but they must cross the electrode–skin interface (Z_{Ei}, v_{Ei}) to reach the input of the amplifier.

and offset values are dependent on several factors, some rather unpredictable, like skin conditions (humidity, temperature, among others) and are highly variable. Electrode impedances range from 10 kΩ to 1 MΩ at 10 Hz (Rossell *et al.*, 1998), whereas DC offsets potential can be up to 300 mV and above.

When dealing with such small amplitudes, a question arises: What is the lowest amplitude that we can sensibly measure using superficial electrodes? The answer lies not within the signal characteristics but in the noise phenomena. A relatively high noise amplitude is produced at the electrode–skin interface. Any biopotential signal lower in amplitude would be masked by this noise, and hence, it would not be measurable. Huigen *et al.* (2002) conducted a study of the noise produced by electrodes on the skin and showed what amplitude and spectral distribution to expect.

One remarkable fact is that the power spectrum seems to follow a $1/f^\alpha$ law with $1 < \alpha < 2$, which means that much of the signal's power is condensed at the low frequencies. Hence, noise from a pair of electrodes placed on the skin tends to look like the dark-gray signal in Fig. 12.2a. Its power spectral density (PSD) is shown in

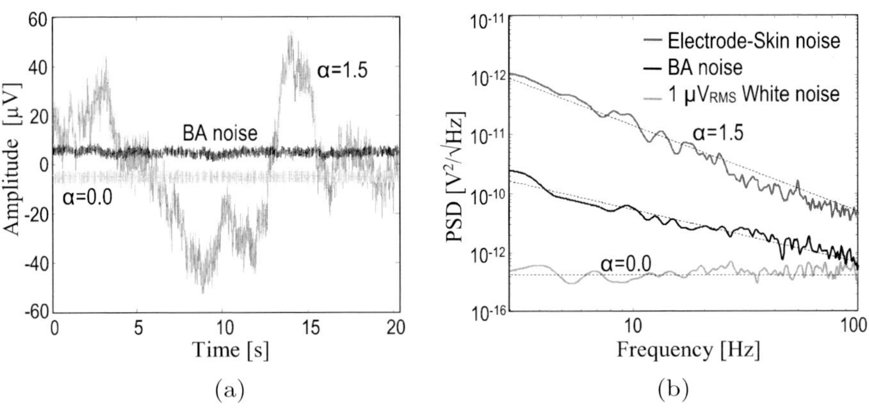

Figure 12.2: Representative noise signals (a) and their respective PSDs (b). White noise is shown in light-gray, a typical BA's noise in black, and noise from electrodes placed on the skin in dark-gray. In (b), $1/f^\alpha$ curves fitted to the experimental data show the α-values typical of each noise source, which translate to the signal behavior seen in (a). Noise from the electrode–skin interface has more power concentrated at low frequencies.

Fig. 12.2b next to a $1/f^{1.5}$ curve (dashed line). For comparison, the light-gray line shows the well-known "white noise" that has a flat spectrum ($\alpha = 0$), and the black line depicts noise typical of a biological amplifier (BA) when its input terminals are short-circuited.

Visual inspection of these curves confirms what the PSD shows, that is, as the α value increases, the power signal becomes more skewed toward low frequencies. Consequently, the noise in biopotential measurements has a marked low-frequency baseline drift that can be filtered out if the requested signal has no content of interest in this band. The black curve of Fig. 12.2a shows, in fact, a noise record of a BA. The resolution of the BA is given by its inherent noise sources, which can all be represented by one unique noise source at the input. We could argue that if the system's noise is of the same order of magnitude or below the value of the electrodes' noise, no further advantage would be gained by enhancing the resolution (i.e., lowering the system's noise). It is noteworthy, as Huigen *et al.* (2002) point out, that the predominant noise in surface biopotential measurements is due to the outer skin layer characteristics, so that it can be significantly diminished by skin preparation.

Biosignal amplitudes vary according to several factors, such as the type and number of recruited excitable cells, the synchronization between them, and the distance from the biopotential source to the electrodes. Typical amplitude and bandwidth (BW) values for the most common biosignals are presented in Table 12.1.

In summary, biopotentials are strong signals inside our body, but we can only reach them through high-value electrode impedances and electrode offset potentials. Figure 12.3 depicts the fundamental

Table 12.1: Typical amplitude and bandwidth values of some biomedical signals.

	Amplitude (typ)	Bandwidth (typ)
EOG	10 μV	0–10 Hz
EEG	100 μV	0.1–100 Hz
EMG	100 μV	10–1000 Hz
ECG	1 mV	0.05–100 Hz

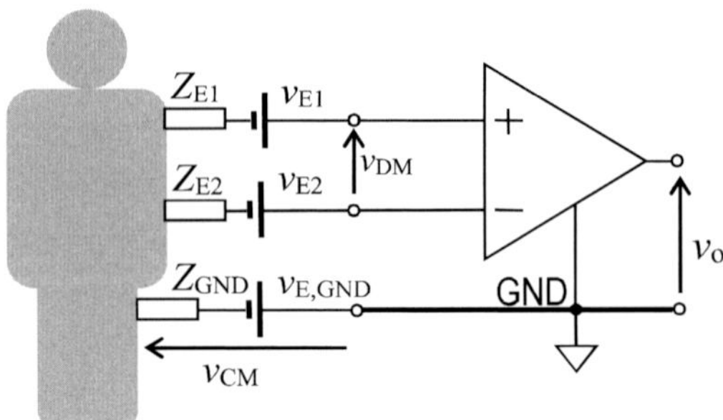

Figure 12.3: Biopotentials are strong signals inside the body, but we can only reach them through widely variable (and sometimes high) impedances and DC potentials.

issue faced by biopotential acquisition systems: To pick-up biosignals through variable and unpredictable impedances, contaminated with spurious potential sources, and immersed in high DC voltages. This looks a bit hopeless, but do not worry, millions of biomedical equipments are acquiring biosignals at this moment! Smile.

12.1.2 *Challenges for biological amplifiers*

In the preceding section, the amplifier was modeled with an ideal gain block. When going beyond this abstraction toward a realistic model, several issues arise that shape the design of the amplifier. The intended use of the biomedical signal dictates the requirements of the amplifier, for instance, a sports-oriented cardiac pulse meter will not have the same requirements as a clinical diagnostics device. However, with a general case in mind, one would desire to present the signal as produced in the body without any modification, presenting no risk to the subject's health, and with certain repeatability. Thus, we can subdivide the requirements within three areas, that is,

Interference rejection: This is the main challenge for biological amplifiers and it is thoroughly discussed in Section 12.2.

Signal accuracy: It must meet the demands outlined in Section 12.1.1. Parameters to be considered and precautions to be taken when translating the signal to the digital domain are given in Section 12.3.

Electrical safety: Electrodes establish low-impedance connections for currents to run through the body; therefore, a primary concern is to keep these currents within safe limits. This implies understanding possible leakage paths and implementing appropriate isolation barriers, as described in Section 12.4.

12.1.3 *Modeling analysis: Differential mode and common mode voltages*

The biological amplifier displays a fairly commonplace scenario in the instrumentation world, that is, to measure the potential difference across two nodes (the electrodes), or to put it in electronic analysis jargon, to measure the *differential mode* (DM) signal v_{DM} indicated in Fig. 12.3. In every such measurement setup, one must consider that while the signal of interest is present in the subtraction of the nodes' voltages (actually the subtraction "weighted" by $1/2$, e.g., $(v_1 - v_2)/2$), these signals also carry a *common mode* voltage (CMV) v_{CM}, defined between the electrodes and the amplifier ground. To reduce v_{CM}, a low-impedance Z_{GND} between patient and the latter ground is desired.[1]

The instrumentation challenge is to acquire v_{DM} rejecting v_{CM} effects, as a *differential amplifier* does. The output v_o of a real differential amplifier depends on both, v_{DM} and v_{CM}, so that,

$$v_o = G_{DD}v_{DM} + G_{DC}v_{CM} \tag{12.1}$$

where G_{DD}, G_{DC} are the amplifier gains for differential mode voltage (DMV) and CMV, respectively[2] (Pallàs-Areny and Webster, 1999). An amplifier figure of merit is its common-mode rejection ratio (CMRR), which describes its capacity to amplify v_{DM} while rejecting v_{CM} (exactly what we need!). It is defined as,

$$\text{CMRR} = \frac{G_{DD}}{G_{DC}} \tag{12.2}$$

and usually expressed in decibels by,

$$\text{CMRR} = 20 \log_{10}\left(\frac{G_{DD}}{G_{DC}}\right) \tag{12.3}$$

---

[1] Although it is also possible to pick-up biopotentials using just two electrodes (without Z_{GND}), this approach is very vulnerable to power-line interference. For details, see Spinelli and Mayosky (2005).

[2] These gains are defined as: $G_{DD} = v_o/v_{DM}|_{v_{CM}=0}$; $G_{DC} = v_o/v_{DM}|_{v_{DM}=0}$

The output of a perfect differential amplifier only depends on v_{DM}, its G_{DC} gain is null and its CMRR infinite.[3]

12.2   Power-Line Interference

Patients are not floating perfectly, untethered in the universe, as Fig. 12.3 suggests. They are tied up to the power line (PL) by electric and magnetic fields. Through these couplings, PL produces v_{CM} and v_{DM} potentials, thus interfering with biopotential signals. PL interference (50/60 Hz and its harmonics) is perhaps the main problem in biopotential measurements. Although there are many methods to "clean" signals up that were contaminated with $f_{PL} = 50/60$ Hz components, this goal is difficult to achieve without some signal distortion or information loss (Levkov *et al.*, 2005). As Pallàs-Areny and Webster (1999) state: *A measuring system requiring a permanent notch filter to remove power-line interference should be inspected and, probably, its front-end redesigned.* Therefore, the requirement of notch filters or digital signal processing techniques for removing PL interference are evidence and consequence of a poor instrumentation design. When interference is too high, it can jeopardize the amplifier input range, leading to signal distortion and even to the amplifier saturation, which makes it impossible to recover the desired signal. Furthermore, high-frequency interference signals, interacting with the circuit's nonlinearities, could produce low-frequency components in the range of the biomedical signal's BW (Van der Horst *et al.*, 1998; De Jager *et al.*, 1996). These problems show the importance of early avoidance and rejection of PL interference. The first step in solving 50/60-Hz interference is to know how it invades a biopotential acquisition system.

12.2.1   *PL interference model*

Given that biopotential measurement system dimensions are much smaller than the PL wavelength ($\lambda_{50\,Hz} \cong 6$ km), the electromagnetic

---

[3]As a rule-of-thumb, one could consider 100–120 dB a very high CMRR value and 60 dB a poor one.

interference (EMI) effects produced by the electric and magnetic fields can be considered separately (Ott, 1988).

In this case, electric-field interference coupling can be modeled by stray capacitances, and even — as a second approximation step — by lumped capacitances as presented in Fig. 12.4. This simple network allows analyzing, and also estimating with reasonable precision, PL interference effects. This circuit describes the main coupling mechanisms reported in Huhta and Webster (1973), Pallàs-Areny and Colominas (1989), Metting van Rijn *et al.* (1990a), Metting van Rijn *et al.* (1991), and Wood *et al.* (1995). In the following analysis, an acquisition system isolated from earth is considered (the only kind that biomedical standards admit). Some elements associated with minor effects are omitted for the sake of clarity. Furthermore, considering that impedance values inside the human body are much lower than the other impedances involved in the model, they were neglected and replaced by short circuits.

Note that no biopotential generators are present in the network of Fig. 12.4; hence, the output voltage $v_{\mathrm{O.EMI}}$ is exclusively due to PL interference. The network also includes the BA input impedances for CMV and DMV (Z_{C} and Z_{D}, respectively). Impedances Z_{C} include capacitances from the wires and those associated with input protection circuits. The elements of the EMI model can be experimentally estimated (Haberman *et al.*, 2011), but they are widely variable depending on several factors, such as geometry, proximity to power cables and to large area objects. Typical values for these parameters are listed in Table 12.2.

12.2.2 *PL interference output voltage*

PL voltage produces currents and potential differences across the impedances depicted in Fig. 12.4. These potentials appear at the BA input as a DMV between the electrodes v_{DM} and as a v_{CM} defined between patient and amplifier ground. Assuming that no biopotential signal is present, the BA output voltage $v_{\mathrm{o.EMI}}$ depends exclusively on the PL voltage, that is,

$$v_{\mathrm{o.EMI}} = v_{\mathrm{DM}}G_{\mathrm{DD}} + \frac{v_{\mathrm{CM}}}{\mathrm{CMRR}}G_{\mathrm{DD}} \tag{12.4}$$

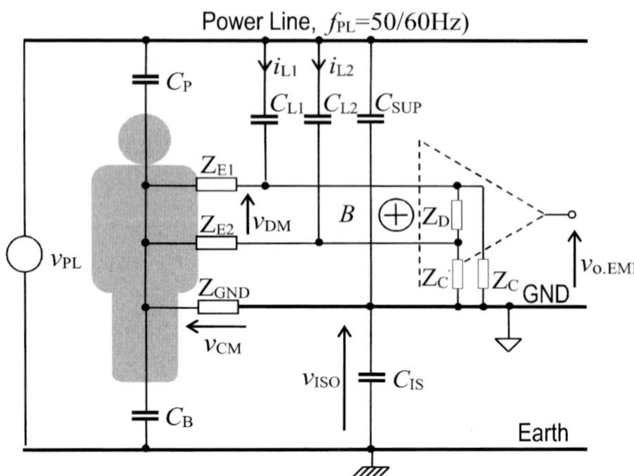

Figure 12.4: Electrical network describing PL coupling to a biopotential acquisition system.

Table 12.2: Main interference model parameters.

C_P	Line-to-patient capacitance (0.05–2 pF) (Haberman *et al.*, 2011)
C_B	Patient-to-earth capacitance (100–300 pF) (Haberman *et al.*, 2011)
C_{Li}	Line-to-patient lead i capacitance ($\approx$0.1 pF/m) (Metting van Rijn *et al.*, 1991)
C_{SUP}	Line-to-ground capacitance (0.03–3 pF) (Haberman *et al.*, 2011)
C_{ISO}	Ground-to-earth capacitance (20–100 pF) (Haberman *et al.*, 2011)
Z_{Ei}	Electrode i impedance (1 kΩ–1 MΩ) (Rossell *et al.*, 1988)
Z_D	DM input impedance
Z_C	CM input impedance
v_{CM}	Amplifier CM input (0–20 mV) (Haberman *et al.*, 2011)
v_{DM}	Amplifier DMV
B	Magnetic field (5–200 nT) (Pallàs-Areny and Colominas, 1989)

where G_{DD} is the BA differential gain and CMRR its common mode rejection ratio. In order to obtain values independent on G_{DD}, the output voltage $v_{o.EMI}$ is referred to the input as $v_{i.EMI}$,

$$v_{i.EMI} = v_{DM} + \frac{v_{CM}}{\text{CMRR}} \tag{12.5}$$

The first term in Eq. (12.5) corresponds to interference components that appear at the BA input as DMVs (as the biopotentials of interest are). The second term represents the contribution of CMVs to $v_{i.EMI}$

due to the limited CMRR of a real amplifier. A "perfect" differential amplifier has an infinite CMRR, and CM input voltages v_{CM} do not affect its output.

The value of $v_{\text{i.EMI}}$ can be calculated from Eq. (12.5) by solving the complete network of Fig. 12.4 for v_{iD} and v_{CM}. The resulting expression, however, would be as intricate as useless, because its complexity would hide the mechanisms through which the PL voltage interferes. Therefore, we will choose a more conceptual method of analysis, that is, the contribution of the different mechanisms of PL interference will be estimated separately, using a model as simple as possible for each case. This approach will not yield precise results, and it will not be able to calculate the total $v_{\text{i.EMi}}$ by addition of voltages from these partial models, but it will provide a clear idea of the order of magnitude of each different contribution, hence, allowing to finding specific solutions for interference reduction.

12.2.3 *Common mode voltage v_{CM}*

The value adopted by v_{CM}, depends on the impedance of the ground electrode Z_{GND} and on the values of stray capacitances C_{P}, C_{B}, C_{SUP}, and C_{ISO} (Fig. 12.5). These parameters vary with the type and size of the biomedical device, and with the patient's conditions. An extreme value for v_{CM} can be found when a multichannel, PL powered acquisition system is used and the patient is in contact with a large metallic area (e.g., a metallic locker); nonetheless, even in this condition, v_{CM} rarely exceeds 10 mV for Z_{GND} as high as 100 kΩ. For well-isolated devices, such as battery-powered ones, v_{CM} reduces to tenths of mV (Haberman *et al.*, 2011).

Considering $v_{\text{CM}} = 10$ mV, and according to Eq. (12.5), a BA with a CMRR of 10,000 (80 dB) reduces the contribution of v_{CM} to 1 μV, which is in the order of the amplifier's noise. Although there is a traditional concern to obtain BAs with ultra-high CMRR values, a CMRR of 80 dB is enough for most biomedical applications. Moreover, as it will be shown later, when active v_{CM} reduction techniques are included, CMRR constraints are even less significant.

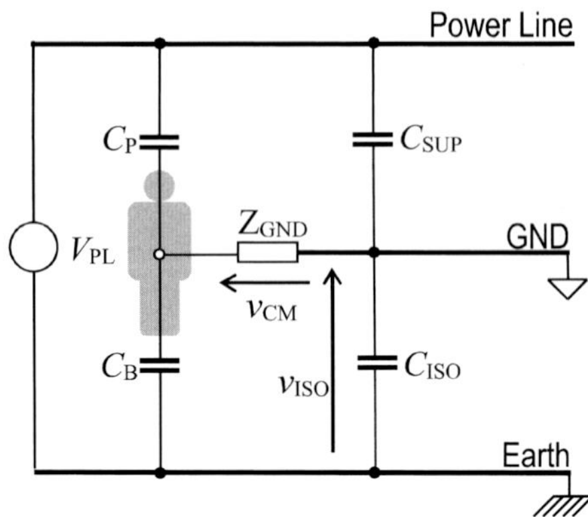

Figure 12.5: Reduced EMI model to estimate the v_{CM}.

12.2.4 *Differential mode voltage v_{DM}*

Some interference mechanisms directly produce DMVs, whereas others produce them by transforming CMVs into differential ones. This *mode transformation* does not occur only inside the BA, as in the case of the CMRR of the BA, but outside it and before signals reach the amplifier's input. These v_{DM} voltages are amplified because they are "mode-indistinguishable" from the biomedical signal. Let us see such cases.

12.2.4.1 *Coupling to the patient leads*

PL voltage produces displacement currents i_{L1}, i_{L2}, which run through the patient leads (Fig. 12.4). Given that the BA has high-input impedances, the above-mentioned currents flow through the electrode's impedances Z_{E1}, Z_{E2}. When electrode impedances are unbalanced, a DMV appears, that is,

$$v_{DM}\big|_{\text{Patient Leads}} = i_{L1} \cdot Z_{E1} - i_{L2} \cdot Z_{E2} \tag{12.6}$$

The displacement currents i_{L1}, i_{L2} are similar because the coupling capacitances to the leads are of the same order, too, that is, $C_{L1} \approx C_{L2} \approx C_L$. The worst condition shows up when the whole PL voltage V_{PL} is applied to these capacitances. In this case, $i_L \approx V_{PL} 2\pi f_{PL} C_L$ resulting in,

$$v_{DM}\big|_{\text{Patient Leads}} \approx V_{PL} 2\pi f_{PL} C_L (Z_{E2} - Z_{E1}) \tag{12.7}$$

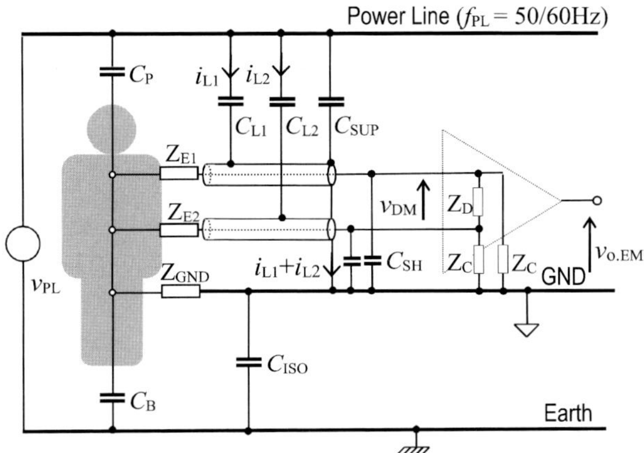

Figure 12.6: Interference caused by displacement currents coupled to the patient leads can be avoided using shielded cables. C_{SH} represents the cables' capacitances, which degrade the amplifier's CM input impedance.

Considering the typical values of Table 12.1, that is, $C_L = 0.1$ pF, $V_{PL} = 220$ V, $f_{PL} = 50$ Hz, $Z_{E1} = 10$ kΩ, and $Z_{E2} = 100$ kΩ, the following can be obtained,

$$v_{DM} = 220 \text{ V} \cdot 2\pi \cdot 50 \text{ Hz} \cdot 0.1 \text{ pF}(100 \text{ k}\Omega - 10 \text{ k}\Omega) \cong 0.6 \text{ mV} \quad (12.8)$$

The latter value, regrettably, is of the order of magnitude of the ECG signal! Facts sometimes do not help, as Murphy's law wisely and laughingly predicts.

12.2.4.2 *Solution to EMI coupling to the patient leads*

This effect produces high PL interference contribution, but it is easy to solve. It suffices to use shielded cables, with their shield connected to the amplifier ground as Fig. 12.6 shows. In this way, the displacement currents i_{L1}, i_{L2} flow to ground without circulating through the electrodes' impedances; thus, they are unable to produce significant interference voltages. It is important to note that cable capacitances between the central conductor and the shield (typically 100–200 pF per meter of cable) reduce the BA CM impedance Z_C. In general, this is not a serious problem, but it must be taken into account. This is a manifestation of the **Principle of Difficult Conservation**: *Coupling to the patient leads is solved at the expense of CM impedance* *degradation.* Engineering work consists in moving the problems to where they are less harmful. Another way to reduce this kind of interference is to assure low Z_{E1}, Z_{E2} values. This approach may require "preparing" the skin of the patient. This can be done by cleaning it, placing an electrolyte (liquid or gel) or even by skin abrasion (a minor invasive practice that must be avoided, because *primum non nocere* remains as a basic medical stand). Biomedical devices, especially those with a large number of channels, provide circuitry to measure the impedance of the electrodes in order to verify if it is low enough to acquire good quality records.

12.2.4.3 *Electromotive force induced in the patient leads loop*

The magnetic field B produced by the PLs is also an interference source. It makes itself apparent through the electromotive forces (EMFs) induced into the patient lead loops, as shown in Fig. 12.7a. These EMFs appear at the BA input as a DMV. Huhta and Webster (1973) obtained the DMV due to this effect assuming that the loop has an effective area[4] S,

$$v_{\mathrm{DM}}|_{\text{Magnetic field}} = B \cdot S \cdot 2\pi f_{\mathrm{PL}} \tag{12.9}$$

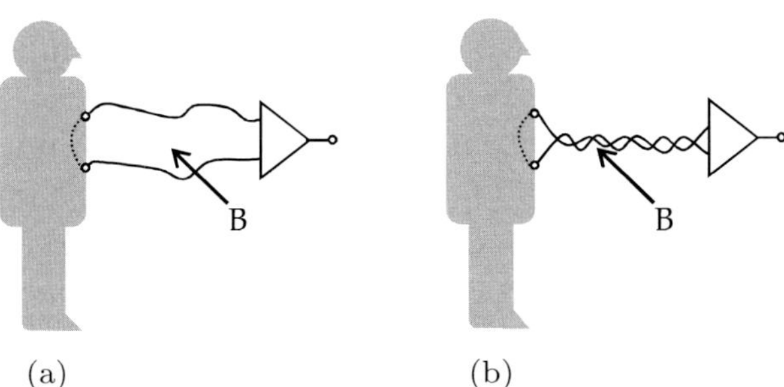

Figure 12.7: Electromotive induced voltage. The magnitude of this effect increases with the patient leads loop area (a). To reduce it, cables can be twisted as in (b). Magnetic field B is perpendicular to this sheet.

---

[4]The "effective area" is smaller than the geometric area because it involves additional factors such as its relative orientation (area normal to the magnetic field) and partial EMF cancellation when multiple loops are defined (e.g., twisted wires, as in Figure 12.7b).

Typical values of B ranges from 5 to 200 nT (Pallàs-Areny and Colominas, 1989). For an effective area $S = 0.1$ m^2, considering $B = 200$ nT, at 50 Hz v_{DM} results in,

$$v_{\text{DM}}|_{\text{Magnetic field}} = 200 \text{ nT} \cdot 0.1 \text{ m}^2 \cdot 2\pi 50 \text{ Hz} = 6\mu\text{V} \qquad (12.10)$$

This value of v_{DM} is not significant for ECG signals, but it could be too high for EEG applications. Fortunately, in EEG acquisition setups, electrodes are closer to the amplifier, so areas are smaller, yielding EMF contributions usually below the amplifier's noise floor.

12.2.4.4 *Solution to EMFs induced in the patient lead loops*

In general, magnetic-field interference does not produce serious problems. This is really good news, because there is not much to do in order to deal with this EMI (Huhta and Webster, 1973). The only feasible measure that can be taken to compensate for this effect is to reduce the effective area S by twisting the leads, as Fig. 12.7b shows. It is possible to implement magnetic-field shielding techniques, but thick ferromagnetic materials must be used, resulting in large, heavy, and expensive shields. This kind of shielding is only necessary for very magnetic-sensitive instruments, like magneto-encephalographers.

12.2.4.5 *Mode transformations*

CM potentials, such as v_{CM}, could be transformed to DMVs before reaching the amplifier's input due to imbalances between the electrode impedances Z_{E1}, and Z_{E2}. The simplified model of Fig. 12.8 is adequate to analyze this effect,[5] which is known as "potential divider effect" (Huhta and Webster, 1973).

The v_{CM} falls across two voltage dividers formed by Z_{E1}, Z_C and Z_{E2}, Z_C, respectively. Assuming that these dividers are not equal (in general, $Z_{E1}/Z_C \neq Z_{E2}/Z_C$), a DM potential appears at the BA

---

[5] For the sake of clarity, the amplifier's DM input impedance Z_D has not been included. However, as demonstrated by Pallàs-Areny and Webster (1999), this impedance does not affect the relationship between v_{DM} and v_{CM}, as predicted in Eq. (12.11).

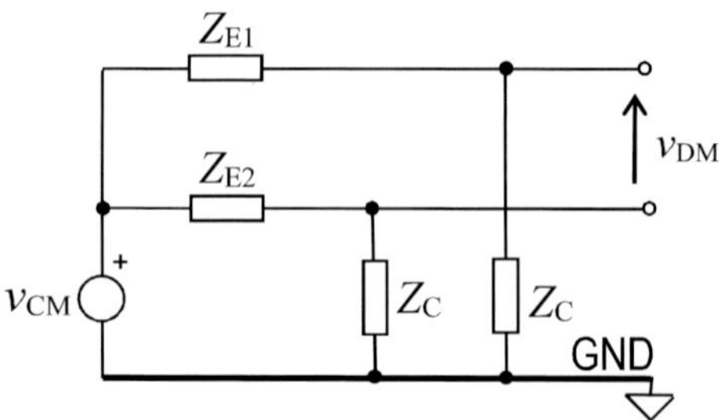

Figure 12.8: Potential divider mode transformation. Amplifier CM impedances are considered equal because their imbalances produce minor effects.

input given by,

$$v_{\mathrm{DM}} = v_{\mathrm{CM}} \left(\frac{Z_{\mathrm{E1}}}{Z_{\mathrm{E1}} + Z_{\mathrm{C}}} - \frac{Z_{\mathrm{E2}}}{Z_{\mathrm{E2}} + Z_{\mathrm{C}}} \right) \tag{12.11}$$

By and large, $Z_{Ei} \ll Z_C$, so that Eq. (12.11) can be approximated by,

$$v_{\mathrm{DM}}\big|_{\text{Potential divider}} \cong v_{\mathrm{CM}} \frac{\Delta Z_{\mathrm{E}}}{Z_{\mathrm{C}}} \tag{12.12}$$

where $\Delta Z_E = Z_{E1} - Z_{E2}$.

Considering $\Delta Z_E = 5k\Omega$ and $Z_C = 50$ MΩ,[6] the transformation factor of v_{CM} into DM $(\Delta Z_E/Z_C)$ is 3×10^{-4}, which corresponds to a CMRR of 80 Db.[7] For $v_{\mathrm{CM}} = 10$ mV, a $v_{\mathrm{DM}} = 1$ μV results.

12.3 Biological Amplifiers

In summary, PL interference imposes requirements on BA amplifiers. To reduce the effects of PL interference, it is important to avoid the circulation of PL displacement current through the electrode impedances, for example, by using shielding cables. Otherwise, special care must be taken to ensure low-electrode impedance values.

PLs also produce CMVs that can become differential voltages (directly superposing to biopotentials) by means of two mechanisms:

---

[6]50 MΩ corresponds to the impedance of a 60-pF capacitance at 50 Hz.

[7]This is further evidence of how meaningless is to look for amplifiers with ultra-high CMRR values. A CMRR of 80 dB is enough to keep the amplifier CMRR effect below other interference contributions.

Before the signal reaches the BA, due to the *potential divider* effect, and in the BA itself due to its limited CMRR. A CMRR of 80 dB is enough to reduce PL interference below the amplifier's noise level. Besides, to reduce the potential divider effect, large Z_C values are needed.

These requirements define a BA: **It is a differential amplifier with a good CMRR and high CM input impedance Z_C.** For low-level signals, such as EEG, low voltage and current noise are also desirable.

12.3.1 *Integrated instrumentation amplifier*

A differential amplifier can be implemented using operational amplifiers (OAs) (Horowitz *et al.*, 1989), but nowadays many differential amplifiers are available as integrated circuits (ICs) with high CMRR and high Z_C. These circuits are called *instrumentation amplifiers* (IA) and allow us to solve almost all instrumentation problems. They present a differential input, adjustable gain, and an additional ground-referred input v_{REF}. The output of an IA is given by,

$$v_{\mathrm{o}} = G \cdot v_{\mathrm{DM}} + v_{\mathrm{REF}} \tag{12.13}$$

where G is a function of an external resistor R_G through an expression like

$$G = 1 + 50 \text{ k}\Omega / R_G \tag{12.13b}$$

The IA amplifies differential voltages v_{DM} by this factor G and allows adding an auxiliary voltage v_{REF} that can be useful to shift the output DC level (Fig. 12.9a). Devices commonly used for BA are the AD620 (from Analog Devices[TM]) and the INA111 (from Texas Instruments[TM]). An IA can be also implemented with discrete OAs, for example, when special features (as operation with low power-supply voltages) are required. A classic and popular circuit is shown in Fig. 12.9b.

IAs and OAs as well, have some parameters that describe how much they part away from the "ideal model" we suppose assuming that Eq. (12.13) holds fully. We can mention a selection of importance to BA design: Besides the already discussed CMRR, the input

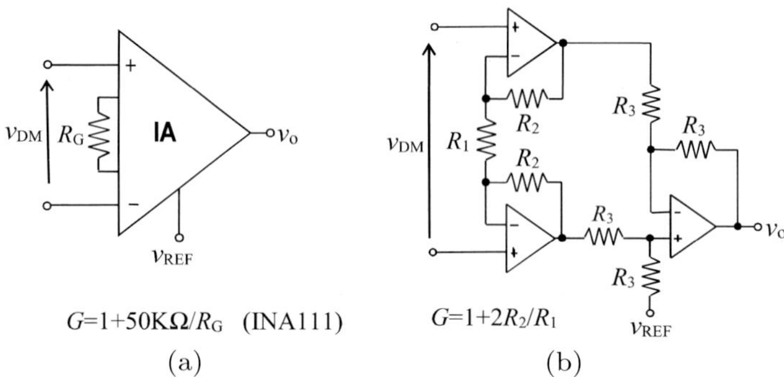

Figure 12.9: Instrumentation amplifiers: (a) Diagram of a commercial IA (INA111); (b) An IA implementation with discrete OAs.

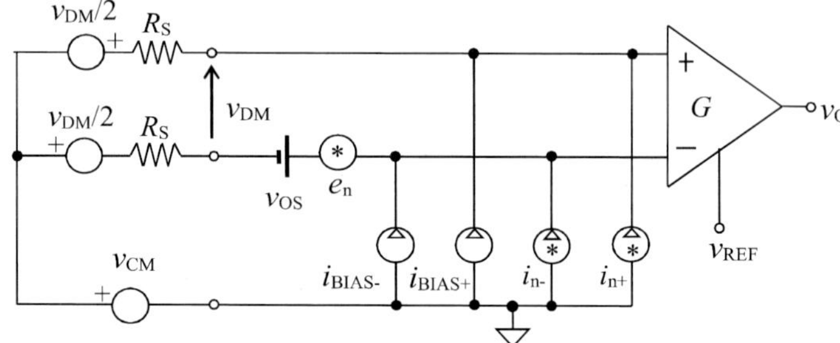

Figure 12.10: A realistic model of an IA. This model is also valid for OAs, omitting the input v_{REF}. For the sake of simplicity, this model assumes resistive source impedances, although we know that electrode impedances are a bit more complex!

offset voltage (v_{OS}), the input bias currents ($i_{\mathrm{BIAS}+}$, $i_{\mathrm{BIAS}-}$), and the voltage and current noise (e_n and i_n, respectively).

The effect of these "nonidealities" can be made apparent by adding the sources indicated in Fig. 12.10. The offset voltage can be modeled by a DC source v_{OS} at the amplifier's input, which for CMOS technology can be up to a few millivolt. The bias currents $i_{\mathrm{BIAS}+}$, $i_{\mathrm{BIAS}-}$, also associated with DC effects, have values up to a few pA for the same technology, and their contribution can be estimated by including two current sources, as indicated in Fig. 12.10. The amplifier noise sources are also represented in Fig. 12.10 by the voltage generator e_n and the current sources i_n. In general, this noise sources are specified by their spectral densities, and typical values are $10\ \mathrm{nV}/\sqrt{f}$ and $1\ \mathrm{fA}/\sqrt{f}$. Taking into account all these additional sources, the output of a real IA does not depends exclusively on v_{DM} and v_{REF} as assumed in Eq. (12.13), but also on v_{CM} and on the other parameters, as shown in,

$$v_{\mathrm{o}} = Gv_{\mathrm{DM}} + v_{\mathrm{REF}} + \mathrm{L}$$

$$\times \mathrm{L} + \frac{G}{\mathrm{CMRR}} v_{\mathrm{CM}} + G(v_{\mathrm{OS}} + (i_{\mathrm{BIAS+}} - i_{\mathrm{BIAS-}})R_{\mathrm{S}}) + G(e_{\mathrm{n}}$$

$$+ (i_{\mathrm{n+}} - i_{\mathrm{n-}})R_{\mathrm{S}}) \tag{12.14}$$

By means of careful design and component selection, it is possible to reduce the effect of the last three terms in Eq. (12.14) and neglect their contributions.

12.3.2 *Analog-to-digital conversion: The biopotential acquisition system*

The convenience and pervasiveness of digital systems have transformed BA design. Amplifiers must be codesigned with analog-to-digital converters (ADCs) to carry the information from the analog signal $s(t)$ to the quantized sequence $s(k)$, in a digital support. The BA characteristics established in the preceding section are catered to the need for interference rejection. An additional set of requirements emerge when the BA is designed to provide the signal conditioning that the ADC demands. The role of the BA is conditioning biopotentials to obtain the appropriate signals for the ADC's input.

Important features of data acquisition systems are their *resolution,* or the minimal variation we wish or we are able to detect, and their *span or measurement range.* The quotient of these parameters (*span/resolution*) defines a quality factor called the *dynamic range* (DR). It is dimensionless and usually expressed in decibels as,

$$\mathrm{DR} = 20 \log \left(\frac{\mathrm{span}}{\mathrm{resolution}} \right) \tag{12.15}$$

The DR can be used for both analog and digital stages. An ideal N-bit ADC has a span of $2^{N} - 1$ levels, and a resolution of 1 level represented by the least significant bit (LSB); thus, its *DR* is related to the number of bits N by,

$$\mathrm{DR} = 20\log(2^N) \approx 6 \cdot N \ [dB] \tag{12.16}$$

As a reference, a 12-bit ADC has a $DR = 72$ dB, resolving 1 part in 4096, whereas a 16-bit ADC shows a $DR = 96$ dB, resolving 1 part in 65536.

Let us consider a 12-bit ADC with a span of 0–5 V. It would have a resolution of approximately 1.2 mV. This means that some kind of signal conditioning is required for the ADC to be useful for acquiring biomedical signals with resolutions below 10 μV. An amplifier with a gain of 1.2 mV/10 μV = 120 times would be necessary. However, as shown in Section 12.1.1, the electrodes on the skin generate a DC offset potential v_E that can be as high as a few hundred millivolt superimposed to the signal. Assuming a power supply of ± 5 V, this potential is not adequate for an amplifier with a gain higher than around 10–15 times, lest it causes saturation.

There are two approaches to deal with v_E: The classic *AC-coupled* approach, where DC components are removed (or blocked before) they reach the ADC, and the modern *DC-coupled* approach, where offset potential and biosignals are digitized together, and low-frequency components are removed by digital signal processing. The requirements for the ADC depend on the signal-conditioning strategy adopted to manage the large DC electrode offset potential, which impacts on the number of bits and, therefore, on the ADC type.

An acceptable resolution for almost all biomedical applications is ± 1 μV$_\mathrm{PP}$, which is below the noise level of a good amplifier. For a DC-coupled system, the span is imposed by the DC component v_E, so it must be of almost ± 300 mV.[8] It is necessary to resolve 1 μV immersed in 300 mV (1 part in 300,000), leading to a DR of around 110 dB; thus, forcing the use of 18–19 bits ADCs (Berry *et al.*, 2009). On the other hand, for an AC-coupled system, a span of ± 5 mV is enough, DR reduces to 74 dB, and 12 bits ADCs can be used (Table 12.3).

---

[8]This value is required by biomedical equipment standards (AAMI, 1998).

Table 12.3: DR and number of bits.

Span	Resolution	Dynamic Range	# bits
± 5 mV	± 1 μv	74 dB	≈ 12 bits
± 300 mV	± 1 μv	110 dB	≈ 19 bits

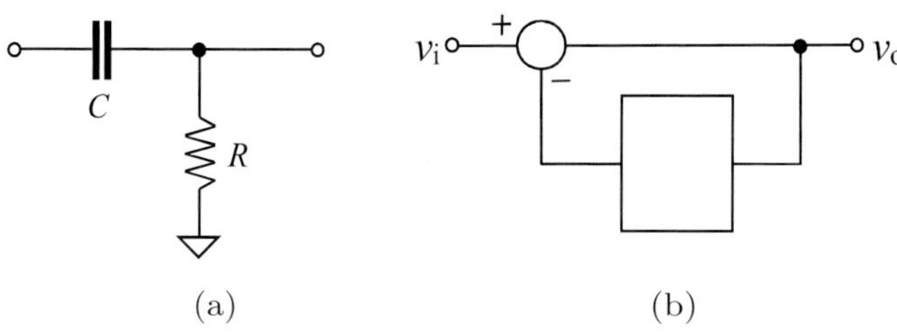

(a)   (b)

Figure 12.11: Two methods to reject a DC voltage: (a) blocking it with a capacitor, and (b) subtracting it using a feedback loop with an integrator.

12.3.3 *AC-coupled biopotential amplifiers*

The goal of rejecting low-frequency components can be achieved in two ways: By a passive network or by and active circuit. The first one consists of blocking DC components with a series capacitor (Fig. 12.11a), whereas in the active approach (Fig. 12.11b), the DC components present at the amplifier output are fed-back to its input by an integrator, so ensuring a null DC component at the output (Pallàs-Areny and Webster, 1999). In both cases, the transfer function verifies a first-order high-pass filter response given by Eq. (12.17),

$$T(s) = \frac{s}{\left(s + \frac{1}{\tau}\right)}; \quad \tau = RC \tag{12.17}$$

The lower cutoff frequency $f_{\mathrm{L}} = 1/(2\pi RC)$ is an important design parameter, for it defines the lowest frequency component that will be acquired, and also affects the transient response of the amplifier (Spinelli *et al.*, 2003) that must meet biomedical standards (AAMI, 1998).

The circuits in Fig. 12.11 show the main alternatives for AC-coupled BAs. They can be applied in an intermediate stage but not at the input, because a differential input is required. Figure 12.12a

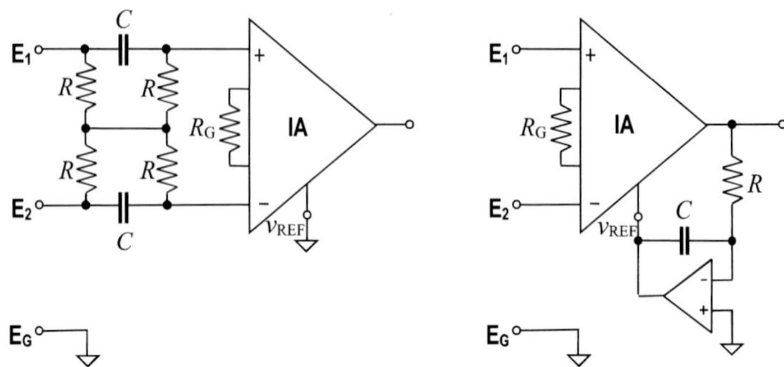

Figure 12.12: Two approaches to reject electrode offset voltages v_E: (a, or left) a passive network and (b, or right), a DC-suppression circuit implemented with an IA that subtracts the DC components through its auxiliary input v_{REF}. Both circuits have the same cutoff frequency $f_L = 1/(2\pi\ RC)$.

depicts a practical passive circuit with differential inputs, which also presents high CM input impedances Z_C, as a BA requires.

The passive circuit approach admits large differential DC potentials independently of the power supply voltage, but some residual DC potentials, due to amplifier offset and polarization currents (see Fig. 12.10), appear at the BA input and, therefore, are amplified, too. On the other hand, the DC suppression circuit of Fig. 12.12b ensures a null mean voltage at the output; however, the maximum admissible DC voltage at the input, $v_{\text{iDC.MAX}}$, depends on its power supply voltage V_{CC} and BA gain (Spinelli *et al.*, 2004) through Eq. (12.18),

$$v_{\text{iDC.MAX}} = \frac{V_{CC}}{G} \tag{12.18}$$

For example, for $v_{\text{iDC.MAX}} = \pm 300\,\text{mV}$ and $V_{CC} = \pm 5$ V, the gain must be lower than 16 times; far away from the $G \approx 1000$ required to amplify 1-mV biopotentials to 1 V for the ADC. This can be easily solved including additional amplification stages. The two AC-coupling approaches can be combined, as it will be show later in Section 12.3.7.

12.3.4 *DC-coupled biopotential amplifier*

The DC-coupled BA is a modern approach. The idea is simple: Electrode offset and biopotentials are amplified and digitized together,

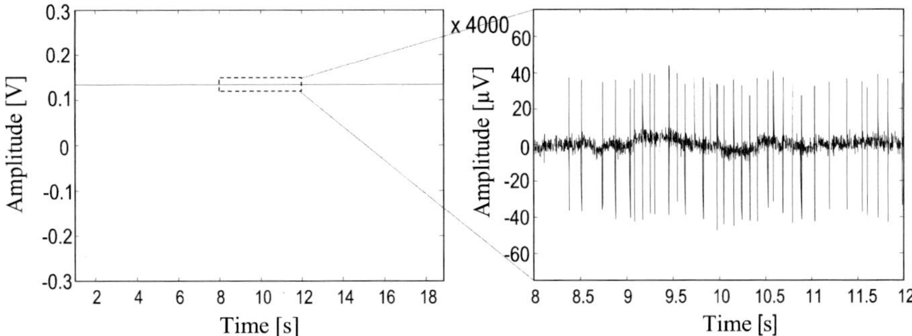

Figure 12.13: DC electrode offset is several orders of magnitude larger than the amplitude of biopotential signals. A 4000-fold magnification only allows to encompass the full range of the muscle activation EMG signals in the figure. This plot also shows the huge resolution of the current 24-bit sigma-delta ADCs!

but the ADC must be able to resolve about ± 1 μV within ± 300 mV, thus implying a DR of 19 bits and above (Fig. 12.13). Such constrains can be met at a reasonable cost by the current 24-bit sigma delta ADCs. These high-resolution ADCs have differential inputs, so for best performance the BA must provide a differential output, and the previously presented general purpose IA is not adequate. A simple circuit with differential input for the measurement of electrodes and differential output to connect to the ADC's input is displayed in Fig. 12.14. Its gain is $G = 1 + 2R_2/R_1$. A gain of 10 is enough to achieve the necessary resolution while the ± 300 mV offset does not exceed ± 3 V at the amplifier's output. The current tendency is toward DC-coupled amplifiers, and semiconductor companies produce ICs with this technology to provide a solution for the complete biopotential acquisition problem. For example, the ADS1298 IC, from Texas Instruments (2010), has a complete 8-channel ECG system, including amplifiers and one 24-bit sigma-delta converter per channel. It only requires connecting the patient leads, electrodes and the digital communication to a microcontroller. This kind of integrated system is likely to become a usual module in biopotential acquisition equipment, as IAs and op-amps are nowadays.

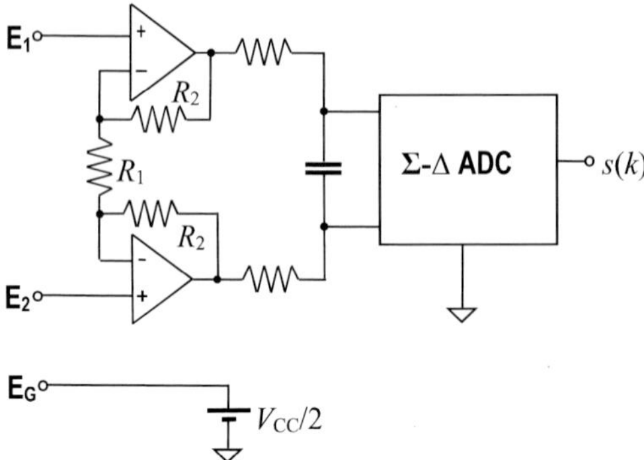

Figure 12.14: A biopotential amplifier with differential input and differential output (FD: fully differential) that also includes an FD low-pass filter. A DC voltage $V_{CC}/2$ was included to ensure a CMV at the ADC input suited to its power supply rails (in general, 0–5 V or 0–3.3 V). Note that the circuit is really simple, as many processing functions have now been moved to the digital domain.

12.3.5 *Antialiasing filter*

It is well known that before digitizing a signal, at sampling rate f_S, it is necessary to limit its BW to $f_S/2$ in order to avoid the aliasing effect (Oppenheim *et al.*, 1997).[9]

Therefore, a low-pass filter must be included in order to reduce signal BW. The filter also reduces noise BW; hence, as a general rule, no more BW than the necessary should be provided. AC-coupled amplifiers usually work with sampling frequencies up to a few kHz. For these applications, active second-order filters are appropriate. When the DC-coupled approach is used, filter constrains are less strict. Sigma-delta ADCs work at very high sampling rates, up to a few megahertz, so there is an almost two-decade long frequency range for the filter roll-off zone. Simple single-pole passive filters may suffice in this case. Remember that the steeper the roll-off, the higher the "Q" or quality factor of the filter.

---

[9]Recall that **aliasing** is an effect that causes different signals to become indistinguishable (or *aliases* of one another) when sampled. It also refers to the distortion, or artifact, which results when the signal reconstructed from samples is different from the original continuous signal.

12.3.6 *Active solutions for PL interference reduction*

As described in Section 12.2, a biopotential amplifier tries to prevent the CM signal v_{CM} produced by the PL to interfere with the differential biopotential signal, but v_{CM} can produce DMV before it reaches the BA (see Section 12.2.4). One appealing solution is to reduce interference using active circuits; for example, the driven-right-leg (DRL) circuit to reduce v_{CM} (Winter and Webster, 1983). Other problems find mitigation with active circuits: Active electrodes can avoid the effects of PL coupling to the patient leads.

12.3.6.1 *DRL circuit*

The DRL circuit senses the CM potential v_{CM}, compares it with a reference value (usually the ground potential or a DC voltage), and amplifies their difference, which is then injected as feedback to the patient through the ground electrode Z_G. In ECG measurements, this electrode is placed on the right leg, thus naming this circuit. The voltage v_{CM} can be estimated averaging out many signal channels (Fig. 12.15b) or with a specific electrode (Fig. 12.15a). This latter alternative is usual in systems with a large number of channels.

The effect the DRL circuit has, reducing v_{CM}, is equivalent to a reduction of the ground electrode impedance at f_{PL} (50/60 Hz) by

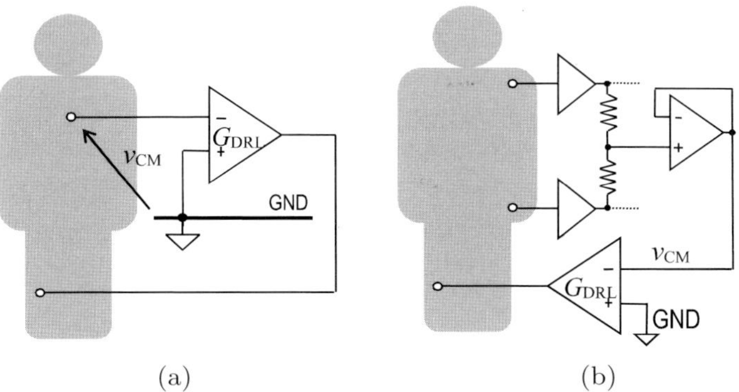

Figure 12.15: Two approaches for the DRL loop implementation. In (a), a single, independent electrode is used to measure the v_{CM} voltage, whereas in (b) v_{CM} is obtained by averaging out the electrodes' potentials.

a factor equal to the gain the DRL has at this frequency (Metting Van Rijn *et al.*, 1990b). Note that in this way we achieve the same as abrading the skin, but without any patient discomfort. A high DRL gain at f_{PL} is desirable, but limited by the loop stability. The dynamics introduced by the electrodes' impedances and the active components transferences need to be compensated for to assure circuit stability. The classic solution is to implement the DRL amplification as an integrator with a gain of around 30 dB at f_{PL} (Winter and Webster, 1983).

12.3.6.2 *Active electrodes*

A clear manifestation of the tendency to the increasingly pervasive use of active techniques is the use of active electrodes. The basic active electrode consists in just a unity gain buffer that is included inside the electrode itself (Fig. 12.16). In this way, displacement currents coupled to the patient leads do not flow through the electrode impedances; they are driven to the power supply's ground by the low-output impedance of the buffer and do not produce a significant potential difference.

Active electrodes make the biopotential acquisition system very robust against PL interference, but there is a cost to pay: The buffer

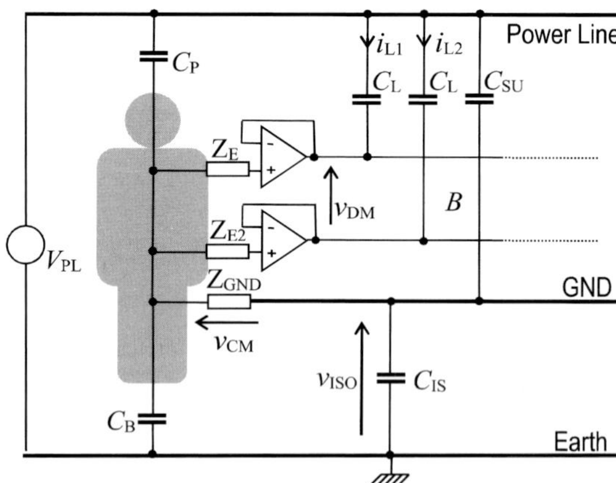

Figure 12.16: Active electrodes solve the problem of interference caused by PL coupling to the patient leads.

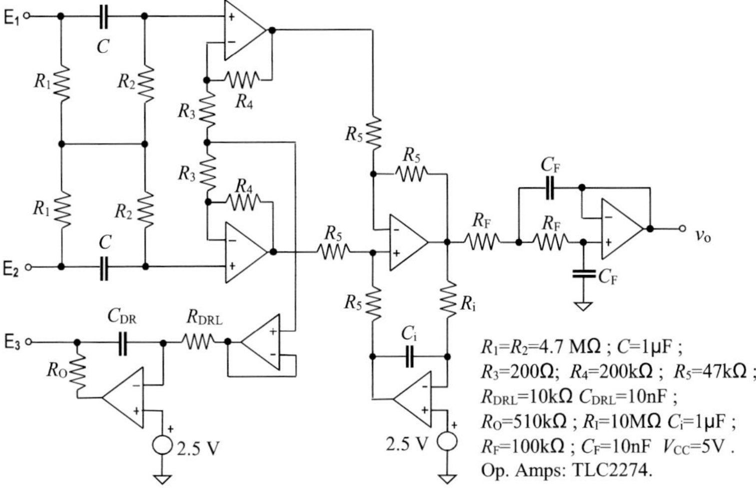

Figure 12.17: A simple but complete AC-coupled biopotential amplifier for single-supply operation.

inside the electrode needs power to work; hence, additional wires are needed to provide it. However, since no shielded cables are required, low-cost flexible ribbon cables are also adequate.

12.3.7 *A complete biopotential amplifier*

The previously described techniques can be integrated in a practical biopotential amplifier as the one showed in Fig. 12.17. It uses a passive AC-coupling network and a second active DC-suppression stage to remove DC voltages produced by the amplifier's offset voltage and bias currents effects. For the IA, a CMOS (low i_{BIAS}) IC must be used and, for a single supply operation, the IA can be implemented by rail-to-rail OAs, as described by Spinelli *et al.* (2003). The circuit includes a low-pass filter to limit BW (Section 12.3.5) and a DRL circuit (Section 12.3.6) that estimates v_{CM} by averaging. In this circuit, v_{CM} is not compared against the GND potential but to a DC voltage, thus driving the patient's potential to a DC offset in order to allow single supply operation (0–5 V). This is another useful function of the DRL circuit: To shift the DC-patient potential.

The circuit of Fig. 12.17 combines the two AC-coupling approaches depicted in Section 12.3.3. A passive AC-coupling circuit at the input to reject electrode offset potentials, and a subsequent active AC-coupling scheme that rejects OA DC effects, such as OA's offset voltage and bias currents.

12.3.8 *Multichannel acquisition systems*

The previously described circuits are intended for *bipolar* channels, for which the biopotential signal of interest is defined as the difference between two electrodes (Fig. 12.18a). Another possibility, generally adopted for systems with a large number of channels, is the *monopolar* topology. In this case, all channels are referred to the same electrode, as shown in Fig. 12.18b. In this figure, it can be observed that the reference electrode potential (E_{REF}) is common to all channels, thus reducing the CM impedance Z_C seen by E_{REF} (all the amplifiers' input impedances appear in parallel). The usual solution is to connect the reference electrode to a unity gain buffer (Fig. 12.18c). In this way, all the channels present the same input impedance Z_{C}.

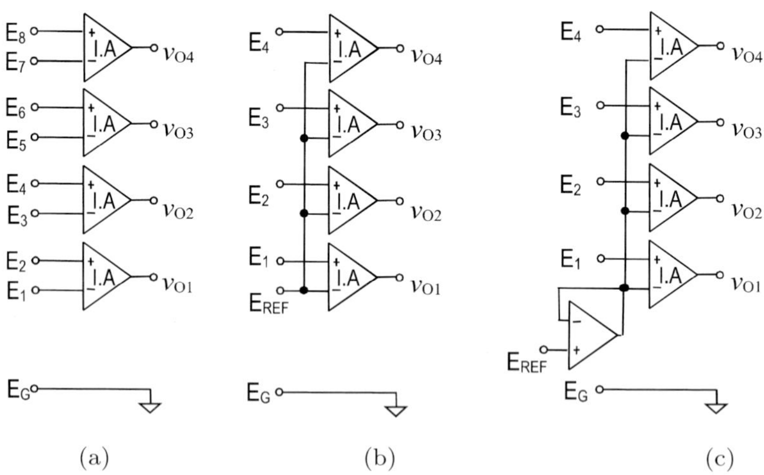

Figure 12.18: Usual topologies for BA. (a) Bipolar, (b) and (c) Monopolar. Network (c) provides the same CM input impedance Z_C for all channels, thus equalizing input impedances.

12.4 The Isolation Barrier

According to biomedical standards, the patient must be electrically isolated from earth and, obviously, from the PL. Mostly, the equipment that receives the biosignals for visualizing, processing, storing them or else (such as a PC), is powered by the PL, and its case is connected to earth through the power cord. Hence, an isolation barrier must be established between the BA and the final equipment, defining the "patient side" and the "grounded side" (Fig. 12.19). Signals and power supply must be transferred through this barrier.

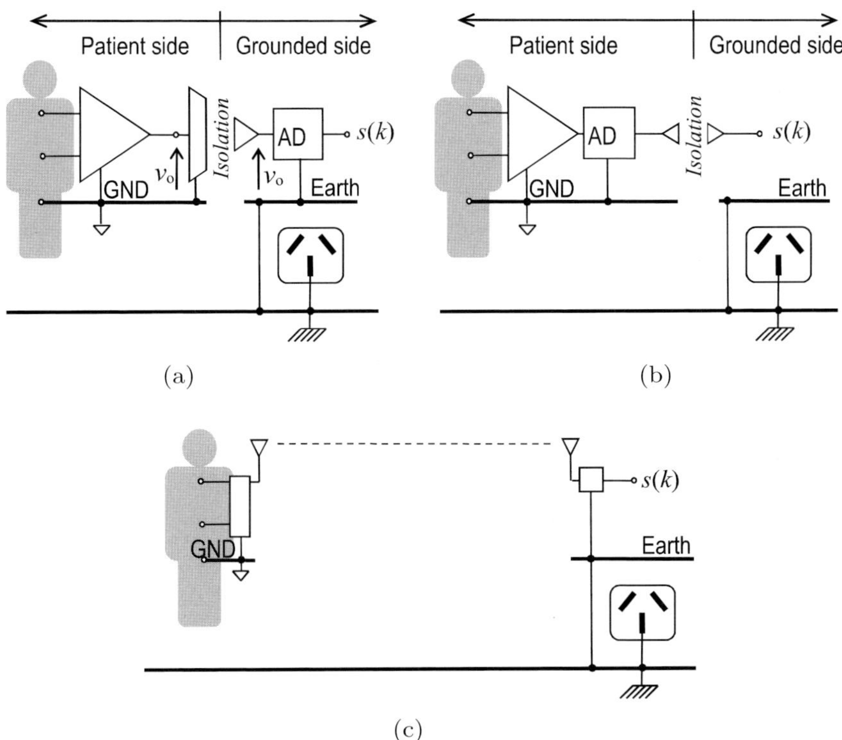

Figure 12.19: Different isolation barrier alternatives. (a) Isolation amplifier with a barrier in the analog signal path; (b) moves the isolation solution to a point where the signal has been transferred to the digital domain; and (c) shows a third option relying entirely on a battery-operated wireless device.

Power is transferred by the so-called medical-grade isolation transformers, sometimes with two or more isolation barriers. Biomedical signals may be carried to the grounded side by an isolation amplifier as in many traditional designs (Fig. 12.19a), but nowadays the ADC is placed on the patient side and signals cross the barrier through less expensive digital isolators (Fig. 12.19b). There are ICs, like the ADuM family from Analog Devices^{TM}, that provide a complete isolation solution both for power and signal transfer. An attractive alternative for small devices is to power them with batteries and relay the data through a wireless link Fig. 12.19c. This approach, because of its ultra-high isolation, results in low C_{ISO} capacitance values and so, as it was described in Section 12.2.3, in low v_{CM} voltages, thus leading to systems with high immunity to PL interference.

12.5 Toward Zero Invasiveness: Capacitively Coupled Electrodes

Biomedical instrumentation engineers must work to measure good quality signals minimizing invasiveness and patient discomfort. This goal has resulted in an instrumentation trend toward electrodes that do not require skin preparation or the use of conductive gel.

Active electrodes, described in Section 12.3.6, can function without the addition of gel (*dry electrodes*), but the "zero invasiveness" challenge is to capture biopotentials without metallic contact with the skin and without any ionic exchange. This can be achieved by using CEs that pick-up biopotentials capacitively through a dielectric film or even clothes (Fig. 12.20a).

The concept behind this technology is really simple: It is just an AC-coupled amplifier (Fig. 12.20b). However, the coupling capacitor is very small, around 100 pF and as low as 10 pF when biopotentials are acquired through clothing (e.g., a T-shirt). These small coupling capacitances demand an ultra-high bias resistor R_{BIAS} to fulfill the lower cutoff frequency required in biomedical applications, calling for R_{BIAS} in the range of tera-ohms (1 TΩ = $10^{12}\Omega$). The resistor R_{BIAS} cannot be omitted because it provides a path for the amplifier bias currents, which would otherwise be integrated by C_{S}, leading

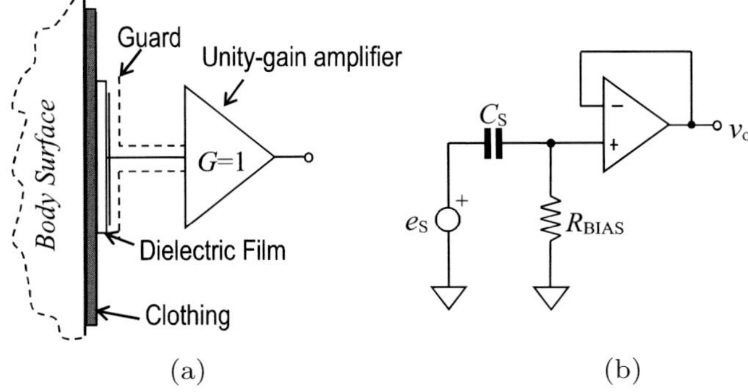

Figure 12.20: (a) General scheme of a CE to acquire biopotentials through clothing, and (b) its equivalent circuit. Note that it reduces to a simple AC-coupled stage.

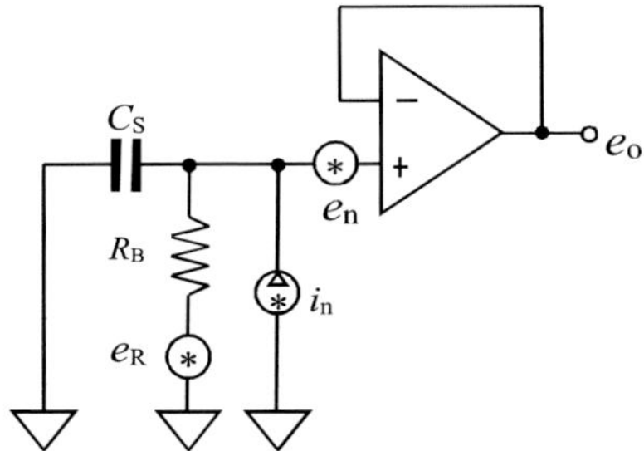

Figure 12.21: Simplified schematic of the front-end circuit including the main noise sources: e_R and R_B stand for the noise source, whereas e_n and i_n correspond to the amplifier voltage and current noises, respectively.

to electrode saturation. The ultra-high impedance values involved in this technology make it vulnerable to electric-field interference, and active electrodes must be used. The front-end circuitry is included inside the electrode itself (Chi *et al.*, 2012).

12.5.1 *Noise analysis*

A simplified scheme of a front end for CEs, including the main noise sources, is shown in Fig. 12.21. Clothes, when working as dielectric layer, also contribute with noise (Chi *et al.*, 2010), but this falls beyond the electronic design domain. The resistor noise PSD is denoted as e_R, whereas e_n and i_n represent, respectively, the voltage and the current noise PSD of the amplifier. Solving this circuit, the

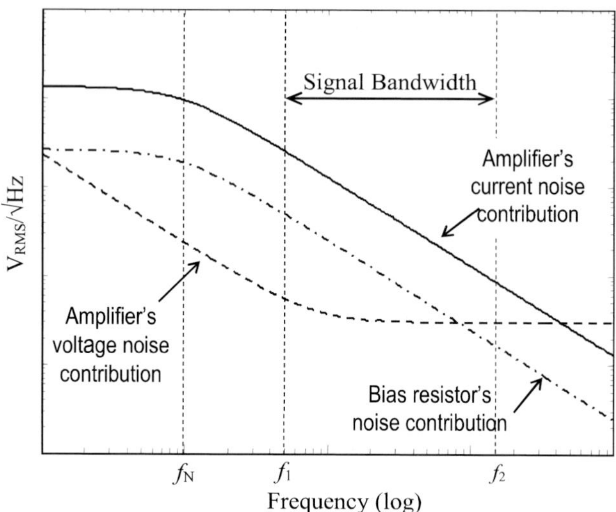

Figure 12.22: Spectral characteristics of CEs' noise sources.

output noise PSD e_o results in

$$e_o^2 = e_R^2 \frac{1}{1 + (f/f_N)^2} + i_n^2 \frac{R_B^2}{1 + (f/f_N)^2} + e_n^2 \tag{12.19}$$

where f_N is the -3-db noise cutoff frequency, defined as,

$$f_N = \frac{1}{2\pi R_B C_S}. \tag{12.20}$$

A typical curve of the overall output noise and the contributions of each term in Eq. (12.19) is shown in Fig. 12.22.

The low-frequency noise values e_R and $i_n R_B$ are very large, around 100 μV/$\sqrt{\text{Hz}}$ for $R = 1$ TΩ, and i_n as low as 0.1 fA/$\sqrt{\text{Hz}}$ (an amplifier for wet electrodes has around 100 nV/$\sqrt{\text{Hz}}$, at low frequencies). In order to reduce these noise contributions to acceptable values, the noise cutoff frequency f_N must be considerably below the signal BW. In this frequency range, where noise PSD decays with f (see Eq. 12.19), the noise voltage can be approximated by,

$$e_o^2 \approx e_R^2 \left(\frac{f_N}{f}\right)^2 + i_n^2 R_B^2 \left(\frac{f_N}{f}\right)^2 + e_n^2. \tag{12.21}$$

Replacing the resistor noise e_R with the Nyquist expression ($e_R^2 = 4\mathrm{k}TR_\mathrm{B}$) and f_N by Eq. (12.20), leads to,

$$e_\mathrm{o}^2 \approx \frac{kT}{(\pi C_\mathrm{S}f)^2}\frac{1}{R_\mathrm{B}} + \frac{i_\mathrm{n}^2}{(2\pi C_\mathrm{S}f)^2} + e_\mathrm{n}^2 \qquad (12.22)$$

$$e_\mathrm{o}^2 \approx \frac{1}{(\pi C_\mathrm{S}f)^2}\left(\frac{kT}{R_\mathrm{B}} + \frac{i_\mathrm{n}^2}{4}\right) + e_\mathrm{n}^2 \qquad (12.23)$$

As observed in Eq. (12.23), noise reduces with C_S, mostly with the electrode area. Therefore, noise can be reduced by increasing electrode size, as it happens when wet electrodes are used (Huigen *et al.*, 2002). Expression (12.23) corresponds to the output noise PSD. The root mean square (RMS) noise of the front end can be obtained by integrating Eq. (12.23) over a given bandwidth BW $= f_1 - f_2$ as,

$$E_\mathrm{o}^2 = \int_{f_1}^{f_2}\left(\frac{kT}{(\pi C_\mathrm{S}f)^2}\frac{1}{R_\mathrm{B}} + \frac{i_\mathrm{n}^2}{(2\pi C_\mathrm{S}f)^2} + e_\mathrm{n}^2\right)df \qquad (12.24)$$

Considering that BW usually covers few decades ($f_2 \gg f_1$), Eq. (12.24) becomes,

$$E_\mathrm{o}^2 \approx \frac{kT}{(\pi C_\mathrm{S})^2}\frac{1}{R_\mathrm{B}f_1} + \frac{i_\mathrm{n}^2}{(2\pi C_\mathrm{S})^2}\frac{1}{f_1} + E_\mathrm{n}^2, \qquad (12.25)$$

where f_1 is the signal lower cutoff frequency. It is important to note that output noise E_o does not depend on the signal BW but on f_1, which clearly shows that noise can be reduced by increasing the low cutoff frequency f_1, a usual practice in CEs (Lim *et al.*, 2006; Baek *et al.*, 2008). Cutoff frequencies up to 8 Hz have been used for heart rate detectors (Wu and Zhang, 2008), but f_1 must be in the order of tenths of Hz if one aims at meeting medical standards, and thus keeping the same diagnostic capacity of wet electrodes (AAMI, 1998).

12.5.2 *PL interference rejection*

As it was stated previously, CEs are very susceptible to PL interference due to the ultra-high impedances involved (Kim and Park,

2008). In order to reduce interference to acceptable levels, the front-end circuit must be placed in the electrode itself. This technique, previously depicted as "active electrodes" and used to improve interference rejection in wet electrodes, is mandatory for CEs. Once the electrode containing the front end is properly shielded, the main interference contribution that remains is due to the v_{CM} imposed by the PL, because of the already depicted "potential divider effect". The CM potential produces a DMV v_{iD} as a result of imbalanced coupling capacitances of different channels, as it happens with standard wet electrodes in the presence of electrode impedance imbalance. In the CE case, expression (12.12) becomes,

$$v_{\mathrm{DM}} \approx v_{\mathrm{CM}} \frac{C_{\mathrm{in}}}{C_{\mathrm{S}}} \frac{\Delta C_{\mathrm{S}}}{C_{\mathrm{S}}} \tag{12.26}$$

where $\Delta C_{\mathrm{S}} = C_{\mathrm{S1}} - C_{\mathrm{S2}}$ denotes the imbalance between the coupling capacitance of each channel and C_{in} is the amplifier's input capacitance. Hence, to reduce this interference, the v_{CM} and the input capacitance C_{in} should be made as low as possible. Coupling capacitance C_{S} imposes the measurement conditions and it cannot be managed by design. The input capacitance C_{in} is imposed by the amplifier, but it can be reduced using *neutralization*, which is a technique that reduces the input capacitance by using a positive feedback scheme (Prance *et al.*, 2000). To reduce v_{CM}, a capacitive version of the DRL circuit can be used (Lim *et al.*, 2010).

As with C_{in}, special care must be taken with all the stray capacitances, and carefully guarding techniques must be used. A typical CE contains DC basic circuit of Fig. 12.20, plus accessory circuits to neutralize C_{in} and to drive guards. A simplified schematic is shown in Fig. 12.23. More details and a noise analysis of this technology can be found in Spinelli and Haberman (2010) and Spinelli *et al.* (2012).

Figure 12.24 shows ECG signals acquired simultaneously by CE placed on the chest over a cotton T-shirt, and by standard wet electrodes (Spinelli and Haberman, 2010). There are no significant differences between ECG signals while the patient is in apnea, but CEs are very vulnerable to motion artifacts, for example, like breathing.

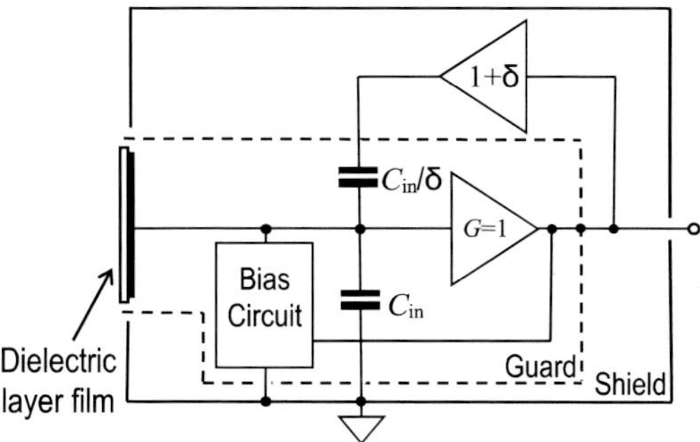

Figure 12.23: Block diagram of the front end used to obtain the experimental data.

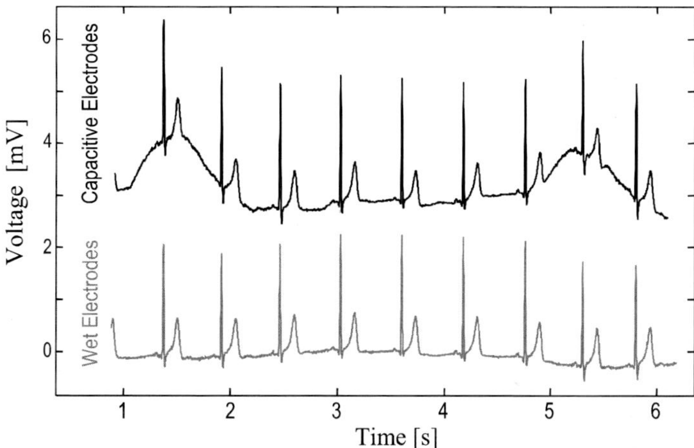

Figure 12.24: Two differential channels were taken at same time in the same location: One using CEs through cotton clothes (signal above) and the other by wet electrodes (signal below). No significant differences are seen on the ECG signal, but CEs are more vulnerable to motion artifacts, like breathing.

12.6 Conclusions

Nowadays, a few commercial ICs can provide a solution for the complete biopotential acquisition signal path, including the isolation barrier. Although these off-the-shelve solutions are very useful and appropriate for the most common situations, special demands such as restrictions in power supply voltage or nonstandard biopotential signals require custom-tailored biopotential amplifiers.

As a consequence, it is important to know, and be able to model, the pathways through which PL interference victimizes our biopotential acquisition system. Moreover, biopotential amplifiers for new applications with challenging constrains will always need to be designed. In other words, technology advances and provides solutions for existing problems, but new problems arise and it is important to know them and analyze their possible solutions. We hope to have contributed to this end.

References

AAMI. ANSI/AAMI EC38: American National Standard for Ambulatory Electro-cardiographs. Association for the Advancement of Medical Instrumentation, Arlington VA, 1998.

Baek H, Kim J, Kim K and Park, K. System for unconstrained ECG measurement on a toilet seat using capacitive coupled electrodes: The efficacy and practicality. *Proceedings of the 30th Annual International Conference IEEE Engineering in Medicine and Biology Society*, pp. 2326–2328, Vancouver, BC, 2008.

Berry D, Duignan, F, Hayes, R. An Investigation of the use of a high resolution ADC as a digital biopotential amplifier. *4th European Conference International Federation Medical and Biological Engineering* 22:142–147, 2009.

Chi Y, Jung T, Cauwenberghs G. Dry-contact and noncontact biopotential electrodes: Methodological review. *IEEE Rev Biomed Eng* 3:106–119, 2010.

Chi Y, Maier C, Cauwenberghs G. Ultra-high input impedance, low noise integrated amplifier for noncontact biopotential sensing. *IEEE J Emerg Sel Topics Circuits Syst* 1:526–535, 2012.

De Jager P, Peper A, Metting Van Rijn A, Grimbergen C. High frequency effects in amplifiers for biopotential recordings. *Proceedings of the 18th Annual International Conference of the Engineering in Medicine and Biology Society IEEE*, pp. 26–27, 1996.

Haberman M, Cassino A, Spinelli E. Estimation of stray coupling capacitances in biopotential measurements. *Med Biol Eng Comput* 49:1067–1071, 2011.

Haberman M, Spinelli E. A multichannel EEG acquisition scheme based on single ended amplifiers and digital DRL. *IEEE Trans Biomed Circuits Syst* 6:614–618, 2012.

Horowitz, P, Hill W, Hayes T. The art of electronics. Cambridge University Press, Cambridge, 1125 pp., 1989. ISBN: 0-521-37095-7.

Huhta, JC, Webster J. 60-Hz interference in electrocardiography. *IEEE Trans Biomed Eng* 2:91–101, 1973.

Huigen, E, Peper A, Grimbergen C. Investigation into the origin of the noise of surface electrodes. *Med Biol Eng Comput* 40:332–338, 2002.

Kim K, Park K. Effective coupling impedance for power line interference in capacitive-coupled ECG measurement system. *Proc. International Conference on Technology and Applications in Biomedicine ITAB2008*, pp. 256–258, 2008.

Levkov C, Mihov G, Ivanov R, Daskalov I, Christov I, Dotsinsky I. Removal of power-line interference from the ECG: A review of the subtraction procedure. *BioMed Eng OnLine* 4, 50, 2005.

Lim Y, Chung G, Park K. Capacitive driven-right-leg grounding in indirect-contact ECG measurement. *Proceedings 32th Annual International Conference IEEE Engineering in Medicine and Biology Society*, pp. 1250–1253, 2010.

Lim Y, Kim K, Park K. ECG measurement on a chair without conductive contact. *IEEE Trans Biomed Eng* 53:956–959, 2006.

Metting van Rijn A, Peper A, Grimbergen C. High-quality recording of bioelectric events: Part 1. Interference reduction, theory and practice. *Med Biol Eng Comput* 28:389–397, 1990a.

Metting van Rijn A, Peper A, Grimbergen C. High-quality recording of bioelectric events: Part II: A low-noise low-power multichannel amplifier design. *Med Biol Eng Comput* 29:433–440, 1990b.

Metting van Rijn A, Peper A, Grimbergen C. The isolation mode rejection ratio in bioelectric amplifiers. *IEEE Trans Biomed Eng* 38:1154–1157, 1991.

Oppenheim A, Willsky A, Nawab S. Signal and systems. 2nd ed. Prentice-Hall, New Jersey, 1997. ISBN 0-138-14757-4.

Ott H. Noise reduction techniques in electronic systems. 2nd ed. John Willey & Sons, New York, (xx + 426) pp., 1988. ISBN 0-471-85068-3.

Pallàs-Areny R, Colominas J. Differential mode interference in biopotential amplifiers. *Proceedings 11th Annual International Conference IEEE Engineering in Medicine & Biology Society*, pp.1721–1722, 1989.

Pallàs-Areny R, Webster J. Analog signal processing. John Willey & Sons, New York, (xv + 586) pp., 1999. ISBN 0-471-12528-8.

Plonsey R, Barr R. Bioelectricity: A quantitative approach. Springer, New York, 305, 2007. ISBN 978-0-387-48865-3.

Prance R, Debray A, Clark T, Prance H, Nock M, Harland C, Clippingdale A. An ultra-low-noise electrical-potential probe for human-body scanning. *Meas Sci Technol* 11:291–297, 2000.

Rossel J, Colominas J, Riu P, Pallàs-Areny R, Webster J. Skin impedance from 1 Hz to 1 MHz. *IEEE Trans Biomed Eng* 35:649–651, 1998.

Spinelli E, Haberman M. Insulating electrodes: A review on biopotential front-ends for dielectric skin-electrode interfaces. *Physiol Meas* 31:183–198, 2010.

Spinelli E, Haberman M, García P, Guerrero F. A capacitive electrode with fast recovery feature. *Physiol Meas* 33:1277–1288, 2012.

Spinelli E, Martinez N, Mayosky M, Pallàs-Areny R. A novel fully-differential biopotential amplifier with DC suppression. *IEEE Trans Biomed Eng* 51:1444–1448, 2004.

Spinelli E, Mayosky M. Two-electrode biopotential measurements: Power line interference analysis. *IEEE Trans Biomed Eng* 52:1436–1442, 2005.

Spinelli E, Pallàs-Areny R, Mayosky M. AC-coupled front-end for biopotential measurements. *IEEE Trans Biomed Eng* 50:391–395, 2003.

Valentinuzzi ME. Understanding the human machine: A primer to bioengineering. World Scientific Publisher, New Jersey & Singapore, (xii + 396) pp. 8 chapters. Series on Bioengineering & Biomed Engineering, vol 4, 2004. ISBN 981-238-930-X & ISBN 981-256-043-2.

Van der Horst M, Metting van Rijn A, Peper A, Grimbergen C. High frequency effects in amplifiers for biopotential recordings. *Proceedings 20th Annual International Conference Engineering in Medicine and Biology Society IEEE*, pp. 3309–3312, 1998.

Winter B, Webster J. Driven right-leg circuit design. *IEEE Trans Biomed Eng* 30:62–66, 1983.

Wood D, Ewins D, Balachandran W. Comparative analysis of power line interference between two- or tree-electrode biopotential amplifiers. *Med Biol Eng Comput* 33:63–68, 1995.

Wu K, Zhang Y. Contactless and continuous monitoring of heart electric activities through clothes on a sleeping bed. *Proceedings International Conference on Technology and Applications in Biomedicine ITAB2008*, pp. 282–285, 2008.

CHAPTER 13

MATHEMATICAL MODELS IN BIOENGINEERING

Max E. Valentinuzzi and Pedro Arini

Nessuna umana investigazione si puo dimandare vera scienza s'essa non passa per le matematiche dimostrazione. Leonardo da Vinci (1452–1519).

Theoretical studies in biology and physiology may sound weird, especially to the young student, as they frequently involve appalling oversimplifications. However, mathematical physics also makes use of amazing oversimplifications. The works of Maxwell and Einstein once were considered purely speculative. Yet it was Maxwell's work that made our present day electrical industry possible. It was Einstein's speculations about four-dimensional space-time that made the atomic age possible. The thousands of ungenerous patents that followed them, mostly buried in oblivion and many giving healthy profits to their inventors and more comfortable life to the users, represent the practical arm. Biology as yet has not had its Maxwell or its Einstein ... but they may come ... perhaps sooner than expected ... perhaps they are in front of these modest lines.

Nicholas Rashevsky (1964)

The muscles of mathematics are connected to the bones of experimental science by the tendons of mathematical modeling. Anonymous.

However ... mathematics does not really exist, for it is a creation of the Human Mind, and in that respect it approaches a Supreme Idea, as some kind of Divine Enlightenment.

Abstract

We intend here to offer an overview of how biology and medicine entered into a process of mathematization, starting with Leonardo until reaching our current days, pointing out some outstanding areas without forgetting essential concepts that remain as theoretical columns or pillars.

13.1 Introduction

One objective of bioengineering is the quantification of the biological and medical sciences, with the goal of improving their exactness and preciseness, always in an attempt to remove as many indeterminations and uncertainties as possible, especially when seeking predictions. The extract at the beginning of the column, which is attributed to Leonardo da Vinci, seems to reflect a very early Renaissance concept in such a direction; this concept is indeed more general because it encompasses all the sciences. The famous Leonardo's Vitruvian Man, so many times reproduced on book covers and posters, appear perhaps as an anticipatory geometrical epitome of that modern recent aim. However, 200 years before Leonardo, Ramón Llull (c.a. 1232–1315), a Majorcan philosopher, member of the Third Order of Saint Francis, and author of important works of Catalan literature, came up with contributions to logic that made him a candidate as pioneer of computation theory. He might also be considered a very early predecessor of biomathematics (Valentinuzzi and Kohen, 2013).

13.2 Early Attempts

After Llull and Leonardo, and more significant in terms of actual influence, Giovanni Alfonso Borelli (1608–1679), an Italian physicist and mathematician, made significant contributions to medicine. In the city of Pisa, Italy, he met Marcello Malpighi (1628–1694), a physician who encouraged him to get into medical subjects (by the way, Malpighi is well known for the discovery and description of the capillaries, thus completing what William Harvey (1578–1657) had anticipated regarding the blood pathway as a closed circuit. As a result, and after studies and experimentation, Borelli came up with an opera magna, *De Motu Animalium* (*On Animal Movement*), which was published two years after his death. The book deals with mechanics and mathematics as applied to medicine. He used the term *iatromechanics* (*iatros* is the ancient Greek word for physician). This root was later used by the medieval medical alchemists who called their practice iatrochemistry, meaning the chemistry of healing (perhaps it should be considered an ancestor of biochemistry). Thus, *iatromechanics* would be the predecessor of our current biomechanics. Somehow, Borelli followed the mechanistic philosophy of René Descartes (1596–1650), that is, believing that living things behave like machines. Descartes sustained the Pitagoric and Platonic ideas according to which the visible world was a mere illusion hiding the mathematical reality of things. For Descartes, then, everything that exists is *res extensa* (extended thing or. corporeal substance), that is, matter amenable of mathematical measurement and known with certainty by the human mind.

Philosophers of science call such a view a *reductionist approach*, that is, attempting to reduce complex biological systems to simpler ones (such as mechanical, hydraulic, or electrical systems).

Perhaps other examples could be found, like the simple numerical thinking of William Harvey, who on the basis of animal experimentation, asserted that blood moves in a closed circuit, and accepting this fact as a biomathematical achievement. Luigi Galvani (1737–1798), an experimental biophysicist and the discoverer of animal electricity, might be included as a forefather of biomathematics for many were the mathematical developments that much later explored electrophysiology. Philosophically, the contributions made by Carl Ludwig's school during the 19th century must be recalled, as well as that of several of his followers and disciples (Valentinuzzi *et al.*, 2012). They certainly gave a significant impulse to the quantification of physiology and medicine; however, the true mathematical model of a biological system had to await more maturation.

13.3 The 20th Century: Verhulst, Lotka, Volterra, Kostitzin, and van der Pol

Perhaps the first contribution starting the mathematization of biology belongs to Pierre Francois Verhulst (1804–1849), who introduced

         M. E. Valentinuzzi & P. Arini

the following equation (Verhulst, 1838),

$$\frac{dN}{dt} = rN\left(1 - \frac{N}{K}\right) \tag{13.1}$$

where $N(t)$ represents the number of individuals at time t, r is the intrinsic growth rate, and K is the carrying capacity or the maximum number of individuals an environment can support. He called its solution the *logistic function* (name still kept), that is,

$$N(t) = \frac{K}{1 + CKe^{-rt}} \tag{13.2}$$

In the latter Equation (13.2), the constant

$$C = \frac{1}{N(0)} - \frac{1}{K} \tag{13.3}$$

is determined by the initial condition $N(0)$. Bacteriologists use daily the logistic equation because it describes the growth of bacteria in an adequate culture broth. Figure 13.1 depicts an experimental bacterial growth curve obtained by the impedance method, which essentially follows the logistic equation and can be easily adjusted to it (Felice *et al.*, 1988).

However, the systematic mathematization of biology with its eventual extension to medicine can be credited to Alfred James Lotka (1880–1949). He is the author of a number of theoretical articles on chemical oscillations during the early decades of the 20th century and of the first book on theoretical biology (Lotka, 1925). This early first book, full of philosophical–epistemological digressions and not an easy read, indeed, goes into difficult areas such as consciousness and the concept of life itself. Lotka is best known for the predator–prey model, now called the Lotka-Volterra model, which he developed independently from Vito Volterra (1860–1940). This model is still the basis of many others, as, for example, in such odd fields as the re-engineering of corporations (Modis, 1997). Vito Volterra, 20 years older than Lotka, was an Italian mathematician and physicist. After World War I, he turned his attention to the application of mathematics to biology, principally reiterating and developing the

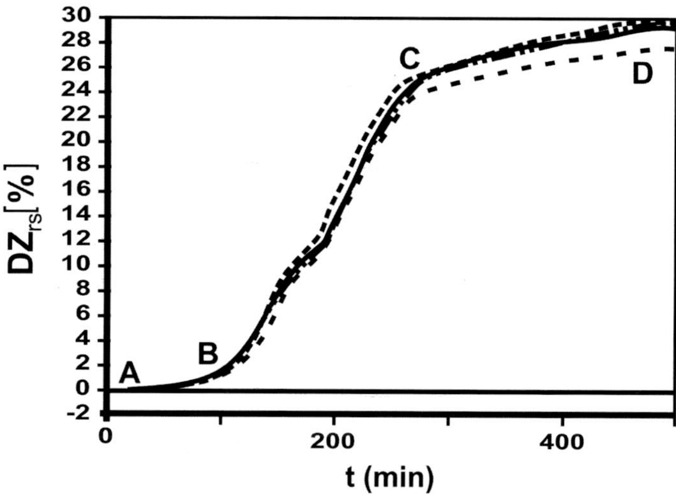

Figure 13.1: Five typical *E. coli* growth curves, at temperature $t = 37.25°$C. The sterile media were inoculated with 10^7 bac/ml. Vertical axis: Relative impedance change, displayed upwards (impedance actually decreases as the number of bacteria goes up), normalized to a reference cell. Horizontal axis: Time in minutes. The curve starts on the left at a steady level (A) getting into a fast growing period (BC) where the number of bacteria proliferated rapidly. Thereafter, it relatively stabilized in the final phase (CD). This curve can be easily mathematically adjusted. From Felice CJ, Clavin OE, Gallo B, Armayor MR, Valentinuzzi ME (1988) *Medical Progress Technology* (discontinued), 14:25–33. Redrawn by Gustavo Idemi.

previous work of Verhulst (Volterra, 1931). The Lotka–Volterra equations, or predator–prey equations, are a pair of first-order, nonlinear differential equations describing the dynamics of certain biological systems. Two species interact: One is a predator and the other is its prey; they can refer to bacteria, insects, or any other kind of animal that responds to such competitive interaction. Something of the type would also be the acetylcholine-acetylcholinesterase kinetics at the myoneural junctions. They evolve in time according to the hypothesis that states that the rate of growth of the prey is proportional to the number of prey units minus a nonlinear term that includes the number of prey units and the number of predator pieces; besides, the rate of growth of predators is accepted as being negatively proportional to the number of predator units plus a nonlinear term including both prey and predator units, that is,

$$\frac{dx}{dt} = x(\propto -\beta y) \tag{13.4}$$

$$\frac{dy}{dt} = -y(\gamma - \delta x) \tag{13.5}$$

where x is the number of prey (say, mosquitoes in a swamp), y is the number of predators (for example, frogs in the same swamp), dy/dt and dx/dt represent the growth of the two populations against time, t stands for time, a, b, c, and d are parameters taking into account the interaction of the two species.

Thereafter, Vladimir Alexandrovitch Kostitzin (1883–1963?), a prominent scientist though perhaps not well recognized, mathematician by training but also a naturalist and closely related to Volterra, produced a second full book devoted to mathematical biology (Kostitzin, 1937). One chapter deals with Verhulst's logistic equation, expanding considerably the subject of growth (Valentinuzzi and Kohen, 2013).

Within a somewhat different context, Balthasar van der Pol (1889–1959), with his collaborator van der Mark, specifically dealt with the heart using a model based on oscillations by relaxation (van der Pol, 1926; van der Pol and van der Mark, 1929). Van der Pol's relaxation paper starts by refreshing the usual sinusoidal oscillations stemmed at a linear second-order differential equation, wherein the squared damping factor α must be much smaller than the squared natural frequency, or

$$\alpha^2 \ll \omega^2 \tag{13.6}$$

The relaxation idea is based on reversing the above-mentioned inequality, that is,

$$\alpha^2 \gg \omega^2 \tag{13.7}$$

In the latter condition, the system tends initially to jump off from zero to a positive value, decreasing gradually, and then suddenly returning to a negative level. This cycle is continued indefinitely. The relaxation period is given by

$$T_{\text{relax}} = \omega^2/\alpha \tag{13.8}$$

Hence, the time period, which is determined by the discharge of a capacitor, is called the *relaxation time*. It was pointed out that these types of oscillations are found in nature. Van der Pol actually constructed a four-element simple electrical circuit (capacitor, neon tube, battery, and resistor) and demonstrated several features related to cardiac activity. Students were given a laboratory exercise which required them to check how the different elements influenced the output signal. All this, no doubt, represented one of the first fully bioengineering accomplishments, as it gathered a rather solid theoretical background along with a hard model aimed at a specific physiological system. Thus, it went beyond biomathematics. Exhaustive information on the subject has been given by Ginoux (2011; 2014).

13.4 Nicolas Rashevsky (1899–1972) and His Chicago School

Nicholas Rashevsky systematized modern mathematical research in the biological and medical sciences. A physicist born in Russia, Rashevsky moved to the United States in 1924, where he shifted his interests toward biology (Valentinuzzi and Kohen, 2013). In 1934, Rashevsky got a position at the University of Chicago, where he later created the outstanding and far reaching Committee on Mathematical Biology. There are innumerable contributions from renowned authors serving as members of the Committee. Several students obtained their doctorate degrees, disseminating the discipline on national and international levels. Besides, he was the founder of the *Bulletin of Mathematical Biology*, which is still an active and prestigious publication. In 1938, Rashevsky produced the first edition of a significant book on mathematical biology, which comprehensively collected well-thought knowledge and set the framework of the emerging discipline (Rashevsky, 1960). One of the many subjects found there refers to excitable tissues, using the so-called two-factor theory, also called the activation–inhibition theory, before potassium, sodium, and calcium had entered the scene and obviously a predecessor of other models developed much later. It is based on two differential equations, or

$$\frac{d\varepsilon}{dt} = KI - k(\varepsilon - \varepsilon_{\mathrm{o}}) \tag{13.9}$$

$$\frac{d\iota}{dt} = KI - k(\varepsilon - \varepsilon_{\mathrm{o}}) \tag{13.10}$$

where K, M, k, and m are constants. In other words, whenever a current flows through an excitable tissue (nerve or muscle), ions are transported, say of the activating type ε and of the inhibiting kind ι. It is assumed that the rate of change of the activating or excitatory type and also of the inhibiting kind is proportional to the current flow I. Besides, any excess of ε or ι over the threshold values ε_o or ι_o, respectively, or any deficiency below those levels tends to decrease at a rate proportional to the excess or deficiency. Relationships with other similar theories are well covered in addition to a good number of predictions and explanations.

A relatively recent text by Allman and Rhodes (2004) needs to be mentioned because it freshens up the biomathematics conceptual framing while it follows in a way the traditional line of thought established by the previously referred to forerunners. It starts with the Malthusian model and difference equations, after Thomas Maltus (1766–1834), who introduced anonymously his Principle of Population in 1798. Very pessimistic, he predicted that the population would eventually outstrip the food supply because the human population grows geometrically, while the food supply only expands arithmetically. Quite a subject to study and discuss under the light of the current knowledge, but we better leave it hanging out to stimulate further research.

13.5 Sampling Theory, Discrete Mathematics, Computers, and Other Modeling Ideas

Steadily and at a growing rate, computers got into the field along with sampling theory and discrete mathematics so reviving old mathematical methods and generating new ones to solve complex and nonlinear systems of equations. However, it must be recognized that the two former concepts, Sampling Theory (ST) and Discrete Mathematics (DM), are essential primary ideas which readily became implemented by computers. To classify and separate models out is for the time being a challenging and not easy task. Herein, we will try to briefly review some of them. Mathematical details must be searched by the reader as they would require an entire text.

In its early days, science consisted of careful observation and experimentation, with a focus on collecting facts. In contrast with science, mathematics is a purely mental discipline. It can be very useful in science, but it has to be connected to science carefully if valid results are to be achieved. All this is very well stated by Glenn Ledder (2013), from Lincoln, Nebraska, USA. This book, after refreshing calculus, proceeds to models underlying the importance of their creation based on biological principles. Probability is also treated as a necessary set of knowledge to handle the natural spread of biological data. Dynamic systems (meaning changing systems, as biological systems are), opposed to static systems (meaning unchanging systems, as a mechanical one), constitute the subject of one part of the book, getting also into the discrete concept and phase analysis, comparing with continuous cases, as for example, with compartmental models. It is stressed that a significant part of the power of mathematics lies in its capacity for generalization. A single symbol can represent a range of numerical values, allowing the mathematical work to be done on a whole class of problems rather than an individual example (Ledzewicz *et al.*, 2013).

The International conference on Mathematical Modeling and High-Performance Computing in Bioinformatics, Biomedicine and Biotechnology (http://conf.nsc.ru/mmhpc2014), organized in Novosibirsk, Russia, proposed the following research areas:

(1) High-performance and distributed computing in systems biology
(2) Data mining methods in natural sciences
(3) Problems of mathematical modeling in life sciences: direct and inverse problems
(4) Mathematical modeling of gene and metabolic networks

(5) Mathematical modeling in biomedicine
(6) Mathematical modeling of biopolymers structure and dynamics
(7) Modeling of DNA–protein and protein–protein interactions
(8) Use of application program packs for problems of bioinformatics and biomedicine
(9) Modeling of pharmacokinetic processes in the body
(10) Mathematical models of immune processes, both direct and inverse
(11) Classification of phase portraits of nonlinear dynamic systems[/NL]

The spread of themes (sometimes not well-defined) is very wide showing also some overlapping, especially because nowadays we should consider ecology, biology at large with its many branches, and medicine, the latter as more narrowly directed to human health, but strongly dependent on the two former fields. Human life cannot proceed without an adequate environment, simple truth not fully recognized by the same human being.

13.6 Compartmental Models

Compartmental models can be considered as a separate family. They are composed of sets of interconnected mixing chambers or stirred tanks. Each component of the system is considered to be homogeneous, instantly mixed, with uniform concentration. The state variables are concentrations or molar amounts of chemical species. Chemical reactions, transmembrane transport, and binding processes, determined by electrochemical driving forces are generally treated using first-order rate equations. Such simplicity makes them easy to compute, since ordinary differential equations are readily solved numerically and often analytically. Although compartmental systems have a reputation for being merely descriptive, they can often provide realistic mechanistic features. Generally, one is considering multi-compartmental systems. Compartments can be used as **black box** operators without explicit internal structure, but in pharmacokinetics compartments are considered as homogeneous pools of particular solutes, with inputs and outputs defined as flows or solute fluxes, and transformations expressed as rate equations. Descriptive models providing no explanation of mechanism are nevertheless useful in modeling of many systems. In **pharmacokinetics** (a discipline that uses mathematical models to describe and predict the time-course of drug concentrations in body fluids), compartmental models are in widespread use for describing the concentration-time curves of a drug following administration. This gives a description of how long it remains available in the body, and is a guide to defining dosage regimens, method of delivery, and expectations for its effects. **Pharmacodynamics** (as the study of the time course and intensity of drug effects on the organism), instead, requires more depth since it focuses on the physiological response to the substance, and therefore stimulates a demand to understand how the drug works on the biological system. Kinetic–dynamic modeling uses mathematical methods to link drug concentrations directly to clinical effects. Many models have been developed over the years to aid in the understanding of these systems. Almost all have solved only ordinary differential equations (Greenblatt *et al.*, 2000; Bassingthwaighte *et al.*, 2012). Compartmental models are quite old, and have been used also in the determination of cardiac output, regional blood flow, and acetylcholine release (Valentinuzzi *et al.*, 1972; Hill *et al.*, 1973; Valentinuzzi *et al.*, 1973; Valentinuzzi and Valentinuzzi, 1973).

13.7 Eukaryotic Cell-Cycle Regulation and More

The dynamic properties of complex regulatory networks cannot be reliably characterized by intuitive reasoning alone. Computers can help us to understand and predict the behavior of such networks, and differential equations provide a convenient language for expressing the meaning of a molecular wiring diagram in computer-readable form. Numerical solutions of the differential equations can be compared with experimental results, in an effort to determine the kinetic rate constants and to confirm the adequacy of the wiring diagram. Eventually the model should give accurate simulations of known experimental results and should be pressed to make verifiable predictions. It is proposed, for example, a protein interaction network for the regulation of DNA synthesis and mitosis that emphasizes the universality of the regulatory system among eukaryotic cells.

Let us recall that a *eukaryote* is any organism whose cells contain a nucleus and other structures (organelles) enclosed within membranes. The defining membrane-bound structure that sets eukaryotic cells apart from prokaryotic cells is the nucleus, within which the genetic material is contained. They have obvious biological importance. The idiosyncrasies of cell cycle regulation in particular organisms can be attributed to specific settings of rate constants in the dynamic network of chemical reactions. The values of these rate constants are determined ultimately by the genetic makeup of an organism. To support these claims, the reaction mechanism is converted into a set of governing kinetic equations and provide parameter values (specific to budding yeast, fission yeast, frog eggs, and mammalian cells) that account for many curious features of cell cycle regulation in these organisms.

Using one-parameter bifurcation diagrams, it is shown how overall cell growth drives progression through the cell cycle, how cell-size homeostasis can be achieved by two different strategies, and how mutations remodel bifurcation diagrams and create unusual cell-division phenotypes. The relation between gene dosage and phenotype can be summarized compactly in two-parameter bifurcation diagrams. Thus, a theoretical framework is given to better understand these complex processes (Csikász-Nagy *et al.*, 2006). Keeping with the progress of the eukaryotic cell, its division cycle is driven by an underlying molecular regulatory network. Cell cycle progression can be considered as a series of irreversible transitions from one steady state to another in the correct order. Bifurcation analysis of a model for the budding yeast cell cycle has identified only two different steady states using cell mass as a bifurcation parameter. By analyzing the same model, using different methods of dynamical systems theory, evidence was produced for transitions among several different steady states during the budding yeast cell cycle.

Hence, by calculating the *eigenvalues* of the Jacobian of kinetic differential equations, the stability of the cell cycle trajectories has been determined using a model. Based on the sign of the real part of the *eigenvalues*, the cell cycle can be divided into excitation and relaxation periods. During an excitation period, the cell cycle control system leaves a formerly stable steady state and, accordingly, excitation periods can be associated with irreversible cell cycle transitions, like entry into mitosis and exit from mitosis. During relaxation periods, the control system asymptotically approaches the new steady state. It was also shown that the dynamical dimension of the model fluctuates by increasing during excitation periods followed by decreasing during relaxation periods. In each relaxation period, the dynamical dimension of the model drops to one, indicating a period where kinetic processes are in steady state and all concentration changes are driven by the increase of cytoplasmic growth. In conclusion, the used methods are more sensitive than the bifurcation analysis used before because they identify those transitions between steady states that are not controlled by a bifurcation parameter.

Therefore, a deeper understanding of the dynamical transitions in the underlying molecular network can be supplied. These authors (mentioned above and below) propose a protein interaction network for the regulation of DNA synthesis and mitosis that emphasizes the universality of the regulatory system among eukaryotic cells. The idiosyncrasies of cell cycle regulation in particular organisms can be attributed to specific settings of rate constants in the dynamic network of chemical reactions. The values of these rate constants are determined ultimately by the genetic makeup of an organism. They convert the reaction mechanism into a set of governing kinetic equations and provide parameter values (specific to budding yeast, fission yeast, frog eggs and mammalian cells) that account for many features of cell cycle regulation in these organisms. Using one-parameter bifurcation diagrams, it is showed how overall cell growth drives progression through the cell cycle, and how mutations remodel bifurcation diagrams and create unusual cell-division phenotypes (Csikasz-Nagy *et al.*, 2006). This group of authors has published several papers in this difficult and attractive area.

It is an exciting time to be a mathematical biologist, or a practitioner of systems biology. The use of mathematical ideas, models, and techniques is rapidly growing and increasingly important throughout the biosciences. New programs have eliminated the division between theory and experiment. The culture of biology is changing with a growing awareness that "to think is to model." Before the rapid development of molecular biology, the field was largely qualitative; today no aspect of biology can afford to disregard quantitative measurement and analysis as too boring, time-consuming, or difficult. Theorists and experimenters are now even working in the same buildings (Schnell *et al.*, 2007).

There is a book, edited by Richard E. Reiss (1964) already over 50 years ago, that deserves to be recalled because it marked several actual lines in mathematical modeling of biological systems, mainly neural systems. Age of a publication does not necessarily mean obsolescence, and this one, in particular, pointed out to several subjects that at that time were opening new roads, such as behavioral models, theoretical significance of dendritic trees, signal processing in axon systems, or systems analysis of visual perception. It also has some fascinating applications in insects so much broadening the spectrum of possibilities.

13.8 Closing Remarks

Little actual math has been given in this chapter for only general orientation was taken as the main line. It is the reader's task to dig into the literature as development of the area proceeds. Several relatively recent titles have been added to the specialized literature, as Barnes and Chu (2010), where these authors underline that in biological modeling, there is no one modeling technique that is suitable for all problems. Instead, different problems call for different approaches. Furthermore, it can be helpful to analyze the same system using a variety of approaches, to be able to exploit the advantages and drawbacks of each. It is often unclear which modeling approaches will be most suitable for a particular biological question.

This is a good place to remember those mathematical discoveries that have had enormous influence in science and technology, an influence that no doubt will continue and that propagated in lower or higher degree to biology and medicine, too. Often, and since such knowledge is usually taken for granted, they are forgotten or disregarded. Let us briefly mention them even though they may seem obvious.

Pythagoras's theorem is perhaps the first one in the list (Pythagoras of Samos, 569–500 BC, Greek philosopher and mathematician considered the first **pure of his kind**). Its applications are countless, including in Relativity Theory (Valentinuzzi and Arini, 2014).

Logarithms come next, introduced by John Napier (1550–1617), in 1614, as a means to simplify calculations. He was a Scottish mathematician who made common use of the decimal point or comma in arithmetic operations. Logarithms were rapidly adopted by navigators, scientists, engineers, and many others to perform computations more easily, using slide rules and logarithm tables (both resources fully outdated nowadays).

Infinitesimal calculus, as the mathematical discipline focused on limits, functions, derivatives, integrals, and infinite series. Decisive concepts independently brought up by Isaac Newton (1642–1727) and Gottfried Leibniz (1646–1716), discovery or invention that went into a long controversy. The former was a British physicist, philosopher, theologian, inventor, alchemist, and mathematician, while the latter was also a philosophist and logician, mathematician, jurist, librarian, and German politician. In those days, intellectuals or wise men, if it sounds better, used to combine knowledge from different disciplines. Nowadays, our current Philosophy Doctor, or PhD scientific degree, remains as residual tenuous memorial to such polymathy.

Newton's law of gravity defines the attractive force between all objects that possess mass. The law offers profound insights into the way the universe functions and it also helps in the understanding of very small particles behavior.

Complex numbers have practical applications in many fields, including physics, chemistry, biology, economics, electrical engineering, and statistics, and their appearance historically preceded Napier's, Newton's, and Leibniz' basic contributions. The Italian mathematician Gerolamo Cardano (1501–1576) conceived complex numbers around 1545, introducing the *Square Root of Minus One* as the imaginary unit i, though his understanding was rudimentary. He wrote more than 200 works on medicine, mathematics, physics, philosophy, religion, and music.

Euler's formula, named after Leonhard Euler (1707–1783), Swiss mathematician and physicist, establishes the relationship between trigonometric functions and the complex exponential function. Euler's formula states that, for any number x,

$$\mathrm{e}^{ix} = \cos(x) + i\sin(x) \tag{13.11}$$

where e is the base of the natural logarithms, i is the imaginary unit, and *cos* and *sin* are the trigonometric functions *cosine* and *sine*, respectively, with the argument x given in radians. Euler's formula is ubiquitous in mathematics, physics, and engineering. The physicist Richard Feynman (1918–1988) called Equation (13.11) *our jewel* and *the most remarkable formula in mathematics.*

The normal distribution, which some historians attribute to Abraham de Moivre (1667–1754), French mathematician, as having been introduced in 1738, at least in a preliminary and not clear form. Its actual discovery, however, belongs better to Carl Friedrich Gauss (1777–1855), in 1809, while several other authors added later on some conceptual improvements. Gauss was a German mathematician, astronomer, geodesist (land surveyor), and physicist.

The wave equation is a second-order linear partial differential equation that describes sound, light and water waves, with significance in fields like acoustics, electromagnetics, and fluid dynamics, and of course also in hemodynamics, as well shown in the traditional and still valid text by Milnor (1982). Historically, in 1746, Jean-Baptiste d'Alembert (1717–1783), mathematician, philosopher, and encyclopedist, discovered the one-dimensional wave equation, and soon thereafter Euler produced the three-dimensional wave equation. He sometimes appears as Jean le Rond D'Alembert

The Navier–Stokes equations, after Claude-Louis Marie Henri Navier (1785–1836), French engineer and physicist, and George Gabriel Stokes (1819–1903), Irish mathematician and physicist, describe the motion of fluid substances, as for example blood. In their full and simplified forms, these equations help with the study of blood flow and many other problems of physics and engineering. Their solution is a *velocity*, not *position*. The Navier–Stokes equations are nonlinear partial differential equations in almost every situation; in the case of turbulence, at the time-dependent chaotic behavior seen in many fluid flows (as blood flow), the equations present even greater difficulties. Turbulences are present all over the cardiovascular system.

Fourier and Laplace transforms and the concept of mathematical transformation. The first one, named after Jean-Baptiste Joseph Fourier (1768–1830), French mathematician and physicist, is a mathematical operation employed to transform signals from one domain (say, time) to another domain (say, frequency) or vice versa. As well-known, it is defined by,

$$\hat{f}(\xi) = \int_{-\infty}^{\infty} f(x)e^{-2\pi ix\xi}dx \tag{13.12}$$

for any real number ξ. The **Laplace Transform**, in turn, is a linear operator of a function $f(t)$ with a real argument t $(t \geq 0)$ that transforms $f(t)$ to a function $F(s)$ with complex argument s,

$$F(s) = \int_{o}^{\infty} f(t)e^{-st}dt. \tag{13.13}$$

It is named after Pierre-Simon Laplace 1749–1827). This transform is related to the Fourier Transform, but whereas the latter expresses a function as a series of modes of vibration (frequencies), the Laplace Transform resolves a function into its *moments*. Like the Fourier transform, the Laplace transform is used for solving differential and integral equations. In mathematics, **Transform Theory** deals with the study of operations that modify the appearance of a function as well as its independent variable. In essence, by a suitable choice of a *kernel* (core) for a vector space a problem may be simplified. The two exponentials in the equations above are kernels and, in fact, we might choose any arbitrary function as long it is an adequate one.

Maxwell's equations, Scottish physicist, have also profound influence in BME, although perhaps not well recognized because of its very basic nature, which makes it as indirect (Valentinuzzi and Kohen, 2013).

The Laws of Thermodynamics also give dramatic background to bioengineering at large. In simple terms, the First Law is a version of the Law of Conservation of Energy, which states that the total energy of an isolated system is constant; energy can be transformed from one form to another, but cannot be created or destroyed. Sadi Carnot (1796–1832), French physicist and engineer, enunciated the Second Law when he stated that it is impossible to extract an amount of heat from a hot reservoir and use it all to do work, as some amount of heat must be exhausted to a cold reservoir. Such fact precludes a perfect heat engine. He introduced it with the concept of a heat engine ideal cycle. The third law says that the entropy, defined as $S = Q/T$ (heat quantity divided by absolute temperature), contained by any pure substance in thermodynamic equilibrium approaches zero as the temperature approaches zero. The entropy of a system at absolute zero is zero (Valentinuzzi and Arini, 2015).

Information theory, introduced by Claude Shannon (1916–2001), North American electronics engineer and mathematician, linked to thermodynamics and also with biology, brings up another avenue of research and many practical consequences. Entropy and information, the two central concepts of Shannon's theory, play transparent roles when applied to symbolic sequences in genomics. Such an approach can distinguish between entropy and information in genes, predict the secondary structure of ribozymes, and detect the covariation between residues in folded proteins. It envisions new tools in the characterization of resistance mutations and in drug design (Adami, 2004). But there were early attempts, as when authors aimed at calculating the amount of information contained in a chemical or biological structure, and, to estimate the energy needed for obtaining an organization unit (Valentinuzzi and Valentinuzzi, 1963).

Chaos theory, about which much has been published, said and expected, but with no demonstrated application yet, at least in the field of BME. An early proponent of this theory was Henri Poincaré

(1854–1912), prestigious French polymath (that is, proficient in several scientific disciplines), in the 1880. This theory belongs to mathematics with apparent applications to several fields, including biology. It studies the behavior of dynamical systems that are highly sensitive to initial conditions. Let us remind that a **dynamical system** is a concept in mathematics where a fixed rule describes the time dependence of a point in a geometrical space, such as the swinging of a clock pendulum, the flow of water in a pipe or the number of fish in a lake (Alligood *et al.*, 2000; Baianu, 2010; Galor, 2011). Interestingly enough, Poincaré posed the still valid question of whether the Solar System would always be stable. Anyhow, do not keep any concern, for if it is not, we *ain't gonna* be here!

Some people may add to this list of essential concepts and theories the **Relativity Theory** (both **Special** and **General**, stated in 1905 and 1915, respectively, by Albert Einstein, 1879–1955) and **Schrödinger Equation**, stated in 1925 by Erwin Schrödinger (1887–1961), but all this would lead us too far away from our central theme. To close this chapter, it should be mentioned that not too long ago (2008), in Houston, Texas, USA, the Center for Mathematical Biosciences was created, jointly between Rice University and the Texas Medical Center (see http://www.mathematics.uh.edu/featured-news/mathematical-medicine/index.php).

References

Adami C. Information theory in molecular biology. *Physics of Life Reviews* 1(1):3–22, 2004.

Alligood KT, Sauer TD, Yorke JA. Chaos: An introduction to dynamical systems, Springer Verlag, New York, 2000. ISBN 0-387-94677-2.

Allman ES, Rhodes JA. Mathematical models in biology: An introduction. 8 Chapters, 6 Appendices, Cambridge University Press, Cambridge, UK, 367 pp., 2004.

Baianu IC. Complexity and dynamics, theories, dynamical systems, and applications to biology. It deals with the study of the behavior of complex systems, dynamic programming and control theory, neurodynamics and psychodynamics, 2010. https://archive.org/details/ComplexityEmergentSystemsLifeAnd Complex BiologicalSystemsComplex.

Barnes DJ, Chu D. Introduction to modelling for biosciences. Springer Verlag, New York, 2010.

Bassingthwaighte JB, Butterworth E, Jardine B, Raymond GM. Compartmental modeling in the analysis of biological systems. *Methods in Molecular Biology* 929:391–438, 2012.

Csikász-Nagy A, Battogtokh D, Chen KC, Novák B, Tyson JJ. Analysis of a generic model of eukaryotic cell-cycle regulation. *Biophysical Journal* 90(12):4361–4379, 2006. http://www.ncbi.nlm.nih.gov/pmc/articles/PMC14 71857/.

Felice CJ, Clavin OE, Gallo B del V, Armayor MR, Spinelli JC, Valentinuzzi ME. Impedancimetric bacterial detection: Theoretical and experimental aspects. *Medical Progress through Technology* 14:25–33, 1988.

Galor O. Discrete Dynamical Systems, Springer Verlag, New York, 2011. ISBN 978-3-642-07185-0.

Ginoux JM. Analyse Mathématique des Phénomènes Oscillatoires Non Linéaires, PhD Dissertation, Université de Paris VI, France, 2011.

Ginoux JM. History of nonlinear oscillations' theory (1880–1940) Laboratoire LSIS, CNRS, UMR 7296, Université de Toulon, Archives Henri Poincaré, CNRS, UMR 7117, Université de Nancy II, March 27, 2014.

Greenblatt DJ, Harmatz JS, von Moltke LL, Shader RI. Pharmacokinetics and Pharmacodynamics. Only in http://www.acnp.org/g4/GN401000085/ CH084.html; Work supported in part by grant MH-34223 from the Department of Health and Human Services. With the collaboration and assistance of Bruce L. Ehrenberg and Lawrence G. Miller, 2000.

Hill DW, Valentinuzzi ME, Pate T, Thompson FD. The use of a compartmental hypothesis for the estimation of cardiac output from dye dilution curves and the analysis of radiocardiograms. Med & Biol Eng 11(1):43–54, 1973.

Kostitzin VA. Biologie mathématique. Libraire Armand Colin, Paris, 223 pp., 1937.

Ledder G. Mathematics for the life sciences: Calculus, modeling, probability, and dynamical systems, Springer Verlag, New York, 2013.

Ledzewicz U, Schättler H, Friedman A, Kashdan E (eds). Mathematical methods and models in biomedicine. Series on lecture notes on mathematical modeling in the life sciences, Springer Verlag, New York, 427 pp., 94 illus, 2013.

Lotka AJ. Elements of physical biology. Williams & Wilkins, Baltimore, 495 pp., 1925. https://archive.org/details/elementsofphysic017171mbp.

Csikasz-Nagy A, Battogtokh D, Chen KC, Novák B, Tyson JJ. Analysis of a generic model of eukaryotic cell cycle regulation. *Biophysical Journal* 90:4361–4379, 2006.

Milnor WR. Hemodynamics, Williams & Wilkins, Baltimore/London, 390 pp., 1982.

Modis T. Genetic re-engineering of corporations. *Technological Forecasting and Social Change* 6:107–118, 1997.

Rashevsky N. Mathematical biophysics: Physico-Mathematical foundations of biology, 3rd ed., vols 1–2, Dover Publications, New York, 1960.

Rashevsky N. Some medical aspects of mathematical biology. Charles C. Thomas Publisher, Springfield, IL, 314 pp., 1964.

Reis RF(ed). Neural theory and modelling. Stanford University Press, Stanford, CA, 426, 1964.

Schnell S, Grima R, Maini PK. Multiscale modeling in biology. *American Scientist* 95:134–142, 2007. http://eprints.maths.ox.ac.uk/567/01/224.pdf.

Valentinuzzi ME, Arini PD. Relativistic effect in history, *IEEE Pulse* 5(4):64–74, 2014.

Valentinuzzi ME, Arini PD. Intracardiac pressure-volume diagrams (PVD) and their links with thermodynamics. *IEEE Pulse* 5(6):48–56, 2015.

Valentinuzzi ME, Beneke K, González GE. Ludwig: The bioengineer, *IEEE Pulse* 3(4):68–78, 2012.

Valentinuzzi ME, Hoff HE, Geddes LA. A two compartment model for the release and action of acetylcholine in the heart. *Circulation Research* 33(5):532–538, 1973.

Valentinuzzi ME, Kohen AJ. The mathematization of biology and medicine: Who, when, how? *IEEE Pulse* 4(1):50–56, 2013.

Valentinuzzi ME, Kohen AJ. James Clerk Maxwell, Kirchoff's laws, and their implications on modeling physiology, *IEEE Pulse* 4(2, March–April):40–46, 2013.

Valentinuzzi M, Valentinuzzi ME. Information content of chemical structures and some possible biological applications. *Bull Mathematical Biophysics* 25(1):11–27, 1963.

Valentinuzzi M, Valentinuzzi ME. Newman's chamber models in the study of blood circulation through the liver and heart. *Bulletin Math Biology* 35:19–29, 1973.

Valentinuzzi M, Valentinuzzi ME, Posey JA. Dilution curve area: Fast estimation by a compartmental procedure. *J Association for the Advancement Medical Instrumentation* 6(5):335–343, 1972.

van der Pol B. On relaxation oscillations, *Jahrbuch der drahtlosen Telegraphie* 28:178, 1926.

van der Pol B, van der Mark J. The heartbeat considered as a relaxation oscillation, and an electrical model of the heart, *Archives des Nederlands de Physiologie de l'Homme et des Animaux* 14:418–443, 1929.

Verhulst PF. Notice sur la loi que la population poursuit dans son accroissement, *Correspondance Mathématique et Physique* (Ghent, Belgium) 10:113–121, 1838. See also Vogels M, Zoeckler R, Stasiw DM, Cerny LC. Verhulst's "Notice sur la loi que la populations suit dans son accroissement" published as showed above. *J Biological Physics* 3(4):183–192, 1975. This paper reports what is perhaps the first mathematical treatment of population statistics. It was written at a time when the ideas of Malthus were popular.

Volterra V. In: Gabay J, ed. Leçons sur la Théorie Mathématique de la Lutte pour la Vie (in French, Lessons on the mathematical theory of the struggle for life). Gauthier-Villars, Paris; reissued in 1990, 1931.

CHAPTER 14

THE FUTURE OF BIOENGINEERING: POSSIBLE NEW AREAS

Max E. Valentinuzzi, Sibel Ertek, and B. Silvano Zanutto

Weißt du, wie viel Sternlein stehen
an dem blauen Himmelszelt?
Weißt du, wie viel Wolken gehen
weithin über alle Welt?
Gott der Herr hat sie gezählet,
dass ihm auch nicht eines fehlet
an der ganzen großen Zahl[1]

Wilhelm Hey (1789–1854), lyrics released in 1837.

The Past is always known, with its accomplishments and mistakes, irretrievably.

The Present is Now and, in fact, as soon as "it is" it becomes "it was", so much that one can eventually feel it did not exist, as a continuum.

The Future remains unknown, the huge endless Hope for the Overall Better that Man for ages has been craving for, always trying to dig in deeper since the beginnings of Times, always trodding rough roads with uncertain destinations.

Free Picture Country Path ID 246410
©Dainis Derics-Dreamstime Stock Photos

---

[1]Do you know how many stars there are up in the blue sky? Do you know how many clouds move around over the entire world? God the Lord has counted them, so that not even one is missing from the big overall number.

523

Let us begin this last chapter by more or less repeating what was said in a previous book (Valentinuzzi, 2010): The last chapter is the most difficult of all for it intends to play the magician, the Nostradamus, by watching in the crystal ball to signal possible new avenues for research, development and applications. During such endeavor, some references not given before will be brought up to better complete the list and partially compensate for ignorance and/or forgetfulness. Thus, the subject order tries to fill up possible gaps left in the main text but...definitely YES...the traditional pending questions are,

- What can we expect?
- How difficult it may be?
- When will it become reality?

and, obviously, we cannot guarantee how close to the truth the answers will be. In such effort, there will be no abstract and sections; subjects will flow out more or less according to their respective significance...or not. Have you ever heard the improvisations of a musician when playing an instrument, as in the Old New Orleans Roaring Twenties Jazz Days? The notes come out as he or she feels them. Let us then freely play the melodies of scientific subjects, unfortunately too often hidden by mundane glistening and flickering shines.

Actual fact is that combining knowledge in cell and molecular biology, physiology with biomaterials, biomechanics, and biotransport phenomena, we find **regenerative medicine**, as a field which aims to understand the mechanical, structural, and biological processes associated with designing and developing systems to repair or replace damaged organs and tissues. Current research activities address the aspects of **cellular engineering**, **biomaterials**, and **tissue engineering**.

Shannon Fischer (2014), a freelance writer of Boston, Massachusetts, USA, claims that 382 million people in the world have **diabetes** while about 175 million do not even are aware they have it. Besides, in the order of over 300 million are at high risk for diabetes. These numbers combined make over 10% of the global population (about 7,500 million, as for April 2016). If now **hyperlipidemia** is added, that is, high lipid levels in blood, the health risk grows dramatically. Hyperlipidemia includes several conditions, but it usually means high low-density lipoprotein (LDL)-cholesterol and high triglyceride concentrations. The former is sometimes called in the daily street talk "bad cholesterol" as opposed to the "good cholesterol" or high density lipoprotein (HDL). Since low levels of HDL cholesterol is another risk for atherosclerosis, the term "hyperlipidemia" is replaced by the term "dyslipidemia" in clinical terminology. Dyslipidemia can speed up the process called atherosclerosis, that is, the formation of plaques or atheromas, along with hardening of the arterial walls (arteriosclerosis). Eventually, enough plaque may build up to reduce arterial blood flow or rupture and occlude the vessel lumen. Atherosclerosis increases the risk of heart disease, stroke, and other vascular diseases, and if the latter situation is immersed in a sweet sea full of sugar, things become truly harsh. Together with new findings at the molecular biology level about these two diseases, new targets and use of new technologies for both treatment and diagnosis have appeared. Since both diseases are important cause of mortality and morbidity with increasing prevalence, besides the chronic nature of them, treatment itself and complications cause economic burden to many countries, especially in Occident. Bioengineering, with its multiple-frame approach, can help quite a bit, eventually bringing successful answers. **Nanotechnology** and **microtechnology** are important, already established, areas with good possibilities for the diagnosis and treatment of these diseases and their related complications. Chapter 11 on biosensors anticipated information regarding such subject.

Let us briefly remind that one micron (μm) is one millionth of a meter, or 10^{-6} (say, a single strand of human hair is about 100μm wide). Similarly, a nanometer equals one billionth of a meter, or 10^{-9} (a single sheet of paper is about 100,000 nm thick). By definition, micro and nano technologies deal with structures of the size of microns (μm) or nanometers (nm) and involve the development of materials and/or devices within those sizes. Nanotechology is revolutionizing today's and tomorrow's products. The U.S. Government has created the National Nanotechnology Initiative (NNI), which provides a multi-agency framework to improve human health and economic well-being (Shatkin, 2013). Nano products such as stain-resistant clothing, coatings on eyeglasses to keep the lenses from scratching, lighter baseball bats and tennis rackets, and other tiny electronic devices are already available. Protective masks, for example, utilize a proprietary nanocoating technology on the outermost layer that not only traps viruses, bacteria, and fungi, but also effectively breaks them down on the cellular level, rendering them harmless (Harris, 2010), and hyperlipidemia can also be helped by these very small technologies. Remember that such pathological condition, in some dyslipidemia types, develops because of the lack of LDL receptors in hepatocytes. Since injected polymeric nanoparticles are quickly taken up by the liver Kupffer cells, it is possible to enhance LDL delivery to the liver through the use of LDL-absorbing nanoparticles. (Maximov *et al.*, 2010) demonstrated the feasibility of the approach *in vitro* using biodegradable and biocompatible polylactide nanoparticles (~ 100 nm in diameter) with an attached covalently antibody to adsorb LDLs at physiologically relevant concentrations. They showed that up to six-fold decreases of LDL levels can be achieved *in vitro* upon treatment of LDL suspensions. The study of the uptake of the antibody–nanoparticle–LDL complexes by cells was performed using a mouse macrophage cell line as a model for liver Kupffer cells. These authors found that macrophages can quickly take up antibody–nanoparticle–LDL complexes and digest them within 24 hrs with no evidence of cytotoxicity. Hence, this is potentially a **nanodevice for hyperlipidemia treatment**.

Inorganic nanoparticles to make **synthetic high density lipoprotein cholesterol** (HDL-C) provide a new strategy to dyslipidemia treatment. A few years ago, Thaxton *et al.*, (2009) and Thaxton *et al.*, (2009) found that colloidal gold nanoparticles could provide an inorganic platform to base it on, that is, the synthesis of HDL biomimetic nanoparticles capable of binding cholesterol. Iron oxide and quantum dots could also be used as core material, as described by Luthi *et al.*, (2010), where they highlighted the most common strategies for treating atherosclerosis using HDL while further detailed potential treatment opportunities utilizing nanotechnology to increase the amount of HDL in circulation. The synthesis of biomimetic HDL nanostructures that replicate the chemical and physical properties of natural HDL provides novel materials for investigating the structure-function relationships of HDL and for potential new therapeutics to combat coronary heart disease (CHD), many times an obesity complication (Luthi *et al.*, 2012). Quite interestingly, Thaxton's group (see references above) first reported the gold nanoparticle bio-assay probe for the detection of prostate specific antigen (PSA) after radical prostatectomy finding that the test was approximately 300 times more sensitive than commercial immunoassays.

What about using **negatively charged liposomes in atheroma plaques?** Phospholipid composition of lipid bilayer could be modified to create liposomes to be used as an impermeable delivery particle. Rhodamine-labelled liposomes were found at highest concentration at lipid-laden atheromas in animal studies. Thus, liposomes to target atheromas in a Watanabe heritable hyperlipidemic (WHHL) rabbit model have been proposed. They were labeled with rhodamine and nanogold and were injected intra-arterially into the descending thoracic aortas of WHHL rabbits. The arterial segments of interest were perfusion-fixed and evaluated with immuno-histochemistry, light microscopy, and electron microscopy. Deconvolution (see Chapters 5, 6, and 7) microscopy showed that rhodamine label was concentrated in the plaque shoulder regions of advanced-stage atheromas; however, rhodamine label was not found in adjacent, non-atherosclerotic aorta. Transmission electron microscopy revealed liposome remnants and the highest concentration of nanogold label in lipid-laden areas of atheromas. Liposomes were concentrated in areas of lipoprotein-associated phospholipase. It was concluded that modified liposomes may be useful therapeutically for targeting metabolically active plaque (Walton *et al.*, 2010).

Another appealing possibility within this very small but potent world refers to **drug-carrying non-conventional systems**. To improve disadvantages of conventional drug-delivery systems (biodistribution, low bioavailability, lack of water solubility, poor delivery to the intended site of action, low therapeutic response despite high dosages, side effects, drug resistance, toxicity, barriers in the body such as the blood brain barrier), drugs are entrapped in small vesicles and injected into the bloodstream, to decrease drug degradation and increase bioavailability at specific tissues. Self-MicroEmulsifying Drug Delivery Systems (SMEDDS), ternary solid nanosuspensions stabilized by surfactants, ion-exchange resins, nanosponge drug delivery, nanoparticle formulation, mucoadhesive microcapsules, buccoadhesive drug delivery system, gastric floating drug delivery systems, and pulsatile drug delivery system are important alternatives to conventional drug delivery methods (Farooq *et al.*, 2013). New nanotechnological drug delivery systems are also promising in the treatment of diabetic complications, such as more efficient drug delivery to the posterior eye with nanoparticles (e.g., liposomes, *dendrimers*,[2] cationic nanoemulsions, lipid and polymeric nanoparticles) (Fangueiro *et al.*, 2014).

Images represent perhaps the best way to know what goes on within the body, and nanotechnology can produce **smart contrast agents and new Imaging Techniques**. Superparamagnetic Iron Oxide (SPIO), Ultrasmall Superparamagnetic Iron Oxide (USPIO) and Monocrystalline Iron Oxide Nanoparticles (MIONS) have been the subjects of extensive research over the past decade (Wang *et al.*, 2001). The iron oxide particle size of these contrast agents varies widely and influences their physicochemical and pharmacokinetic properties. Superparamagnetic agents enhance both relaxation times. It should be recalled that return to equilibrium of net magnetization is called *relaxation*. During relaxation, electromagnetic energy is retransmitted and this radio frequency (RF) emission is called the Nuclear Magnetic Resonance or NMR signal. Relaxation combines two different mechanisms, longitudinal relaxation corresponds

---

[2]*Dendrimers* are highly branched, star-shaped macromolecules with nanometer-scale dimensions. Dendrimers are defined by three components (Figure 14.1):

- a central core;

- an interior dendritic structure (the branches); and

- an exterior surface with functional surface groups.

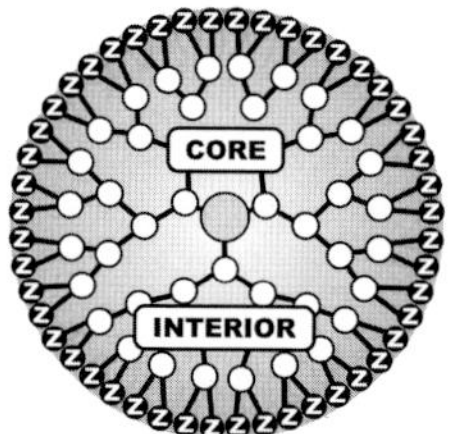
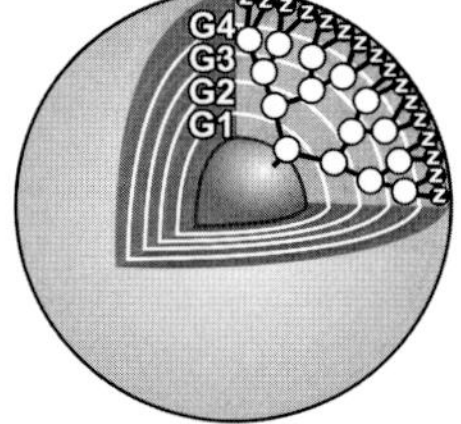

Figure 14.1: Dedrimer. Downloaded from Google, public domain website; enter with the word "dendrimer".

The varied combination of these components yields products of different shapes and sizes with shielded interior cores that are ideal candidates for applications in both biological and materials sciences. While the attached surface groups affect the solubility and chelation ability, the varied cores impart unique properties to the cavity size, absorption capacity, and capture–release characteristics. Applications highlighted in recent literature include drug delivery, gene transfection, catalysis, energy harvesting, photo activity, molecular weight and size determination, rheology modification, and nanoscale science and technology. Monodisperse dendrimers are synthesized by stepwise chemical methods to give distinct generations of molecules with narrow molecular weight distribution, uniform size and shape, and multiple surface groups.

to longitudinal magnetization recovery, and transverse relaxation corresponds to transverse magnetization decay. Large iron oxide particles are used for bowel contrast and liver/spleen imaging while smaller iron oxide particles are selected for lymph node imaging, bone marrow imaging, perfusion imaging, and Magnetic Resonance (MR) angiography. Even smaller MIONS are under research for receptor-directed MR imaging and magnetically labeled cell probe MR imaging. Iron oxide particles for bowel contrast are coated with insoluble material, and all iron oxide particles for intravenous injection are biodegradable. Superparamagnetic agents open up an important field for research in MR imaging, surely forecasting newer advances. The technique may allow direct visualization of atheromatic plaques and estimation of plaque stability, which is correlated with macrophage content (Ruehm *et al.*,). Many new methods for diagnostic purposes could also have therapeutic applications in the near future, as anticipated by Wickline and Lanza (2003). These authors emphasize that

     M. E. Valentinuzzi, S. Ertek & B. S. Zanutto

molecular imaging is not a substitute for the traditional process of image formation and interpretation, but is intended to improve diagnostic accuracy and sensitivity by providing an *in vivo* analog of immuno-cytochemistry or *in situ* hybridization. It is less concerned with native image contrast or resolution, which is a key for depicting the effects of the disease on surrounding normal tissues, but rather, it seeks to enhance the conspicuity of microscopic pathologies by targeting the molecular components or processes that represent actual mechanisms of disease. Moreover, imaging will become crucial for *in vivo* characterization of the complex behaviors of disease in time and space that will clarify question such as, where the process is, how big it is, how fast it is developing, how many molecular processes are contributing simultaneously, what to treat it with, how it is responding to therapy, and how it is changing. Perhaps it should be mentioned that Michel Zanca is with the Nuclear Medicine & Functional MRI at University Hospital, in Montpellier, France. In turn, Ansary and Faddah (2010) reviewed the use of **nanoparticles as biochemical sensors**, since glucose and triglyceride nanosensors could open important area for more sensitive measurements compared to classical methods.

Mathematical models are especially important in clinical evaluation of diabetic patients. Currently, mean plasma glucose levels are a significant index for macro- and micro-vascular diabetic complications. HbA1c (glycohemoglobin)[3] measurement is the most available, feasible, and standard method to evaluate nearly 120 days mean glucose levels, which are related to mean red blood cell life span (Thomas *et al.*, 2009). The problem is that changes in red blood cell lifespan and many chronic diseases (such as renal failure or hemoglobinopathies) affect it and, especially, in some

---

[3]A glycohemoglobin test, or hemoglobin A1c, is a blood test that checks the amount of sugar (glucose) bound to the hemoglobin in the red blood cells (when hemoglobin and glucose bond, a coat of sugar forms on the hemoglobin). That coat gets thicker when there is more sugar in blood. A1c tests measure how thick that coat has been over the past 3 months, which is how long a red blood cell lives. People who have diabetes or other conditions that increase their blood glucose levels have more glycohemoglobin.

patients needing shorter estimations of near-past glycemia (such as gestational diabetics or diabetics with new therapeutic arrangements). In continuous blood glucose monitoring systems, many new markers and data could be available to arrange and evaluate glycemia. Besides, marker of "mean" blood glucose leads to the question of how to evaluate. The authors referred to above make use of the so called *glucose pentagon*. Connecting the values of these five parameters provided an enclosed area of a given size. For a patient with diabetes, these parameters and the connected area describe how the glycemic condition was during the monitoring period. The area of the glucose pentagon for a patient with diabetes, divided by the standard area of healthy subjects, yields a non-dimensional characteristic value defined as the glycemic risk parameter. They assume that this risk parameter provides a more meaningful overall description of metabolic control than the HbA(1c) alone. In addition, it might also allow a better assessment of a patient's risk for developing diabetes-related late complications in comparison to the HbA(1c) alone. Adequate long-term clinical studies are still needed.

Nanosensors and nanomaterials could also be used for glucose monitoring and each year they become more sensitive, more stable, smaller, and with expected longer lifetime (Cash and Clark, 2010). The latter paper discusses developments in the past several years on both nanosensors that directly measure glucose and nanomaterials that improve glucose sensor function. They also consider challenges that must be overcome to apply these developments in the clinic. There are other reports regarding similar trends, like one by Viviana Scognamiglio (2013), who is a researcher at the Institute of Crystallography in Rome, Italy, belonging to the National Research Council (CNR) of that country.

Because glycemic fluctuations are also related with diabetic complications, often associated with activation of oxidative stress, and they may also affect HbA1c values, we need better parameters to follow glycemic stability and mean plasma glucose (Derr *et al.*, 2003; Wu *et al.*, 2013; Cheng *et al.*, 2014). Variability of blood glucose (rate of change), low blood glucose and high blood glucose indices are also being used (McCall *et al.*, 2006). Formulas to postulate mean plasma glucose from HbA1c values could be used (Rohlfing *et al.*, 2002), but HbA1c itself has an interesting kinetics; according to a study carried out by Monier *et al.*, (2003), the contribution of postprandial glucose levels to HbA1c predominate in fairly controlled diabetic patients, whereas well-controlled patients have preprandial glucose levels correlated with HbA1c values (Bonora *et al.*, 2001). Therefore, mathematical models to analyze HbA1c behavior are needed (Kahrom, 2010) and are important for the prevention and prediction of acute and chronic complications (Kilpatrick *et al.*, 2007). Another paper to mention is that by Monnier *et al.*, (2006) where oxidative stress triggered by glucose fluctuations is analyzed. By the way, these authors belong to a well-established and productive group at Lapeyronie Hospital, in Montpellier, France.

There are two main practical problems faced with diabetic patients, that is,

(1) Glucose monitoring systems require needles which tend to decrease the patients' own follow-up of glycemia;
(2) Insulin treatment itself is invasive, decreasing the patient's compliance and leading, in the long term, to serious chronic disease-related depression.

Nanotechnology and microtecnology also try to find new methods for insulin administration. Since insulin is structurally a protein, oral administration is not successful because of gastric digestion. Newly developed nanostructure of liposomes that contain insulin molecules coated with sugar molecules are attached to each other by Concanavalin A (Con A) and used by inhalation to prevent gastric destruction of insulin. When blood sugar is high, the Con A detaches from sugars covering liposomes and latch on to the blood sugar, making the bundle unstable. Thereafter, insulin in liposomes is released to the bloodstream. The system has been tested in rats leading to a 6-hour insulin effect after inhalation. However, Con A should be replaced when tested in humans because it is known to be inflammatory (Karathanasis *et al.*, 2007).

Insulin delivery systems based on hydrogels to measure glucose levels are slow and mechanically weak, causing leak of insulin.

Gu *et al.*, (2013) relied on a special gel-like substance made of nanoparticles containing spheres of dextran loaded with an enzyme that converts glucose to gluconic acid. Glucose diffuses freely through the gel; in hyperglycemia, that enzyme produces more gluconic acid which leads to local pH decrease. This acid microenvironment causes dextran spheres to disintegrate and release insulin proportional to glucose levels. In tests with Type 1 diabetes in mice models, single injection caused normoglycemia for 10 days. Since the particles are composed of polysaccharides, they are eventually degraded in the body (Balaconis and Clark, 2013). Agrawal *et al.*, (2014) reported folic acid functionalized insulin loaded stable liposomes with improved bioavailability following oral administration. Liposomes were stabilized by alternating coating of negatively charged poly(acrylic acid) and positively charged poly(allyl amine) hydrochloride over liposomes. Furthermore, folic acid was appended as targeting ligand by synthesizing folic acid-poly(allyl amine) hydrochloride conjugate. The insulin entrapped within the freeze-dried formulation was found stable both chemically as well as conformationally and developed formulation exhibited excellent stability in simulated biological fluids. Overall the proposed strategy is expected to contribute significantly in the field of designing ligand-anchored, polyelectrolyte-based stable systems in drug delivery. Besides, micro and nanosystems could also be used for incretin-based oral treatments in diabetics (Araujo *et al.*, 2014).[4]

Microjet injectors could also be used for insulin injections as non-painful. They have smaller diameter than conventional jet injectors (50–100 micrometers compared to 75–200 micrometers), contain less fluid (2–15 nanoliters compared to 30,000–100,000 nanoliters), show easier control of penetration, and can be as small as 200 micrometres, compared to 2–20 millimeters in the conventional jet injector. Injection is activated by pulsed electricity obtained from a piezoelectric source. It was tested in rats and it safely gives the required amount of

---

[4] *Incretins* are a group of hormones that stimulate a decrease in blood glucose levels. They cause an increase in the amount of insulin released from the pancreatic beta cells (Figure 14.2).

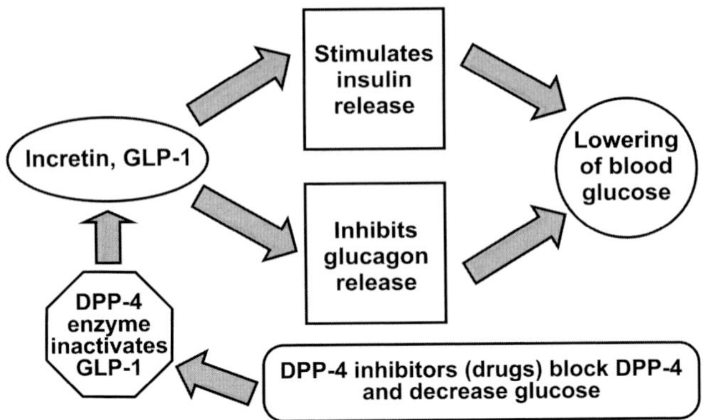

Figure 14.2: Incretins and their effects. Public domain figure downloaded from the WEB.

Incretins are gut hormones that are secreted from enteroendocrine cells into the blood within minutes after eating. One of their many physiological roles is to regulate the amount of insulin that is secreted after eating. There are two incretins, known as *Glucose-dependent Insulinotropic Peptide* (GIP) and *Glucagon-Like Peptide*-1 (GLP-1), that share many common actions in the pancreas but have distinct actions outside of the pancreas. Both incretins are rapidly deactivated by an enzyme called dipeptidyl peptidase 4 (DPP-4). Short-term studies have shown that incretins/incretin-based therapies protect β-cells (by enhancing cell proliferation and differentiation and inhibiting apoptosis) and stimulate their function (by recruiting β-cells to the secretory process and increasing insulin biosynthesis/secretion) (Kim,W and Egan JM (2008).

insulin subcutaneously so replacing common injections (Arora *et al.*, 2007).

Micropumps are important new delivery modalities for injectable drugs (Laser and Santiago, 2004). Electroosmotic micropumps provide high flow rates and pressures, whereas dynamic micropumps based on electrohydrodynamic and magnetohydrodynamic effects have also been produced. More than 60 prototype silicon electroosmotic micropumps have been produced and have nearly 1.5 cm^3 package sizes generate flow rates up to 170 μL $\times$ min^{-1} and pressures of nearly 10 kPa, and new technology continues to improve their flow rates and depth of injections. The use of micropumps in the clinical area is new and promising (Selam *et al.*, 1992; Gin *et al.*, 2003; Schaepelynck *et al.*, 2011).

Acidity (pH) sensitive nano-valves also appear as useful tools in nano-targeted drug delivery systems and probable subject of new developments for insulin delivery units (Angelos *et al.*, 2008). A Swiss company has developed a 65-mm membrane-based insulin pump that injects constant amount of liquid on every beat. Although not at nanoscale, it has high accuracy and reliability, despite different temperature and pressures. It uses silicon technology, similar to computer chips.

Elevated levels of low-density-lipoprotein cholesterol (LDL-C) in plasma are a well-established risk factor for the development of CHD. Plasma LDL-C levels are in part determined by the rate at which LDL particles are removed from the bloodstream by hepatic uptake. The uptake of LDL by mammalian liver cells occurs mainly *via* receptor-mediated endocytosis, a process which entails the binding of these particles to specific receptors in specialized areas of the cell surface, the subsequent internalization of the receptor–lipoprotein complex, and ultimately the degradation and release of the ingested lipoproteins constituent parts. Pearson *et al.*, (2009) formulated a mathematical model to study the binding and internalization (endocytosis) of LDL and very low-density lipoprotein (VLDL) particles by hepatocytes in culture. The system of ordinary differential equations, which includes a cholesterol-dependent pit production term representing feedback regulation of surface receptors in response to intracellular cholesterol levels, was analyzed using numerical simulations and steady-state analysis. Numerical results showed good agreement with *in vitro* experimental data describing LDL uptake by cultured hepatocytes following delivery of a single bolus of lipoprotein.

Alterations in removal rate of LDL-C and changed cholesterol absorption by age could also be formulated by mathematical models (McAuley *et al.*, 2012). These authors presented a quite elaborated and well-organized system of equations to describe and anticipate cholesterol behavior. In turn, Cilla *et al.*, (2014) proposed a mathematical model to reproduce atheroma plaque growth in coronary arteries using the Navier-Stokes equations and Darcy's law for fluid dynamics, convection-diffusion-reaction equations for modelling the mass balance in the lumen and intima, and the Kedem–Katchalsky equations for the interfacial coupling at membranes, that is, the endothelium (Jarzyńska M, 2008). Their results showed that the mathematical model is able to qualitatively capture the atheroma plaque development observed in the intima layer. Eussen *et al.*, (2011) constructed a mathematical model to simulate reductions in LDL cholesterol in persons who combine the use of statins with a high intake of phytosterols/-stanols, e.g., by the use of functional foods. The model includes the cholesterol pool size in the liver and serum levels of VLDL cholesterol to simulate the reduction in LDL cholesterol after separate and combined intake of statins and functional foods acting on intestinal (re)absorption of cholesterol or bile acids in humans. Zhang *et al.*, (2013) studied theoretically the problem of foam cell formation in atherosclerosis.

In a near future, bioengineering has new challenges, not only because of the great development of the field and new technologies, as neurorobotics, but also because individual genetic screening would be an easily available test. It would start a new era of personalized medicine, leading perhaps to finding possible genetic disorders along with bioengineering solutions (as tissue engineering) to improve or replace biological functions. The latter requires new developments on bioinformatics for genomics and proteomics. Nanorobots will be a great help in Bioengineering; they can be used to detect blockages in blood vessels, destroy kidney stones, and dissolve blood clots. They also would detect foreign bodies (bacteria, viruses, toxins) to help produce antibodies and detect cancer cells for diagnosis of diseases, and even its treatment. Nanorobots could be placed on damage tissue to assure proper repair, for example, on fractures or damaged organs, to ensure proper repair. Besides, they can be used in combination with stem cells, say, to recognize the place where to put the cells. On top of that, nanorobots could synthesize hormones and enzymes from amino acids. Novel biomedical technologies and computational capacities would improve Brain Computer Interphase applications, as the control of Exoskeletons. Another growing area refers to surgical planning, and for guiding or performing surgical interventions (Computer Assisted Surgery). No doubt, bioengineers must be prepared to face amazing theories if they have to understand better the biological processes and to give new solutions as we enter into a renewed era for the biomedical sciences. No doubt, the tools and conceptual frames brought about by Biomedical Engineering (BME) constitute elements opening new gates of fascinating research (Fantini *et al.*, 2011).

However, surprising news show up here and there, sometimes unexpected, or eventually as a cascade. Animal and two-dimensional cell culture models have had a profound impact on, not only lung research but also medical research at large, despite inherent flaws and differences when compared with *in vivo* and clinical observations. Three-dimensional (3D) tissue models are a natural progression and extension of existing techniques that seek to plug the gaps and mitigate the drawbacks of two-dimensional and animal technologies. A recent review, describes the transition of historic models to contemporary 3D cell and organoid models, the varieties of current 3D cell and tissue culture modalities, the common methods for imaging these models, and finally, the applications of these models and imaging techniques to lung research. One such example is the idea of a model system that provides a platform for studying and testing novel methods in a controlled and replicable environment. Many models in which various aspects of a disease can be reproduced were designed and rigorously tested. These models range from small and large animals, cell and tissue cultures, and the advent of tissue-engineered organoids (Konar D *et al.*, (2016).

We honestly do not know whether a future vision was offered in these last pages, perhaps we did not, or perhaps it was barely a timid peeping out. None the less, admitting such weakness, there is something we feel that must at least be mentioned: WORLD POVERTY. The reader is invited to visit https://www.dosomething.org/us/facts/11-facts-about-global-poverty, where he/she will learn that,

1. Nearly 1/2 of the world's population — more than 3 billion people — live on less than \$2.50 a day. More than 1.3 billion live in extreme poverty — less than \$1.25 a day.
2. One billion children worldwide are living in poverty. According to UNICEF, 22,000 children die each day due to poverty.

3. About 805 million people worldwide do not have enough food to
   eat.
4. More than 750 million people lack adequate access to clean drinking water. Diarrhea caused by inadequate drinking water, sanitation, and no hand hygiene kills an estimated 842,000 people every
   year globally, or approximately 2,300 people per day.
5. In 2011, 165 million children under the age 5 were stunted
   (reduced rate of growth and development) due to chronic mal-
   nutrition.

**Is there something bioengineering can do to alleviate such
tremendous situation? Think it over, please!** Often we scientists become blinded or glared by new technologies, gadgets of all
sorts, ever more complex cell phones, fancy cars, drones . . . you name
it! How many can reach them? **And violence, wars, and terror-
ism . . . ?**

On the other side of the coin, if you make a quick visit to
http://www.forbes.com/billionaires/, it will show that 1,810 bil-
lionaires hold an aggregate net worth **$6.5 trillion** . . . By checking
in http://www.dictionary.com/browse/trillion, we realized that one
trillion is a cardinal number represented in the United States by
1 followed by 12 zeros, and in Great Britain by 1 followed by 18
zeros. Big numbers, indeed! World population is about 7,500 million
people, considering the number given above, the percentage holding
a huge amount of money, we repeat, a handsome **$6.5 trillion**, is
negligible, that is, $\mathbf{1,810/7,500 = 24.1 \times 10^{-6}\%}$. Bioengineering
or science at large appears as a non-entity within such frame work,
but **we are still human beings . . . or so we think!**

References

Agrawal AK, Harde H, Thanki K, Jain S. Improved stability and antidiabetic
    potential of insulin containing folic acid functionalized polymer stabilized
    multilayered liposomes following oral administration. *Biomacromolecules*
    15(1):350–60, 2014.
Angelos S, Yang YW, Patel K, Stoddart F, Zink JI. pH-Responsive supramolecu-
    lar nanovalves based on cucurbit[6]uril pseudorotaxanes. *Angewandte Chemie*

47(12):2222–2226, 2008. *Cucurbiturils* are macrocyclic molecules made of gly-coluril ($=C_4H_2N_4O_2$ =) monomers linked by methylene bridges ($–CH_2–$). The name is derived from the resemblance of this molecule with a pumpkin of the family of Cucurbitaceae. Cucurbiturils are commonly written as cucurbit[n]uril, where n is the number of glycoluril units. *Pseudorotaxane* is a supramolecular species consisting of a linear molecular component encircled by a macrocyclic component.

Ansary A, Faddah LM. Nanoparticles as biochemical sensors. *Nanotechnology, Sci Appl* 3:65–76, 2010.

Araujo F, Shrestha N, Granja PL, Hirvonen J, Santos HA, Sarmento B. Antihyperglycemic potential of incretins orally delivered via nano and microsystems and subsequent glucoregulatory effects. *Current Pharmaceutical Biotechnol* 15(7):609–19, 2014.

Arora A, Hakim I, Baxter J, Rathnasingham R, Ravi Srinivasan R, Fletcher DA, Samir Mitragotri S. Needle-free delivery of macromolecules across the skin by nanoliter-volume pulsed microjets. *Proc Natl Acad Sci USA* 104(11):4255–4260, 2007.

Balaconis MK, Clark HA. Gel encapsulation of glucose nanosensors for prolonged *in vivo* lifetime. *J Diabetes Sci Technol* 7(1):53–61, 2013.

Bonora E, Calcaterra F, Lombardi S, Bonfante N, Formentini G, Bonadonna RC, Muggeo M. Plasma glucose levels throughout the day and HbA(1c) interrelationships in Type 2 diabetes: Implications for treatment and monitoring of metabolic control. *Diabetes Care* 24(12):2023–2029, 2001.

Cash KJ, Clark HA. Nanosensors and nanomaterials for monitoring glucose in diabetes. *Trends Mol Med* 16(12):584–93, 2010.

Cheng D, Fei Y, Liu Y, Li, J, Xue Q, Wang X, Wang N. HbA1C variability and the risk of renal status progression in diabetes mellitus: A meta-analysis. *PLOS* 9(12):e115509, 2014.

Cilla M, Pena E, Martinez MA. Mathematical modelling of atheroma plaque formation and development in coronary arteries. *J. Royal Soc Interface* 11(90):20130866, 2014.

Derr R, Garrett E, Stacy GA, Saudek CD. Is HbA(1c) affected by glycemic instability? *Diabetes Care* 26(10):2728–2733, 2003.

Eussen SR, Rompelberg CJ, Klungel OH, van Eijkeren JC. Modelling approach to simulate reductions in LDL cholesterol levels after combined intake of statins and phytosterols/-stanols in humans. *Lipids in Health & Disease* 10:187, 2011. http://www.ncbi.nlm.nih.gov/pmc/articles/PMC3229468/.

Fangueiro JF, Silva AM, Garcia ML, Souto E. Current nanotechnology approaches for the treatment and management of diabetic retinopathy. *European J Pharmacology and Biopharmacology* 96:307–322, 2014.

Fantini S, Bennis C, Kaplan D. Biomedical engineering continues to make the future. *IEEE Pulse Magazine* 2(4):54–59, 2011.

Farooq SA, Saini V, Singh R, Kaur K. Application of novel drug delivery system in the pharmacotherapy of hyperlipidemia. *J Chemical and Pharmaceutical Sciences* 6:138–146, 2013.

Fischer S. The sugar spectre. *IEEE Pulse* 5(3):12–17, 2014.

Gin H, Renard E, Melki V, Boivin S, Schaepelynck-Bélicar P, Guerci B, Selam JL, Brun JM, Riveline JP, Estour B, Catargi B or EVADIAC Study Group. Combined improvements in implantable pump technology and insulin stability allow safe and effective long term intraperitoneal insulin delivery in Type 1 diabetic patients: The EVADIAC experience. *Diabetes Metab* 29(6):602–607, 2003.

Gu Z, Dang TT, Ma ML, Tang BC, Cheng H, Jiang S, Dong YZ, Zhang Y, Anderson DG. Glucose-responsive microgels integrated with enzyme nanocapsules for closed-loop insulin delivery. *ACS Nano* 7:6758, 2013.

Harris P. Microtechnology vs Nanotechnology. *Occupational Health & Safety* 79(1):36–37, Jan 2010. http://ohsonline.com/Articles/2010/01/01/Micro-technology-vs-Nanotechnology.aspx.

Jarzyńska M. Derivation of practical Kedem–Katchalsky equations for membrane substance transport. *Concepts of Physics* 5(3):459–474, 2008. doi 10.2478/v10005-007-0041-8.

Kahrom M. An innovative mathematical model: A key to the riddle of HbA(1c). *Int J Endocrinol*, August 29, 2010. doi: 10.1155/2010/481326. http://www.ncbi.nlm.nih.gov/pmc/articles/PMC2939406/.

Karathanasis E, Bhavane R, Annapragada AV. Glucose sensing pulmonary delivery of human insulin to systemic circulation of rats. *Int J Nanomedicine* 2:501–513, 2007.

Kilpatrick ES, Rigby AS, Atkin SL. Variability in the relationship between mean plasma glucose and HbA1c: Implications for the assessment of glycemic control. *Clin Chemistry* 53(5):897–901, 2007.

Kim,W, Egan JM. The role of incretins in glucose homeostasis and diabetes treatment. *Pharmacol Review* 60(4):470–512, 2008.

Konar D, Devarasetty M, Yildiz DV, Atala A, Murphy SV. Lung-on-a-chip technologies for disease modeling and drug development. *Biomed Eng Computational Biology* (Suppl 1):17–27; review published Apr 20, 2016. http://www.la-press.com/lung-on-a-chip-technologies-for-disease- modeling-and-drug-developments-article-a5547-abstract?

Laser DJ, Santiago JG. A review of micropumps. *J Micromech Microeng* 14:R35–R64, 2004.

Luthi AJ, Patel PC, Ko CH, Mutharasan RK, Mirkin CA, Thaxton CS. Nanotechnology for synthetic high-density lipoproteins. *Trends Mol Med* 16(12):553–560, 2010.

Luthi AJ, Zhang H, Kim D, Giljohann DA, Mirkin CA, Thaxton CS. Tailoring of biomimetic High-Density Lipoprotein (HDL) nanostructures changes cholesterol binding and efflux. *ACS Nano* 6(1):276–285, 2012. doi: 10.1021/nn2035457.

Maximov VD, Reukov VV, Barry JN, Cochrane C, Vertegel AA. Protein-nanoparticle conjugates as potential therapeutic agents for the treatment of hyperlipidemia. *Nanotechnology* 21(26):265103, 2010.

McAuley MT, Wilkinson DJ, Jonas JJL, Kirkwood TBL. A whole-body mathematical model of cholesterol metabolism and its age-associated dysregulation. *BMC Syst Biol* 6(1):845–880, 2012.

McCall AL, Cox DJ, Crean J, Gloster M, Kovatchev BP. A novel analytical method for assessing glucose variability: Using CGMS in type 1 diabetes mellitus. *Diabetes Technol Ther* 8(6):644–653, 2006.

Monnier L, Lapinski H, Colette C. Contributions of fasting and postprandial plasma glucose increments to the overall diurnal hyperglycemia of type 2 diabetic patients: Variations with increasing levels of HbA(1c). *Diabetes Care* 26(3):881–885, 2003.

Monnier L, Mas E, Ginet C, Michel F, Villon L, Cristol JP, Colette C. Activation of oxidative stress by acute glucose fluctuations compared with sustained chronic hyperglycemia in patients with type 2 diabetes. JAMA 295(14):1681–1687, 2006.

Pearson T, Wattis JAD, O'Malley B, Pickersgill L, Blackburn H, Jackson KG, Byrne HM. Mathematical modelling of competitive LDL/VLDL binding and uptake by hepatocytes. J Math Biol 58(6):845–880, 2009.

Rohlfing CL, Wiedmeyer HM, Little RR, England JD, Tennill A, Goldstein DE. Defining the relationship between plasma glucose and HbA(1c): Analysis of glucose profiles and HbA(1c) in the diabetes control and complications trial. *Diabetes Care* 25(2):275–278, 2002.

Ruehm SG, Corot C, Vogt P, Kolb S, Debatin JF. Magnetic resonance imaging of atherosclerotic plaque with ultrasmall superparamagnetic particles of iron oxide in hyperlipidemic rabbits. *Circulation* 103(3):415–422, 2001. doi: 10.1161/01.circ.103.3.415.

Schaepelynck P, Darmon P, Molines L, Jannot-Lamotte MF, Treglia C, Raccah D. Advances in pump technology: Insulin patch pumps, combined pumps and glucose sensors, and implanted pumps. *Diabetes Metab* 37(Suppl 4):S85–S93, 2011.

Scognamiglio V. Nanotechnology in glucose monitoring: Advances and challenges in the last 10 years. *Biosens Bioelectron* 15:47:12–25, 2013.

Selam JL, Micossi P, Dunn FL, Nathan DM. Clinical trial of programmable implantable insulin pump for type I diabetes. *Diabetes Care* 15(7):877–885, 1992.

Shatkin JA. Nanotechnology: Health and environmental risks, 2nd ed, CRC Press, Boca Raton, Fl, 2013.

Thaxton CS, Daniel WL, Giljohann DA, Thomas AD, Mirkin CA. Templated spherical high density lipoprotein nanoparticles. *J American Chem Society* 131(4):1384–1385, 2009.

Thaxton CS, Elghanian R, Thomas AD, Stoeva SI, Lee JS, Smith ND, Schaeffer AJ, Klocker H, Horninger W, Bartsch G, Mirkin CA. Nanoparticle-based bio-barcode assay redefines "undetectable" PSA and biochemical recurrence after radical prostatectomy. *Proc Natl Acad Sci USA*, 106(44):18437–18442, 2009.

Thomas A, Schönauer M, Achermann F, Schnell O, Hanefeld M, Ziegelasch HJ, Mastrototaro J, Heinemann L. The "glucose pentagon": Assessing glycemic control of patients with diabetes mellitus by a model integrating different parameters from glucose profiles. *Diabetes Technol Ther* 11(6):399–409, 2009.

Valentinuzzi ME. Fibrillation-Defibrillation: Clinical and Engineering Aspects, 8 chapters. World Scientific Publisher (WSP), New Jersey & Singapore, (xxii + 279) pp., 2010. Series on Bioengineering & Biomedical Engineering, vol 6; ISBN 978-981-4293-63-1///981-4293-63-6.

Walton BL, Leja M, Vickers KC, Estevez-Fernandez M, Sanguino A, Wang E, Clubb Jr FJ, Morrisett J, Lopez-Berestein G. Delivery of negatively charged liposomes into the atheromas of Watanabe heritable hyperlipidemic rabbits. *Vasc Med* 15(4):307–313, 2010. BWalton@heart.thi.tmc.edu.

Wang YX, Hussain SM, Krestin GP. Superparamagnetic iron oxide contrast agents: physicochemical characteristics and applications in MR imaging. *Eur Radiol* 11(11):2319–31, 2001.

Wickline SA, Lanza GM. Nanotechnology for molecular imaging and targeted therapy. *Circulation* 107:1092–1095, 2003. http://circ.ahajournals.org/ content/107/8/1092.full.

Wu D, Gong CX, Meng X, Yang QL. Correlation between blood glucose fluctuations and activation of oxidative stress in Type 1 diabetic children during the acute metabolic disturbance period. *Chin Med J* 126(21):4019–4022, 2013.

Zhang S, Ritter LR, Ibragimov AI. Foam cell formation in atherosclerosis: HDL and macrophage reverse cholesterol transport, *Discrete and Continuous Dynamical Systems* Supplement, 825–835, 2013. www.aimSciences.org.

INDEX